Micro Instrumentation

Edited by Melvin V. Koch,
Kurt M. VandenBussche,
and Ray W. Chrisman

1807–2007 Knowledge for Generations

Each generation has its unique needs and aspirations. When Charles Wiley first opened his small printing shop in lower Manhattan in 1807, it was a generation of boundless potential searching for an identity. And we were there, helping to define a new American literary tradition. Over half a century later, in the midst of the Second Industrial Revolution, it was a generation focused on building the future. Once again, we were there, supplying the critical scientific, technical, and engineering knowledge that helped frame the world. Throughout the 20th Century, and into the new millennium, nations began to reach out beyond their own borders and a new international community was born. Wiley was there, expanding its operations around the world to enable a global exchange of ideas, opinions, and know-how.

For 200 years, Wiley has been an integral part of each generation's journey, enabling the flow of information and understanding necessary to meet their needs and fulfill their aspirations. Today, bold new technologies are changing the way we live and learn. Wiley will be there, providing you the must-have knowledge you need to imagine new worlds, new possibilities, and new opportunities.

Generations come and go, but you can always count on Wiley to provide you the knowledge you need, when and where you need it!

William J. Pesce
President and Chief Executive Officer

Peter Booth Wiley
Chairman of the Board

Micro Instrumentation

for High Throughput Experimentation
and Process Intensification – a Tool for PAT

Edited by
Melvin V. Koch, Kurt M. VandenBussche,
and Ray W. Chrisman

WILEY-VCH Verlag GmbH & Co. KGaA

The Editors

Dr. Melvin V. Koch
University of Washington
Center for Process Analytical Chemistry
Box 351700
Seattle, WA 98195-1700
USA

Dr. Kurt M. VandenBussche
UOP Process Technology and Equipment
25 East Algonquin Rd.
Des Plaines, IL 60017
USA

Dr. Ray W. Chrisman
University of Washington
Visiting Scholar
Center for Process Analytical Chemistry
Box 351700
Seattle, WA 98195-2326
USA

All books published by Wiley-VCH are carefully produced. Nevertheless, authors, editors, and publisher do not warrant the information contained in these books, including this book, to be free of errors. Readers are advised to keep in mind that statements, data, illustrations, procedural details or other items may inadvertently be inaccurate.

Library of Congress Card No.: applied for

British Library Cataloguing-in-Publication Data:
A catalogue record for this book is available from the British Library.

Bibliographic information published by the Deutsche Nationalbibliothek
The Deutsche Nationalbibliothek lists this publication in the Deutsche Nationalbibliografie; detailed bibliographic data are available in the Internet at http://dnb.d-nb.de.

© 2007 WILEY-VCH Verlag GmbH & Co. KGaA, Weinheim

All rights reserved (including those of translation into other languages). No part of this book may be reproduced in any form – by photoprinting, microfilm, or any other means – nor transmitted or translated into a machine language without written permission from the publishers. Registered names, trademarks, etc. used in this book, even when not specifically marked as such, are not to be considered unprotected by law.

Composition   Manuela Treindl, Laaber
Printing   betz-druck GmbH, Darmstadt
Binding   Litges & Dopf GmbH, Heppenheim
Cover   Grafik-Design Schulz, Fußgönheim

Wiley Bicentennial Logo: Richard J. Pacifico

Printed in the Federal Republic of Germany
Printed on acid-free paper

ISBN: 978-3-527-31425-6

Contents

Preface *XVII*

List of Contributors *XIX*

Part I Introducing the Concepts *1*

1 Introduction *3*
Melvin V. Koch

1.1 Background *3*
1.2 Analytical Tools for use in PAT *7*
1.3 The Center for Process Analytical Chemistry (CPAC) and the Summer Institute *9*
1.4 Topics covered by Previous CPAC Summer Institutes *14*
1.5 Recent Emphasis of CPAC Summer Institutes: High Throughput Experimentation and Process Intensification *16*
1.6 Conclusion *20*
References *20*

2 Macro to Micro ... The Evolution of Process Analytical Systems *23*
Wayne W. Blaser and Ray W. Chrisman

2.1 Introduction *23*
2.1.2 Early Developments *23*
2.1.3 Developments since 1980 *25*
2.1.4 Sampling Systems *28*
2.1.4.1 Filtration *30*
2.1.4.2 The Fast Loop–Analytical Loop Strategy *31*
2.1.4.3 NeSSI *32*
2.1.5 General Reviews *32*
2.2 Chromatography *33*
2.2.1 Gas Chromatography *33*
2.2.2 Liquid Chromatography *35*
2.2.3 On-line Spectroscopy *36*

2.2.4   On-line Mass Spectrometry   37
2.2.5   Microflow Techniques   38
          References   42

3   Process Intensification   43
          Kurt M. VandenBussche

3.1   Introduction, Scope and Definitions   43
3.2   Process Intensification in the Field of Reaction Engineering   44
3.3   Process Intensification through Micro-structured Unit Operations   49
3.3.1   Gas Phase Mass Transfer   51
3.3.2   Liquid–Liquid Mass Transfer: Mixing and Emulsions   52
3.3.3   Gas–Liquid Mass Transfer   53
3.3.4   Mass Transfer in Gas–Solid Systems   54
3.3.5   Heat Transfer   56
3.4   Case Studies   57
3.4.1   Distributed Production of Methanol   57
3.4.2   Distributed Production of Hydrogen   60
3.5   Conclusions   64
          References   64

4   High Throughput Research   67
          Ray W. Chrisman

4.1   Introduction   67
4.2   Description of Terms   67
4.3   Concept of a Research Process   68
4.4   High Throughput Analytical   70
4.5   Extracting Information from the Process   71
4.6   Process Development becomes the Next Bottleneck   73
4.7   Use of High Throughput Concepts for Process Development   74
4.8   Microreactors for Process Development   75
4.9   Current Barriers and Limitations to Microscale Reaction Characterization   76
4.10   Conclusion   77
          References   77

Part II   Technology Developments and Case Studies   79

5   Introduction   81
          Melvin V. Koch, Ray W. Chrisman, and Kurt M. VandenBussche

6   Microreactor Concepts and Processing   85
          Volker Hessel, Patrick Löb, Holger Löwe, and Gunther Kolb

6.1   Introduction   85

6.2	Microreactor Technology – Interfacing and Discipline Cross-boundary Research *85*	
6.3	Microstructured Mixer-reactors for Pilot and Production Range and Scale-out Issues *88*	
6.3.1	Caterpillar Microstructured Mixer-reactors *88*	
6.3.2	StarLam Microstructured Mixer-reactors *91*	
6.3.3	Microstructured Heat Exchanger-reactors *94*	
6.4	Fine-chemical Microreactor Plants *96*	
6.4.1	Laboratory-range Plants *96*	
6.4.2	Pilot-range Plants *98*	
6.5	Industrial Microreactor Process Development for Fine and Functional Chemistry *100*	
6.5.1	Phenyl Boronic Acid Synthesis (Scheme 6.1) (Clariant/Frankfurt + IMM) *100*	
6.5.1.1	Process Development Issue *100*	
6.5.1.2	Microreactor Plant and Processing Solution *101*	
6.5.2	Azo Pigment Yellow 12 Manufacture (Scheme 6.2) (Trust Chem/Hangzhou + IMM) *102*	
6.5.2.1	Process Development Issue *102*	
6.5.2.2	Microreactor Plant and Processing Solution *103*	
6.5.3	Hydrogen Peroxide Synthesis (UOP/Chicago + IMM) *104*	
6.5.3.1	Process Development Issues *104*	
6.5.3.2	Microreactor Plant and Processing Solution *104*	
6.5.4	(S)-2-Acetyl Tetrahydrofuran Synthesis (SK Corporation/Daejeon; IMM Tools) *105*	
6.5.4.1	Process Development Issues *105*	
6.5.4.2	Microreactor Plant and Processing Solution *105*	
6.5.5	Synthesis of Intermediate for Quinolone Antibiotic Drug (LG Chem/Daejeon; IMM Tools) *106*	
6.5.5.1	Process Development Issues *106*	
6.5.5.2	Microreactor Plant and Processing Solution *106*	
6.6	Industrial Production in Fine Chemistry *107*	
6.6.1	Nitroglycerine Production Plant for Acute Cardiac Infarction (Xi'an Huian Industrial Group/Xi'an + IMM) *107*	
6.6.2	Production-oriented Development for LED Materials (MERCK-COVION/Frankfurt + IMM and Other Partners) *107*	
6.6.3	Pilot Plant for MMA Manufacture (Idemitsu Kosan, Chiba/Japan; MCPT Works) *108*	
6.6.4	Production Operation for High-value Polymer Intermediate Product (DSM Linz/Austria + FZK) *109*	
6.7	Microreactor Laboratory-scale Process Developments for Future Industrial Use *110*	
6.7.1	Michael Addition – Amine Addition to α,β-Unsaturated Compounds (Fresenius + IMM) *110*	
6.7.1.1	Process Development Issues *110*	

6.7.1.2 Microreactor Plant and Processing Solution *110*
6.7.2 Kolbe–Schmitt Synthesis (IMM) *112*
6.7.2.1 Process Development Issues *112*
6.7.2.2 Microreactor Plant and Processing Solution *113*
6.7.3 Hydrogenation of Nitrobenzene (UCL + IMM) *115*
6.7.3.1 Process Development Issues *115*
6.7.3.2 Microreactor Plant and Processing Solution *116*
6.7.4 Brominations of *m*-Nitrotoluene (IMM) *118*
6.7.4.1 Process Development Issues *118*
6.7.4.2 Microreactor Plant and Processing Solution *118*
6.7.5 Brominations of Thiophene (Fresenius + IMM) *119*
6.7.5.1 Process Development Issues *119*
6.7.5.2 Microreactor Plant and Processing Solution *120*
6.8 Free-radical Polymerizations (Uni Strasbourg, IMM) *121*
6.8.1 Process Development Issues *121*
6.8.2 Microreactor Plant and Processing Solution *122*
6.8.2.1 Process Development Issues *122*
6.9 Future Directions – Establishing a Novel Chemistry by Enabling Function *124*
6.9.1 Traditional Chemistry – Processes Follow Limitations of Reactors *124*
6.9.2 Novel Chemistry – Tailoring Protocols to Increase Microreactor Performance *124*
6.9.2.1 High-p,T Processing *124*
6.9.2.2 Solvent-free Processes – Contacting "All-at-once" *124*
6.9.2.3 Use of Hazardous Elements and Building Blocks – Direct Routes *125*
6.9.2.4 Routes in the Explosive Regime *125*
6.9.2.5 Simplified Protocols *125*
6.10 Summary *126*
 References *127*

7 Non-reactor Micro-component Development *131*
Daniel R. Palo, Victoria S. Stenkamp, Jamie D. Holladay, Paul H. Humble, Robert A. Dagle, and Kriston P. Brooks

7.1 Overview *131*
7.2 Introduction *131*
7.3 Heat Exchange *132*
7.4 Mixing *135*
7.4.1 Active Micromixers *136*
7.4.1.1 Periodic Flow Switching *136*
7.4.1.2 Electric Field Mixing *137*
7.4.1.3 Ultrasound/Piezoelectric Membranes *137*
7.4.1.4 Microimpellers *138*
7.4.1.5 Acoustic Bubble Shaking *139*

7.4.2	Passive Micromixers	*139*
7.4.2.1	T and Y Type Micromixers	*139*
7.4.2.2	Multi-laminating Mixers	*140*
7.4.2.3	Split and Recombination	*141*
7.4.2.4	Chaotic Flow	*142*
7.4.2.5	Recirculation Mixers	*142*
7.4.2.6	Structured Packing	*142*
7.4.2.7	Colliding Jet Mixers	*142*
7.4.2.8	Moving Droplet Mixers	*143*
7.5	Microchannel Emulsification	*143*
7.6	Phase Separation	*148*
7.7	Phase Transfer Processes	*149*
7.7.1	Adsorption	*149*
7.7.1.1	Microchannel Chromatography	*149*
7.7.1.2	Microchannel Adsorption for Component Collection	*150*
7.7.2	Extraction	*151*
7.7.3	Microchannel Electrophoresis	*154*
7.7.4	Absorption	*158*
7.7.5	Desorption	*158*
7.7.6	Pervaporation	*159*
7.7.7	Distillation	*160*
7.7.8	Heat Pump Systems	*162*
7.8	Biological Processes	*165*
7.8.1	Cell Testing with Microchannel Systems	*165*
7.8.2	Hemodialysis	*165*
7.9	Body Force Driven Processes	*167*
7.9.1	Electric Fields	*167*
7.9.1.1	Electroosmotic Flow (EOF)	*167*
7.9.1.2	Electric Field Actuated Valves	*169*
7.9.1.3	Electromagnets	*169*
7.9.2	Acoustic/Ultrasonic Forces	*169*
7.10	Summary and Future Directions	*171*
	References	*172*

8	**Microcomponent Flow Characterization**	*181*
	Bruce A. Finlayson, Pawel W. Drapala, Matt Gebhardt, Michael D. Harrison, Bryan Johnson, Marlina Lukman, Suwimol Kunaridtipol, Trevor Plaisted, Zachary Tyree, Jeremy VanBuren, and Albert Witarsa	
8.1	Introduction	*181*
8.2	Pressure Drop	*182*
8.2.1	Friction Factor for Slow Flows	*182*
8.2.2	Mechanical Energy Balance for Turbulent Flow	*183*
8.3	Dimensionless Mechanical Energy Balance	*184*

8.3.1	Mechanical Energy Balance for Laminar Flow *186*
8.3.2	Pressure Drop for Flow Disturbances *187*
8.3.3	Pressure Drop for Contractions and Expansions *189*
8.3.4	Manifolds *192*
8.4	Entry Lengths *194*
8.4.1	Contraction Flows *194*
8.5	Diffusion *196*
8.5.1	Characterization of Mixing *198*
8.5.2	Average Concentration along an Optical Path *199*
8.5.3	Peclet Number *199*
8.5.4	Diffusion in a Rectangular Channel *201*
8.5.5	Diffusion in a T-Sensor [13, 14] *202*
8.5.6	Serpentine Mixer *204*
8.5.7	Reactor System *205*
8.6	Conclusion *206*
	Nomenclature *207*
	References 208

9 Selected Developments in Micro-analytical Technology *209*

9.1 Introduction *209*
Melvin V. Koch

9.2 Application of On-line Raman Spectroscopy to Characterize and Optimize a Continuous Microreactor *211*
Brian Marquardt

9.2.1	Introduction *211*
9.2.2	Optical Sampling *212*
9.2.3	Continuous Reaction Monitoring *213*
9.2.4	Improved Reactor Sampling Using NeSSI Components *219*
	References 220

9.3 Developments in Ultra Micro Gas Analyzers *222*
Ulrich Bonne, Clark T. Nguyen, and Dennis E. Polla

9.3.1	Overview *222*
9.3.2	MGA Performance Goals *223*
9.3.3	PHASED Air-based Analyzer *223*
9.3.4	Sandia's Hydrogen-based Analyzer *228*
9.3.5	Preliminary Results *230*
9.3.5.1	Preconcentration *230*
9.3.5.2	Separation *231*
9.3.5.3	Detection *232*
9.3.5.4	False Alarm Rate Metrics *234*
9.3.6	Conclusions *238*
	References 239

9.4	Nuclear Magnetic Resonance Spectroscopy *241*	
	Michael McCarthy	
	References 245	
9.5	Surface Plasmon Resonance (SPR) Sensors *246*	
	Clement E. Furlong, Timothy Chinowsky, and Scott Soelberg	
9.5.1	Introduction *246*	
9.5.2	Monitoring Refractive Index, Fundamentals of SPR *246*	
9.5.3	Instrumentation *247*	
9.5.3.1	Spreeta Sensor Elements *247*	
9.5.4	Automated Hand-held 24-Channel SPR Instrument with Modular Fluidics *247*	
9.5.5	Examples of Specific Applications *249*	
9.5.5.1	Analysis of Small Organics *249*	
9.5.5.2	Analysis of Larger Analytes such as Production Proteins or Contaminants *250*	
9.5.5.3	Monitoring of Larger Analytes such as Viruses, Whole Cells or Spores *251*	
	References 251	
9.6	Dielectric Spectroscopy: Choosing the Right Approach *253*	
	Alexander Mamishev	
9.6.1	Introduction *253*	
9.6.2	Dielectric Spectroscopy in Comparison with Other Techniques *253*	
9.6.3	Electrode Patterns *254*	
9.6.3.1	Control of Spatial Wavelength *254*	
9.6.3.2	Control of Signal Strength *255*	
9.6.3.3	Imaging Capability *255*	
9.6.4	Direct Sensing vs. Pre-concentrators *256*	
9.6.5	Frequency of Excitation *256*	
9.6.6	Conclusions *256*	
	References 257	
9.7	The Future of Liquid-phase Microseparation Devices in Process Analytical Technology *258*	
	Scott E. Gilbert	
9.7.1	Introduction *258*	
9.7.2	Historical Perspectives of Microseparation Device Technology *259*	
9.7.2.1	Development of Chip-based CE and CEC *261*	
9.7.2.2	Development of Chip-based LC *264*	
9.7.3	Applications of Liquid Microseparation Devices for Process Stream Sampling and Coupling to Microreactors *281*	
9.7.3.1	Continuous-sampling CE Chips for Process Applications *281*	
9.7.3.2	Continuous Sampling LC Chips for Process Applications *292*	

9.7.3.3 Coupling of Microseparation Devices to Microreactors 294
9.7.4 Potential Applications in Process Analytical Technology 295
9.7.4.1 Relevance of Micro-scale Device Technology 295
9.7.4.2 Potential Applications in Industrial Biotechnology and Biopharmaceutical Production 296
9.7.4.3 Sample Pretreatment 296
9.7.4.4 Potential Applications in the Fields of Process Development and Microreactor Technology 297
9.7.5 Future Perspectives 297
References 298

9.8 Grating Light Reflection Spectroscopy: A Tool for Monitoring the Properties of Heterogeneous Matrices 305
Lloyd W. Burgess

9.8.1 Overview 305
9.8.2 Introduction 305
9.8.3 Theory 307
9.8.4 Results and Discussion 309
9.8.5 Significance/Conclusions 313
References 313

10 New Platform for Sampling and Sensor Initiative (NeSSI) 315
David J. Veltkamp

10.1 Introduction 315
10.2 What is NeSSI™? 316
10.3 NeSSI™ Background 320
10.4 NeSSI™ Physical Layer (Generation I NeSSI™) 323
10.4.1 The Swagelok® MPC System 325
10.4.2 The CIRCOR Tech µMS3™ (Micro Modular Substrate Sampling System) System 330
10.4.3 The Parker-Hannifin Intraflow™ System 334
10.4.4 General Characteristics 338
10.5 NeSSI™ Electrical Layer (Generation II NeSSI™) 339
10.5.1 The NeSSI™ Bus 340
10.5.2 The NeSSI™ Sensor Actuator Manager (SAM) 345
10.6 Advantages of NeSSI™ in Laboratory Applications 346
10.6.1 Provide a Consistent Physical Environment for making Analytical Measurements 347
10.6.2 Easily Constructed and Configured Interfaces for Analytical Instrumentation 347
10.6.3 Facilitate Reduced Sensor or Analyzer Development Costs 348
10.6.4 Increased Ability to Integrate Multiple Sensing Elements 349
10.6.5 Ease Adoption of Developed Analytical Solutions to Industrial Applications 349

10.6.6	Applications to Other Laboratory Tasks 350
10.6.7	Cost, Convenience, and Safety 350
10.7	Conclusions 351
	References 352

| 11 | **Catalyst Characterization for Gas Phase Processes** 353 |
| | *Michelle J. Cohn and Douglas B. Galloway* |

11.1	Introduction 353
11.2	Characterizing Reactivity, Reaction Mechanisms, and Diffusion 354
11.2.1	Reactivity Testing 355
11.2.2	Kinetic Tools 356
11.2.3	Diffusion Tools 359
11.3	Summary 361
	References 362

12	**Integrated Microreactor System for Gas Phase Reactions** 363
	David J. Quiram, Klavs F. Jensen, Martin A. Schmidt, Patrick L. Mills,
	James F. Ryley, Mark D. Wetzel, and Daniel J. Kraus

12.1	Overview 363
12.2	Introduction 364
12.3	Microreactor Packaging 367
12.3.1	Microreactor Packaging Hardware 367
12.3.2	DieMate™ Electrical Interconnect Details 367
12.3.3	DieMate™ Fluidic Interconnect Details 368
12.3.4	DieMate™ Manifold Heating 370
12.3.5	DieMate™ Manifold Fabrication 370
12.3.6	DieMate™ Manifold Insulation 370
12.4	Microreactor System Design 371
12.4.1	System Chassis 371
12.4.2	Primary Feed Gas Manifold 373
12.4.3	Reactor Feed Gas Manifold 374
12.4.4	Reactor Board 374
12.4.5	Transport Tube Heating 376
12.4.6	Temperature Controller Board 376
12.4.7	Heater Driver Circuit Boards 377
12.4.8	Testing Procedures 378
12.5	System Control and Process Monitoring 378
12.5.1	System Hardware 378
12.5.2	Process Control Software 381
12.5.3	Process Control Interface 383
12.5.3.1	Run Time (RT) Engine tab 384
12.5.3.2	Microreactors Tab 384
12.5.4	Automation Testing Procedures 386
12.6	Microreactor Process Safety 386

12.6.1 Overview 386
12.6.2 Process Hazards 387
12.6.3 AIMS Safety Features 388
12.7 Microreactor Experimental Methods 389
12.7.1 Overview 389
12.7.2 Experimental Protocol 390
12.7.3 GC Method 391
12.8 AIMS Testing Results and Discussion 391
12.8.1 Overview 391
12.8.2 AIMS vs. MARS Setups 392
12.8.2.1 System Size 392
12.8.2.2 Gas Flow Manifold 393
12.8.2.3 Process Automation 393
12.8.2.4 Electrical Wiring 394
12.8.2.5 Reactor Failure and Replacement 395
12.8.2.6 Redwood Microvalve Operation 395
12.8.2.7 Redwood Mass Flow Controllers 395
12.8.2.8 Other Process Components 396
12.8.3 Automation and Hardware Evaluations 396
12.8.3.1 Temperature Sensors 396
12.8.3.2 Temperature Controllers 397
12.8.3.3 Process Control Loops 397
12.8.4 Reaction Performance Results 397
12.8.4.1 Methane Oxidation 398
12.8.4.2 Ammonia Oxidation 401
12.9 Summary and Conclusions 404
References 404

13 Liquid Phase Process Characterization 407
Daniel A. Hickman and Daniel D. Sobeck

13.1 Overview 407
13.2 Background 407
13.3 System Design Basis 409
13.3.1 Heat Transfer in Laminar Pipe Flow 411
13.3.2 Mixing at the Reactor Inlet Tee 412
13.3.2.1 Characteristic Mixing Time Analysis 412
13.3.2.2 Two-dimensional Simulation 413
13.3.3 Pressure Drop in Laminar Pipe Flow 415
13.3.4 Axial Dispersion in Laminar Pipe Flow 415
13.3.4.1 Plug Flow Criterion 416
13.3.5 Injector Loop Sizing 417
13.4 System Capabilities 417
13.4.1 System Overview 417
13.4.1.1 Feed System 417

13.4.1.2	Robotics Feed Preparation 418
13.4.1.3	Reactor 418
13.4.1.4	On-line GC Analysis 419
13.4.1.5	Process Control 419
13.4.2	Experimental Evaluation 419
13.4.2.1	Materials of Construction Effects 422
13.4.2.2	Summary 424
13.4.3	Quantitative Measurements of Reaction Kinetics 425
13.5	Conclusions 427
	Nomenclature 427
	References 428

14 Novel Systems for New Chemistry Exploration 431
Paul Watts

14.1	Introduction 431
14.2	Chemical Synthesis in Microreactors 436
14.2.1	Synthesis of Pyrazoles 436
14.2.2	Peptide Synthesis 437
14.2.3	Reaction Optimization 439
14.2.4	Stereochemistry 440
14.3	Chemical Synthesis in Flow Reactors 442
14.3.1	Large-scale Manufacture 446
14.4	Conclusions 447
	References 447

15 Going from Laboratory to Pilot Plant to Production using Microreactors 449
Michael Grund, Michael Häberl, Dirk Schmalz, and Hanns Wurziger

15.1	Introduction 449
15.2	Nitration 449
15.2.1	General Remarks 449
15.2.2	Orienting Laboratory Nitrations 451
15.2.2.1	Acetic Acid 451
15.2.2.2	Acetic Anhydride 452
15.2.2.3	Laboratory Optimization 452
15.3	Microreaction System "MICROTAUROS" 454
15.3.1	First Prototype Silicon Micromixer 454
15.3.2	Second Prototype Silicon Micromixer with Connector 455
15.3.3	Optimized Connection System 456
15.3.4	First Summary 457
15.3.5	Optimized Micromixer with an Advanced Connection System 458
15.4	Automated Reaction Optimization 459
15.4.1	The Principle 459
15.4.2	First Prototype Reaction Optimizer 460
15.4.3	Sampling System without Cross-contamination 461

15.5	Upscale in Larger Laboratory Scale 462
15.5.1	Laboratory Modules for Process Development 462
15.5.2	Nitration with 65% Nitric Acid in Concentrated Sulfuric Acid 463
15.5.3	Nitration with Neat 65% Nitric Acid 464
15.6	Upscale in a Pilot Plant 466
15.7	A Concept for the Future 466
15.8	Conclusion 466
	References 467

Part III A Summary and Path Forward 469

16 Concluding Remarks 471
Melvin V. Koch, Ray W. Chrisman, and Kurt M. VandenBussche

| 16.1 | Summary 471 |
| 16.2 | The Path Forward 473 |

Subject Index 475

Preface

This book recognizes the important role that advances in miniaturization technology have had in improvements in the field of micro-instrumentation as it pertains to Process Analytical Technology (PAT).

The Center for Process Analytical Chemistry (CPAC) at the University of Washington in Seattle, Washington, has been pursuing developments in real-time measurement and control since 1984. In 1996, CPAC developed a strategic objective structured around micro-analytical developments as the future for achieving effective monitoring for process control. To validate this objective CPAC conducted a Summer Institute that year with that as the theme of the gathering. It was a successful event and the Summer Institute has been repeated each summer since, building on that theme of topics surrounding micro-instrumentation.

The concept of the Development Cycle, where there is a reactor product that must be analyzed and the data from the analysis evaluated before the next reaction is run, is presented

The possibility to scale-down engineering unit operations and subsequently measurement techniques has shown value in the field of High Throughput Experimentation, particularly in the discovery phase of new materials and product development.

There is growing interest in Process Intensification, where studies to obtain a basic understanding of unit operations – often via miniaturization approaches – is important to reducing capital and operating costs. This approach has resulted in the application of micro-instrumentation to the areas of process development and process optimization.

Based on developments from recent CPAC Summer Institutes, this book describes and ties various fields together in order to highlight the importance and impact of advances in micro-instrumentation on High Throughput Experimentation and Process Intensification.

Seattle, December 2006 *Melvin V. Koch*

List of Contributors

Wayne W. Blaser
Tek-Creations, L.L.C.
2700 Longfellow Lane
Midland, MI 48640
USA

Ulrich Bonne
Honeywell Laboratories
4936 Shady Oak Road
Hopkins, MN 55343
USA

Kriston P. Brooks
Pacific Northwest National Laboratory
902 Battelle Blvd., K6-28
Richland, WA 99354
USA

Lloyd W. Burgess
University of Washington
Center for Process Analytical Chemistry
Box 351700
Seattle, WA 98195
USA

Timothy Chinowsky
Verathon Inc.
21222 30th Drive SE, Suite 120
Bothell, WA 98021
USA

Ray W. Chrisman
University of Washington
Visiting Scholar
Center for Process Analytical Chemistry
Box 351700
Seattle, WA 98195-2326
USA

Michelle J. Cohn
UOP LLC
Advanced Characterization Group
50 E. Algonquin Rd
Des Plaines, IL 60017
USA

Robert A. Dagle
Pacific Northwest National Laboratory
902 Battelle Blvd., K8-93
Richland, WA 99354
USA

Pawel W. Drapala
University of Washington
Department of Chemical Engineering
Box 351750
Seattle, WA 98195-1750
USA

List of Contributors

Bruce A. Finlayson
University of Washington
Chemical Engineering
Box 351750
Seattle, WA 98195-1750
USA

Clement E. Furlong
University of Washington
Departments of Medicine (Division of Medical Genetics) and Genome Science
Box 357720
Seattle, WA 98195
USA

Douglas B. Galloway
UOP LLC
50 East Algonquin Rd.
PO Box 5016
Des Plaines, IL 60017
USA

Matt Gebhardt
University of Washington
Department of Chemical Engineering
Box 351750
Seattle, WA 98195-1750
USA

Scott E. Gilbert
Crystal Vision Microsystems, LLC
65 Pine St. Unit 201
Edmonds, WA 98020
USA

Michael Grund
Merck KGaA
Frankfurter Str. 250
64293 Darmstadt
Germany

Michael Häberl
Merck KGaA
Frankfurter Str. 250
64293 Darmstadt
Germany

Michael D. Harrison
University of Washington
Department of Chemical Engineering
Box 351750
Seattle, WA 98195-1750
USA

Volker Hessel
Institut für Mikrotechnik Mainz GmbH
Chemical Process Technology Department
Carl-Zeiss-Straße 18–20
55129 Mainz
Germany

Daniel A. Hickman
The Dow Chemical Company
Engineering & Process Sciences
1776 Building
Midland, MI 48674
USA

Jamie D. Holladay
Pacific Northwest National Laboratory
902 Battelle Blvd., K6-28
Richland, WA 99354
USA

Paul H. Humble
Pacific Northwest National Laboratory
902 Battelle Blvd., K6-28
Richland, WA 99354
USA

Klavs F. Jensen
Massachusetts Institute of
Technology
Department of Chemical Engineering
77 Massachusetts Ave, MIT 66-566
Cambridge, MA 02139-4307
USA

Bryan Johnson
University of Washington
Department of Chemical Engineering
Box 351750
Seattle, WA 98195-1750
USA

Melvin V. Koch
University of Washington
Center for Process Analytical
Chemistry
Box 351700
Seattle, WA 98195-1700
USA

Gunther Kolb
Institut für Mikrotechnik Mainz
GmbH
Carl-Zeiss-Str. 18–20
55129 Mainz
Germany

Daniel J. Kraus
DuPont Company
Central Research & Development
Experimental Station
Wilmington, DE 19880-0357
USA

Suwimol Kunaridtipol
University of Washington
Department of Chemical Engineering
Box 351750
Seattle, WA 98195-1750
USA

Patrick Löb
Institut für Mikrotechnik Mainz
GmbH
Carl-Zeiss-Str. 18–20
55129 Mainz
Germany

Holger Löwe
Johannes-Gutenberg-Universität
Mainz
Department of Chemistry
Pharmaceutics and Earth Sciences
Institute of Organic Chemistry
Duesbergweg 10–14
55128 Mainz
Germany

Marlina Lukman
University of Washington
Department of Chemical Engineering
Box 351750
Seattle, WA 98195-1750
USA

Alexander Mamishev
University of Washington
Electrical Engineering
Box 352500
Seattle, WA 98195-2500
USA

Brian J. Marquardt
University of Washington
Center for Process Analytical
Chemistry
Box 351700
Seattle, WA 98195-1700
USA

Michael J. McCarthy
University of California
Department of Food Science and
Technology
Davis, CA 95616-8598
USA

Patrick L. Mills
Texas A&M University-Kingsville
Department of Chemical and Natural
Gas Engineering
700 University Blvd, MSC 188
Kingsville, TX 78363-8202
USA

Clark T. Nguyen
University of Michigan
Ann Arbor, MI
USA

Daniel R. Palo
Pacific Northwest National Laboratory
Microproducts Breakthrough
Institute
PO Box 2330
Corvallis, OR 97339
USA

Trevor Plaisted
University of Washington
Department of Chemical Engineering
Box 351750
Seattle, WA 98195-1750
USA

Dennis E. Polla
DARPA/Microsystems Technology
Office
3701 North Fairfax Drive
Arlington, VA 22203-1714
USA

David J. Quiram
TXU Energy
Strategic Planning Department
1601 Bryan Street, Suite 7-060B
Dallas, TX 75201
USA

James F. Ryley
DuPont Company
Central Research Development
Experimental Station, P.O. Box 0357
Wilmington, DE 19880-0357
USA

Dirk Schmalz
Merck KGaA
Frankfurter Str. 250
64293 Darmstadt
Germany

Martin A. Schmidt
Massachusetts Institute of
Technology
Department of Electrical Engineering
and Computer Science
77 Massachusetts Ave, MIT 39-521
Cambridge, MA 02139-4307
USA

Daniel D. Sobeck
The Dow Chemical Company
Engineering & Process Sciences
1710 Building
Midland, MI 48674
USA

Scott Soelberg
University of Washington
Departments of Medicine (Division
of Medical Genetics) and Genome
Science
Box 357720
Seattle, WA 98195
USA

Victoria S. Stenkamp
Pacific Northwest National Laboratory
902 Battelle Blvd., K6-28
Richland, WA 99354
USA

Zachary Tyree
University of Washington
Department of Chemical Engineering
Box 351750
Seattle, WA 98195-1750
USA

Jeremy VanBuren
University of Washington
Department of Chemical Engineering
Box 351750
Seattle, WA 98195-1750
USA

Kurt M. VandenBussche
UOP Process Technology and
Equipment
25 East Algonquin Rd
Des Plaines, IL 60017
USA

David J. Veltkamp
University of Washington
Center for Process Analytical
Chemistry
Box 351700
Seattle, WA 98195-1700
USA

Paul Watts
The University of Hull
Department of Chemistry
Cottingham Road
Hull, HU6 7RX
UK

Mark D. Wetzel
DuPont Company
Central Research & Development
Experimental Station, P.O. Box 0323
Wilmington, DE 19880-0323
USA

Albert Witarsa
University of Washington
Department of Chemical Engineering
Box 351750
Seattle, WA 98195-1750
USA

Hanns Wurziger
Merck KGaA
Frankfurter Str. 250
64293 Darmstadt
Germany

Part I
Introducing the Concepts

1
Introduction

Melvin V. Koch

1.1
Background

Mankind is continually investigating ways to improve their surroundings and to open new horizons, often creating new areas for research. Developing improvements in the ability to effectively manufacture goods is one of the areas to benefit from these investigations, as it achieves better product quality, less negative impact on the environment, and lower costs of production. Monitoring and control of manufacturing processes is a measure of how well these improvements are being achieved. A basic tool in this monitoring and control is Process Analytical Technology (PAT).

PAT has been an important part of industrial manufacturing for many years in the application of measurement science technology within a production environment. The first examples of using sophisticated measurement technology rather than the physical property tools of temperature, pressure, and flow to monitor a process were within the German chemical industry in the late 1930s [1, 2]. Although the ability to monitor processes with measurement technology is not new, the recognition that it will enhance productivity, quality, and environmental impact has grown markedly over the past 30 years. The beginning of PAT was driven by problem solving reasons to understand the composition of molecules of interest in the process. These measurements were often a part of a trouble-shooting effort of a process research group that was trying to optimize reaction conditions and/or understand underlying factors that influenced the success of the reaction. Initially, process analysis was carried out by taking samples from the process in the production plant and transporting them to the analytical laboratories. It often took 8 to 24 hours before receiving the results of the analysis and was it relatively expensive and inaccurate, as the sample could change during the transport and conditioning steps. With time, many analytical laboratory measurements were moved to laboratory sites located in the production plants. This approach reduced the time (10 to 60 min) for receiving the results, but it was still somewhat inaccurate due to the sampling and conditioning techniques.

Micro Instrumentation for High Throughput Experimentation and Process Intensification – a Tool for PAT
Edited by M. V. Koch, K. M. VandenBussche and R. W. Chrisman
Copyright © 2007 WILEY-VCH Verlag GmbH & Co. KGaA, Weinheim
ISBN: 978-3-527-31425-6

The early, traditional approach to implementing PAT was to modify selected laboratory analytical technology to create instrumentation suitable to the production plant environment. Significant efforts were made to obtain a representative sample from the process in such a manner that it would interface with the analytical instrumentation. The history of process analyzers is best represented by process gas chromatographs (GC). The first generation of process GCs were relatively accurate and maintainable, and normally gave analytical results in 10 to 60 min. The recent generation GCs, as with various other process analyzers today, have improved accuracy and increased speed. The next generation of GCs is expected to approach real-time response, almost like a sensor, with an expected high level of accuracy and reliability.

PAT has evolved as the proven usefulness of many measurement technologies in laboratory studies has been moving to applications of these technologies to industrial processes. PAT has also become a field of measurement science that involves several contributions from various technical disciplines, including chemistry, engineering, biology, data handling, and control strategies. The outcome has been an ability to achieve an increasing amount of knowledge about the process being operated.

Today, there is growing interest in duplicating analytical instrumentation that has been proven in the laboratory into a process environment. This is logical, since this instrumentation had been used to originally characterize the product that is being manufactured, as well as raw materials and the reaction intermediates. Typically, a series of analytical laboratory measurements are made on a new compound or new formulation to characterize the product as part of its registration or product specification summary. Thus, many analytical technologies that performed this "analytical profile" that defined the registered product and its ultimate specifications were considered as candidate technologies to be taken to the production process streams. Historically, these technologies have, primarily, included the analytical science technologies of spectroscopy, chromatography, and thermal techniques. The sample studied was primarily

Fig. 1.1 Converting a laboratory instrument into a process instrument. Courtesy Dow Chemical.

vapor or liquid, though there was some solid material handling measurements of polymeric and inorganic materials. When taken to the production environment these laboratory instruments were packaged in specially designed containment cabinets (Fig. 1.1) that were purged to maintain an inert atmosphere in order to meet the requirements for safe operation in a manufacturing environment. Subsequent transportation (Fig. 1.2) and installation of the "process-ready" cabinet into a production location (Fig. 1.3) was often a significant event, as production operations were interrupted for many hours while the equipment was transferred into the production plant. The fact that production needed to be halted was a significant inhibitor to installation of process analytical tools, in that the value of lost production had to be realized in the production planning from a business point of view. However, once the analytical instrumentation was installed and functioning, the value of the data from the instrument proved that the effort was worth the inconvenience of the cost of installation.

Fig. 1.2 Transportation of the process instrument. Courtesy Dow Chemical.

Fig. 1.3 FTIR in a production environment. Courtesy Dow Chemical.

This traditional approach of placing laboratory based instrumentation into a production environment has shown great value to the chemical and petrochemical industries for several years. It was able to provide an insight into the process that gave information on real-time composition and concentrations. It is also a very valuable way to study reaction pathways, and reaction kinetics for the purpose of understanding the reaction mechanisms. This has proven to be a very important to developing an effective process control system. It was also often possible to obtain valuable data on relevant environmental and hazard evaluation issues related to the process. Real-time analysis almost always shows something of unexpected value compared with taking a sample from a process stream and transporting it to an analytical laboratory for analysis. As mentioned earlier, such transportation often results in temperature and matrix changing conditions that alter the original sample make up, thus missing something important that is in the process stream. Similar things can also happen during conditioning the sample for analysis in a laboratory setting.

In real-time analysis, it is often possible to see fleeting intermediates and impurity build-up, thus being able to construct a realistic reaction profile (Fig. 1.4). Again, it is possible to monitor for indicators of safety and environment concern that may be missed in the additional time and sample preparation techniques required when taking the sample to the analytical laboratory.

Instrumentation that can provide data from which a reaction profile of the process under production conditions can be developed is invaluable. This helps to define the capabilities of the process and to gather data to locate critical process parameters. Once critical process parameters (including indicators of composition,

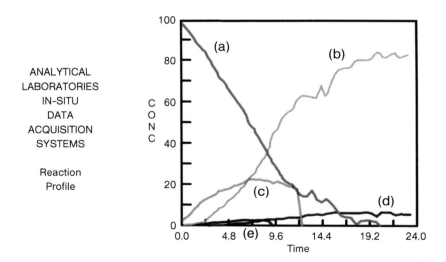

Fig. 1.4 Reaction profile data.
(a) raw material, (b) product, (c) intermediate, (d) impurity,
(e) intermediate, previously unknown.
Courtesy Dow Chemical.

physical properties, and interfacial characteristics) are identified and monitored, process models can be developed or modified with ultimate improvements in process control. PAT becomes a combination of applying process analytical tools for monitoring process unit operations and developing feedback process control strategies based on information management tools or process optimization studies for the purpose of achieving process understanding.

1.2
Analytical Tools for use in PAT

Developments in measurement science have been largely influenced by advances in the technical fields driven by needs in the military, communication, and the computer related industries. These developments have resulted in improvements in techniques for miniaturization, as reflected in MEMS (micro-electrical mechanical systems) technology.

Also, the development of new materials and new electro-optics has afforded improved instrumentation, particularly in detector systems, due to the ability to employ new construction and heat transfer techniques. There has also been valuable use of advanced data handling tools that are tied to increased computing power capabilities. These improvements have been incorporated into many new analytical instruments. Now, measurement technologies that were discovered and demonstrated years ago, but were underutilized because of instability or data gathering limitations, have been enhanced by these advances and are subsequently playing a major role in analysis today. An example of this is in Raman spectroscopy, where improvement in material development has resulted in more stable lasers, and computing advances have realized better data interpretation. With new technical capabilities being developed within analytical sciences it is important to consider how to prioritize them relative to their potential use in PAT. In considering what measurements are appropriate for PAT it needs to be recognized that any technique that gives more relevant data than is presently being obtained is appropriate for PAT. The constraints of cost and reliability of the data will enter into the decision of what instrumentation can be effectively used, but it must be recognized that more measurement points usually allow for better models and this results in better control strategies.

As the market for process analytical tools expands, the analytical instrument vendors will need to develop and market instruments that are conducive to responding to the goals of PAT. The instruments should be flexible, modular, and have the capability to be integrated into analyzer systems. Often the term analyzer system is thought to be a device that responds to the analyte of interest. However, the term analyzer system needs to be defined and understood in a process related environment, as it is much broader than just being able to detect the analyte. The analyzer system needs to incorporate sampling, sample conditioning, detection, data gathering, and information generation. Measurement technologies applied to processes also need to expand from compositional characterization to also include more physical characterization tools. This is increasingly important as it is necessary to also develop measurements that correlate with, or predict, final product properties. This approach, called "inferential analysis", has become a key processing and marketing parameter. Properties such as material science (rheology) characterization, physical properties, and interfacial characteristics are being used increasingly for this purpose. This has led to an emphasis on measurement science to characterize product properties differently, and often beyond those laboratory technologies that are traditionally used to determine the "analytical profile" for product definition and submission for registration.

The field of Process Analysis, and "inferential analysis" in particular, has received a huge boost in importance by the emphasis that the Center for Drug Evaluation and Research (CDER) within the US Food and Drug Agency (FDA) has given to the use of PAT in the pharmaceutical and biotechnology industries [3]. The FDA realizes that using PAT tools under the recently reissued Current Good Manufacturing Practice (CGMP) guidelines [4] will help to achieve process understanding and result in improved process control strategies for good quality and cost effective pharmaceutical products.

In the FDA Guidance on PAT and in resulting global efforts (ICH Q8 and the ASTM E55 committee), a definition of PAT has been formulated. Presently it is: Process Analytical Technology – a system for designing, analyzing, and controlling manufacturing through timely measurements (i.e., during processing) of critical quality and performance attributes of raw and in-process materials and processes with the goal of assuring final product quality.

Actually, PAT has been used selectively within the pharmaceutical industry for over 25 years [5]. This activity has been largely for API (active pharmaceutical ingredient) production and monitoring separation operations, drawing from the technical successes experienced by the chemical related industries in monitoring chemical and petrochemical production. However, with the FDA emphasis on the use of PAT in the pharmaceutical industry is expanding rapidly into the unit operations associated with formulated pharmaceutical production (including milling, granulation, mixing, blending, drying, and compaction) and in operations within the biotechnology industry. Much of the present PAT related discussion across the pharmaceutical and biotechnology industries involves learning about what new measurement technologies are available and evaluating these technologies versus prioritized measurement needs. An important, and often overlooked, subject with PAT is the cost of installation and the subsequent cost of ownership of the instrumentation versus the benefit of the measurement. As the regulatory agencies emphasize the importance of PAT, PAT understanding and education are needed within each organization that will practice the approach. This includes not only the functions that will develop the analytical methods, install, and use the data that results from the measurements, but often more importantly, the other industrial functions of corporate and site management, QC, QA, and regulatory affairs personnel. PAT has become a business driven initiative within the pharmaceutical and biotechnology industries, with a growing appreciation of how important it can be to cost, quality, and regulatory concerns.

The types of instrumentation presently being evaluated and implemented in PAT for pharmaceutical and biotechnology applications involve a broad range of analytical tools, including vibrational and scattering spectroscopies, chromatographies, mass spectrometry, acoustics, chemical imaging, light-induced fluorescence, and light scattering. Many other technologies like Raman spectroscopy, optical coherence, optical scattering, dielectric spectroscopies, and NMR are showing increased value today, as improvements in the stability of optics, new materials, and computing strength have been incorporated into the instrumentation. As this list of possible PAT tools grows there will be expanded career opportunities for analytical chemists and engineers associated with modifying, implementing, and interpreting these technologies.

1.3
The Center for Process Analytical Chemistry (CPAC) and the Summer Institute

It is very difficult for any one corporate organization to remain abreast of all of the technical developments in PAT that are occurring across research operations in industry, government, and academia. Limited resources of funding and time have caused organizations to look outside their company for help. The concept of using a consortium approach has value in that the members of a consortium can combine their resources to fund various research projects. This leveraged approach allows them to evaluate a broad set of technologies and then select if and when to concentrate on those projects that meet their immediate needs. Several consortia have been established to address the areas that are of interest to the field of PAT. One such consortium that is concentrating on PAT has been the Center for Process Analytical Chemistry (CPAC) at the University of Washington in Seattle, Washington (www.cpac.washington.edu). CPAC was formed in 1984 for the purpose of highlighting PAT by taking advantage of the availability of low cost microprocessors that presented a unique opportunity to transform and/or upgrade the analytical and control functions in industrial processes. CPAC has provided valuable services and resources to sponsor organizations (multi-industry and government groups) with an interest in real-time analysis and control. These services have included university research projects in selected technical fields (Fig. 1.5), training of graduate students and postdoctoral research associates, semi-annual project review meetings for the members, the CPAC Summer Institute for special topics in PAT, technical workshops, and a visiting scientist program (Fig. 1.6). In its first 20 years CPAC graduated 128 PhDs and published in excess of 400 scientific papers. A key benefit of the way CPAC is structured is its fostering of communication by bringing together individuals from across a wide spectrum of industries with an interest in process analysis (Fig. 1.7). CPAC was initially influenced by the interests of the existing base of process monitoring strength in the chemical and petrochemical industries, but, with time, it was realized that the process analysis needs of these industries were actually mirrored by the process analysis needs in other industries such as materials, pharmaceuticals, food, and biotechnology.

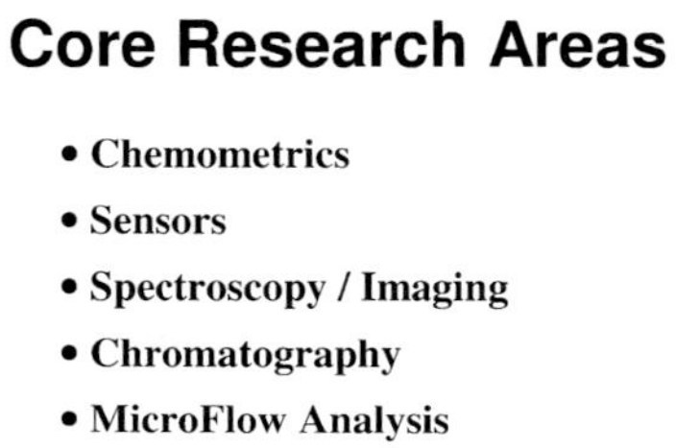

Core Research Areas

- Chemometrics
- Sensors
- Spectroscopy / Imaging
- Chromatography
- MicroFlow Analysis

Fig. 1.5 Technical fields covered by CPAC.

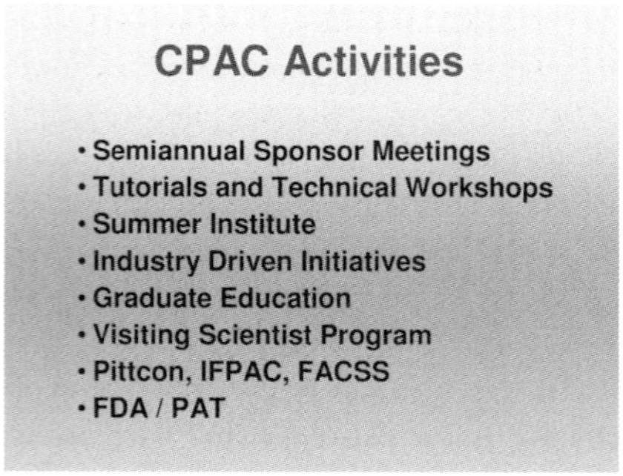

CPAC Activities

- Semiannual Sponsor Meetings
- Tutorials and Technical Workshops
- Summer Institute
- Industry Driven Initiatives
- Graduate Education
- Visiting Scientist Program
- Pittcon, IFPAC, FACSS
- FDA / PAT

Fig. 1.6 List of CPAC activities.

Sponsor Focus Groups

- Oil & Petrochemicals
- Chemicals & Materials
- Food & Consumer Products
- Bio Tech / Pharmaceuticals
- National Labs & Agencies
- Instrument Companies

Fig. 1.7 CPAC fosters communication by bringing together individuals from a wide spectrum of industries with an interest in process analysis.

At CPAC member meetings an atmosphere has been established that allows for peer communication, where researchers interested in real-time measurement that are from different industries can discuss how various technology developments can impact on common problems. For example, a discussion may occur on a new technology that can characterize the degree of uniform mixing of materials. One industrial researcher may wish to apply the newly developed technology to the manufacturing of a polymer system while another might be interested in meeting regulatory requirements in a pharmaceutical formulation. The needs, ideas, and technology in PAT often span the products of many industries but the researchers in a given industry tend not to – until they become involved with a consortium like CPAC. CPAC has sustained itself by providing meaningful university developed research results and creating a forum for communication among its members, while evolving with the needs of industry. During the past several years CPAC has averaged 30–35 industrial sponsors and a significant number (about one-third) of these are involved with the pharmaceutical/biotechnology industry sector. CPAC personnel have participated in FDA committees associated with the FDA Guidance on PAT and in training of the FDA reviewers and inspectors who concentrate on PAT submissions to the FDA. This increased involvement from the pharmaceutical sponsors and government agencies has encouraged new research initiatives in the development of the micro-instrumentation for high throughput experimentation and in fermentation monitoring and control.

In 1995 CPAC developed strategic objectives to address the key problem areas of measurement science in industry. The problem areas identified were: (a) surface and interfacial characterization and control (largely a solids handling concern), and (b) multicomponent materials characterization and control (involving composition and spatial concerns). The approach identified to address these problem areas for future CPAC research was Versatile Micro-analytical Systems (VMAS) for Process Monitoring, Modeling and Control. This area of research concentration was selected largely due to the impact, as mentioned earlier, that miniaturization advances has had on developments in the computing and communications fields. As a way to validate the choice of emphasizing developments using VMAS, it was decided to bring together a group of multidiscipline researchers in the summer of 1996 in an informal setting that was called the CPAC Summer Institute. The results of that meeting showed a strong support for the concept of VMAS and a keen interest to further define it.

From that beginning in the summer of 1996, CPAC has conducted a Summer Institute each summer and has focused it around the theme of the value of miniaturization of analytical systems. This theme was adopted and embellished as it was believed that micro-systems would provide the modularity and flexibility needed to improve laboratory and process control operations. The CPAC Summer Institute has evolved into a venue for gathering engineers, measurement scientists, micro-instrumentation vendors, and data handling specialists for brainstorming on how to merge micro-instrumentation developments with measurement science and engineering needs. The brainstorming atmosphere includes multidisciplinary faculty from the University of Washington campus (including chemistry, electrical engineering, chemical engineering, mechanical engineering, bioengineering, forestry, and medical genetics). There are also speakers and participants from industry, national laboratories, international institutes, and other academic institutions.

The CPAC Summer Institute has proven to be a very productive approach to validating technology concepts and discussing how these concepts could assist in solving industrial problems. The atmosphere of the Summer Institute is informal. The gathering is structured over a two and one-half day time period. The first day has a few plenary talks on the title topic(s) and a series of shorter presentations (Fig. 1.8). The presentations are state-of-the-art and often leading edge research. By mid-afternoon a discussion topic has been outlined and the total audience is encouraged to participate in a roundtable, brainstorming fashion (Fig. 1.9). The first day meetings are held on the campus of the University of Washington, and broad student and faculty participation is encouraged. The academic participants benefit from hearing industrial and government developments along with a selection of technical concerns being described. It has been extremely valuable

Fig. 1.8 A presentation on the first day of the Summer Institute.

Fig. 1.9 Round-table debate on a chosen topic.

for students to hear the various viewpoints of industry and to interact with global leaders in these fields. It has been a useful resource to industrial participants as a form of continuing education and as a vehicle for benchmarking. Instrument vendors have seen the Summer Institutes as a way to learn of evolving instrumental needs and technologies that could be solutions to those needs.

For the past several years the second day of the Summer Institute has been held at a mountain retreat in the nearby Cascade Mountains (Fig. 1.10). A series of short technical presentations are delivered (Fig. 1.11) and by mid-afternoon an informal "flip chart" session is conducted, where participants add their comments to a series of discussion topics that have been developed during the first and second days of the CPAC Summer Institute.

Fig. 1.10 View from a mountain retreat in the Cascade Mountains – location of the second day of the Summer Institute.

Fig. 1.11 One of a series of short technical presentations.

The participants interact freely in the open-air atmosphere followed by an interactive barbeque dinner (Fig. 1.12). The multidisciplinary make-up of the group is critical. The disciplines represented are normally those of chemists (organic, analytical, and physical), engineers (process control, reaction engineering, mechanical, electrical, and biological), health science researchers, and computer science and information technology. It is quite unusual and very productive to bring together such a technically diverse group of scientists in a problem solving, brainstorming forum.

The morning of the third and final day of the Summer Institute involves some selected plenary technical talks, a summary of the "flip charts" comments, development of action plans to work on during the next year, and a roundtable critique of the value of that Summer Institute event (Fig. 1.13).

Fig. 1.12 An opportunity for informal interaction

Fig. 1.13 Scene from third and final day of the Summer Institute.

1.4
Topics covered by Previous CPAC Summer Institutes

The Summer Institute has been an excellent way for CPAC to demonstrate the value of peer communication and of multi-industry, multidiscipline project involvement. It is also a very good outlet for CPAC researchers to describe their research in a setting focused on a selected meeting theme. Several companies have subsequently joined the CPAC consortium as a result of participation in the Summer Institute and observing the value of the atmosphere for peer communication.

The early CPAC Summer Institutes (1996–1997) highlighted a multidisciplinary problem that was often not recognized, and for which there is no easy solution. Engineering curriculum does not stress measurements to define concentration or composition of a process stream, but rather primarily the measurements of flow, pressure, and temperature. Measurement scientists and chemists often are developing tools to measure composition and pushing to have production personnel use them for process monitoring. As part of the discussion at the 1997 CPAC Summer Institute, Professor Charlie Moore, of the University of Tennessee's Chemical Engineering Department, pointed out that engineering education does not emphasize analytical instrumentation that measures composition, but rather how to use basic measurements of flow, pressure, and temperature in a production plant setting. It was rationalized that this is because engineers have valves in their plants to control flow, pressure, and temperature – and usually no valve for controlling composition. Often, compositional and concentration measurements are made for quality, alarm, and/or archival purposes. It was furthered emphasized that very few, if any, algorithms have been developed for converting the results from compositional measurements into what the changes need to be made in that which the engineers can control (flow, pressure, and temperature). It was very enlightening for the chemists and measurement scientists who participated in this discussion to realize the constraints that plant engineers have in dealing with compositional control issues.

The realization that the engineering curriculum did not emphasize the importance of compositional measurements, nor did it provide a base for developing algorithms for connecting the results of this compositional data to control parameters (flow, pressure, and temperature), has catalyzed a next generation of PAT. One example of this is where a fiber optic based insertion probe was developed for observing onset of crystallization in a production environment. This observation evolved into the development of algorithms that controlled the parameters of the crystallizers, resulting in control of the unit operation and subsequent production of the desired crystal product characteristics. Further refinement of the data generated by this probe has allowed its use in following the dynamics of processes where the particles are changing in size and shape during the process – as is observed in a fermentation process [6]. The next generation of PAT will incorporate more ability to utilize feed-back and feed-forward approaches to process control.

In the late 1990s (1998–1999), the CPAC Summer Institute focused on the advances in miniaturization that were influencing areas that are related to microanalytical and micro-instrumentation developments. These included presentations and discussions on developments in micro-unit operations (microfluidics, micropumps, microfilters, etc.). A concern that continues to arise in these discussions was the question as to how one would sample and subsequently monitor these micro-operations. The topic of sampling has always been an area of concern to the traditional macro-operations as well. Sampling seemed to be a major problem in almost every PAT application discussed. An action item at these Summer Institutes was to address potential sampling system improvements that could work across industries.

In the 2000–2002 CPAC Summer Institutes the concepts of microreactors and how they were influencing the field of high throughput experimentation was presented and discussed. The discussion expanded to how the microreactors could assist in process optimization studies. In addition, there was much discussion surrounding the concept of using microreactors in a number up mode. This mode is where the microreactors are duplicated in a parallel processing approach, as compared with the traditional scale-up approach. These discussions led to the establishment of an action item at the end of SI 2002 of "what do the engineers need in order to design, operate, and optimize processes". This was the reason for introducing the concept of process intensification into the 2003 Summer Institute [7]. During this same period of 2000–2002 a new sampling concept was elucidated. In 1999, CPAC industrial focus groups established improvement in sampling technology as one of industry's key needs for enhancing developments in PAT. As a result, CPAC has served as a focal point for the development of the New Sampling and Sensor Initiative (NeSSI) that is now ANSI/ISA standard 76.00.02 [8], which defines the footprint and standardizes the flow geometry of components and interconnects that make up the platform. The use of NeSSI as a sampling system for traditional process analyzers and in process development activities has already begun to show value. Chapter 10 describes the development of the NeSSI platform for process sampling and its evolution into a platform for micro-analytical devices.

1.5
Recent Emphasis of CPAC Summer Institutes: High Throughput Experimentation and Process Intensification

The recent emphasis (2003–2006) at the CPAC Summer Institute has been on how micro-instrumentation can impact high throughput experimentation and process intensification. It was the unique way that the multi-industry, multidiscipline approach at the CPAC Summer Institute to this topic that captured the attention of John Wiley Publishing and the subsequent request to put a book together on this complementary compilation of the relevant subjects. High throughput experimentation is now being used for more than combinatorial synthesis purposes (Fig. 1.14). It is demonstrating value in several functions.

Process intensification is, preferably, based on a sound understanding of the underlying fundamentals of the process and it has gained importance as a means

High Throughput Experimenation is Being Used For:
- Discovery and screening
- Process optimazation
- Process development
- Process control
- Production

Fig. 1.14 Uses of high throughput experimentation.

to reduce capital and operating expenditure, as described in Chapter 3. Process intensification techniques have achieved drastic reductions in resource use and waste generation while maintaining productivity as well as the quality of the desired product [7]. Microtechnology (microreactors, etc.) has significantly impacted this field and this will be described further in Chapter 6.

Both high throughput experimentation and process intensification are fields that have benefited from technical advances in hardware and related automation technology and now are generating large numbers of small volume samples for product development and process optimization studies. Each of these areas will be described in the following chapters. There have been many significant developments recently in the technology for microreactor hardware that is being used for high throughput experimentation and process optimization purposes. Various organizations, including IMM, mikroglas, Cellular Process Chemistry (CPC), Argonaut, and Symyx, now offer commercial systems. In addition, many companies within various industries, government laboratories, and academic groups have also developed microreactor systems for their internal use. Each system has unique fluidic and inter-connection approaches. Since there were particular advantages to some of the unit operations within each system, an effort was started by MicroChemTec, a group of organizations within Dechema, [9, 10] to develop a "backbone" (Fig. 1.15) that will result in a way to standardize the connections of several microreactor and related unit operation components, illustrated below.

However, these advances in micro-hardware devices often have not addressed how the various unit operations will be monitored and controlled. This void is

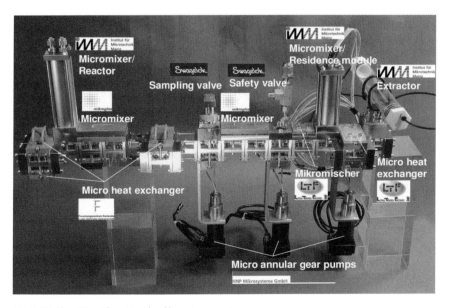

Fig. 1.15 The MicroChemTec "backbone" concept.

creating a demand for miniaturized measurement devices, scanning sensor devices, and rapid data handling that will grow with the need to analyze an increasing number of samples. These analytical tools need to mesh with high throughput experimentation platforms as well as microreactors and other MEMS based developments.

CPAC has responded to the need for micro-analytical measurement devices by focusing much of its present research effort in this direction. Over the past decade CPAC has funded developments in the miniaturization of several technologies (Fig. 1.16) and has sponsored workshops and tutorial sessions on developments both inside and outside of CPAC member meetings.

The CPAC sponsored workshops and tutorials on the topics surrounding the needs and developments in micro-analytical technology have resulted in new initiatives being pursued at CPAC (Fig. 1.17).

Some relevant CPAC research efforts in mini-instrumentation

- Micro-LC
- Surface tension detection
- Gas chromatography
- Fringe field sensors
- Vapochromic sensors
- Raman
- Grated light reflective spectroscopy
- Reflectometry
- Flow-through particle analyzer
- Surface plasmon resonance
- Mini-NMR
- Ultrasound
- Data handling

Fig. 1.16 Some relevant CPAC research efforts in mini-instrumentation.

CPAC INITIATIVES

OPEN
- *NEW SAMPLING AND SENSOR INITIATIVE (NeSSI)*
- *CHEMOMETRICS ON-LINE INITIATIVE*
- *DEGRADATION MONITORING AND CONTROL*
- *FOOD SAFETY, SECURITY, AND NUTRITION*

CLOSED
- *MICRO-INSTRUMENTATION FOR HIGH THROUGHPUT EXPERIMENTATION PLATFORM*
- *FERMENTATION MONITORING AND CONTROL PLATFORM*
- *SUSTAINABLE CHEMISTRY VIA BIO-BASED RESOURCES*

Fig. 1.17 CPAC initiatives.

The initiatives are normally lead by industry and serve as platforms or a base on which to direct research activity. These platforms are then focus areas for the research and development of new measurement technologies at CPAC.

Chapter 9 covers several of the projects indicated in Fig. 1.16. CPAC's Summer Institute and workshops at CPAC member meetings have focused on micro-analytical device research, as well as highlighting those micro-analytical devices that have been commercialized, creating a forum for discussing how well the needs of industry are being met by these developments.

As miniature analytical devices have been developed it has not always been obvious how they would be used or even demonstrated in laboratory or process applications. A test bed for demonstrating these small instruments was needed. Often, the successful commercialization of new mini-analytical concepts was delayed due to a void in the commercialization process. This void occurred when a proof-of-concept on a new micro-analytical measurement technique was demonstrated at a laboratory level, but then these results were difficult to compare with those from the traditional measurements in a production setting. However, a pivotal observation/development occurred. In addition to serving as an improvement to traditional sampling systems, the NeSSI platform (introduced earlier and described in Chapter 10) was shown to be an ideal template for demonstrating new micro-analytical devices. It is becoming apparent that the process analyzer does not need to be a large or laboratory level modified instrument. Besides interfacing with a process stream, a NeSSI platform containing analytical instrumentation (Fig. 1.18) appears to be capable of being used in several valuable laboratory based applications. As analytical technologies are developed and/or adapted to be NeSSI compatible, the number of potential applications for a flexible/portable analytical laboratory on a NeSSI platform will increase.

Over the years, most comments by the CPAC Summer Institute participants have been centered on the value of the unique nature of the discussion that occurs in a multidiscipline, multi-industry group. The subjects are meaningful and the past successes resulting from the CPAC Summer Institute are impressive. Figure 1.19 summarizes some of these discussion topics.

The CPAC Summer Institute has been an ideal forum for presenting and discussing these valuable concepts.

Fig. 1.18 NeSSI platform with sampling components and sensors.

- Micro-analytical devices and sensors for the future
- The New Sampling an Sensor Initiative (NeSSI) platform for demonstrating and deployment of micro-analytical devices
- Using micro-analytical devices to monitor micro-reactors
- Process intensification

Fig. 1.19 Successful discussion topics at CPAC Summer Institutes.

1.6 Conclusion

There have been many key technical developments in the field of Process Analytical Technology (PAT) over the six decades that it has been practiced in industry. As the number and importance of these developments increase, it is necessary to understand and prioritize them in a timely manner. The use of academic consortia to leverage resources and to serve as a benchmarking and peer communication vehicle is proving to be a cost-effective manner to accomplish this in a resource challenged research environment. The CPAC Summer Institute has proven to be a valuable forum for discussion of industry concerns and new technology concepts. It has become a successful annual event at the University or Washington that explores the impact of advances in micro-instrumentation on high throughput experimentation. The recent CPAC Summer Institute theme of "Micro-instrumentation for High Throughput Experimentation and Process Intensification" is a dynamic topic that is showing continued technical advances and improved positive impact on PAT.

References

1 T. McMahon, E. L. Wright, *Analytical Instrumentation: A Practical Guide for Measurement and Control*, ed. R. E. Sherman, L. J. Rhodes, Instrument Society of America, Research Triangle Park, N.C., **1996**.

2 C. H. Gregory (Team Leader), H. B. Appleton, A. P. Lowes, F. C. Whalen, *Instrumentation & Control in the German Chemical Industry*, British Intelligence Operations Subcommittee Report #1007 (12 June 1946) (per discussion with Terry McMahon).

3 *Guidance for Industry, PAT – A Framework for Innovative Pharmaceutical Development, Manufacturing, and Quality Assurance*, U.S. Department of Health and Human Services, Food and Drug Administration, Center for Drug Evaluation and Research (CDER), Center for Veterinary, Medicine (CVM), Office of Regulatory Affairs (ORA), Pharmaceutical CGMPs, September **2004**. http://www.fda.gov/cder/guidance/6419fnl.htm

4 *Pharmaceutical CGMPs for the 21st Century – A Risk-Based Approach*, Final Report, September 29, **2004**. http://www.fda.gov/cder

5 R. E. Cooley, C. Egan, The impact of process analytical technology (PAT) on pharmaceutical manufacturing, *Am. Pharm. Rev.* **2004**, 7(1), 62.

6 C. A. Snook, Dow Chemical, presentation, Council for Chemical Research workshop on PAT, December 13, **2005**.

7 K. Vanden Bussche, UOP, presentation, CPAC Summer Institute 2004 report, (www.cpac.washington.edu).
8 R. D'Aquino, The incredible shrinking sampling system, *Chem. Eng. Prog.*, December **2003**, 9.
9 A. Müller, V. Cominos, B. Horn, A. Ziogas, K. Jähnisch, V. Grosser, V. Hillmann, K. A. Jam, A. Bazzanella, G. Rinke, M. Kraut, *Chem. Eng. J.* **2005**, 107, 1–3, 205–214.
10 A. Müller, V. Cominos, V. Hessel, B. Horn, J. Schürer, A. Ziogas, K. Jähnisch, V. Hillmann, V. Großer, K. A. Jam, A. Bazanella, G. Rinke, M. Kraut, *Chem. Ing. Technik* **2004**, 76(5), 641–651.

2
Macro to Micro ... The Evolution of Process Analytical Systems

Wayne W. Blaser and Ray W. Chrisman

2.1
Introduction

Analytical instruments have been tied to chemical processes for decades. Process engineers were quick to realize that "eyes into the process" were key for following process conditions and to quickly determine when process irregularities were occurring. Fortunately, to support this growing realization there were several important technical innovations that completely revolutionized process analytical technology (PAT). Just as past technology innovations have had an impact on the growth of process analytical instrumentation, new technology developments that are now evolving in the micro-instrumentation area are expected to have a significant impact on the growth of the field into the future. These past innovations and their impact on the field are outlined in this chapter. In addition some key developments in the area of microflow path devices and their potential relationship to process analytical will be introduced and lay the groundwork for more detailed discussions in subsequent chapters.

2.1.2
Early Developments

The limitations of early instrumentation were extreme. Engineers were first able to get on-line measurements for only the simplest process parameters – initially only temperature and pressure. As interest in on-line measurements grew, various manufacturers developed instrumentation for chemical process streams. Soon, instrumentation was available for simple infrared measurements, water determination, simple ultraviolet measurements, density, and viscosity. The spectroscopic instruments (IR, UV, etc) of this era were simplistic devices similar to colorimeters that used filters to provide the appropriate frequency or range of frequencies to quantitate one or more components of the process stream.

A major step forward was achieved in the late 1960s when the gas chromatograph became a laboratory workhorse and was modified for process areas. The petroleum industry was the primary beneficiary of this technology since many of their process streams were gaseous, easily vaporized or of mixed phase, which could be easily separated and the gas phase generated readily analyzed. Reliable sampling of chemical streams has always been the limiting factor of most on-line instrumentation systems and was the most challenging factor in the installation of early process analytical systems. Early process gas chromatographic systems had detectors that were limited to variations of the flame ionization detector and the thermal conductivity detector. When detection sensitivity was not a major concern, thermal conductivity was the detection system of choice. This detector required two fewer gases and operated in an environment that was simpler to house for Class I, Div 2 classified areas (potential of a flammable gas in the area), which covered most of the areas of a petrochemical based process. This detector became increasingly popular when dual filaments were utilized and the sensitivity was considerably increased due to the increased stability of the compensated filament system. In many cases the need for calibration was minimal and area percent of the peak areas was close enough to weight percent to follow process conditions.

In the late 1960s and early 1970s chemical engineers increasingly turned to computer control to tighten parameters on chemical and petrochemical processes. This strategy proved useful for both batch and continuous processes. However, to achieve maximum benefit from the computer control, real-time feed back on process parameters was a critical need. Thus, a high priority for on-line analytical devices developed in the chemical and petrochemical industries. The same computer revolution that changed the nature of chemical processes also impacted and changed the face of process analytical systems. Computers were first implemented in laboratory instruments. One of the first to benefit from this technology revolution was Fourier-transform infrared spectroscopy. The instruments themselves were not significantly more accurate and definitely not more rugged. It was the interfacing with computers that allowed much better data transformation and utilization of the raw spectroscopic data. The computer revolution led to the development of more sophisticated and reliable analytical devices. Improvements included improved gas chromatographic instruments, detectors with higher sensitivity and stability, more sophisticated spectroscopic instruments, and hardened mass spectrometers. Mass spectrometers proved especially useful in process areas containing streams that were predominantly gaseous in nature (or very volatile liquids). A second area where on-line mass spectrometers proved especially valuable was in the analysis of process streams that existed at sub-ambient pressures. Streams that previously had been difficult to sample properly without contamination of the process stream itself were now easily sampled by the high vacuum existing in the mass spectrometer. A third area, which lent itself very nicely to analysis by the mass spectrometer, was environmental air monitoring. Many of the existing chemical processes utilized toxic volatile intermediates or generated toxic products. To safeguard the workers and the environment, in general, many companies utilized mass spectrometers as pseudo-on-line area monitors. The result was a very sensitive, specific analytical technique for low level volatile organic materials.

2.1.3
Developments since 1980

The early 1980s brought a new impetus to on-line analytical technology. These were the years in which the prices of energy and oil feedstocks began skyrocketing. Since the chemical industry is a cyclic business, and the early 1980s were the start of the low ebb of the cycle, many chemical companies began scrambling to fine tune their processes to obtain the maximum yield of prime product and minimize their waste and off-prime production. On-line analytical data was crucial to fully understand a process and to provide insight when parameters of the process were changed. Since many of these studies were designed for short-term investigation and were to be terminated when the optimum process conditions were developed, a new type of on-line analytical instrument was developed for these studies. As commercial instrument manufacturers were slow to respond to this changing need, several chemical companies began developing their own short-term on-line analytical instruments to better understand their process issues (see Fig. 1.2 in the preceding chapter).

One of the leaders in this area was the Dow Chemical Company, which developed the technology and built many different types of analytical instrument systems for this purpose. Many of these short-term studies simplified the reliability needs usually required for on-line systems. Normally, when an on-line system was designed for permanent installation on a process, the sampling system was developed to maximize reliability with minimal routine maintenance. Now, under the short-term study needs, this high reliability was desired, but not a critical need. Since the process investigation usually involved several research chemists and process engineers to define and vary critical process parameters, these people were intimately involved in the studies and were more tolerant of maintenance over several hours or days to keep the sampling systems functioning. The critical need in this case was to ensure that the sample being obtained was representative of the process being studied. Long-term reliability was not a major issue. Another advantage of this type of short-term study was that new and innovative approaches to sampling systems could be tried and evaluated. Thus, a rare opportunity to implement sampling approaches to obtain experimental data allowed these sampling strategies to be evaluated for longer-term application in other process areas.

Several categories of analytical instruments were modified for continuous analyses in a process area. This usually involved interfacing the primary analytical instrument with a computer system, which would control the instrument, obtain the raw data, apply any calibration factors necessary and finally provide the chemist or engineer with the pertinent data in graphical and tabular form. A few of these instruments, such as gas chromatography, were already well accepted in the process areas and required only minimal development to capture and present the data in the desired form. Other instruments had rarely, if ever, been introduced in the process area and required extensive development for appropriate packaging as well as computer control, data acquisition, and graphical presentation. Examples of these technologies included Fourier-transform infrared spectroscopy, liquid chromatography, near-infrared spectroscopy, mass spectroscopy, and Raman spectroscopy.

The challenge with most of these analytical techniques was to develop a process interface that would take an appropriate sample and condition it for introduction into the analytical device. In a few cases, the sampling components required were commercially available and merely required assembly in the appropriate configuration. This was true for most chromatography and mass spectrometry process interfaces. In other cases there were no reliable leak-free sampling components commercially available that could be introduced in the process area. In several such cases, the design of laboratory grade sampling components had to be modified to ensure a leak-free reliable operation in plant environments, and these hardened components had to be fabricated, many times out of exotic materials such as titanium or tantalum, before the instrument could be utilized. In addition to hardening the sampling interface components, the analytical instrument itself was typically originally designed and sold for laboratory use and, prior to its introduction in the process area, a suitable purged enclosure was designed and fabricated. The enclosure served several purposes: (a) to allow the instrument to be taken into a process area where volatile organic materials could or were typically present. The purged enclosure would prevent the instrument from producing any sparks or hot surfaces, which could conceivably ignite the vapors present in the environment; (b) to protect the analytical instrument from any corrosive materials, which would damage the electronics or sensitive optical components; and (c) to provide stability for vibration or temperature sensitive instruments. The preferred purge gas is dry instrument air, which is typically present in most process areas. In some cases, where the system was sensitive to oxygen, the purge gas employed was dry instrument grade nitrogen. In many cases the prototype process instruments provided such valuable data and were so reliable that they were utilized for permanent process analytical systems after short-term studies. In a few instances the prototype was marginal only because of the design of the sampling system. After a few iterations of redesign, the shortcomings of the sampling system were resolved and the entire system was reliable enough for permanent installation.

The emphasis on process optimization generated another opportunity for the utilization of short-term on-line analytical instrumentation. In many cases, the optimization strategies were not well suited for investigation in the full-scale process equipment. Many of these experiments, particularly those where the raw materials used were very expensive or those where the experimental design had the likelihood for producing large amounts of off-spec and unusable product, were better suited for smaller scale equipment such as that found in a pilot plants or research laboratory environments. These smaller scale experiments still required real-time analytical data for optimum results. The short-term on-line systems were ideally suited for these applications as well, yielding valuable process data in real time (or pseudo-real-time) to aid researchers in making critical experimental decisions. The only equipment modifications usually necessary on going from plant scale to pilot plant or research laboratory scale was the size of the sample taken. Laboratory experiments in small process reactors required samples sufficiently small so as not to alter the experimental conditions due to continuous withdrawal of the material being sampled.

Typically it was a rather simple project to scale down the sampling equipment size to maintain experimental integrity. (R. A. Bredeweg, Dow Chemical Company, described the in-situ process analytical chemistry techniques for both process monitoring and laboratory/pilot plant monitoring at several of the spring and fall annual meetings of the Center for Process Analytical Chemistry at the University of Washington from 1993 through 1996.)

Chemometric research and developments in the late 1980s and early 1990s brought about significant improvements for data mining from process analytical systems. Dr. Bruce Kowalski at the Center For Process Analytical Chemistry at the University of Washington was a leader in the implementation of statistical and chemometric techniques to deconvolute more useful information from spectroscopic data. Previously, techniques such as near-infrared (NIR) had been used for some on-line applications, but their utility was rather limited due to the rather nondescript nature of the spectra obtained. By applying chemometric techniques to these spectra, Kowalski demonstrated that these on-line instruments could have much broader application, yielding useful data.

These findings created an increased market for this type of spectroscopic instrument and spawned an interest in the commercial production of several on-line process NIR instruments. [For example, the revolutionary PIONIR(TM) 1024 Process Near Infrared Analyzer provided new levels of accuracy, intelligence, speed, stability, industrial ruggedness, and reproducibility in on-line measurements. It was developed from concept to reality for the petroleum and chemical industries through collaborations among Perkin-Elmer, the Center for Process Analytical Chemistry (CPAC), and a major refining and petrochemical corporation.] These instruments were popular with the petrochemical industry due to their simple construction and capability for obtaining a broad range of data. These instruments likewise gained popularity by the development of several fiber optic probe assemblies that allowed an extremely simple process–instrument interface that was leak free and reliable. During this time the availability of high quality glass fibers increased dramatically and many researchers devoted significant effort to develop probe assemblies using these fibers. In some cases, the probe assemblies were used to transmit the source light to the reactor or sample site and additional fibers were used to collect the light transformed by the sample. Interfaced to the appropriate spectrometer, useful analytical data was obtained with the simple interface. These probe assemblies were classified as intrinsically safe and could transmit light over relatively long distances. Thus, the probe assembly could be installed in a process area classified as hazardous and the fibers could be strung to an area in a nearby control room or laboratory that was not hazardous. Thus, the spectrometer/light source portion of the instrument avoided the necessity of hazardous area packaging while the benefit of the instrument was realized in the hazardous classified area. In other applications, those process reactors utilizing strong light sources for reaction catalyzation did not require fibers bringing light to the reactor and only required one or more fibers to collect light from the reactor. In these cases, typically, the attenuation of the original light in the reactor was calibrated to reflect the concentration of one or more of the reactants or products. Again, these process analytical devices proved to be very reliable, requiring minimal maintenance for cleaning or repair.

The advent and development of microfabrication techniques in the mid-1990s and early 2000s was another milestone affecting the nature of on-line process analytical devices. This led to a much better understanding of microflow technology and how it could be utilized in microreactors and micro-analytical devices. Although the micro-analytical devices have not been fully embraced and installed in many on-line applications, the ground work is being laid and it is fully expected that these devices will be sufficiently hardened to function reliably in the process environment. This technology will be discussed more fully at the end of this chapter and in greater detail in Chapter 9.

2.1.4
Sampling Systems

The heart of any on-line process analytical system is the interface with the process, i.e., the sampling system. These are basically the analytical components that transform an aliquot of process stream from its native state to a form that will promote long-term stability and reliability with the analytical instrument or instruments installed. Each of the components is connected to other components of the system with tubing of appropriate material. The tubing and components are typically all rigidly fastened to a metal or plastic plate so that the components and connecting tubing are protected from exterior trauma. This rigid plate is then mounted in a convenient location either close to the process or to the analytical instruments being utilized. The location is strongly determined by the nature of the process stream and whether the integrity of the components is better preserved when the sampling system is near the process or the instrument. Each occurrence has to be extensively investigated prior to installation to determine the best location.

The sampling panel is typically protected in one of two manners. In cases where the routine maintenance can be conveniently carried out in the outdoor environment, the sample panel is typically mounted within a "box enclosure" of appropriate size. Several manufacturers sell boxes expressly designed to house sampling panels. These are typically fabricated from fiberglass or custom manufactured from metal, usually mild steel that is then protected by epoxy paint coatings. These boxes are generally purged with a clean gas for one of several purposes. Most typically, the boxes are purged with instrument air or plant nitrogen to keep the components clean and minimize, if not prevent, corrosion caused by contaminants in the plant environment. Obviously, if nitrogen is used, the working environment around the purged box must be monitored to ensure that there is not a nitrogen build-up in the area that would make working in the area dangerous for maintenance personnel. This is most conveniently accomplished with an oxygen sensor mounted in the instrument box area. A second reason for purging the sampling system box is to meet the electrical classification for the process area in which the box is mounted. If one or more of the components used for the sampling system is not intrinsically safe and the box is housed in an area classified as hazardous, the boxes are typically purged with plant nitrogen to ensure that the system can not act as an ignition source in the area. This is generally done with the box enclosing the analytical instrument, but it is sometimes done for the sampling system as well. Normally, the sampling system is isolated from the analytical instrument (separate, unconnected enclosures) to prevent an inadvertent leak in the sampling system causing damage to the analytical instrument.

A second commonly used mode to house the sampling system is to mount the instrument panel in an instrument house. There are several manufacturers that sell a wide variety of instrument houses in fabricated concrete as well as metal. These are typically relatively large structures capable of housing multiple sampling systems and analytical instruments. The advantage of the house over the box is that the house provides a more hospitable environment for maintenance personnel to provide their servicing, particularly in the northern areas where winters are rather harsh, and in southern areas where summers are typically inhospitable. In addition to providing a more reasonable work area, the houses are typically temperature conditioned so that the sample system as well as the analytical instrument are protected from wide temperature variations. The air supply to the house is typically filtered and heated/cooled so that environmental contaminants are removed, thereby protecting both worker health and instrument.

The boxes or instrument houses also protect the analytical equipment and sampling systems from other sources that have unexpectedly caused problems in past industrial installations. Normally, one does not anticipate that the wildlife (insects, rodents, birds) in an installation area would contribute to maintenance problems. However, all of these have caused instrument failure, particularly in environmental monitoring stations where the sampling points were open to the environment. Many of these sites have been invaded by insects building homes and, consequently, plugging the sample lines. Rodents have caused damage to power supply and data lines through incessant chewing of the cable insulation.

The ideal situation for a process analyzer is to have the detector/electrical components in an area physically isolated from the sample stream and any liquid components required for calibration or instrument operation (e.g., solvents for liquid chromatography or gel-permeation chromatography). In this arrangement, isolation of the electrical components ensures that a spark source or hot surface (such as a spectroscopic light source) will not cause a problem with flammable components (both sample stream components and solvents). Also, this isolation will limit instrument damage in case of a sample system leak or solvent leak. Although the ideal situation is often not realized in a practical application, one should make every effort to protect the delicate sections of the instrument. In one application of the authors, a fiber optic probe having a quartz window sealed on the end of the probe was installed in the reactor of a process that contained compounds that were extremely corrosive and toxic. The seal on the window (a gold seal) was calculated to isolate and protect the instrument, which was connected to a spectrometer a considerable distance away through an appropriate length of fiber optic cable. The authors were shocked one day shortly after operation initiation to find that the probe seal had leaked and that, by normal process pressure, corrosive and toxic chemicals had been forced through the probe and the fiber optic cable and had filled the spectrometer. After washing down the entire system with copious amounts of water, the authors found that the corrosive material had destroyed the bulk of the system. As a result the components had to be replaced. The experience proved to be an expensive lesson: sample components should be isolated by multiple means from the instrument portion of the system. This example illustrates the more general case that with seals, valves and safety devices a good design must insure redundancy since the failure rate of any component tends to be too high to rely on one device as the sole means of protection.

One of the main functions of the process interface (sampling system) is to condition the sample stream so that the components can be accurately quantitated over an extended period of time with only routine maintenance. Considering the broad spectrum of process streams being sampled, this is not a trivial task. Actually when there is failure of a process analytical system, more than likely it is a malfunction of the sampling system. Process analytical experts report that 70–80% of process analyzer outages can be traced to the sample system [1]. There are several commercially available sampling components that aid in the sample conditioning process. A primary concern of the conditioning process is temperature control. As mentioned above, many times this is accomplished by the housing of the sample panel – either in a sample system box or the more elaborate instrument house. If the process stream temperature is more extreme than can be tempered with the environment of the box or house, there are a myriad of heating and cooling devices available to the system designer. For heating, the most popular choices are the use of steam tracing or resistance heating of the lines. In most chemical process areas, steam tracing is preferable since it is not a source of ignition as the resistance tracing could potentially be. For cooling there are a wide range of devices based either on refrigeration or expansion of a gas source. Refrigeration units are a proven commodity for temperature reduction but are expensive to purchase initially and require a considerable amount of electricity to operate. In contrast, Vortec coolers are considerably less expensive to purchase initially, but they do require large amounts of compressed air, are noisy when operating and do not have the cooling capability of a large refrigeration unit. If available, industrial cooling water can also be used as a refrigerant. Although not as flexible as refrigeration units or Vortec coolers, it does provide a degree of cooling and is relatively inexpensive.

2.1.4.1 Filtration

Filtration is one of the most challenging yet most important functions of the sampling systems. Most commercial process streams have some type of particulate matter in the process stream. For most analytical chromatography systems downstream there is a small orifice valve that samples the stream and injects the sample into the instrument. Without removing the particulate matter upstream, these small orifices will get plugged, rendering the system unusable. For most optical systems, small diameter tubes lead from the sample to an optical cell. Without removing the particles upstream, the small diameter tubing will ultimately become plugged or the optical cell windows may become coated, depending upon the "stickiness" of the particulate matter. There are a wide variety of filters commercially available. The materials of construction vary to the extent that almost any liquid stream can be effectively filtered without degradation of the filter material or filter holder. As one might expect, a device designed to remove something from a process stream will at some period in its product lifetime become coated with the substance being removed, rendering the filter ineffective. Although for some applications this situation occurs at a lifetime that is acceptable to the application designer, several strategies have been developed over the years to rejuvenate the filter element and extend the useful lifetime. One strategy has been the backflushing of the filter element at either periodic intervals or on-demand when the process operator feels that either the backpressure has risen to an unacceptable level or the filtration rate has been reduced to an unacceptable level. This strategy is most effective when the backflush solvent is either the process stream itself (which can be re-introduced to the reactor) or of a compatible process solvent that again can be reused in the process. If the flush solvent cannot be reused in the process, this will be the source of a waste stream that will require disposal, thereby increasing the complexity of the installation. A second strategy is to use a cross-flow filter. In this device, the raw process stream is forced over the surface of the filter element and is returned to the process. A portion of the stream going over the filter surface actually penetrates the filter element and is directed to the analytical instrument. The raw product flowing over the surface of the filter tends to clean the particulates off the filter surface, thereby extending the useful element lifetime. Again, these elements and filter holders are available in a wide variety of materials so that a material compatible with the process stream can be utilized.

2.1.4.2 The Fast Loop–Analytical Loop Strategy

Most continuous process owners do not want an analytical system directly interfaced to their process. The reason is that no matter how rugged the analyzer it will fail or need maintenance when the process is running and they are not willing to stop the process to repair an analyzer. This rule is somewhat relaxed for batch processes but is often times still followed. Thus most sample systems make use of the "fast loop–analytical loop" strategy, which has been utilized for decades in process analytical systems. In review, the fast loop consists of a relatively large diameter (typically not small diameter tubing) pipe system that is connected to the process equipment to be sampled through a double block and bleed valve system (a key redundancy system design) to insure safe system maintenance. Sample material flows rapidly through the pipe system and is returned to the process, either at the same point at which it was removed or at a point in the process of reduced pressure. If the material is returned to the same point in the process, some means to move the material (e.g., a mechanical pump) is utilized. If the material flows to a point of reduced pressure, the process pressure is typically sufficient to keep the process fluid moving and no pump is needed. This approach can often be used when there is a recycle loop and the sample is taken just after the process pump and returned just before the pump. The purpose of this circuit is to provide a continually renewing source of sample to the sample system. This is the best means to assure that a sample representative of the process is available for analysis without a large waste of process fluid. At some point in the fast loop, an outlet to the analytical loop is established. This is located at a point that is relatively close to the analytical instrument so that the sample being analyzed at any time is representative of the process without a significant time lag. For example, if a cross-flow filter is being used, then the fast loop would be used to flow over the surface of the filter element and the material, which passes through the element, would then be in the analytical loop. Pressures and flows in both the fast and analytical loops are held at the desired parameters by the use of flow and/or pressure controllers. A wide variety of these devices are commercially available in several ranges of pressure and flows. Again, these devices are manufactured in several materials of construction to be suitable in any process stream. If a cross-flow filter is not being used, a tee is inserted into the fast loop so that an analytical loop can be established. Suitable small diameter tubing is connected to the "T" and then the pressure and flow is regulated using appropriate flow and pressure controllers. The temperature of this analytical line is regulated if necessary and the stream is directed to the analytical instrument. The choice of analytical technique will determine the quantity and conditioning of the sample being directed toward it. These will be covered more completely in the sections below.

2.1.4.3 NeSSI

A new approach has recently been introduced to standardize the sampling system. It is based on a modular concept where each active component fits into a standardized flow path. This system, called NeSSI (New Sampling/Sensor Initiative), is described in much more detail in Chapter 10.

2.1.5
General Reviews

While we give a general overview here of the current state of analytical instruments the reader is encouraged to read the process analytical instrument reviews that appear every two years in the journal *Analytical Chemistry* for more details. These reviews have been published every two years since June 1993 through June 2005 (there was no process analytical chemistry review in 1997, but the 1999 review covered all major developments from late 1994 through 1999). This is undoubtedly the most current comprehensive compilation of process analytical technology available [2].

2.2 Chromatography

2.2.1 Gas Chromatography

Gas chromatography was one of the first analytical techniques used successfully on-line. Since there have been, and continue to be, such a large number of gas chromatographic installations in both the chemical and petroleum industries both the instrumentation and the systems for sample introduction have had extensive development efforts. As a result, the gas chromatographic equipment commercially available is the most "turn-key" ready of any analytical instrument for numerous applications. Several of these applications have been exhaustively described in both the scientific and product literature. This is undoubtedly the easiest technique to apply, with almost assured success. There are several techniques with highly developed commercially available equipment for moving the sample from the analytical loop to the front of the chromatographic column. Most of these depend on some valve action to remove a small aliquot of the sample from the analytical loop. In one design, there are a series of plates with small sample orifices in which the center plate moves. When the center plate is in one position, the sample flows from the analytical loop through the small slot in the center plate and then back out of the valve. The center plate is then slid to a second position, moving the sample aliquot into the gas stream of the gas chromatograph. The valve is heated sufficiently so that when the sample is moved into the gas stream it is quickly volatilized and swept to the front of the column where the component separation begins. A second valve design uses a moving rod with a groove or dimple in place of the sliding plate. The advantage of this design is that the groove is less likely to foul with the passage of the sample stream than is the sliding plate with the small slot. The size of the groove or dimple determines the sample volume injected into the chromatograph. The disadvantage of the moving rod is that the groove edge tends to degrade the seal surface as it moves from position to position. Thus, the moving rod injector potentially will require more frequent seal maintenance than the moving plate. A third design utilizes a multiport rotary valve. This valve uses one or more sample loops to capture a sample from the sample stream. In this design, the valve is turned to the load position and the sample flows through the appropriate valve ports connected to the sample loop. The size of the loop determines the quantity of material injected into the GC. The rotary valve is turned to a second position where the loaded sample loop is now in line with the carrier gas. Since the valve is heated above the vaporization point of the components, the sample is quickly volatilized and swept unto the front of the analytical column. The advantage of this design is that high quality high temperature valves are available that can be incorporated into a highly reliable sampling system. In addition, several valve configurations are available. If necessary, a calibration standard can be incorporated into the same sampling valve system, which can be used to calibrate the system on a periodic basis. In many systems, however, absolute accuracy is not as important to the process operators as a highly repeatable analysis, though long-term drift can go unnoticed if care is not used in the implementation of the approach. In many cases the process computers can be tuned to how the current process parameters compare with previous batches of material. This information can be just as useful for controlling the process as having absolute quantification of the components present. In many cases this allows a quicker, simpler analysis strategy than going for absolute values. All of these valve designs are readily automated, with air actuators being the best for handling flammable materials so that the entire analysis can run automatically.

The columns and injectors developed for on-line gas chromatography have improved dramatically in the past several years. Now, most of the separations required can be achieved on thermally stable high-resolution capillary columns. The column length is adjusted so that the component resolution is achieved in a minimum time. Thus, a great deal of pseudo-real time data can be obtained for process control. Although the modern process gas chromatograph can be readily temperature programmed, most process analytical designers choose to run the chromatograph isothermally if the required component resolution can be obtained in a reasonable time frame. Although temperature programming can significantly speed up the chromatographic analysis, it adds to the complexity of the overall analysis. At the top of the temperature program, after all of the components have eluted, the oven has to recycle to the initial temperature and stabilize at that temperature before a second analysis can begin. Getting rid of excess heat in a process area can be difficult due to concerns about flammable materials in the environment. If the exact starting temperature and temperature program is not repeated for each analysis, there will be elution instability for the components of interest. When components do not elute at exactly the same time as in calibration runs, it is sometimes difficult for the integration/quantitation system to correctly identify the eluting components as well as to correctly quantitate the peak from the last calibration run. These problems are usually not encountered when isothermal operation is utilized.

Several stable detectors for on-line gas chromatography have been developed. By far the most popular is flame ionization, which will detect most organic compounds in the low part per million range to 100% with appropriate sample split ratios or sample dilution. The second most popular detector is thermal conductivity, which will detect most volatile compounds with roughly the same response factor. Other detectors used for special applications include electron capture detection for low level organic compounds containing chlorine, nitrogen-phosphorus detection for low level detection of nitrogen- or phosphorous-containing organic compounds, and sulfur detectors for low level detection of sulfur compounds. Bolt-on photo-ionization detectors can also be utilized for special applications. The advantage in using these specialty detectors is that they discriminate against other compounds in favor of those that contain the heteroatoms they detect. For example, an electron capture detector will accurately detect ppm levels of a chlorine-containing compound even if it co-elutes with a large amount of a nonchloride containing compound. This allows accurate quantitation without the need for extremely high-resolution columns.

2.2.2
Liquid Chromatography

Liquid chromatography is the complementary analytical technique to gas chromatography. The basic difference between the two techniques is that in liquid chromatography the sample components partition between the column packing and a mobile liquid phase whereas in gas chromatography the components partition between the column packing and a mobile gas phase. The major advantage of liquid chromatography is that higher molecular weight or highly polar compounds can be separated and detected. The technique is also valuable for the separation and quantitation of thermally labile compounds. A limiting factor for gas chromatography is that all components to be separated must be volatile compounds. Liquid chromatography utilizes the same basic sampling system and injection systems that are applicable to gas chromatography; although all of the injection valve types described previously could be utilized, the best choice for most applications is the multiport rotary valve. This valve allows easy change of the sample size injected by changing the sample loop. This flexibility, along with the ability to easily incorporate calibration standards in the same valve, makes this the sampling valve of choice for most system designers.

The four dominant detectors used in LC analysis are the UV detector (fixed and variable wavelength), the electrical conductivity detector, the fluorescence detector and the refractive index detector. These detectors are employed in over 95% of all LC analytical applications. In contrast to the availability of on-line gas chromatographs, there are relatively few commercial on-line liquid chromatographs. When the application calls for the use of liquid chromatography, many analytical process designers opt to buy a laboratory instrument and convert it for the on-line application. This involves repackaging the pump and column components in a box, which can be purged to eliminate ignition sources in the process area. As mentioned previously, the sampling and inlet system can be the same rotary valve that is commonly used for on-line gas chromatography applications. A liquid chromatography application is typically easier than gas chromatography because in most instances the sample is a liquid stream, which does not require heating to volatilize it before or during injection into the instrument. On-line liquid chromatography has found its major usage in pharmaceutical applications where the compounds of interest are either water soluble or higher molecular weight compounds that are not readily volatilized for gas chromatography application, or the compounds are thermally sensitive and would not survive gas chromatography application. The major disadvantage of on-line liquid chromatography, which has severely limited its on-line uses, is that the solvent(s) used for the separation are a source of waste that must be dealt with. Although the volumes are relatively small at ~1 mL min^{-1}, the effluent from the detector must be captured, safely stored until disposal, and then disposed of in an environmentally acceptable manner.

The high solvent usage also tends to require more maintenance to continually replenish the solvent reservoir, and while solvent recycling can be used when analytes are not present the solvent usage is still a problem.

2.2.3
On-line Spectroscopy

Various forms of spectroscopy have been used on-line for years. Several commercial instruments utilizing infrared and near-infrared spectroscopy as well as ultraviolet spectroscopy have been used as fixed wavelength detectors. Although somewhat limited in analytical capability, these on-line instruments were valuable in monitoring various process and environmental streams. In their simplicity, they proved to be highly reliable and trouble-free. The weakest parts of these instruments for long-term performance were the cell and cell windows followed by the light source. In many cases, exotic materials such as gold plating were used to produce a cell that could withstand the corrosive nature of the chemicals they were subjected to. A limiting factor in many cases was the fouling of the cell windows, which decreased the light intensity through the instrument or changed the characteristics for quantitation. LEDs have very positively impacted the lifetime of the light sources where they are applicable.

As the instrument manufacturers have become more sophisticated in their on-line offerings, the spectroscopic instruments available for on-line use have begun to approach the capabilities available in the research laboratory. For example, Fourier-transform infrared instruments have become readily available for on-line implementation. Computer controlled FT-IR instruments have made this a viable option for on-line use. Scans are accomplished very readily, the transform is done quickly and either single- or multiple-scan results are passed to a second computer that controls the overall on-line system. With the sophistication of this instrument, multiple components can quickly be quantitated in the process stream. Again, the cell is the instrument component most likely to fail or cause quantitation problems. Much effort has been expended by both instrument manufacturers and companies utilizing the equipment to design process cells. The most critical aspect of the process cell is the ability to withstand the process stream parameters without leakage. Many companies have developed elaborate cell body designs using exotic materials to prevent sample stream leakage. In addition, many have a multiple seal arrangement to hold the cell windows securely to the cell body.

A second area of on-line spectroscopy that has experienced breakthrough technology improvements is near-infrared. Three areas of instrumentation have seen this technology breakthrough: (a) computer controlled scanning and detection capabilities, (b) development of fiber optic probes, and (c) implementation of chemometric techniques. As in the FT-IR instrumentation, near-IR has benefited immensely from computer control to either scan the entire spectrum or to monitor specific areas. Equally important from a process perspective is the ability to use optical fibers to transmit the near-IR light between the process and the instrument. This allows the instrument to be placed in an area that does not require the conventional instrument packaging necessary in most process areas. As a consequence, lower cost, higher capability state-of-the-art laboratory instruments can be used for process areas without additional process packaging. There has also been significant development in the probe end of the fiber optic bundle. There are several varieties available, including those with windows physically bonded to the probe body and those with spring-loaded seals against the probe window. The latter are available commercially in many exotic materials, including stainless steel, titanium and tantalum. Several of the spring-loaded probes have been installed in very corrosive process streams and have been fully functional without leaks for several years. The final breakthrough has been the implementation of chemometric techniques to glean more analytical information from the data obtained. Chemometric techniques are very helpful in pulling relevant chemical data from the normally convoluted spectral information.

2.2.4
On-line Mass Spectrometry

Mass spectrometry has been utilized for on-line measurements for many years. The technique has been widely used in the petroleum industry as well as the chemical industry where the process stream contains primarily gases. This type of sample is the most compatible with the mass spectrometer and requires little sample preparation before introduction into the instrument. Early instruments designed for process monitoring did not have full scanning capability, but rather had several detectors at fixed position to monitor specific ions. Most of these instruments had the capability to monitor 5–7 components in the gas stream. These instruments were also widely used as leak detectors since they had the requisite sensitivity for low level detection and the speed to detect leaks very quickly.

Many companies now use high capability laboratory instruments and are installing them at site labs near the process area, where the samples can be directed to the instrument but the instrument is not required to possess a hazardous area packaging. Liquid process samples can be analyzed by injecting a small quantity (microliters) in a non-reactive container and vaporizing the sample prior to injecting into the mass spectrometer. A salient feature of mass spectrometry is the large library of characteristic fragment patterns of organic compounds. Not only can process stream components be routinely quantitated, but unknowns that arise in the process stream can quickly be identified and quantitated.

Several techniques used in process analytical chemistry have not been discussed here. They are rather specialized, such as size exclusion chromatography, ion chromatography, atomic absorption, etc. Rather than employing commercially available equipment, most of these installations rely on heavily modified laboratory equipment adapted to the process analytical application. Again, the reader is referred to the Process Analytical Reviews [2] for complete details of these applications.

2.2.5
Microflow Techniques

The advent and development of micromachining techniques, which enabled microflow based systems, has opened a potentially very important new area for development. The small size and potential for low cost means that new chemical production concepts such as point of use manufacturing could be realized. Microfabrication techniques also offer the potential to imbed analytical measurement within the flow path for close coupled control of the production process. Finally, new chemistries, which are difficult or just dangerous to practice in conventional equipment, are much more easily controlled in microscale equipment. However, given the conservative nature of the chemical processing industry this technology, like all new innovations, will have a slow introduction into mainstream manufacturing.

Microscale equipment has high surface-to-volume ratios, which facilitates designs with high heat and mass transfer rates. Hessel, Hardt and Lowe have written a complete review of the current state of microscale processing equipment [3]. The typical types of processing equipment can be broken into three general classes: mixers, heat exchangers and reactors. The micromixer can achieve very high mixing rates with low energy input. In the simple case a multilayer approach is used. The idea is that at the molecular level all mixing is by diffusion. With turbulent flow mixing, the goal is to reduce the size of the individual domains to shorten the diffusion distances since the rate of mixing goes up in proportion to the square of the distance the molecules have to travel. However, most of the energy is used achieving bulk flow as opposed to going directly into the mixing process. An example is the mixing process in a stirred tank. With a micromixer, stream A is broken into multiple sub-streams and interleaved with multiple sub-streams from stream B. To reduce the distance even further the linear velocity is increased by transitioning into a smaller size channel through a funnel like transition. The effect is to have micron size layers, which then means that there is very short, millisecond, mixing times. An additional feature of micromixers is that the very small layers can lead to very controlled particles sizes, in the micron range, with low energy input. One area of use has been the formation of emulsions with reduced surfactant levels.

The heat exchanger takes advantage of the high surface-to-volume ratios of the channels, which enables the rapid transfer of heat into or out of the system. The absolute rate of energy transfer is a function of many parameters, such as channel size, materials of construction, wall thickness and flow rates of process and heat exchange fluid, but the key is that very high heat transfer rates can be achieved in a small area. While the general area of heat exchangers is less directly related to the concept of this book, when they are combined with micromixers a microreactor is a useful result. A microreactor, which incorporates rapid mixing and high heat transfer, can be used to carry out very exothermic reactions. For example, ring nitration reactions can be performed with much more control of product distributions, and Grignard reactions afford much cleaner reaction products at higher temperatures and higher concentrations. These components are covered at length later in this book (Chapter 6).

Even though the conservative nature of the bulk chemical processing industry will make the introduction of this technology slow in that field, microreactors are generating considerable excitement among specialty chemical manufacturers who consider this technology to solve several reaction control problems currently encountered in several industrial sectors. In addition, they also have, less obvious, features that can solve some of the problems facing the more traditional chemical processing industry. While microscale equipment applications will be covered in more detail later in the book, we list some of these features and the analysis implications in the next few paragraphs.

A significant problem with current process technology and equipment is the capital outlay and time required to fabricate a world-scale plant for both chemical and petrochemical products. The problem is that with a new process for a new compound the owner may be required to build a very large plant to get the low capital cost per pound needed to ultimately succeed in the marketplace. However, this large investment may be required long before the market has grown large enough to absorb a significant portion of the new capacity. This problem can lead to a very large negative cash flow while waiting for the market to develop. Preliminary data indicates that microreactor technology can be duplicated in as many parallel channels as required to reach needed early production levels at substantial capital savings since capacity can be sized more closely to the expected early demand. The basic idea is that the economy of scale savings will come from mass produced reactors as opposed to a larger custom-built reactor. In addition to the potential cost benefits from microreactors there are technical benefits that suggest microreactors may offer improved processes with higher conversion and less waste. In fact the ability to run more efficient intensified processes will be required in most cases for microreaction technology to be truly cost competitive with traditional manufacturing methods.

The technical benefits come from the small channel size with large surface-to-volume ratios. The channels are usually less than a millimeter and typically in the range of 100s of microns. As mentioned above, designs with small diffusion distances improve mass and heat transfer characteristics. In addition, with no potential for turbulent flow there is less potential for back mixing, which can reduce impurity problems in chemistries where products can react with starting materials. Typically, the rate that energy intensive processes can be safely operated is controlled by the rate that heat can be moved in or out of the process. For less energy intensive processes the rate is usually controlled by the rate of mixing rather than by the inherent rate of reaction. The improved heat and mass transfer work together, enabling the intensification of the reaction step. This allows processes to be operated at the underlying chemistry rate rather than at rates dictated by the processing equipment. Thus, if heat flow can be controlled, then many reactions can also be run at higher temperatures to increase the rates. An additional gain can also be realized over the traditional way of running a reaction in that often reactants must be diluted to control the rate of reaction. The result is not only a slower reaction but also the need to ultimately remove the added solvent.

Another key aspect of microreaction technology is that the reaction is implemented as a continuous process. This can be a disadvantage for implementation for some owners such as batch process operators in the specialty chemicals area that are not used to working with continuous processes. However, continuous processes can be more cost effective to operate since recycling is more easily implemented and they usually require less labor to operate. Notably, at this early stage there is not enough data to clearly demonstrate that these two advantages inherent in large-scale continuous processing will be true with microreaction technology implementations.

Notably, a key problem with the use of microreactors is that, historically, many reactions have been developed to be driven to completion by relying on the formation of either small molecules or precipitates; if small molecules are one of the reaction products, they would be boiled off to drive the reaction. Both mechanisms can be a problem in microreactors. Consequently, work is underway to couple membranes and other separation concepts to enable the small molecules to be extracted from the reaction mix and drive the reaction. With solids formation, the requirement is that the solids exit the system without plugging. While this has been demonstrated it is not clear whether solids formation will be a major problem when dealing with these types of reactions.

An additional factor in the use of microreaction technology is that the process that is developed in the laboratory is essentially the same process that is commercialized. This is significant when dealing with complex chemistries where the reaction product needs to be qualified for use in an application. With traditional approaches there is always the concern that as the process is scaled up the product characteristics will change as the reaction runs with differing heat and mass transfer characteristics. This leads to the requirement that, for scale-up, reactions often need to be run in ever larger test equipment (usually a process needs to be tested for every 10–100-fold increase in volume) to be sure that some unforeseen change is not occurring like the formation of unwanted byproducts. A related problem can be that the production process must be locked in at a very early stage before much process optimization, which is often done in larger scale equipment, can be done. It is locked in to insure no changes in product quality for materials with long qualifications times. These very real potential problems lead to much higher scale-up costs, increased time to market, and the potential for the production of materials with less than optimum processes.

An unexpected key benefit from the use of microreactors has been the detailed flow information that is available from their construction. To build optimized channel structures detailed models of the devices have been developed. These models offer the potential to use the information as the basis for reactor modeling. The result is that with detailed reactor models and precise composition measurements from on-line analysis the rate expressions for the chemistry can be extracted from the laboratory data. The potential of this is significant in that general laboratory data only gives a sense of what is happening whereas now the possibility exists of obtaining precise chemistry models from early stage lab work. In fact UOP has indicated in several Summer Institute meetings that they now can accurately scale lab data by 10^6 or more. To be safe they still make pilots runs but the time has been reduced to confirming lab reaction rates as opposed to developing reaction rates. The general concept of using microreactors coupled with on-line analysis to generate a detailed understanding of the chemistry will be discussed in Chapter 13.

There are several impacts of this technology on the on-line analysis field. First, the flow information is offering new ways of sampling that takes advantage of some of the properties of laminar flow. One concept is that in a wide channel that is equally fed with a sample stream and a solvent stream, molecules will move out of the sample half and across the solvent stream based on their diffusion rate, which is a function of molecular size. Also, particles like blood cells will not move out of the sample stream. This is one way to separate small molecules from blood cells without filtering. This general concept of particle and molecule separation is expected to find more uses for sample clean up and preparation. Small intersecting streams can form the basis of very precise low volume sample injection systems needed for micro-scale chromatography and has been effectively used for electro-driven chromatography. In addition, the benefits of using coated wall micro-scale LC for real time analysis were also discussed at the Summer Institute meeting. This technology could be a significant advantage for on-line analysis due to the low solvent usage and the rapid analysis times. In much the same manner as the micro-scale processing equipment benefits from short diffusion distances the key advantage for micro-scale chromatography stems from the short (between 1 and 10 μm in this case) and uniform diffusions distances. Other facets of the technology have not yet found their way into routine use. However, the expectation is that since reactions can be rapidly and efficiently performed on a micro-scale the potential for more chemistry based analysis schemes, such as is currently done for derivatization approaches for enhanced detection, will become more common.

More generally, as the microflow equipment becomes increasingly available there will continue to be advancements in the general area of sample preparation for analysis. As stated earlier the key failures in an on-line analysis system tend to be in the sampling system. Thus, approaches that improve the ability to manipulate a process stream will be significant for process sampling as well. Clearly, most analyzers need very small amounts of material for analysis; this means that if smaller amounts of material are processed to present to the analyzer then it is often easier to prepare the sample. For example, it is much easier to adjust temperature on a small sample than a large one, and there is much less load on a filter when cleaning a small amount of material. Working in small volume equipment makes dealing with pressure and corrosive materials much easier as well. The NeSSI equipment that will be described in more detail later in Chapter 10 is a key step in taking advantage of these concepts to reduce the cost and improve the reliability of process analysis.

References

1 Center for Process Analytical Chemistry, *New Sampling/Sensor Initiative (NeSSI) Information Package*, August 1, **2000**.
2 W. W. Blaser et al., *Process Analytical Chemistry*, American Chemical Society, **1995**, 47R–70R;
J. Workman, Jr., D. J. Veltkamp, S. Doherty, B. B. Anderson, K. E. Creasy, M. V. Koch, J. F. Tatera, A. L. Robinson, L. Bond, L. W. Burgess, G. N. Bokerman, A. H. Ullmann, G. P. Darsey, F. Mozayeni, J. A. Bamberger, M. Stautberg Greenwood, *Anal. Chem.*, **1999**, *71* (12), 121–180;
J. Workman, Jr., M. Koch, D. J. Veltkamp, *Anal. Chem.* (Review), **2003**, *75* (12), 2859–2876; J. Workman, M. V. Koch, D. J. Veltkamp, *Anal. Chem.*, **2005**, *77*, 3789–3806.
3 V. Hessel, S. Hardt, H. Lowe, *Chemical Micro Process Engineering*; Wiley-VCH, Weinheim, **2004**.

3
Process Intensification

Kurt M. VandenBussche

This chapter introduces the concept of process intensification. Several examples of industrial relevance are given. In addition, the field of microtechnology is situated and its importance for process intensification assessed. Two case studies are presented, showing how this concept can impact the processing industries. A third case study is presented in Part II of this book, applying again the concepts of process intensification and illustrating how microtechnology can enable novel synthesis routes.

3.1
Introduction, Scope and Definitions

In its broadest sense, process intensification can be defined as a series of methodologies aimed at reducing the capital cost associated with chemical processing, by removing existing "limitations".

Capital expenditure typically makes up over 20% of the overall cost of production for bulk-chemicals. A significant reduction of the size of the equipment or a reduction of the number of units, for instance by coupling several functions in one piece of equipment, can result in substantial production cost savings. Interestingly, process intensification may have further reaching implications than mere reduction of capex. It may enable alternative processing routes or may lead to distributed production of chemicals. In addition, the lower cost may render the industry more nimble, accelerating the response to market changes, facilitating scale-up and providing the basis for rapid development of new products and processes.

Within this definition, the field of process intensification is very broad. It can not be the objective of this chapter to provide an exhaustive overview of the area. Whole conferences are dedicated to process intensification technology every year and networks of experts have been formed, see, e.g., Ref. [1]. Disciplines like mechanical engineering, process engineering, materials science and others all bring exciting new angles and ideas. Rather than attempting to condense the vast literature into a single chapter and not doing it justice, we focus here on how process intensification has influenced the thinking for reaction engineering and how microtechnology can contribute to this.

3.2
Process Intensification in the Field of Reaction Engineering

The field of reaction engineering attempts to quantify several physical and chemical phenomena, the interplay of which determines what happens in the conversion process. It is important to gain a sufficient understanding of these to quantify them and recognize which is/are limiting the intensity of the process. These phenomena are:

1. The chemical mechanism of the conversion. This includes the determination of reaction intermediates, the rate-determining step in the mechanism, the nature of the transition state (i.e., the high energy transient state that dictates the activation energy). For catalytic systems, one needs to examine the role and nature of adsorption and desorption of feed and product on the catalyst surface, and the occurrence of physical changes or solid state reactions in the catalyst under process conditions (oxidation/reduction, sintering, carbon deposition, etc.).
2. The transfer of molecules from the bulk phase to the location where reaction occurs (catalytically or thermally) may significantly influence the overall reaction rate. When the transfer from the bulk fluid phase to the catalyst is limiting, this is referred to as external mass transfer limitation. If diffusion of reactants or products in the pores of the catalyst is slow, it is termed internal diffusion limitation (Fig. 3.1). Both of these effects commonly occur under relevant conditions.
3. Analogously, resistances to heat transfer can occur between bulk and catalyst as well as within the catalyst pellet. The first of these, "external heat transfer limitation" is observed occasionally, particularly in stagnant or slow flowing liquids. Significant resistance to heat transfer within a pellet "internal resistance to heat transfer" is only observed for highly endo- or exothermic reactions. This is because the heat is more easily distributed across the pellet by means of conduction through the solid, than by convection in the pores.
4. Proper description of hydrodynamic effects and the momentum balance is often neglected in reactor modeling today. Assumptions of "plug flow" or "perfectly mixed" are common and simplify the calculations tremendously. A trend towards full calculation of flow and momentum profiles is starting to take shape in the literature, but it is still hampered by excessive computing times.
5. Phase behavior in multiphase environments. Many reaction systems either occur in a multiphase environment or form products that will exist in a different phase. The description of how species are divided up into different phases is a key element of reaction engineering and often the more challenging one.

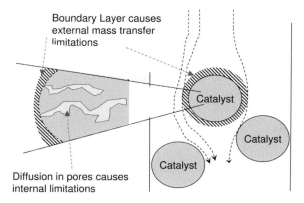

Fig. 3.1 Depiction of internal and external resistances to mass transfer.

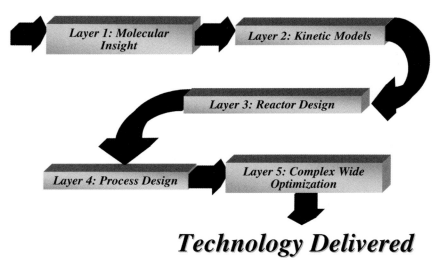

Fig. 3.2 Schematic representation of multiscale modeling.

These phenomena, and their interplay, are usually described by a set of coupled differential equations [2]. A hierarchical approach, often referred to as "multiscale modeling", is used to structure the interaction between the various effects [3, 4]; see also Fig. 3.2.

At the lowest level, reaction mechanisms are translated into a set of kinetic equations, describing the rate and its functional dependence on species concentration, temperature and pressure. This is also the level where any change in physical state of the catalyst (loss of oxidizing agents, deactivation by carbon build up, poisoning, etc.) is described.

One step up from the kinetic equations and for a catalytic system, the reaction–diffusion equation describes the interaction of internal mass transfer and reaction kinetics within the catalyst. An energy equation for the solid determines the temperature gradient over the catalyst pellet, if applicable. These equations are coupled through the rate of the reactions and the heat they generate.

Moving up into the reactor level, effects of convection, dispersion and generation are described in the conservation equations for mass and energy. The momentum balance describes the behavior of pressure. The interface between the reactor and the catalyst level is described by the external mass transfer conditions, most often represented in a Fickian format, i.e., a linear dependence of the rate of mass transfer on the concentration gradient. In cases where an explicit description of mixing and hydrodynamic patterns is required, the simultaneous integration of the Navier–Stokes equations is also conducted at this level. If the reaction proceeds thermally, the conversion of mass and the temperature effect as a result of it are described here as well.

The reactor level can subsequently be absorbed in levels for process design and plant wide optimization. These are, strictly speaking, not part of the reaction engineering framework but are tightly interwoven with it. The impact of the latter is illustrated briefly in Section 3.4.

Once a thorough and systematic description of the various phenomena is in place, it is relatively straightforward to identify the limitations and intensify the process by removing them. The following thought-experiment, starting from a real-life situation, is useful to gauge the impact of process intensification.

Consider a dehydrogenation process, currently commercially operated at an intensity of "3", as shown in Fig. 3.3. This intensity can be understood as an indication of the throughput of the unit, or, in other words, the amount of product that can be made per unit of time for a fixed capital investment. As the intensity increases, the capital contribution to the overall cost decreases inversely. The

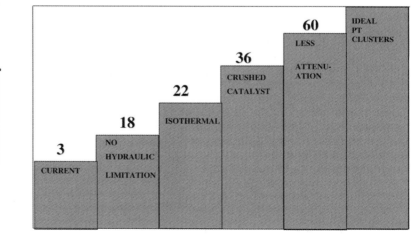

Fig. 3.3 Potential impact of process intensification on a commercial dehydrogenation process.

current process, at its intensity of "3", benefits from running at low pressure, as dehydrogenation reactions cause an increase in the number of moles. It is limited by pressure drop over the reactor. If one could somehow remove this hydraulic limitation, one could increase the intensity to a level of "18" before the next limitation manifests itself. This next limitation lies in the amount of heat that can be supplied to the vessel from the outside. There is a maximum temperature for the catalyst in this dehydrogenation system, above which deactivation by coking rapidly occurs. Given this limitation, at the intensity level of "18", it becomes impossible to provide a sufficient heat flux into the reactor to fuel the endothermic reaction system without causing excessive deactivation close to the walls.

If one could keep the bed isothermal, by somehow removing the heat transfer limitation, one could raise the intensity to "22", upon which the diffusion in the catalyst pellet (internal mass transfer) becomes limiting. Crushing the catalyst pellets to a diameter of 400 µm, again under the absence of hydraulic limitation, can raise the intensity to "36". At this level, the inherent activity of the active site becomes limiting. As the catalyst activity is currently attenuated in this system to avoid over-reaction, the removal of the attenuator allows the process to run at an intensity of "60", now limited by the intrinsic activity of the noble metal clusters on the catalyst support. One can further hypothesize the existence of perfect noble metal sites and what kind of intensity these would allow, but the point to be gained from this is that the identification and successive removal of limitations (hydrodynamic, heat and mass) can have significant impact on the capital intensity of a typical process.

In the example of the dehydrogenation reaction above, the mass flows are assumed invariant, their composition is not disturbed. Pressure and temperature are dictated by the thermodynamics of the system to attain a certain conversion of the feed. An alternative form of PI (process intensification) can be seen when one selectively removes one of the reaction products to shift the equilibrium and intensify the process. The combination of reaction and separation is a key example of PI. The literature abounds with schemes to accomplish this. Its commercial use, however, is limited to a small number of cases. Following are examples, successful and not so successful, of this mode of operation:

1. Membrane reactors are the object of intense study in the literature, but have not revolutionized reactor technology so far. In many cases, their application is hampered by the need to run the reactor at relatively low pressure (for reasons of reaction equilibrium or strength of the membrane wall), while simultaneously requiring enough of a partial pressure in the reactor for the permeant to create a sufficient flux through the membrane. Where re-use of the permeate gas is needed, this system often fails, as a sweep gas is typically required to reduce the partial pressure of the permeate on the other side of the membrane to maintain a driving force. This sweep gas then often dilutes the permeate to a partial pressure where it is of little value.
For these and other reasons, membrane reactors play only a minor role in the bulk processing industry today. An area for future application, however, may

lie in the distributed production of hydrogen, where often only low partial pressures of the product are required. For further references to this specific application, see, e.g., Refs. [5–7].
2. Catalytic distillation is a good example of process intensification in the field of reaction engineering. Here, the difference in boiling point between feed and product is exploited to remove a product from the reaction and hence alleviate the equilibrium limitation. In addition, the reaction heat can be taken up by the boiling liquid, offering an elegant integration of reaction and heat transfer. This technique has also been considered to improve the selectivity of reaction systems, by quickly taking the product away from the feed or the catalyst, avoiding secondary reactions. For a good reference work on this topic see Ref. [8].
3. Selective removal of a product to shift the equilibrium of the reaction can also be performed through a chemical conversion. In the Lummus/UOP Styrene Monomer Advanced Reheating Technology (SMART) technology, the hydrogen produced in the dehydrogenation of ethylbenzene to styrene is burned between successive reactor beds, re-instating the driving force in the equilibrium limited system, as well as providing heat to overcome the endothermicity of the reaction [9].
4. Kuczynski and Westerterp have proposed an elegant means of removing the reaction products from a system [10, 11]. In the area of methanol synthesis, where the conventional process is limited in conversion per pass to about 30%, they proposed to adsorb the product onto an amorphous silica-alumina powder that trickles through the solid catalyst packed bed. At the bottom of the bed, this solid absorbent is collected and depressurized to yield the methanol product. Even though the process looked economically viable, it has, to the best of our knowledge, not seen any commercial application.

Other facets have been explored to achieve process intensification in reactor technology. A first one, the interplay between reaction time constant and catalyst deactivation time constant, has led to several instances of PI since the 1930s. Figure 3.4 shows the various reactor types practiced for catalytic processes on an industrial scale today. Early process technology used fixed bed reactors, with catalyst lifetimes expressed in years. Later, it became apparent that much higher product yields could be reached per unit of time if one could use certain catalysts that exhibited higher deactivation rates. This resulted in the invention of moving bed reactors, where catalyst is gradually and continuously moving from inlet to exit and is regenerated offline to be re-entered later. This whole process occurs on a timescale of days. The ultimate in reactor intensification can be found in riser technologies, where catalyst and feed are shot up into a tube at high velocity, with reaction times in the order of a second and catalyst deactivation times of 5 s.

In yet another effort, the combination of regenerative heat exchange and reaction has been the objective of a long-term development effort at Novosibirsk and elsewhere [12–14]. In this "Reverse Flow" technology, for an exothermic reaction, a fixed bed catalytic reactor is initially heated to the desired reaction temperature.

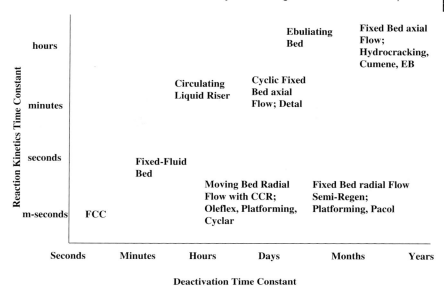

Fig. 3.4 Reactor classification according to time constant for reaction and catalyst deactivation.

A colder feed is brought in and causes an energy wave to form as it cools down the bed. Before the feed gets the chance to push all of the energy out of the bed, the flow direction is reversed and the feed now pushes the wave to the other side. Upon a few reversals, a steady bell shaped temperature profile is established that moves back and forth through the bed at a steady pace. At the front of the bell, the feed is preheated. It reacts once it gets sufficiently hot, releasing the energy at a higher exergetic level. For equilibrium limited reactions, the decrease of the bed temperature on the downstream side of the bell shape allows higher conversions than those achievable in an adiabatic conventional fixed bed. The technology is commercially applied for oxidation of SO_2 [15], as well as for the combustion of volatile organic compounds [16], under the trade names Dynacycle [17] and Regenox [18].

3.3
Process Intensification through Micro-structured Unit Operations

The use of miniaturization of equipment to avoid limitations is relatively recent. It originated in Germany in the early 1990s. It is inspired by the fact that the intensity of heat and mass transfer is determined by the characteristic distance of the system, through the Nusselt and Sherwood formalisms respectively (for further elaboration see, e.g., Ref. [2]). Hence, miniaturization can lead to an increase of heat and mass transfer coefficients of several orders of magnitude of magnitude, be it at the expense of higher pressure drop.

The intensified heat transfer has spawned a myriad of studies, particularly for reactions where precise temperature profile control is crucial. Examples are given in Refs. [19–21].

The better mass transfer and the highly uniform flow path allows for very precise control of the residence time distribution in the reactor system as well. An example of the potential of this feature is given in Section 3.4.

The use of small units also allows for the safe processing of hazardous materials, through minimization of inventory. In addition, explosive reagents can be safely mixed and converted with high yields into the desired product.

Even though the advantages of "miniaturized" process equipment appear clear, the most significant drawback is the increased capital cost of production [22] at the small scale, given in by the famous "0.6 rule". This rule, first published by Chilton et al. [23], teaches that the capital cost of production typically increases with production volume to the power of 0.6 in the processing industries. In the world of microtechnology however, scale up is typically achieved by numbering up or parallelizing, with a scale up factor close to 1.0. To survive this disadvantage during scale up and remain competitive in capital cost, the microtechnology has to be cheaper per unit of production.

The lines in Fig. 3.5, adapted from Ref. [22], show the relationship between scale down ratio and required capital intensification factor for scale up exponents of 0.5, 0.6 and 0.7. If, for example, a new microtechnology is considered, with production volumes smaller than the conventional technology by a factor of 1000 and the conventional scale up rule is ~0.6, the capital cost of this technology will have to be

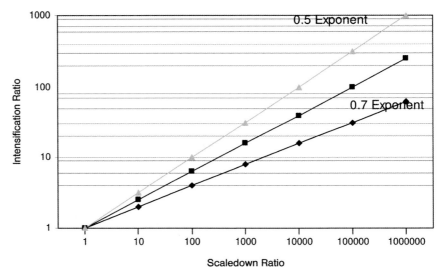

Fig. 3.5 Required capital intensification for various scaledown ratios and various cost exponents (0.5, 0.6, 0.7). (Adapted from Ref. [22].)

some 16 times less than that of the conventional method, per unit of production volume. In other words, an intensification factor of 16 will be required.

Having established a high level methodology to identify intensification factors required for success, the next five sub-sections will to go through the various fundamental reaction engineering processes by means of illustrative examples and discuss differences between a mega- and a micro-scale operation.

Where appropriate, an actual measure of possible process intensification will be given. In some cases, a reaction system will be used to show the potential of the improved process.

The transfer of momentum will not be addressed. The literature abounds with information on micro-devices that accomplish this but, to the best of our knowledge, no efficiency comparisons between mega- and micro-scale have been published.

3.3.1
Gas Phase Mass Transfer

Generally, one would consider the use of microsystems for gas/gas mixing excessive. Large-scale turbulent mixers handle this task with relatively high efficiency at a far lower cost. There does not appear to be a need for further intensification.

In some cases, however, the time scale observed in the large-scale equipment might not be short enough. Cases of this are found where very fast, often homogeneous, competing reactions occur and imperfect mixing affects the desired selectivity or where the mixture goes through an explosion envelope as a function of composition to end up as a non-combustible mixture.

An example of the latter is the direct synthesis of hydrogen peroxide from hydrogen and oxygen. This technology has received significant attention over the last 10 years [24–26], to replace the cumbersome and inefficient anthraquinone route to hydrogen peroxide. All of this work has focused on a range of H_2 contents below 3.5%, staying away from the explosion limits. One would expect H_2/O_2 ratios in the explosive range or even at stoichiometric ratios to be preferred from reaction chemistry, yield and economics point of view. However, the process is inoperable at this ratio because of safety considerations.

By using micro-scale equipment, the inventory can be greatly reduced, leading to an intrinsically safe operation. In addition, the gases can be mixed in a characteristic time lower than the induction time of the reaction. For example, in a first approximation, safe mixing of hydrogen and oxygen will occur at characteristic distances below 100 μm [22].

This relatively simplistic approach only gives a first approximation for a safe mixer. Similar numbers are obtained using the more sophisticated theories of Frank-Kametzki and Semenov [27].

The safe mixing and reaction of hydrogen and oxygen in microsystems has been shown by Veser et al. [28–30] and Inoue [20]. Similar considerations were made for the combination of oxygen and ethylene, as demonstrated by Kestenbaum et al. [31] and Hoenicke and Kursawe [32].

3.3.2
Liquid–Liquid Mass Transfer: Mixing and Emulsions

The fast mixing of liquids is more difficult than that of gasses, due to the lower diffusivities in the environment. Distinction can be made between miscible and non-miscible systems.

For both miscible and immiscible cases, mixing is typically conducted in the turbulent regime when operated in the large scale. If the laminar regime is chosen, other means must be employed to reduce the characteristic distance over which the molecules must diffuse to mix.

To quantify mixing quality in miscible systems, Loewe et al. [33] adapted a mixing characterization technique developed by Villermaux [34]. Using UV/VIS spectroscopy, they presented the data shown in Table 3.1. When comparing macro and micro- cases in the laminar regime, they observed an intensification factor of 40. When comparing laminar micromixing to a macromixer, operated in the turbulent regime, an intensification factor of 10 was found.

More recently, Ushijima and Okamoto [35] compared micromixing with turbulent mechanically induced mixing and found the former to be significantly more homogeneous and efficient.

Yoshida et al. [36] have studied the effect of fast mixing of miscible systems by measuring the selectivity towards mono-alkylation in the Friedel–Crafts alkylation of aromatics. They observed a 20-fold increase in the relative selectivity of the mono-alkylate over the di-alkylated system when using a micromixer instead of a conventional batch reactor. In the cycloaddition of the N-acyliminium ion to styrene [36], 50–80% of the cycloadduct is typically lost towards polymeric by-products. Using an interdigital micromixer, the yield to the cycloadduct increased from 20–50% to almost 80%.

For a further extensive review of micromixing technology in miscible systems, see Hessel et al. [37].

Table 3.1 Comparison of mixing efficiency for various test set-ups. (Data from Ref. [33].)

Mixing equipment	UV absorption (a.u.)
Mixing-tee 3-mm ID, $Re = 177$	1.85
Batch vessel No stirring	1.5
Batch vessel Heavy stirring	0.4
Mixing-tee 1-mm ID, $Re = 531$	0.3
Micromixer array $Re = 19$	< 0.1

In the contacting of non-miscible liquids, mixers become emulsion generators. In the chemical industry (on the mega- as well as the micro-scale) fine emulsions have many useful applications in, e.g., extraction processes or phase transfer catalysis. Additionally, they are of interest for the pharmaceutical and cosmetic industry for the preparation of creams and ointments. Micromixers based on the principle of multilamination have been found to be particularly suitable for the generation of emulsions with narrow size distributions [33]. Haverkamp et al. showed the use of micromixers for the production of fine emulsions with well-defined droplet diameters for dermal applications [38]. Bayer et al. [39] reported on a study of silicon oil and water emulsion in micromixers and compared the results with those obtained in a stirred tank. They found similar droplet size distributions for both systems. However, the specific energy required to achieve a certain Sauter mean diameter was 3–10× larger for the macrotool at diameters exceeding 100 μm. In addition, the micromixer was able to produce distributions with a mean as low as 3 μm, whereas the turbine stirrer ended up with around 30 μm. Based on energy considerations, the intensification factor for the micro-stirrer appears to be 3–10.

3.3.3
Gas–Liquid Mass Transfer

Extremely widely practiced in industry, this type of mass transfer is at the basis of the absorption and distillation unit operations, two or three phase reaction systems and others. The magnitude of the gas–liquid interfacial area is, of course, of prime importance and this is where the process intensification opportunities for microtechnology mostly lie. Table 3.2, adapted from Ref. [40], shows the potential

Table 3.2 Specific interfacial areas of selected conventional and miniaturized reactor types. (Data from Ref. [40].)

Type of conventional reactor	Specific interface ($m^2\ m^{-3}$)	Type of microreactor	Specific interface ($m^2\ m^{-3}$)
Packed column Countercurrent flow; Co-flow	10–350 10–1700	Micro bubble column (1100 × 170 μm) Isopropanol (observation)	5100
Bubble columns	50–600	Micro bubble column (300 × 100 μm)	9800
Spray columns	10–100	Micro bubble column (50 × 50 μm)	14 800
Mechanically stirred bubble columns	100–2000	Falling film microreactor (300 × 100 μm)	27 000
Impinging jets	90–2050		

of miniaturization for various types of gas–liquid interfacing technology. The intensification factor in many cases exceeds 100, and sometimes even 1000.

Cypes and Engstrom [41] reported on the performance of a microstripping column relative to a packed tower. They found an improvement by a factor of at least 10 in the mass transfer capacity coefficient, over a broad range of mass fluxes.

In another example, Hessel et al. [42] and Loewe et al. [43] studied the direct fluorination of toluene with F_2 in a microsystem consisting of reaction channels as well as mass transfer and heat transfer in close proximity. They claim a change in the reaction mechanism from radical in nature to an electrophilic substitution, through careful control of the reaction conditions. This type of microreactor is well described in the literature and is commonly known as a falling film microreactor [44, 45].

3.3.4
Mass Transfer in Gas–Solid Systems

Applications of mass transfer in gas–solid systems are found in gas phase catalytic reaction or adsorption systems as well as three phase reaction systems where the gas phase is continuous (sprays, trickle beds).

Separation by adsorption as a miniaturized unit operation has not yet extensively been discussed in the literature, but the combination of superb hydrodynamics and sharp residence time distributions will clearly lead to extremely sharp mass transfer fronts, enhancing the effectiveness of the adsorbent significantly. It opens perspectives for miniaturized TSA (temperature swing adsorption) technologies, where the components of a mixture are separated from each other by differences in their adsorption isotherms, in a swing-bed semi-continuous fashion.

Extreme control of the hydrodynamics and the residence time distribution has been shown to lead to higher yields in reaction systems on various occasions. The work by Hoenicke et al. [46, 47] on partial hydrogenation of various aromatics is classic in this area. In the system studied here, the hydrogenation of cyclododecatriene to the mono-olefin would open up a new route to the synthesis of Nylon-12. In practice, however, the hydrogenation of the mono-olefin to cyclododecane is far faster, and the reaction is hard to stop at the intermediate product.

Hoenicke et al. constructed a special reactor type to demonstrate the potential of extreme control of reaction conditions. They started by growing hexagonal alumina channels through anodization of aluminum. The pore size, pore length and pore density of this hexagonally shaped nanotube is tightly controlled under these conditions. They subsequently deposited these regular nano-surfaces on a support that contained a strict pattern of etched microchannels (Fig. 3.6). These micro-etched support plates were finally stacked into a reactor block. Through this method of construction, Hoenicke ensured a nearly perfect control of the residence time distribution. Wherever a molecule enters the reactor system, it sees an essentially identical environment, with the same diffusional limitations, the same reaction site density and reactivity and the same convective flows. As a result, the operator can very tightly control the progress of the reaction. Figure

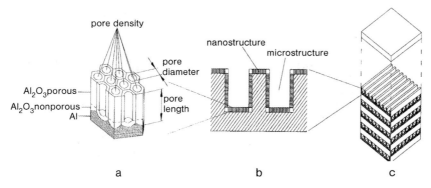

Fig. 3.6 Construction of a microreactor support for extreme control of residence time distribution. (Adapted from Refs. [46, 47].)

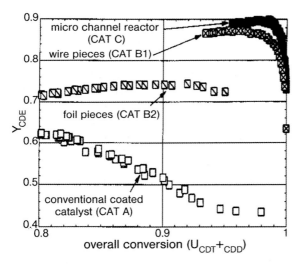

Fig. 3.7 Yield pattern for various reactor types, for the selective hydrogenation of cyclododecatriene to the mono-olefin. (Adapted from Refs. [46, 47].)

3.7 gives a summary of their final yield results. It shows the evolution of the yield of the mono-olefin against the conversion of the di- and tri-ene. Curve A (CAT A) shows the performance of a conventional hydrogenation catalyst. At conversions around 80%, the yield is about 62%, but it quickly drops as the conversion increases further. Curves B1 (CAT B1) and B2 (CAT B2) show how the use of uniform nano-structures enhances the yield. Curve C (CAT C) finally shows the result of the micro-structured reactor block, achieving yields of 90% at conversions as high as 98%.

3.3.5
Heat Transfer

Whereas microsystems are typically expensive heat-exchange devices, they open up a region of specific interfacial area (m^2 of heat exchange area per m^3 of equipment volume) that was previously not attainable (Table 3.3). Numerous papers have been published on this topic. Among them, Brandner et al. [48] studied cross flow heat exchangers with different micro-structures. They found a microcolumn structure to be highest in heat transfer efficiency, with overall heat transfer coefficients of 54 000 W m^{-2} K^{-1}, compared with a conventional coefficient of heat exchange in a forced convection environment of 50 W m^{-2} K^{-1}, or 200 W m^{-2} K^{-1} in a conventional packed bed.

Hardt et al. [49] reached a similar conclusion, using computational fluid dynamics to study the effect of various fin configurations, as well as the effect of heat exchanger material conductivity on the overall performance of the microstructured equipment.

The ability to precisely control the temperature by means of high surface area and high heat transfer coefficients can be very valuable in highly endo- or exothermic systems. An example of the potential impact for an endothermic system has been given in Section 3.2. In the above-mentioned work by Hardt et al. [49], an unnamed endothermic reaction was shown to be run at throughputs exceeding conventional operation by a factor of 2–3.

Hiemer et al. [50] studied the hydroxylation of benzene with nitrous oxide. This highly endothermic reaction proceeded with higher selectivities at significantly higher space time yields than was possible in the conventional tubular reactor.

Increased heat transfer has also been found to be a major benefit in the conversion of hydrocarbons into hydrogen. This topic has received tremendous attention for fuel cell applications (see also Section 3.4). Some of the leading micro-contributions can be found in Refs. [37, 51–55] and those on a slightly larger scale by Heatric [56].

Table 3.3 Overview of heat exchanger types and their specific surface area. (Adapted from Ref. [22].)

Heat exchanger type	Typical surface area (sqft)
Brazed	10–10 000
Shell and tube	100–10 000
Micro exchangers	0.2–100
Double pipe	5–120
Explosion formed plates	3000–100 000

3.4
Case Studies

Two case studies are presented below to illustrate the impact of process intensification. They show how distributed production of hydrogen-age fuels can compete with a larger scale centralized production followed by transportation, provided one can invoke logistical circumstances as well as very intense capital utilization.

3.4.1
Distributed Production of Methanol

Methanol is produced, from natural gas, in very large quantities. Plants produce up to 5000 Mton per day and are usually located close to large gas reserves. The synthesis of methanol from natural gas is a two-step process (Fig. 3.8). In the first step, natural gas is converted catalytically into the carbon oxides and hydrogen ("synthesis gas or syngas") in the presence of steam, in a highly endothermic process called steam reforming. This system sees a significant increase in the number of moles during the reaction. Though thermodynamically more challenging as a result of this, the reaction is conducted at fairly elevated pressures (25 bar), to reduce downstream compression cost. This, in turn, increases the outlet temperature required to ensure sufficient natural gas or methane conversion, to 800 °C.

For the conversion of syngas into methanol, the pressure is raised from 25 to 100 bar. The gas mixture is catalytically converted at temperatures between 250 and 300 °C. The synthesis of methanol is exothermic and strongly equilibrium limited. In conventional operation, the conversion of hydrogen typically does not exceed 30%. In the large-scale plant, methanol and the by-product, water, are flashed off and the unconverted gas is recompressed and recycled to extinction.

The combined plant is a significant steam exporter and is typically highly integrated in the chemical complex surrounding it.

Methanol has received attention as a means to carry hydrogen on board a fuel cell powered vehicle, see e.g., Refs. [57–59]. Indeed, the reverse of the 2nd step mentioned above, the conversion of methanol to synthesis gas, is facile at low

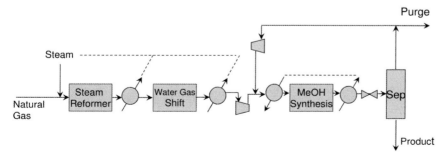

Fig. 3.8 Flow diagram for conventional methanol synthesis.

pressures and temperatures around 250 °C. As such, it may be preferred over the steam reforming or partial oxidation of naphtha/gasoline, the main alternative for indirect hydrogen storage on board.

An energy scenario where methanol is used as a vehicle fuel would require significant logistical changes in the fuel infrastructure. It would also necessitate an increase in the production of methanol by two orders of magnitude. One could consider a continuation of the current methanol model of centralized production and subsequent distribution, or a distributed production model could be followed. In the latter, the fuel will be made locally, tapping into an existing natural gas pipeline. In the latter case though, the synthesis technology needs to be re-evaluated, as there are far fewer integration options for the waste energy now. Typical communities are not set up to make use of steam at various pressures. For this reason, the alternative flowscheme below generates power as a side product; the plant does not have an export of steam.

Figure 3.9 shows the proposed flowscheme. The natural gas, sourced from the pipeline, is desulfurized, preheated and led into a reactor where it is reformed to synthesis gas. The "temperature controlled reactor" continuously combines reaction and heat exchange, to maximize the utilization of the energy contained in the waste stream (discussed later). The synthesis gas is compressed and led into the methanol reactor. The methanol reactor is preferably of the "temperature controlled" type, again to maximize the heat transfer and maximize the intensity of the catalytic operation. The partially converted stream is subsequently flashed to remove methanol and water. Rather than recycling and recompressing the stream, it is expanded in a gas turbine, producing power.

The temperature at the outlet of the turbine is still very high (800 °C) and can now be used to provide the energy for the steam reforming step mentioned above. The waste energy from the flue gas can then be further used either directly to heat the desulfurizer and the local facilities at the power station (in winter) or,

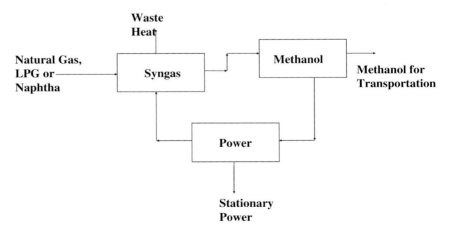

Fig. 3.9 Alternative methanol synthesis flowscheme.

indirectly, in an adsorption cooling cycle (in summer) for air conditioning of the site. The base case design for a process integrated in this way is the production of 200 ton of methanol per day as well as 20 MW worth of power. This would provide energy for some 8000 households, a small to medium size community. By adjusting the conversion in the methanol synthesis reactor, the balance between power and methanol can be adjusted to suit the local needs or to adjust between summer and winter type operations.

Figure 3.10 shows the heat integration for the base case in a conventional temperature–enthalpy plot used for pinch analysis.

For reasons of confidentiality, exact details of the reactor designs have to remain undisclosed in this discussion. The final economic analysis, however, incorporates them in a ±20% estimate of the major equipment cost. The installation factor, defined as the ratio of the final cost over the cost of major equipment items, is projected to be in the order of 2 here, as foundations, piping, general infrastructure will be far less extensive in this compact and highly physically integrated plant.

Table 3.4 compares the cost of methanol produced for three cases:

1. A US case where the (older) plants are somewhat smaller at 2300 tons per day, but where the plant has been depreciated and no further capital cost is incurred.
2. A Middle East case, where new world-scale plants are built, for 5000 tons per day.
3. The distributed methanol case under discussion here, for 195 tons per day.

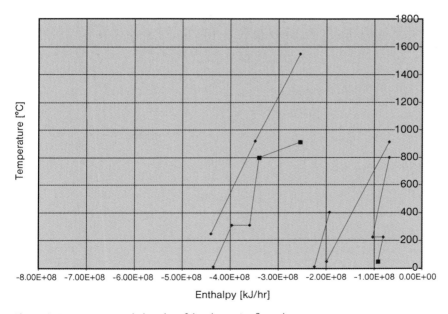

Fig. 3.10 Temperature–enthalpy plot of the alternative flowscheme.

Table 3.4 Cost breakdown for various methanol production scenarios.

Parameter	Means of methanol production		
	Conventional US	Large Middle East	Distributed
Capacity (Tons day^{-1})	2326	5000	195
Capital charge (MM US$)	0	490	35
Methanol production cost ($ gal^{-1})	0.53	0.15	1
ROI ($ gal^{-1})	0	0.12	0.22
Power credit ($ gal^{-1})	0	0	−0.62
Shipping/storage/insurance ($ gal^{-1})	0.14	0.26	0.1
Total ($ gal^{-1})	0.67	0.53	0.7
($ Mton^{-1})	224	178	235

The first thing to note in Table 3.4 is the relative ratio of the capital costs between cases 2 and 3. With reference to Fig. 3.5, to be competitive on a capital cost basis, with a ratio of production of 25, the capital would have to be intensified by a factor of 3–4 to achieve a similar capital cost contribution. Despite the close coupling of the units and the combination of heat exchange and reaction, this is not the case here; the capital charge in $ gal^{-1} is about twice as high in the distributed case, as can be seen in the ROI line. In the latter scenario, however, use is made of the fact that there is a co-production of power to far offset the higher capital cost of production.

The variable cost of production is higher in case 3 as well, since the natural gas feedstock is some 10–20× more expensive when taken from the pipeline in the US than at the well in the Middle East. In addition, no steam credit can be taken in case 3. However shipping, storage and insurance are cheaper for the distributed case.

Adding up all the contributions, it becomes apparent that the economics of the distributed case are equivalent to those for production in large-scale US plants, the capital of which has been paid for, which is remarkable. However, they can not beat the world-scale plant cost structure in the Middle East, primarily due to the differential in the feedstock cost. Based on this evaluation, it is not unreasonable to project some use of this distributed technology if methanol should pan out as a hydrogen carrier.

3.4.2
Distributed Production of Hydrogen

In the same context of a future multi-source energy scenario, the distributed production of hydrogen has received a lot of attention in the literature. It was initially primarily aimed at providing fuel for fuel cells in the 1–5 kW range, for stationary residential applications. However, when the capital cost of the fuel cell did not decrease as quickly as projected, it became clear that this option was not truly cost competitive in the short term. The effort then became mostly focused on larger units for uninterrupted power supply (UPS) or on-demand hydrogen, where "premium use" can support a higher cost structure.

The reduction of capital cost for these units, making some 100 Nm^{-3} h^{-1} at the most, starting from 120 000 Nm^{-3} h^{-1} in today's large-scale hydrogen production plants, is another example of the impact of process intensification.

The cost comparison can not be readily made in this case, as there is intense competition between several players in this area who tightly guard their cost information, but it is interesting to see the process technologies that have emerged, given the request for a smaller scale of cheap production. Two lines of thought can be discerned in the literature. Most have focused on intensifying the various unit operations in the flowscheme, leaving the process scheme identical to that of the large scale version.

Others have taken a broader look and have questioned the integration of the various steps prior to implementation of novel equipment. Examples of the intensified equipment for hydrogen generation have been given above. In this section, we focus on describing the changes in the process flow scheme that have been proposed to intensify the operation.

The conversion of natural gas into hydrogen is conventionally achieved through the generation of syngas, followed by its purification. The conversion step can be done through:

1. catalytic steam reforming (described briefly in Section 3.4.1), which is strongly endothermic [240 kJ $(mol\text{-}CH_4)^{-1}$]:

    $CH_4 + H_2O = 3\ H_2 + CO$

2. thermal or catalytic partial oxidation, which is exothermic [–800 kJ $(mol\text{-}CH_4)^{-1}$]:

    $CH_4 + \frac{1}{2}\ O_2 = 2\ H_2 + CO$

3. catalytic autothermal reforming, which is the combination of (1) and (2), such that the sum of reactions becomes energy neutral.

The choice of technology depends on the desired composition of the produced synthesis gas (preferred carbon-to-hydrogen ratio, presence/absence of nitrogen). If the end-goal is the actual production of hydrogen, steam reforming is preferred, as it produces more hydrogen per unit of methane fed than the other two routes.

The synthesis step is usually followed by a water-gas shift step, where the reducing power of CO is used to convert water into hydrogen:

$CO + H_2O = CO_2 + H_2$

Table 3.5 Comparison of steam reforming and partial oxidation.

Feature	Process	
	Steam reforming	Partial oxidation
Intrinsic safety	+	−
H_2 yield per CH_4	+	−
H_2 conc in product	+	−
Rate of startup	−	+
Load following	−	+

This catalyzed reaction is mildly exothermic at -42 kJ mol^{-1}, and is often executed at different temperature levels (high, medium, low), to optimize reactor volume against the approach to equilibrium.

The purification of the hydrogen is most commonly done via pressure swing adsorption, where the stronger adsorption of the carbon oxides and water, compared with that of hydrogen over certain materials, is used to remove them from the stream.

The choice of a reaction system is not as straightforward if one wants to scale down and mass manufacture a unit that will be at least somewhat accessible to the untrained public. Table 3.5 compares the main features of steam reforming and partial oxidation in the context of small-scale production. Safety, the higher hydrogen content of the gas and the higher energy efficiency are major pros for the steam reforming step. The direct heat transfer and, somewhat related, the more rapid startup/better response are points that speak for an autothermal reforming technology. It is hard to choose the superior solution between these, as both have valuable characteristics. There are currently several technologies making it to the market based on either of these approaches.

Another approach, combining the best features of both, has been termed "Hybrid Autothermal Technology", and is discussed in Ref. [60]. Its optimized heat integration was derived through systematic pinch analysis; it can be understood as follows.

Figure 3.11 shows the standard flowscheme for a hydrogen generator connected to a fuel cell. Natural gas is desulfurized (200–400 °C), preheated (400–700 °C) and mixed with steam and air to be partially converted into synthesis gas (600–800 °C). The effluent of the syngas generator is water-gas shifted in one or two steps (450 and 250 °C).

The CO content of the feed to the Proton Exchange Membrane (PEM) fuel cell can not exceed 10–40 ppmv. After the water-gas shift mentioned above, it typically contains in the order of 800–1200 ppmv CO. This CO is most often removed by preferential oxidation (100 °C) or, less practiced, selective methanation (100–300 °C). The gas is then led to the anode of the fuel cell where it is partially converted (70–90 °C). The unconverted gas contains hydrogen and methane and

3.4 Case Studies | 63

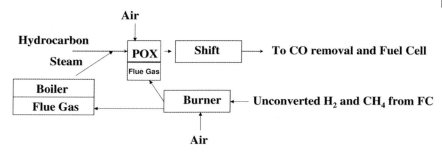

Fig. 3.11 Conventional hydrogen generator flowscheme for fuel cell applications.

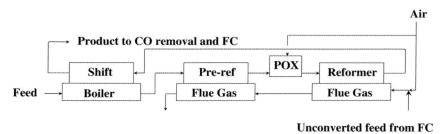

Fig. 3.12 Flowscheme for the "hybrid autothermal technology". (Adapted from Ref. [60].)

can be burned in an "anode waste gas" burner (400–800 °C). Heat integration between all of these units varies, depending on the provider.

The Hybrid Autothermal Technology (Fig. 3.12) contains the same processing steps – its higher intensity lies in the optimized heat exchange and its ability to operate in various degrees of steam reforming and autothermal operation. The feed gas enters the unit through the desulfurizer, operating at 200 °C and heat integrated with the tail end of the anode waste gas heat exchange. The feed gas, now at 200 °C, is preheated in a "pre-reformer block", the second half of which contains a reforming catalyst. Since the reforming step is endothermic, the stream can pick up significantly more energy from the anode waste gas than just the sensible heat. Leaving the pre-reformer at 500–550 °C, it then enters the main reformer, where air is added so that the temperature at the outlet is maintained at 650 °C. This control scheme allows the unit to operate as a partial oxidizer upon startup and gradually migrate to a steam reformer as the equipment gets up to temperature. Indeed, at startup, the unit is cold and the gas is not preheated in the pre-reformer. It enters the main reformer around 200 °C (the desulfurizer has an electric back up heater to keep the catalyst clean), and large quantities of air are needed to reach the 650 °C outlet temperature. As soon as the anode waste gas heats up the pre-reformer, the air injection can be reduced and the unit makes more hydrogen at a higher concentration. Note that the water-gas shift cooling is optimally accomplished by interfacing it with the steam-boiler. In some cases, an electric steam generator may be required at startup for reasons of catalyst stability.

Analysis of the flowscheme results in an intensified process, which will run more efficiently and most likely have a lower capital expenditure.

3.5 Conclusions

This chapter has introduced the concept of process intensification (PI) as a collection of methodologies aimed at reducing the capital cost of chemical processing. Examples have been given of successful and less successful strategies. The area of microreaction technology has also been situated; for some cases, an attempt has been made to quantify the degree of PI achievable.

Two case studies have been presented, looking somewhat ahead to a scenario of multiple energy sources. In the first, the distributed co-production of power and methanol has been found to compete with local US centralized methanol production. In the second case, the intensification of distributed hydrogen production has been discussed, through better process integration and judicious choice of equipment.

Acknowledgments

The contents of this chapter reflect certain aspects of a long-term corporate interest in this matter at UOP. Stimulating discussions with Gavin P. Towler, Anil R. Oroskar, Stanley A. Gembicki and Suheil F. Abdo are gratefully acknowledged.

References

1 www.pinetwork.org
2 G. F. Froment, K. Bischoff, *Chemical Reactor Analysis and Design*, Wiley, 2nd Edition, **1990**.
3 S. A. Gembicki et al., *Chem. Eng. Sci.*, **2003**, 58, 549–555.
4 J. Lerou, K. M. Ng, *Chem. Eng. Sci.*, **1996**, 51, 1595–1614.
5 C. Chen, Annual DOE review, Project DE-FC26-97FT96052, May 24, **2004**.
6 J. Smit et al., *Int. J. Chem. Eng.*, **2003**, 1, A54.
7 P. P. A. C. Pex et al., Proceedings 6th International Conference on Catalysis in Membrane Reactors, Lahnstein, Germany, July 7–9, **2004**.
8 K. Sundmacher, A. Kienle (Eds), *Reactive Distillation, Status and Future Directions*, Wiley-VCH, Weinheim, **2005**.
9 www.chemsystems.com; www.abb.com; www.uop.com
10 K. R. Westerterp, M. A. Kuczynski, *Hydrocarbon Processing*, November **1986**.
11 K. R. Westerterp, M. A. Kuczynski, *Chem. Eng. Sci.*, **1987**, 42, 1539–1559.
12 G. K. Boroskov, Yu. Sh. Matros, *Catal. Rev. Sci. Eng.*, **1983**, 25, 551–590.
13 Yu. Sh. Matros, *Can. J. Chem. Eng.*, **1996**, 74, 566–579.
14 www.matrostech.com
15 Yu. Sh. Matros, Catalytic processes under unsteady state conditions, *Studies Surf. Sci. Catal.*, **1989**, vol. 43.
16 Yu. Matros et al., Preprints Env. Ind. Catal., 1st Eur. Workshop meeting, Louvain-la-Neuve, **1992**.
17 www.water-leau.com
18 www.haldortopsoe.com

19 M A. Schneider, F. Stoessel, *Chem. Eng. J.*, **2004**, 101(1–3), 241.
20 T. Inoue et al., *Proc. IMRET-7*, Dechema e.V., **2003**.
21 E. Dietzch et al., *Proc. IMRET-7*, Dechema e.V., **2003**.
22 A. R. Oroskar et al., Proceedings of Microtec **2000**, Hannover.
23 C. H. Chilton, *Cost Engineering in the Process Industries*, McGraw-Hill, **1960**.
24 Gosser et al., Method for the production of hydrogen peroxide, US 5135731.
25 L. Gosser, Catalytic preparation of hydrogen peroxide from hydrogen and oxygen. US 506893.
26 Chuang et al., Process for the direct synthesis of hydrogen peroxide, US5082647.
27 Warnatz, Maas, *Technische Verbrennung*, Springer Verlag, Berlin, **1998**.
28 G. Veser et al., *Proc. IMRET-3*, Springer-Verlag, Berlin, **2000**.
29 S. Chattopadhyay, G. Veser, Proc. Chem. Conn.-2001, Chennai, India, Dec. **2001**.
30 G. Veser, *Chem. Eng. Sci.*, **2001**, 56, 1265–1273.
31 L. Kestenbaum et al., *Proc. IMRET-3*, Springer-Verlag, Berlin, **2000**.
32 A. Kursawe, D. Hoenicke, *Proc. IMRET-4*, **2000**.
33 H. Loewe et al., *Proc. IMRET-4*, **2000**, pp. 31–47.
34 Villermaux et al., *AIChE Symp. Ser.*, **1991**, 88, 289.
35 T. Ushijima, H. Okamoto, *Proc. IMRET-7*, Dechema e.V., **2003**, p. 97.
36 J. Yoshida et al., *Proc. IMRET-7*, Dechema e.V., **2003**, p.1–4.
37 V. Hessel et al., *Chemical Micro Process Engineering*, Wiley-VCH, Weinheim, **2005**.
38 Schiewe et al., *Proc. IMRET-4*, **2000**.
39 T. Bayer et al., *Proc. MRET-4*, **2000**, pp. 167–173.
40 W. Ehrfeld et al., *Proc. IMRET-4*, **2000**, pp. 3–20.
41 S. H. Cypes, J. R. Engstrom, *Chem. Eng. J.*, **2004**, 101(1–3), 49.
42 V. Hessel et al., *Proc. IMRET-3*, Springer-Verlag, Berlin, **2000**.
43 H. Loewe et al., *Pure Appl. Chem.*, **2002**, 74(12), 2271–2276.
44 K. P. Moelmann et al., Inframation **2004** Proc., **2004**.
45 K. Jaehnish et al., *J. Fluorine Chem.*, **2000**, 105, 117.
46 D. Hoenicke et al., *Proc. IMRET-4*, **2000**, pp. 89–99.
47 L. Wiessmeier, D. Hoenicke, *Proc. IMRET-2*, **1997**.
48 J. Brandner et al., *Proc. IMRET-4*, **2000**, pp. 244–249.
49 Hardt et al., *Chem. Eng.Commun.*, **2003**, 190(4), 540–559.
50 U. Hiemer et al., *Chem. Eng. J.*, **2004**, 101(1–3), 17.
51 G. Pak et al., *Chem. Eng. J.*, **2004**, 101 (1–3), 87.
52 W. E. TeGrotenhuis et al., *Proc. IMRET-7*, **2003**, p. 19.
53 P. Reuse, A. Renken, *Proc. IMRET-7*, **2003**, p. 22.
54 E. Delsman et al., *Chem. Eng. J.*, **2004**, 101(1–3), 123.
55 I. Aartun et al., *Chem. Eng. J.*, **2004**, 101(1–3), 93.
56 B. Haynes, A. M. Johnston, Proc. AICHE Spring meeting, March **2002**.
57 www.navc.org; www.hydrogen.org; www.crest.org
58 Y. Jamal, M. L.Wyszynski, *Int. J. Hydrogen Energy*, **1994**, 19(7) 557–572.
59 M. Prigent, *Oil Gas Sci. Technol.*, **1997**, 52(3), 349.
60 G. P. Towler et al., Process for providing a pure hydrogen stream for use with fuel cells, US 6299994, **2001**.

4
High Throughput Research

Ray W. Chrisman

4.1
Introduction

In one sense the reason for doing reaction characterization by high throughput methods would seem obvious given the high number of variables that should be evaluated for almost any reaction. However, reaction characterization by high throughput methods is still very much the exception rather than the rule. Thus in this chapter we explore the lessons learned in the early stages of high throughput research for the discovery phase of research, make the case for why it is important for process development, discuss how micro-instrumentation makes this concept much more reasonable and then discuss some of the current barriers and limitations. In the Part II of this book, groups that are beginning to put the concepts into practice will give examples.

4.2
Description of Terms

As one begins to look into high throughput research the first issue that comes up is the relationship between combinatorial research, high throughput research and the newer term high content research. The lines between the first two are somewhat blurred and often they are used interchangeably, but combinatorial research does have a slightly different connotation and is usually focused on the discovery process for bioactive compounds. The IUPAC definition for combinatorial chemistry is "using a combinatorial process to prepare sets of compounds from sets of building blocks." While speed does not appear in this description, it is usually a hallmark of the concept since processing is performed in sets. Early efforts focused on automated parallel methods for the synthesis of polypeptides and polynucleotides and rapidly expanded to include the production of complex molecules. This was done by reacting the starting material with a series of small molecules that contain the same functional group with a range of modifications in the non-reacting part of the structure. The rapid screening of these new materials for bioactivity is a key part of the process.

High throughput research is a broader term and is less about making new materials and is more about being faster than traditional methods of doing research. It usually is related to high degrees of automation and doing experimentation in parallel. However, a less common approach, which is rapid serial, is also considered a part of high throughput research. The concept also applies to the automation of the analysis of the new materials and includes the various data analysis steps. However, the collection of data into databases and its analysis is described by informatics and is often an enabler of high throughput research. Thus computer analysis of the data is a key part of the high throughput process but, unlike synthesis or analytical analysis, it is usually not preceded by the descriptor high throughput.

High content research is a new term that is not as common. It tends to be used for the analysis part of the process. The basic idea is that a single screen may well miss a new active compound. As an example a traditional feeding study would look at the whole animal for any impact while combinatorial would tend to use only one screen. High content tries to couple both concepts such that more screens are used to detect broader impacts. The multiple screen approach can take the form of using cell cultures, simpler life forms such as fruit flies, or multiple targets per test. In the catalyst area this would be similar to screening a catalyst for multiple reactions instead of just one reaction. While a powerful concept, high content research will probably stay focused on the discovery side of high throughput research techniques.

4.3
Concept of a Research Process

Early work demonstrated fairly quickly that stable work processes were needed for the methodology to work. This meant that for new materials to be made the synthesis steps needed to be stable and reliable for automation. The normal tinkering that was common in traditional serial laboratory experimentation would not work in pre-programmed hardware. Materials could be made rapidly if each step of the synthesis could be counted on to give good yields of desired products. Thus, the synthesis of polypeptides was an ideal place to start since by just changing the starting amino acid in each step-wise addition of monomer whole new polypeptide sequences could be produced.

The rapid synthesis of these new polypeptide structures is a good example of the combinatorial concept owing to the high number of potential materials, all of which could have differing biological activity. An example given by Loch and Crabtree [1] illustrates the point. If one position of a polypeptide varies by only one amino acid of the 20 possible then there are 20 different compounds. However, the numbers of different compounds goes up dramatically if two, three, or four positions are varied. Two positions gives 8000 different compounds, three gives 160 000, and four positions gives 3.2×10^6 different compounds.

The successful combinatorial process for new drug candidates rapidly caught the attention of researchers working with catalysts. The quest for new catalysts has many analogies to the search for bioactive materials in that large numbers of catalyst candidates can be made with somewhat standard automatable synthesis protocols starting with simple building blocks. It is, of course, more complex than this but the analogy was close enough for research groups to begin trying the approach and achieving promising results.

Another insight that evolved from the ability to rapidly make large numbers of new polypeptides was that it got researchers thinking about the whole work process. The work needed to find a new biologically active material, for example, included the steps of design of experiments (DOE), synthesis, testing and then analysis of the resulting data. Figure 4.1 demonstrates the work process. This requirement to think about the whole research process is perhaps one of the more powerful concepts to arise from the efforts. The basic idea is that, to speed up the whole effort, bottlenecks needed to be found and eliminated. As a simple example, if large numbers of new polypeptides were made but it would then take years to evaluate them using historical analysis methods then no useful improvement in speed was achieved.

While the debottlenecking concept has long been practiced on production processes, the idea that research was a process that could be debottlenecked was a somewhat unexpected outcome. Clearly, R&D is a much less well-defined process than a production process but the idea that multiple steps must work together and operate at somewhat the same rate helped focus attention on areas in need of improvement.

While it is not clear if Symyx originated the phrase, they were very good at repeating it in presentations, and it is a good summary of the concept: "One must analyze in a day what is made in a day". Thus, to rapidly go around the discovery cycle it was quickly realized that just as stable processes for the synthesis steps were needed to enable automation, stable processes for the rest of the steps of the cycle were needed.

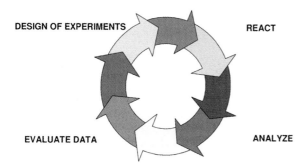

Fig. 4.1 Work process for the discovery of new materials.

Consequently, tools for rapid screening were needed to keep up with the automated methods of making materials. Most of the early work used robotic methods with multi-well sample racks that were manually transported from the synthetic workstation to the analysis workstation. The number of wells in standard equipment rapidly multiplied, as it was clear that doubling the number of wells could double the speed. In general, most work was done in parallel in a batch format, which was a mirror of the standard laboratory methods but now highly automated with parallel processing.

4.4
High Throughput Analytical

Much early analytical effort was directed towards improving the speed of techniques or searching for new techniques, to keep up with the increases in the number of materials being produced. While it was cost-effective to add more wells it was often not so cost-effective to just add more mass spectrometers. In many cases the discovery phase was broken into two stages with primary screening looking for any indication of activity and secondary screening looking for more specific indicators of activity.

In catalyst screening the primary screen could be to look for the evolution of heat as an indicator of activity. All catalysts of interest will produce heat when subjected to the reactants but so will catalysts that are just destroying the starting materials and may in fact look more active by this screen since decomposition could produce more heat. In other words, the primary screen will produce many false positives but it will be fast and will eliminate many non-active catalysts. The slower secondary screen would employ something like mass spectrometry set to detect the product of interest.

In general, in the catalyst area the analytical effort was designed to improve selectivity and quantitative accuracy while maintaining the speed of analysis. Part of the reason for this was that not all catalysts were easy to make, which meant that there were fewer catalysts to evaluate, but more data was desired to understand the performance of the material that was made. In addition, for more complex catalysts, data was needed to confirm what catalyst was made. Thus, analysis was needed on the catalyst composition as well as its performance.

The catalyst area also generated an additional set of needs in that many catalysts have a life cycle, which includes activation and performance over some time interval. To understand catalyst performance therefore required the need for continuous operation, as opposed to batch operation. This also brought the technology of on-line analysis into the high throughput research process, as it could be used to follow product formation as a function of time.

4.5
Extracting Information from the Process

The next bottleneck in the process was the collection, storage and analysis of the data. The whole data manipulation area, known as informatics, became a major source of research into the storage and analysis of data. It was also clear that the presentation of data was a problem since complete data sets often contained results as a function of a whole series of variables, which meant trying to find trends or an optimum of multidimensional data sets. To represent this data requires displays as surfaces or even more complex representations, which are certainly difficult for the unaccustomed to interpret and understand. Consequently, data-mining tools were employed and further developed to help scan the data sets for relationships.

The goal is to have the results of the first pass around the discovery cycle provide information to improve additional passes. While this is a desirable goal it is often difficult to achieve in such a dynamic process. An example would be to have the system cycle around to find the optimum catalyst composition or structure or to go looking for new structures. However, in most cases the DOE is done before any experimentation. While using a static DOE to find new catalysts in high throughput research is much faster than the traditional single experiments and provides a careful evaluation of variables, the ability to detect and follow synergistic variable interactions in a dynamic fashion would be a helpful addition.

One unfortunate, somewhat unexpected, problem arose during the early stages of combinatorial chemistry development. One general reason for combinatorial chemistry development was that the existing theories had a difficult time to predict which molecules would be active and when they could the process was very slow. It became clear that it was faster to do the experiment than it was to model the experiment, and it gave better results. The faster the process became the more mechanical it became or at least the more mechanical it appeared. Consequently, many people developed the attitude that the approach was not science but was rather strictly Edisonian research or, in other words, just highly developed trial and error discovery. This perception somewhat limited the participation of academic groups in the process.

While it is difficult to totally refute the opinion that high throughput research contains a lot of trial and error experiments, this new approach, which is now sometimes called the inductive research approach instead of the deductive approach, clearly cannot possibly evaluate all possible combinations. For example, given the potential number of variables to be explored in the search for new materials, nowhere near all the random possibilities can be made and tested. As an example, Scheidtmann et al. [2] point out that the potential number of organic molecules based on the known structural diversity of five known elements carbon, nitrogen, oxygen, hydrogen and sulfur has been estimated to be 10^{63}. To demonstrate how big this number is they go on to say that if just 1 milligram of each compound was prepared the weight would be 10^{60} g and they then compare it to the sun with an estimated weight of 2×10^{33} g and in fact the total universe is only estimated to weight 10^{58} g, i.e., 100× less. Another way to compare this large quantity of potential compounds is that 1 mg of the known organic compounds would only weigh 18 kg. If one then thinks about catalysts and the potential of using up to 75 elements, some with multiple oxidation states, the number of potential compounds to explore is indeed very large.

Thus, while it is difficult to test all combinations the approach can generate extremely large data sets of information that maps out the variable space. However, the problem can be somewhat like undersea treasure hunting in that, unless logic is used to choose the area in which to search, a lot of sea floor can be explored with no useful results. Powerful correlations can be developed only, therefore, if data from information rich areas that have a fundamental relationship to the properties of interest are mined. The point in all this is that without logical scientific thought to plan these experiments in regions of maximum potential a lot of useless information can be collected.

One example of mining in the wrong area in synthesis is that the high throughput concept tends to force the concept of making a lot of compounds with a known reaction to be able to keep the instrumentation busy. As such it is easy to make many similar compounds based on a common molecular scaffold that is not active. While there is a small potential of success in that one in a long series of test compounds will work because a new molecular pathway is exposed the probability is low and time is wasted. If the short-term goal of making and testing compounds becomes too strong then the number of tests looks good. However, since the real ultimate goal is to find new active compounds the time is wasted. Thus, it is often interesting to see whether a high throughput group is more excited about the number of compounds run or the number of new hits found.

On the other hand, exploring molecular variations of an active scaffold can be very productive. It is not uncommon for an active molecular scaffold to have variations that can be much more active than the original compound tested. In addition, variations can be even more useful in that they can provide data to molecular models to further refine the understanding of the mode of activity and often lead to even more fruitful areas to explore. Again, high throughput research is most effective when exploring in information rich areas.

Another point that has surfaced as the high throughput concept has been practiced is related to data quality. Initially, there was a concern about all the old data and how to assimilate it into the growing new databases. Indeed, closer inspection of much of the old data showed that assimilation was going to be difficult. It seems that since so few experiments were performed researchers would often change multiple variables before successive runs. The changes were often straightforward and logical, such as increasing the temperature a few degrees or changing the agitation rate or a mixing time, but they made comparing data from one experiment to the next very difficult.

A related quality issue was that data precision in historic data was usually unknown and often was not very high. In contrast, the highly automated experiments, which required stable work processes, tended to have a more reproducible precision and often much higher precision. Also, since the experiments were run with more consistent variable changes this also improved data comparability. In addition, when large data sets are collected it is often easy to detect questionable data in the line or surface variable plots of all the related experiments. The improved data quality made it easier to extract subtle trends from the data for improved understanding of the system under study to enable better models to be built.

The ability to extract much more understanding from high quality large experimental data sets focused even more attention on the need for better data quality. Jim Cawse [3] presented a Six Sigma study of a high throughput research process and was able to demonstrate that even more variability could be taken out of the process. This was possible because the repeatable steps could be studied to understand their contribution to the variability in the overall processes. The result was significantly better data quality, which enabled more powerful uses of the data sets.

4.6
Process Development becomes the Next Bottleneck

The rate at which new materials could be discovered soon began to cause another problem, which was that the bottleneck had begun to move down stream of the discovery process. As the road to commercialization requires the efforts of many different groups and corporate organizations the bottleneck on the journey began to appear in other areas. One immediate area that had problems was in process development.

Since the catalyst area was already looking at early stage catalyst lifecycle information it was not difficult for them to make the leap into gathering more data for the process development phase of research. In fact VandenBussche [4] has stated that by careful modeling of the microscale gas phase laboratory reactor, initially developed for their discovery operation, it is possible to obtain data from the system that enables scaling by more than a factor of 10^6. He also stated that they have parallel reactors, which provides them this high quality data on new catalysts under study in a high throughput mode. While he stressed that this was partly possible because they have spent many years studying these types of reaction systems, he said that this approach should not be limited to their type of materials.

Ian Maxwell who left Shell to set up Avantium used to suggest in his presentations that "one should scale down, to scale up". One reason to do this is that it becomes easier to deal with extreme conditions of corrosion, temperature, and pressure. In addition, since the mass is small, changing conditions can be rapid. Finally, much less material is needed both for the catalyst and for reactants. Another very powerful aspect of this concept is that as reactors get smaller it becomes possible to understand and model the reactor's performance so well that its impact on the chemistry being measured can be removed and the intrinsic chemistry can be determined. This is what enables VandenBussche to scale his data. He has been able to remove the impact of the laboratory reactor on his measurements such that he can now go to any scale reactor whose performance is known and by suitable manipulation of the data predict performance at large scale.

This is a very important addition in that not only can the laboratory process characterization data be gathered more quickly by high throughput research concepts but, if done in the manner described by VandenBussche, the data can eliminate the need for extensive intermediate plant scale testing. VandenBussche is quick to point out that at this stage of comfort with the approach they still confirm their results at intermediate scale but the time needed to do this is greatly reduced.

4.7
Use of High Throughput Concepts for Process Development

The concept of high throughput for process development gained momentum for several reasons. First, in the pharmaceutical area, as mentioned the bottleneck had moved from the discovery area to the process development. Thus it was fairly rapidly recognized that to get new materials to the market improvements were needed in the process development area. In addition, a more subtle reason that is evolving from process analytical technology (PAT) thrust by the US FDA is the concept that with a better understanding of the process higher quality products could be produced. One way this is achieved is by operating a process in plateau regions of variables as opposed to operating in regions where minor condition changes can significantly change concentrations and reaction rates. A simple example of this is using a buffer to set pH instead of operating with an unbuffered pH. Temperature and its relation to products and impurities is another simple example. To find these stable process variable regions requires a better understanding of the underlying chemistry and its sensitivity to conditions. To develop this understanding requires careful exploration of the variable space. However, given the short time often allotted in the overall path to market for the process development phase of pharma research this enhanced understanding of the process can probably only be achieved with high throughput concepts.

However, in the petrochemical area, other drivers, which also inspired the concept of process intensification, appeared at about the same time. These drivers were based on the fact that new more efficient, lower cost, processes were needed due to shrinking margins related to increasing energy and raw materials costs as well as increasing waste disposal costs. While these drivers are somewhat different than the pharmaceutical and specialty chemical drivers the need is somewhat the same, which is a much better understanding of the chemical process in less time and preferably at lower cost. A better understanding of the chemical process clearly requires more data and this need logically led to the introduction of high throughput concepts into the process development area.

As mentioned above, researchers in the area of catalyst discovery such as VandenBussche at UOP made the move from discovery to process development very quickly since the equipment required for discovery lent itself very well to longer-term physical variable studies. The gas phase catalyst discovery research was generally carried out with very carefully designed microscale metal laboratory reactors operating in a continuous mode. As mentioned previously, it was recognized at a very early stage that by scaling down not only could speed be improved but also many of the hardware limitations such as heat transfer and mass transfer could be reduced, enabling access to the intrinsic chemistry rates.

In the pharmaceutical and specialty chemical research areas, process characterization research was more commonly carried out by using parallel glass batch reactors. While this approach generates useful process development information, it is often difficult to get the intrinsic reaction rates since it is often hard to have an accurate understanding of the heat and mass transfer characteristics of the laboratory scale reaction equipment. Thus, it is important to test the reaction characteristics at every 10–100-fold increase in production volume to insure no unwanted byproducts are beginning to appear. Finally, this approach makes it somewhat more difficult to explore new reaction regimes such as operating above the boiling point of the solvent, which can be done in pressure equipment, or the use of very reactive reagents such as in the direct fluorination of organics, which can be done in falling film reactors.

4.8
Microreactors for Process Development

While the many features of microreactors such as rapid heat and mass transfer as well as safety and low material usage would make them a reasonable choice for high throughput process development, the potential for modeling offers the greatest opportunity for significant new capabilities. The basic concept is that as more and more data is developed there is a need to convert the information into more useable knowledge about the process. For process development one of the more powerful ways to summarize information is to determine a rate expression for the process. From this information the chemical engineer can begin to design an efficient and highly controllable process. To develop a true rate expression the intrinsic chemistry must be deconvoluted from the measured chemistry by removing the impact of the laboratory equipment characteristics, which is easier with good models of that equipment. In addition, while not as easily done, the potential to explore a wide range of variables offers the possibility that the mechanism of the reaction can be determined. This information would be very useful to the process organic chemist if the reaction needs to be altered to be optimized.

A flow reactor tends to be easy to automate and high precision measurements of conditions are easily achieved by measuring before and after a variable change. High precision is important when trends are being determined that can have fairly subtle impacts on the chemistry. One method for liquid phase reactions that will be described in Chapter 13 by Hickman and Sobeck is the use of a pulse reactant injection technique. This approach uses very small amounts of reagents and has high precision as the measurement can be made before, during and after the reaction to eliminate any system drift. The basic concept of their approach is to use an LC type injection valve in both feed streams to their microreactor. By using synchronized valve injections and controlling flow path volumes and flow rates they can insure that both reagent plugs reach the reactor at the same time. In addition they have control of the reactor temperature, pressure (pressure control enables the study of reactions above the normal boiling point of the solvent), and residence time. Also, they can vary the reagent concentrations by using a robot to load the valves with desired reagents of known concentrations. To follow their chemistry in real time they have integrated an on-line analysis system into their flow path. Their fully automated system gives them the ability to explore a wide range of variables to develop the sought for rate expressions.

4.9
Current Barriers and Limitations to Microscale Reaction Characterization

While microreactors tend to offer advantages for reaction control, which facilitates careful study, not all reactions are equally easy to perform in the systems. General purpose equipment exists to study gas phase reactions as well as gas phase heterogeneous catalyst reactions. However, while liquid phase reactions are also easily studied, specialized microreactor systems are usually built to study liquid phase heterogeneous catalyst reactions. Consequently, these systems are much more specialized and are often tuned to the specific chemistry under study.

In a general sense particles are a major concern for reactions in microreactors due to the potential for plugging. Therefore, not only is it a problem with heterogeneous catalysts but it is a major limitation as many reactions use precipitate formation to drive the reaction to completion. A related problem is encountered when studying reactions that are driven by the removal of small molecules from the reaction mixture as the small channels make this more difficult.

As Hickman and Sobeck point out, it is important to characterize the microreactor before using a system for reaction characterization. As the current generation of microreactors were not designed for use as reaction characterization systems, the performance may not be usable for this purpose. As they describe, care must be taken to understand how the reactor works so that it can be modeled to extract the intrinsic chemistry rates of the reactions. In addition, they point out that given the high surface-to-volume ratio of the reactors care must also be taken to demonstrate that there are not unexpected interactions of the materials with the reactor walls.

4.10
Conclusion

Many existing chemistries can be studied over very wide variable ranges efficiently, safely and rapidly in microreactor equipment, especially when the reactor is interfaced to on-line analysis equipment. Unfortunately, a significant number of reactions are still difficult to study in this equipment. However, notably, since microreactors can also be numbered up to provide commercial scale production, new chemical routes to product production may be possible. Thus, reactions that would not have been deemed reasonable for large volume production in traditional processing equipment due to problems with heat and mass transfer issues may now be possible in microscale reactors. The good news is that it should be possible to explore these new chemistries much more rapidly.

References

1 J. A. Loch, R. H. Crabtree, Rapid screening and combinatorial methods in homogeneous organometallic catalysis, *Pure Appl. Chem.*, **2001**, 73(1) 119–128.

2 J. Scheidtmann, P. A. Weiss, W. F. Maier, Hunting for better catalysts and materials – combinatorial chemistry and high throughput technology, *Appl. Catal. A* **2001**, 222, 79–89.

3 J. Cawse, Application of six sigma quality tools to high throughput and combinatorial materials development, Presented at AIChE National Meeting, November 15, **2000**.

4 K. VandenBussche, Reaction engineering aspects of the Sintef-UOP combinatorial chemistry effort, CPAC Summer Institute meeting July **2001**.

Part II
Technology Developments and Case Studies

5
Introduction

Melvin V. Koch, Ray W. Chrisman, and Kurt M. VandenBussche

In Part I, the origin of this book was explored and the areas of process intensification, microtechnology and high throughput techniques have been elaborated. These seemingly disjointed areas of knowledge are starting to interact synergistically in the field of the design of new chemical conversion and separation processes and are expected to spread to the rapidly expanding field of bioprocessing.

This trend will grow in importance as calculation speed increases, experience expands, and a larger body of scientists graduate with these tools as part of their knowledge base. The interaction of the three disciplines is graphically represented in Fig. 4.1 in the preceding chapter.

This figure shows how the collection of data (through high throughput techniques or similar approaches) using advanced analytical techniques leads to the design of equipment (lab scale or even commercial) which, in its turn, using proper data-acquisition, leads to further suggestions for data generation. Using a combination of thorough data collection and analysis, scale up factors upwards of 10^6 have been reported, i.e., based on data taken on 10 cm^3 of catalyst, decisions are made on reloading a vessel with 10 to 100 m^3 of catalyst.

Applying proper data collection and analysis is a broad topic that encompasses, among other things, exact knowledge of flow rates, composition, three-dimensional temperature profiles, and pressures. It also requires understanding of the typical mass and heat transfer limitations and flow distribution effects. Accurate and representative sampling of the product or an intermediary is critical, as is analysis under representative conditions. In general, the importance of in-situ measurement is often underestimated. Advances in ways to combine effective sampling and sample conditioning with monitoring tools is indeed showing value and will be described in Part II.

An elegant means of gathering intrinsic data is the use of microtechnology, where heat and mass transfer as well as pressure drop can be characterized with great accuracy. Combined with proper resources (like catalyst particle size), microreactors provide direct access to the underlying mechanisms and the kinetics of the various steps.

Micro Instrumentation for High Throughput Experimentation and Process Intensification – a Tool for PAT
Edited by M. V. Koch, K. M. VandenBussche and R. W. Chrisman
Copyright © 2007 WILEY-VCH Verlag GmbH & Co. KGaA, Weinheim
ISBN: 978-3-527-31425-6

The scale-up of microtechnology towards world-scale production is controversial. The conventional mantra of numbering up or "scaling out" by massively parallelizing micro-equipment is often questioned due to issues of plugging, flow distribution and associated cost and will probably require years of demonstration before being fully embraced by the processing industry. We feel microtechnology is more likely to show near term value where it is incorporated into macro-equipment to fulfill a critical role. This concept is referred to as multi-scale design and can be applied to various areas where precise control of conditions (mixing, temperature, pressure) is crucial.

Part II elaborates on some of the topics presented and discussed in recent CPAC Summer Institute sessions at the University of Washington. The impact of microinstrumentation on high throughput experimentation and process intensification will be highlighted in these chapters as a valuable and emerging field of technology. The initial chapters describe fundamentals and hardware developments and some techniques to monitor and characterize them. These are followed by a series of chapters that emphasize the applications and impact of the microdevices.

Professor Volker Hessel of the Institute for Micro-Technology at Mainz (IMM) in Mainz, Germany and professor at Eindhoven University of Technology (The Netherlands) describes the status of microtechnology developments and their characterization (Chapter 6). Dr. Daniel Palo and colleagues at the Pacific Northwest National Laboratories (PNNL) in Richland, Washington, discuss the development of non-reactor micro-components (Chapter 7). Professor Bruce Finlayson of the University of Washington's Chemical Engineering Department addresses the status of micro-component flow characterization, a valuable tool for understanding the operation of microdevices (Chapter 8). This chapter is constructed from a project that was undertaken by a group of Chemical Engineering students and they are listed as his co-authors.

Chapter 9 is devoted to a selection of analytical technologies that have been miniaturized for use in detection, monitoring and characterizing microdevices, as well as being deployed in process and product applications. These include Raman Spectroscopy by Dr. Brian Marquardt of CPAC (Chapter 9.2), Bioassay via Surface Plasmon Resonance Sensors by Dr. Clem Furlong (Chapter 9.5), NMR technology by Professor Michael McCarthy of the University of California at Davis (Chapter 9.4), Grated Light Reflective Spectroscopy by Dr. Lloyd Burgess of CPAC (Chapter 9.8), Ultra Micro-gas Analyzers by Dr. Ulrich Bonne of Honeywell and colleagues at DARPA and the University of Michigan, USA (Chapter 9.3), Fringe Field Sensors by Professor Alex Mamishev of the University of Washington's Electrical Engineering Department (Chapter 9.6), and Micro-liquid Chromatography by Dr. Scott Gilbert of Crystal Vision (Chapter 9.7). Many of these developments have been associated with CPAC researchers.

Dr. David Veltkamp of CPAC describes in Chapter 10 the developments and present status of NeSSI (New Sampling and Sensor Initiative) and how it could enhance the use of microsystems. This initiative has progressed from primarily a standard for process sampling to a potential base for studies involving microreactors and analytical tools.

One of the most successful applications of microtechnology has been in the area of catalyst characterization for gas phase processes. This is described in Chapter 11, prepared by Dr. Michelle Cohn and Doug Galloway of UOP LLC of Des Plaines, Illinois, USA. The ability to use integrated microtechnology systems for characterizing gas phase reactions is presented in Chapter 12, prepared by Dr. Patrick Mills of Texas A&M and colleagues at DuPont de Nemours, Wilmington, Delaware, USA and at MIT, Cambridge, Massachusetts, USA. An example where microtechnology has been demonstrated with liquid systems is presented in Chapter 13, written by Dr. Daniel Hickman and Daniel Sobeck of The Dow Chemical Company, Midland, Michigan, USA.

Part II then concludes with a section where some examples are presented that show recent successes with microtechnology systems for studying and controlling chemical reactions. Dr. Paul Watts of Hull University in Hull, United Kingdom, describes novel micro-systems for carrying out new routes to exploring chemical reactions in Chapter 14. Dr. Hanns Wurziger of Merck KGaA, Darmstadt, Germany then describes an example where laboratory developments of a microreactor system are taken into a pilot plant and subsequently successfully demonstrated at a production plant level in a number up mode (Chapter 15).

6
Microreactor Concepts and Processing

Volker Hessel, Patrick Löb, Holger Löwe, and Gunther Kolb

6.1
Introduction

Today, microreaction technology is considered to deliver novel innovative tools for the chemical processing industry [1–13] and to change also the way of chemical processing [14]. The technique is now mature for industry and has entered nearly all the R&D laboratories of large chemical and pharmaceutical companies and also some of the SMEs in the field. By virtue of approaching the upper limits of what can be considered to still result in structured micro-flows with all their beneficial characteristics, and by combination within existing technology ("plant upgrading"), the technology is nowadays performed on a production scale for selected applications at industrial sites. Thus, commercial business has begun and the first idea of profitability, margins, and amortization has begin to circulate. To achieve this it was essential that either the pure microreactor device supply was supplemented by planning, constructing, and operating microreactor plants or to show how to implement the microstructured reactors into traditional plant architecture. It was also required to change somewhat the ways of chemical processing and to develop devices that industry is willing to use within their production regimes. Thus, the establishment of microreactor-suited and -tailored process development and processing was a milestone in getting the technology accepted. This is shown in this chapter, mainly by the examples of collaborations of industry with the Institut für Mikrotechnik Mainz GmbH (IMM) and in-house developments at IMM.

6.2
Microreactor Technology – Interfacing and Discipline Cross-boundary Research

At the beginning of microreactor research, many papers focused on establishing suitable microfabrication and assembly techniques and having system integration, following the success stories of MEMS fabrication and packaging [15]. The main

Micro Instrumentation for High Throughput Experimentation and Process Intensification – a Tool for PAT
Edited by M. V. Koch, K. M. VandenBussche and R. W. Chrisman
Copyright © 2007 WILEY-VCH Verlag GmbH & Co. KGaA, Weinheim
ISBN: 978-3-527-31425-6

achievements at that time were to supplement the first-hour microfluidic chip materials, silicon and polymers, with ones having other mechanical, chemical or optical features, such as glass, steels and their special alloys, and ceramics. Later, specialty materials like titanium were introduced as well. In this way, the materials' palette familiar to the chemical engineer was available – along with some more new specialty materials.

The interconnection and arrangement of many microreactor units on one fluidic platform format was the next issue covered (see Refs. [16, 17]). Researchers also felt the need to show some demonstration examples to interest the scientific community and, especially, the chemical industry [2–4, 8, 13]. Initially, the state-of-the-art at that time remained at the premature and scouting stage, at least when compared with similar conventional testing. Nonetheless, such functional tests served to validate and revise reactor engineering, which, in turn, rose to quite a high standard during the first five years or so of development (assuming the 1995 as starting point of world-wide extended, communicated and co-operative efforts).

In about 2000, many microreactors were thus available, some even commercially and as off-the-shelf products. That these tools can improve reaction engineering was evident from more and more papers. Therefore, the raw potential was clear and anybody was free to use it. Nonetheless, academia and industry did not, largely, take this chance; at least not for high-level science or commercial business. Development seemed to be stuck and relevant mainly for a small circle of specially interested scientists and developers.

The missing items were soon recognized, which is to minimize fouling, to solve scale-up, to have interconnection to industrial plant environments, to develop

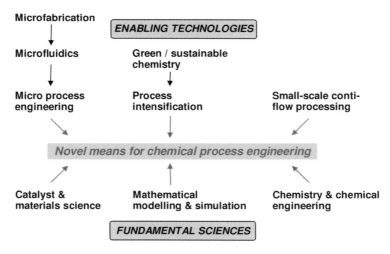

Fig. 6.1 Microreactor technology needs, as an enabling technology, stimulus from the fundamental sciences. (Source: IMM.)
Modern trends in chemical engineering and chemistry are often successful when crossing discipline boundaries.

devices not to the limits of what is technically feasible but to what is needed, to provide complete plant packages, and – probably most important – to "upgrade" microreactor development as an enabling technology by the fundamental sciences (Fig. 6.1) [18].

The most important fundamental sciences here are catalyst & material science, simulation & modeling, and all the special chemical and chemical engineering research fields such as organic chemistry, polymer chemistry, fuel processing, and many more. In this way, the requests (the market view) were addressed by the fundamental sciences (Fig. 6.2) and this changed the way enabling microreactor technology was treated. Technically, this was, for example, achieved by including microreactor sessions and topics in all leading conferences, thus giving platforms for researchers from both sides. Now, and still ongoing, microreactor rigs are used by an increasing number of researchers world-wide and they have almost become a routine development tool. This is a truly interdisciplinary affair, similar to that in other modern areas such as nanotechnology or biomedical science.

Three major impacts were achieved thereby:

- Microstructured reactors were upgraded by better reaction engineering and functionalization [19] (e.g., with new materials, sensors, etc.).
- Plants or means for interconnection and process control were developed to enable non-specialists in the field to also use the technology [1].
- Microreactor processing was revised by comparison with traditional means at a laboratory level [11, 14].

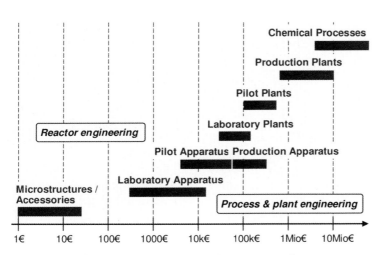

Fig. 6.2 Costs of typical microreactor items and services, given here for fine-chemical scenarios. (Source: IMM.) For bulk processes, deviations to the upper limit are to be expected. The costs at the lower edge and medium range are known today; the costs at the upper edge (production plants and chemical processes) were estimated from first evidence in respective first-hour contracts.

In this way, more and better data were gathered. The ever existing industrial interest now became stronger, due also to the success of industry's in-house developments performed in parallel with the efforts of academia. Now, a new demand was raised for:

- Production-type microstructured reactors and plants and how to operate them (see, e.g., Refs. [12, 20–22]).

Thus, in summary, the development of processing and plants [1], where needed supplemented by tool development, is among the most remarkable developments in recent years – initially for laboratory level and then for pilots and production – and this is reported in the following.

6.3
Microstructured Mixer-reactors for Pilot and Production Range and Scale-out Issues

Presently, most commercial production-oriented microfluidic efforts are made for fine chemistry, since here one typically relies on simple liquid-phase processing in favorable temperature ranges (< 200 °C). A reasonable share of such reactions are mixing sensitive and/or suffer from hot spots due to reaction heat releases [3, 4, 8, 11, 13].

Mixing with microstructured mixers is an operation of very short residence time and thus allows, on its own, high throughputs. The bottleneck, however, is typically the heat-transfer and reaction section, which has a much longer flow path. However, practical solutions have been proposed here as well. It is often enough to diminish the hot spot in the initial interval of the reaction time, i.e., at the start of the flow path. This implies that elements with reasonably short residence time (and usually correspondingly low pressure drop) can be used, while maintaining the advantages of microreaction technology. If needed, residence time can be prolonged by having elements with larger internal dimensions added ("multi-scale technology", see Ref. [23]), having less pressure drop. For all these reasons, mixer-reactors and associated mixer-tube reactors (sometimes even "micro"-sized tubes on their own) were among the very first microfluidic devices to enter industry's production scale. The term mixer-reactor will be used in the following when the reaction has considerably started already in the mixing element, otherwise the function will be restricted to pre-mixing and an elongated delay-loop element will be added.

6.3.1
Caterpillar Microstructured Mixer-reactors

Caterpillar microstructured mixer-reactors owe their name to their total microchannel shape, characterized by alternately up- and down-lifting ramps at the floor and ceiling of the fluid path, which as a whole resemble the fringes along the body of a caterpillar [1, 9, 24–26] (see also Refs. [1, 9]). In a micromixing definition, such devices belong to the class of bas-relief mixers, which are available in a large design choice, e.g., for operation in low, medium, and high Reynolds (Re) number ranges. At high Re, convective mixing by flow recirculation is induced, stretching interfaces up to exponential growth, which may result in a chaotic flow. For specialty applications at low Re, e.g., mixing of viscous media, caterpillar mixer-reactors can be equipped with a splitting plane so that a nearly perfect serial multi-lamination pattern is induced, solely using diffusion mixing by shortening diffusion distances.

Besides the well-understood fluid dynamics, the caterpillar concept profits from robustness (large internal diameter; easy cleaning by screwed two-plate design) and first scale-out strategies, allowing one easily to approach the several-100 L h^{-1}-range with one device only [27]. Certainly, such concepts are still more governed by fabrication capabilities than by chemical engineering logics and the engineering documentation for such scale-out is lagging behind. However, the experimental characterization of mixing efficiency for caterpillar devices of various throughput has begun, see below, and will give more rationale in future.

The caterpillar micromixer consists of a number of serial oriented unit cells that repeat and complete the same type of mixing process. Eight such cells are serially combined in the standard version that is commercially available. Dependent on the mixing problem, however, more or less units may be appropriate, which, especially for production, needs to be optimized to reduce the pressure drop to the limit really needed and for efficient power dissipation. For this reason, caterpillar devices (600 μm width and depth) with 0, 2, 4, 6, and 8 mixing cells have been manufactured to test mixing efficiency by a standardized protocol (Fig. 6.3) [27].

There are two ways for scale-out with caterpillar mixers. One is to enlarge the mixing unit cell, i.e., the corresponding channel width, depth and length, while keeping the width–depth length ratio constant, i.e., maintaining the type of ramp

Fig. 6.3 Caterpillar mixers (600 μm inner channel width and depth, all) with 0, 2, 4, 8, or 12 mixing units, connected in a serial manner.
For better visualization, the end of the units is marked by a black line [27]. From there, a simple unstructured channel follows until the outlet connector. The combined length of structured and unstructured sections is always kept the same. Dotted lines show the respective increase in mixing length for the mixer devices from left to right.

Fig. 6.4 Caterpillar mixers (eight mixing units, all) with 800, 1200, or 2400 μm channel width and depth [27].

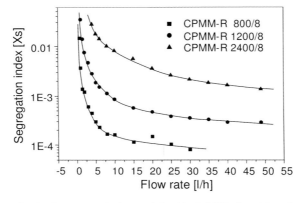

Fig. 6.5 Segregation index, as defined in Ref. [28], for mixing of aqueous solutions in caterpillar mixers with 800, 1200, or 2400 μm channel width and depth [27]. The smaller the segregation index, the better is the mixing and the shorter is the mixing time.

structure [27]. In this way, three caterpillar mixers (8 mixing units, all) with 800, 1200, or 2400 μm channel width and depth have been manufactured (Fig. 6.4).

These mixers have been tested for their mixing efficiency (Fig. 6.5) [27].

Evidently, mixing efficiency improves strongly with miniaturization of the mixing elements for the caterpillar micromixer [27]. For all three mixers tested, mixing efficiency improves with increasing flow rate, which is due to more intense recirculation patterns and thus interfacial stretching. The slope is steeper for the smaller caterpillar micromixers, i.e., the 800-μm device shows a more pronounced increase of mixing efficiency with flow rate.

The second way for scale-out refers to having, in parallel, several caterpillar channels with, ahead, a flow-distribution unit (internal numbering-up) or to simply combine a few of the devices with an external flow manifold. This has not yet been realized. The first route with a slight increase in internal dimensions into the meso range was preferred to avoid problems with flow distribution.

6.3.2
StarLam Microstructured Mixer-reactors

StarLam microstructured mixers consist of disk-like platelets with a hollow star-shaped feeding zone where the fluid is re-directed into an inner central circular mixing zone (Fig. 6.6) [21] (see also Refs. [1, 9]). Many such platelets form a complete, rod-shaped inner mixing channel into which the several streams are injected, virtually via small slit-like nozzles. For efficient mixing, injection in the turbulent range is advised. The flow properties at small Re are less well known. The low mixing efficiencies determined here point to segregated structures, i.e., flow from the "nozzles" is followed by segregated fluid compartments.

This concept has attracted increasing attention from industrial clients, since the StarLam mixer can easily be cleaned, enables large throughputs (up to 1000 L h^{-1}) at a compact size of no larger than a fist, and can be simply attached in existing plants via industrially used flange connectors (see Fig. 6.6) (alternatively, the common Swagelok connectors may be used) [21].

StarLam mixers use the internal numbering-up ("equaling-up") concept for throughput enhancement, by simply adding more platelets into the stack (Fig. 6.7) [27].

Mixing efficiency studies using the segregation index show that best mixing is achieved at small stacks (Fig. 6.8) [27]. The longer the stacks, the more demanding seem to be the issues of flow equipartition. Furthermore, the span in residence time in the mixing element becomes larger. The mixing efficiency is, therefore, less than for typical laboratory micromixers such as the IMM interdigital and caterpillar mixers. Accordingly, the increase in throughput within the StarLam series is paid for by a decrease in mixing efficiency. Nonetheless, the segregation indices are

Fig. 6.6 Generic assembly of StarLam mixers. A two-set series of platelets with star-shaped and circular breakthroughs is stacked and oriented by two guiding pins in a recess of a three-piece housing. By screwing, the platelets are pressed into contact. Industrial flange connectors make insertion of these devices into bypasses of production lines feasible. At laboratory scale, Swagelok connectors may be used. A thorn (not visible) held from the bottom plate and placed in the centre of the mixing channel compensates for the increasing volume flow injected, thus keeping the flow rate along the channel constant.

Fig. 6.7 Equaling-up for the StarLam mixer series by varying the number of microstructured platelets [27]. To have the same volume filled within the recess of the housing, a "dummy" block with inner channel is placed on top of the platelet stack for compensation. The images from the upper left to the upper right to the same sequence of the lower images corresponds to 25, 50, 75, and 100% platelet load (respectively, 75, 50, 25, and 0% replacement by the "dummy" block, see indexing also in Figs. 6.8 and 6.10).

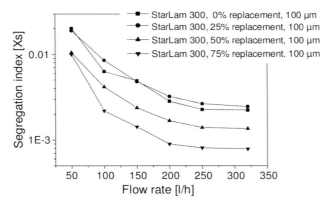

Fig. 6.8 Segregation index, as defined in Ref. [28], for mixing of aqueous solutions in a StarLam300 mixer with a varied number of microstructured platelets, all 100 μm thin [27]. The curves correspond to 25, 50, 75, and 100% platelet load, respectively, 75, 50, 25, and 0% replacement by the "dummy" block. The smaller the segregation index, the better the mixing and the shorter is the mixing time.

still so low that considerable improvement is to be expected compared with the performance of conventional mixing equipment such as impellers. This becomes more apparent when calculating the mixing time from the mixing-efficiency experiments. StarLam mixers have mixing times of less than 10 ms.

There is also another scale-out issue similar to that practiced for the caterpillar mixer. By slight enlargement of internal dimensions, throughput may be increased, while keeping the number of platelets constant [27]. The two dimensions of interest are the thickness (50, 100, or 250 μm) and diameter of the platelet, which means changing the mixing channel diameter. Typically, both specs are changed concurrently. For this reason, the StarLam3000 mixer also has larger outer (housing) dimensions than its StarLam300 and StarLam30 counterparts (Fig. 6.9).

For constant platelet load, the mixing efficiency improves with decreasing platelet diameter (which means increasing pressure drop and throughput) (Fig. 6.10) [27].

Fig. 6.9 StarLam300 (left) and StarLam3000 (right, back) microstructured mixers, with Swagelok connectors [29].

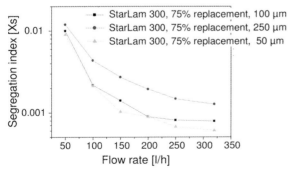

Fig. 6.10 Segregation index, as defined in Ref. [28], for mixing of aqueous solutions in a StarLam300 mixer with varied thickness (50, 100, 250 μm) of the microstructured platelets at constant platelet number [27]. The smaller the segregation index, the better the mixing and the shorter is the mixing time.

6.3.3
Microstructured Heat Exchanger-reactors

For many reactions, the microstructured mixers are simply followed by tube reactors, e.g., double-mantled tubes or shell-in-tube heat exchangers. The benefit from mixing in combination with the benefits of the continuous-flow process through the tube give process intensification, while providing reliable operation, especially when for larger volume flow ranges.

If fast pre-heating is required, larger heats are released, or if quenching of the hot reaction fluid is needed (or for any other more demanding heat-transfer task), cross- or counter-flow microstructured heat exchangers may be the appropriate solution (Fig. 6.11). They typically have a stacked plate-type design, similar to conventional analogues.

For fluids that are liable to fouling, a straight channel routing within the device at not too small an internal diameter is advised. Such microstructured a heat exchanger has been developed with the channels arranged outside of a tube, i.e., on a circle (Fig. 6.12). Inside the tube, a heating cartridge provides electrical heating. This microstructured tube is placed tightly into a hollow cylinder, acting as housing to ceil the microchannels.

At present, virtually all microreactor processing demands can be fulfilled by the tools mentioned above, the conventional or microstructured heat exchangers. Pilot operation with microstructured reactors has been performed up to about 100 L h^{-1} (the microstructured mixers being operated up to about 4000 L h^{-1}, see above). However, microstructured units with a larger capacity for "heavy pilots" may be required in the future, at least in selected cases. The technological basis for this is rooted in the development of such large-capacity microstructured heat exchangers for gas/gas heat exchange (up to 10 kW) for fuel processing for fuel cells (Fig. 6.13). While the design may be different, when applied for liquid heat exchange, the microfabrication and assembly technique is in place.

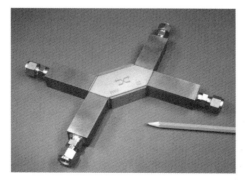 

Fig. 6.11 Microstructured counter-flow heat exchanger with stacked plate design (left); microstructured cross-flow heat exchanger with stacked plate design (right). (Courtesy of the American Chemical Society.)

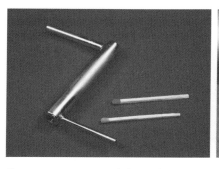
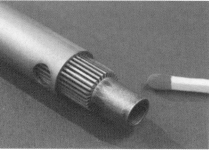

Fig. 6.12 Microstructured electrical heat exchanger with channel-on-a-circle design. Left: assembled device; right: microstructured inner cylinder, partly out of the hollow cylinder housing. (Courtesy of the American Chemical Society.) The recess inside the inner cylinder serves for carrying the heating cartridge.

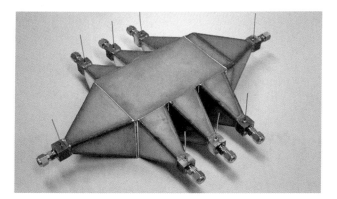

Fig. 6.13 Example of a "heavy-pilot" microstructured heat exchanger: integrated 5 kW$_{el}$ reactor–heat exchanger system for selective oxidation to achieve gas purification of H$_2$-rich reformer gas for fuel cells. (Source: IMM.)
Similar designs may be used for future fine-chemical pilot and production plants operated at much higher throughputs than today.

6.4
Fine-chemical Microreactor Plants

6.4.1
Laboratory-range Plants

The specific nature of microreactors at the laboratory scale, e.g., due to their small size, continuous way of operation, and their interconnectors, raises the need for compact tailored platforms, and indeed suppliers have provided such complete systems, i.e., as bench-scale microreactor plants [2]. Meanwhile, a choice of modular multi-purpose or dedicated microreactor laboratory plants is available on the market, approaching the pilot-scale level [2]. Still, suitable tools for downstream processing are missing to satisfy all customers' needs.

One example of such laboratory plants is the table-top sized fine-chemical plant shown in Figs. 6.14 and 6.15. In a typical version, such plant consists of microstructured heat exchangers for heating up, microstructured mixers, a set of parallel delay loops switched into operation by a multi-port valve, and a microstructured heat exchanger for cooling down (see flow scheme in Fig. 6.14). It can variably be equipped with various microstructured components for either laboratory or pilot-scale applications. The versions are customized according to the process specs of the user, and so the individual flow sheets may differ in the choice of microstructured components and their connection. The complete plant is immersed in a thermostat upside-down, providing one temperature level throughout the whole tools and tubing (Fig. 6.15).

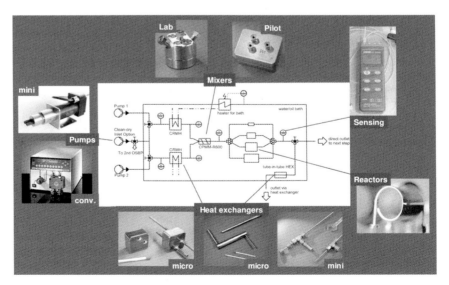

Fig. 6.14 Flow sheet of a fine chemical continuous-flow microreactor plant of table-top size for throughputs from 1 to 100 L h^{-1}. (Source: IMM.)

6.4 Fine-chemical Microreactor Plants

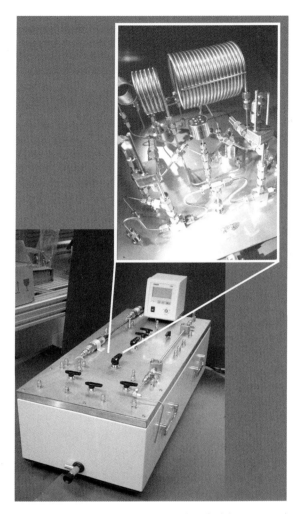

Fig. 6.15 Fine chemical microreactor plant for laboratory- and pilot-range operation, completely assembled (lower image) and microreactor-system core (upper image). (Source: IMM.)

Using a similar construction approach, a cream manufacturing plant has been developed (Fig. 6.16).

Part of the plant (e.g., the pumps and inlets for solids) is above a thermostat bath level, part (e.g., the melting pots for solids, micromixer, all tubing) is immersed into it to be exposed to a hot environment. Solid waxy materials like palmitic or stearic acid are molten in the pots encompassed by a water or oil bath. Being liquefied, these materials are pumped into an eight-component caterpillar micromixer where water and other ingredients are added (Fig. 6.16). In a second stage, another micromixer may feed a further component that is difficult to

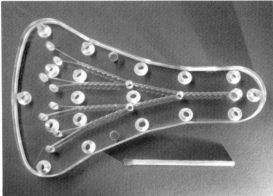

Fig. 6.16 Cream manufacturing plant (top) and its core element, the microstructured eight-component caterpillar mixer (bottom). (Source: IMM.) For better visualization of the flow path, a polymeric mixer is shown. In the plant, however, steel mixers are typically used.

introduce in the first mixing step, e.g., a flagrance of very small volume. The liquid multi-component mixture is either directly introduced into a reservoir with subsequent slow cooling and solidifying or cooled rapidly by using a micro heat exchanger. The cream manufacturing machine can produce about 50 samples of different composition per hour for screening purposes. In terms of production, it can make about 50 L h^{-1} in the current version; larger volumes would need modification of components.

6.4.2
Pilot-range Plants

Microreactor pilot plants consist of, at their core, very similar flow sheets to the laboratory plants and thus in their choice of components and connection. The

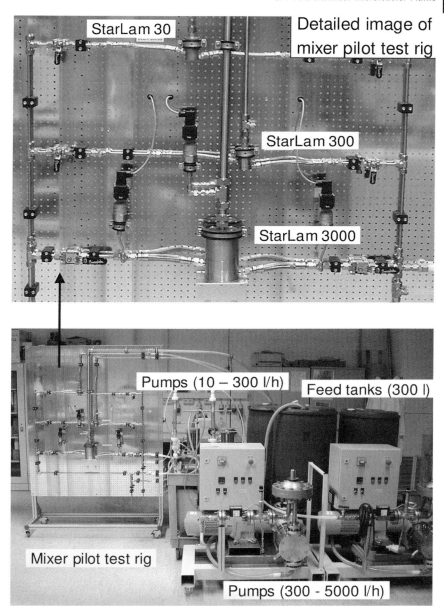

Fig. 6.17 A mixing test station for pilot-sized microstructured mixers up to the 5000-l/h range. Shown above are three StarLam microstructured mixers.
(Source: IMM.)

scaled-out microstructured mixers discussed in Section 6.2 are used here. Conventional (double-mantled) tubes are often used as heat exchangers and reactors. For large heat releases, plate-type microstructured heat exchangers are added instead or in front of the conventional tube-in-tube heat exchangers. Where needed, micro heat exchanger elements are even integrated into the microstructured mixer to withdraw heat already at the initial place of reaction. Around the core of the pilot plants a traditional flow and process control architecture is provided, exceeding the level of functions and system complexity of laboratory versions. Separation tools may be added as well. The few very first such pilot plants cannot be shown here as they are proprietary designs made for industry and operated on-site. However, an in-house pilot rig for testing of pilot-sized microstructured mixers may give an idea of how such fine-chemical plants look (Fig. 6.17).

6.5
Industrial Microreactor Process Development for Fine and Functional Chemistry

The most convincing tests of the new tool microstructured reactor and new type of processing, named chemical micro process engineering, are real-life applications. Some new examples are given below, either with IMM involved as research entity or with IMM tools being used. The next subsection gives the first examples of industrial case studies for chemical production, either with IMM or other suppliers' tools. There are certainly more such examples, some of which are known to the authors; however, these have to remain confidential at present, although some may be made public in the near future. Several subsequent subsections present IMM in-house process developments that were made to be launched to clients.

6.5.1
Phenyl Boronic Acid Synthesis (Scheme 6.1) (Clariant/Frankfurt + IMM)

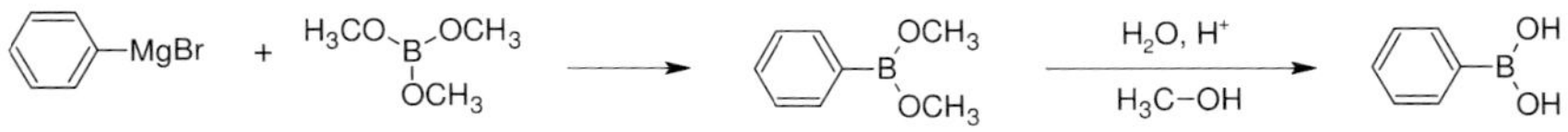

Scheme 6.1

6.5.1.1 Process Development Issue

Many organometallic reactions suffer from insufficient mixing, since these reactions (at room temperature and below) are often faster than the mixing times of many conventional mixer equipments [30]. Thus, the reactions are performed under non-stoichiometric and even temporally and spatially changing concentration profiles, which promotes consecutive reactions. The excessive processing time also allows side reactions such as oxidations and hydrolysis to proceed too far. As a consequence, organometallic reactions are typically carried out under cryogenic conditions to get acceptable selectivity, i.e., the reaction is slowed down. This requires capital investment for cooling utilities and makes the process energy consumptive.

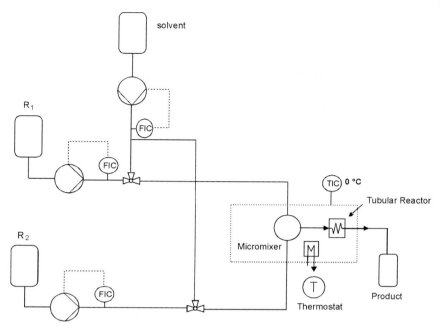

Fig. 6.18 Flow sheet of the laboratory-scale microreactor configuration of the phenyl boronic acid microreactor process, equipped with an interdigital micromixer. (Courtesy of the American Chemical Society [30].)

6.5.1.2 Microreactor Plant and Processing Solution

Micromixer-tube reactor plants have been employed both at laboratory and pilot-scale [30]. For initial process development, a triangular ("focusing") interdigital micromixer was used (Fig. 6.18), while for the pilot scale-out a caterpillar mixer was connected to four tubes of different hydraulic diameter by a five-port valve.

Processing in the microreactor was performed at room temperature instead of using cryogenic temperatures. The micromixer and the tube(s) were simply immersed in a thermostated water bath and no further integrated cooling was involved. Nonetheless, it can be assumed that heat transfer was improved owing to the large heat exchange interfaces and the large reactor mass-to-fluid volume ratio. Mixing was intensified by efficient diffusive and convective mixing with micromixers.

Process Development Results

Phenyl magnesium bromide and boronic acid trimethyl ester react to give phenyl boronic acid with high selectivity (about 90%) even at room temperature, which saves energy costs and the respective CAPEX investment [30]. Process intensification was also achieved by raising the yield by about 25% compared with industrial batch production. The purity of the crude product could be enhanced by about 10%, thereby allowing purification by favorable crystallization only and

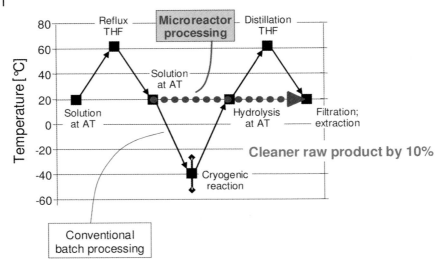

Fig. 6.19 Temperature profile of phenyl boronic acid synthesis along the major steps of the process flow scheme. The difference in temperatures of the conventional batch and microreactor processes stand for the reduction in energy consumption and respective heat-transfer equipment when using the latter. (Courtesy of the American Chemical Society [30].)

rendering the former energy-consumptive distillation step unnecessary (Fig. 6.19). Thus, the higher selectivity not only affected the reaction itself, but also had an impact downstream.

6.5.2
Azo Pigment Yellow 12 Manufacture (Scheme 6.2) (Trust Chem/Hangzhou + IMM)

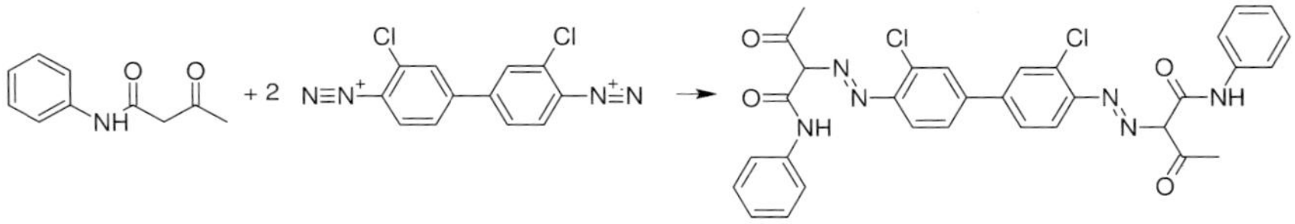

Scheme 6.2

6.5.2.1 Process Development Issue

Micromixers provide very fast mixing, down to milliseconds or even below [9, 26]. The concentration profiles are uniform and there is good agreement between experiment and theory, which may make process development more predictable. Results at the laboratory scale may be maintained throughout the scale-out steps, since the fluid dynamics are not changed or at least shift by analogy in a known manner. This is particularly valid for very fast precipitations such as with Yellow 12 manufacture. Here, the seed formation and agglomeration process is impacted so that narrow-sized, morphologically uniform and (where given) chemically selective crystals result. However, fouling of the mixer's tiny structures may plug the whole system and lead to plant shut down. A careful matching of processing conditions is needed, which may restrict the operational range. The choice of micromixer can diminish fouling problems as not all micromixers have long, narrow flow paths.

6.5.2.2 Microreactor Plant and Processing Solution

Process development was carried out at laboratory scale using a slit-type interdigital micromixer-reactor [31]. The whole flow guidance was provided by the micromixer and the outcoming solution was collected in a vessel.

Process Development Results

For the microreactor synthesis and precipitation, finer particles with more uniform size distribution were yielded for the commercial azo pigment Yellow 12 [31]. This gives an improvement in the optical properties of the particles, such as the glossiness and transparency (Table 6.1). The dye manufacturing is unchanged, as the tinctorial power is the same. Since, therefore, the micromixer-made pigments can be adhered to the product matrix (e.g., wool stuff) with more intense color, lower contents of the costly raw material in the commercial dye products can be employed, which increases the profitability of the pigment manufacture.

Table 6.1 Glossiness of the imprinted color (GU = glossiness units) for micromixer-based and conventional Yellow 12 pigment manufacture [31]. (By courtesy of the American Chemical Society.)

No.	Micromixer (10 mL min^{-1}) (GU)	Micromixer (30 mL min^{-1}) (GU)	Micromixer (50 mL min^{-1}) (GU)	Yellow 12 standard (GU)
1	38.5	49.1	51.9	31.7
2	41.3	47.9	51.1	27.6
3	39.8	45.3	51.0	30.4
4	42.3	46.5	52.0	29.8
5	42.7	46.7	48.8	27.7
Mean (GE)	40.9	47.1	51.0	29.4
σ (GE)	1.8	1.4	1.3	1.8

6.5.3
Hydrogen Peroxide Synthesis (UOP/Chicago + IMM)

6.5.3.1 Process Development Issues

Processing in the explosive envelope is often beneficial, e.g., in terms of space–time yield, since elevated temperatures, pressures, and reactant concentrations are utilized. For some reactant mixtures such as the hydrogen/oxygen case (Scheme 6.3), virtually the whole concentration and operational range is in the explosive regime. Microchannel reactors are said to be inherently safe, since the two major reasons for explosions are not found in microreactors, i.e., thermal runaway and uncontrolled, self-accelerating radical chain propagation [32]. The latter refers to the flame-arrestor effect that quenches chain growth by chain termination at the channel walls due to the large specific interfaces at short diffusion distances. For this catalytic reaction, microreactors also allow rapid catalyst testing and have the option of using the same type of catalyst, e.g., wash-coated, throughout the whole development cycle, up to pilot and production. Finally, mass transfer is improved through the increase in gas–liquid interfaces, and thermal control is also better due to larger exchange surfaces.

$$H_2 + O_2 \longrightarrow H_2O_2$$

Scheme 6.3

6.5.3.2 Microreactor Plant and Processing Solution

Process development was performed at laboratory scale using a slit-type interdigital micromixer-reactor, which was later replaced by a special high-pressure interdigital micromixer [12]. The mixers are followed by a cartridge filled with catalyst, which acts as mini-trickle bed. Many further plant details have been developed for the safe operation of the hazardous process gas mixture, but remain undisclosed.

The new process is realized by direct contacting of hydrogen and oxygen (without inert gas) using a micromixer, in the presence of a heterogeneous catalyst in a trickle-bed mode [12]. The key to high selectivity is to have a noble metal catalyst in a partially oxidized state. Otherwise, either only water is formed or no reaction is achieved.

Process Development Results

Using UOP process specs, a space–time yield of 2 g hydrogen peroxide per g-catalyst-h was achieved, which exceeds literature values. In addition, operation at only 20 bar is considerably lower than for the published processes, and the use of smaller oxygen/hydrogen ratios saves valuable raw materials. Selectivity as high as 85% at 90% conversion has been achieved (at an oxygen/hydrogen ratio of 1.5–3). These laboratory experiments at IMM were followed by pilot tests at UOP, resulting in a basic engineering design for the production of about 150 000-t hydrogen peroxide per year by UOP [12]. This demonstrates that the performance of microstructured units within a plant and process may not necessarily be restricted to small-capacity production.

6.5.4
(S)-2-Acetyl Tetrahydrofuran Synthesis (SK Corporation/Daejeon; IMM Tools)

6.5.4.1 Process Development Issues
In the synthesis of (S)-2-acetyl tetrahydrofuran (Scheme 6.4), the Grignard reagent MeMgCl is very reactive and not easy to handle on a large scale [33]. This results in safety and hazard issues at an industrial scale. Secondly, selectivity issues result from over-alkylation to the tertiary alcohol (Scheme 6.5), an undesired consecutive reaction. This impurity level must be kept to < 0.2%.

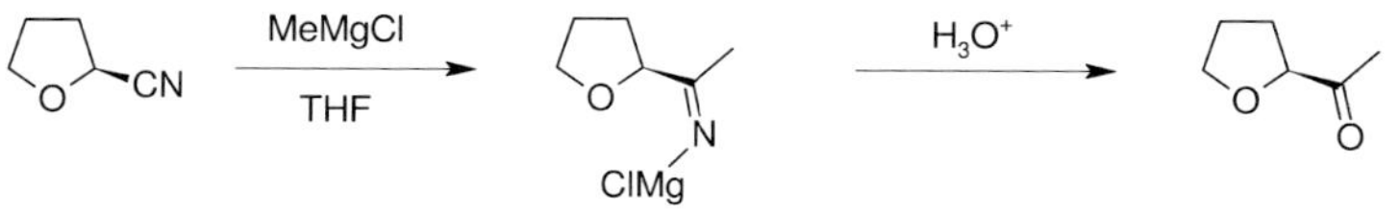

Scheme 6.4

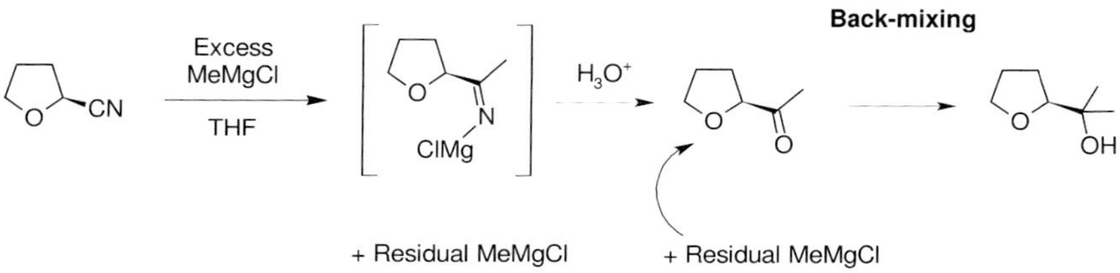

Scheme 6.5

Thirdly, chirality conservation must be maintained, because the α-hydrogen of the reactant is unstable under basic reaction conditions. Therefore, a small degree of racemization occurs, which needs to be minimized.

6.5.4.2 Microreactor Plant and Processing Solution
Details of the devices used and the type of plant as well as on the processing were not given [33].

Process Development Results
The impurity by over-alkylation was 0.18%, while the batch impurity was 1.56% (Table 6.2) [33]. This was explained by the lower back-mixing in the micro-flow system. The optical purity of the microreactor product was 98.4% as compared with 97.9% at batch level.

Table 6.2 Selectivity (individual impurity) and optical purity of the batch and microreactor processes for the (S)-2-acetyltetrahydrofuran synthesis (Scheme 6.4) [34].

Process	Individual impurity (%)	Optical purity (%)
Batch	1.56	97.9
Microreactor	0.18	98.4

6.5.5
Synthesis of Intermediate for Quinolone Antibiotic Drug
(LG Chem/Daejeon; IMM Tools)

6.5.5.1 Process Development Issues

Gemifloxacin (FACTIVE™) is a quinolone antibiotic drug with enhanced activity against Gram positive bacteria [34]. It maintains excellent activity against Gram negative bacteria and atypical strains as well as excellent activity against major respiratory pathogens. During the multi-step synthesis of this compound, an enamine moiety has to be protected with a t-Boc group (Scheme 6.6). The reaction is fast and highly exothermic ($\Delta H = -213$ kJ mol^{-1}). A consecutive reaction occurs with the rate constant $k'_0 = 6.6 \times 10^2$ min^{-1} and activation energy $E'_a = 8.82 \times 10^3$ J mol^{-1}, whereas the main reaction proceeds with a $k_0 = 9.1 \times 10^7$ min^{-1} and $E_a = 3.74 \times 10^4$ J mol^{-1}. Consequently, the side reaction sets in at temperatures > 25 °C. Accordingly, improving heat removal is essential. Also, contact between product and the KOH feed solution should be avoided. Narrowing the residence time distribution is another means of reducing impurities.

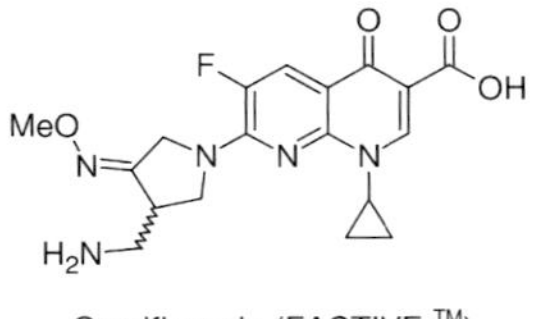

Gemifloxacin (FACTIVE ™)

Structure A

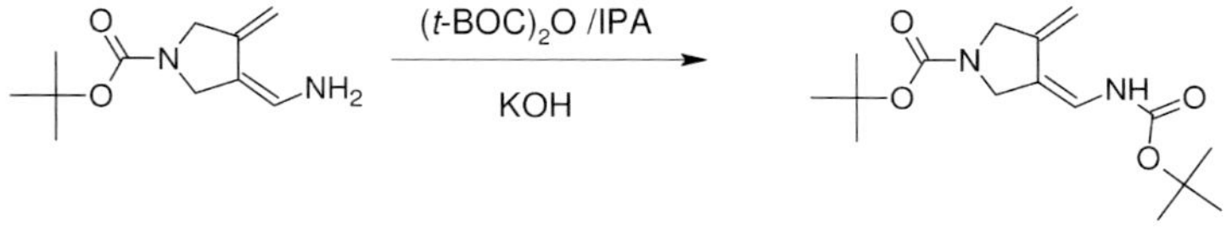

Scheme 6.6

6.5.5.2 Microreactor Plant and Processing Solution

A slit-type interdigital micromixer has been used [34]. Details on the type of plant as well as on the processing were not given.

Process Development Results

Different types of reactors – a tubular reactor, five combinations of Kenics mixers connected in series, and the microstructured reactor – were compared for the t-Boc protection step [34].

For the tube reactor only 27% conversion was achieved [34]. A fast flow rate was required due to phase separation and thus, at $Re > 2000$, about 2 km of tube is needed to have a 5 min residence time, which is impractical. With five Kenics static mixers connected in series a conversion of 97% is achieved. In the microreactor, the heat of reaction was completely removed so that there was no formation of byproducts at conversions as high as 96%. While this microreactor result equals that of the static mixers, the latter need to be operated at 0 °C or –20 °C to avoid side reactions, whereas in the microreactor the reaction temperature was isothermally controlled at 15 °C.

6.6 Industrial Production in Fine Chemistry

6.6.1 Nitroglycerine Production Plant for Acute Cardiac Infarction (Xi'an Huian Industrial Group/Xi'an + IMM)

A plant producing 15 kg h^{-1} of nitroglycerine (Scheme 6.7) has been developed and installed by IMM at the Xi'an Huian industrial group site in China and has been successfully set into operation [35]. The manufactured nitroglycerine is used as medicine for acute cardiac infarction. Therefore, the product quality must be of the highest grade; plant start-up tests already demonstrated higher selectivity and purity by the microreactor operation. The plant is foreseen to operate safely and fully automated. As a second step, a process unit for downstream purification by washing and drying, of notably larger size and complexity than the reactor plant, is going to be developed. Environmental pollution should be excluded by advanced waste water treatment and a closed water cycle. In a final stage, further process units will be added for formulation and packaging to give tablets with nitroglycerine.

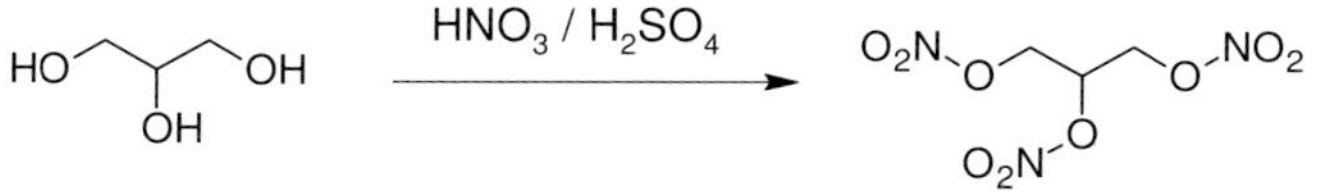

HO–CH(OH)–CH$_2$OH →[HNO$_3$ / H$_2$SO$_4$] O$_2$N–O–CH$_2$–CH(O–NO$_2$)–CH$_2$–O–NO$_2$

Scheme 6.7

6.6.2 Production-oriented Development for LED Materials (MERCK-COVION/Frankfurt + IMM and Other Partners)

A microreactor based plant is currently being designed for the synthesis of polymeric, light-emitting semiconductors for use in displays (see internet description of the German project POKOMI, started in spring 2005, under www.mstonline.de/foerderung/projektliste/detail_html?vb_nr=V3MVT016, 2005 #5166). It will be equipped with process control and on-line analysis. This plant is being developed by IMM and mikroglas chemtec, and will be finally installed at and operated by the MERCK-COVION company, at their Frankfurt/Hoechst site.

Using the example of the Suzuki-coupling reaction as a common synthesis route for polymeric semiconductors, the economic and ecologic improvements for the chemical industry are going to be demonstrated. The aim is to transfer microreactor engineering into chemical production. Generic trends towards reliability, operation times, and controllability of microreactor plants are going to be deduced, in particular concerning the integration of microstructured reactors into existing plants ("plant upgrading"; "multi-scale approach"). This also involves considerations of cost analysis with profitability and amortization time and analyzing corresponding process development time and process safety.

6.6.3
Pilot Plant for MMA Manufacture (Idemitsu Kosan, Chiba/Japan; MCPT Works)

An industrially-suited pilot plant for the free-radical polymerization of methyl methacrylate (MMA) with a capacity of 10 t a^{-1} has been designed and constructed by the Japanese MCPT (Micro Chemical Process Technology) team and has been installed and operated at the industrial site of the Idemitsu Kosan Company in Japan (Fig. 6.20) [36, 37]. Eight microreactor blocks are arranged in a parallel, each consisting of three micro-tube reactors (500 mm internal diameter, 2 m length) in series. The polydispersity index, yield and average number-based molecular weight of the polymer made by the pilot plant resemble the performance of a single microreactor tube, which validates the numbering-up concept applied.

Fig. 6.20 Industrial pilot plant (10 t a^{-1}) with three microreactor blocks for radical polymerization at the Idemitsu site in Japan [37].

6.6.4
Production Operation for High-value Polymer Intermediate Product (DSM Linz/Austria + FZK)

A plant with a customized microstructured reactor has been announced to produce a high-value product for the polymer industry (see internet press release under www.fzk.de/fzk/idcplg?IdcService=FZK&node=2374&document=ID_050927, 2005 #5168). The design and fabrication of this reactor was carried out at the Institut für Mikroverfahrenstechnik (IMVT) in the Forschungszentrum Karlsruhe (FZK), and the installation and operation was performed at DSM Fine Chemicals GmbH in Linz/Austria. During a ten-week production campaign over 300 tons of a precursor for a polymer product was produced by the Ritter reaction. The microstructured reactor (65 cm long; 290 kg in weight, special Nickel alloy, several tens of thousands of micro channels) was operated at a throughput of 1700 kg liquid chemicals per hour and the reaction heat was removed within seconds (Fig. 6.21).

In this way, a central reaction route at DSM was replaced that was previously conducted in a very large reactor tank encasing several tons of explosive and corrosive chemicals. The yield of the microreactor upgraded plant exceeds that of the former route; the process safety for handling the corrosive chemicals was still enhanced when using the microreactor process. The use of raw materials and the waste streams were reduced, improving the cost analysis and eco-efficiency of the process.

Fig. 6.21 Production-type microstructured reactor for throughput at 1700 kg h^{-1} and transfer of a power of 100 kW. This apparatus was used to manufacture a high-value precursor product for the plastics industry at DSM in Linz, Austria [37].

6.7
Microreactor Laboratory-scale Process Developments for Future Industrial Use

6.7.1
Michael Addition – Amine Addition to α,β-Unsaturated Compounds (Fresenius + IMM)

6.7.1.1 Process Development Issues

The batch procedure for the addition of secondary amines to α,β-unsaturated carbonyl compounds (Michael addition; Scheme 6.8), such as acrylic acids, acrylic acid esters or acrylonitrile, gives good yields, in some cases even exceeding 85% [38]. However, the high conversion rates given by the intrinsic kinetics for some such reactions cannot be utilized, since this would be associated with large heat releases due to their great exothermicity. To enable safe removal of the reaction heat created, the α,β-unsaturated compound must be added quite slowly to the amine, which demands long reaction times, typically 17 to 24 h. In addition, the reaction is of an equilibrium type, with the reverse reaction, amine elimination, dominating at high temperature.

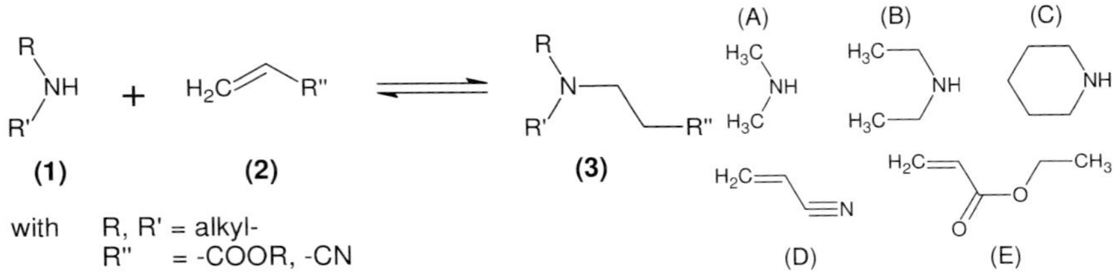

Scheme 6.8

Microstructured reactors enable one to profit from the fast kinetics of the reaction while ensuring efficient heat removal and avoiding thermal overshooting.

6.7.1.2 Microreactor Plant and Processing Solution

Process development was carried out using either a slit-type interdigital micromixer-reactor or a caterpillar micromixer-reactor, connected to a tubular reactor (Fig. 6.22) [38]. The feed solutions were pumped by two syringe pumps and pre-heated by heating capillaries and introduced into the micromixer. The whole flow guidance was provided by the micromixer and the outcoming solution was quenched by a cross-flow micro heat exchanger before being collected in a vessel. A water bath was used for reaction temperatures approaching 100 °C, an oil bath for still higher temperatures. Pressure gauges measure the pressure drop and, thereby, monitor channel blocking. A needle valve adjusts the total system pressure, which is measured by a further pressure gauge.

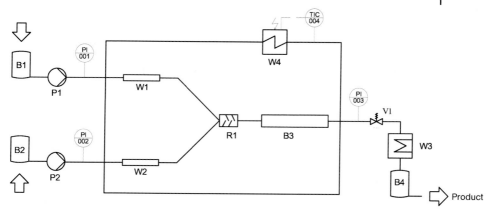

Fig. 6.22 Flow scheme of the experimental setup for performing the Michael addition continuously in a microstructured reactor [38]. B1, B2, supply vessels; P1, P2, syringe pumps; W1, W2, preheating capillaries; R1, microstructured reactor/mixer; B3, tubular reactor; W4, water bath; V1, valve; P001, P002, P003, pressure gauges; W3, cooler/heat exchanger; B4, storage vessel.

Process Development Results

Yields of 95–100% were achieved for four out of the six Michael additions investigated in the microreactor rig, which is higher than for batch operation [38]. For the two diethylamine-related reactions the yields were distinctively lower than for the other microreactor processes and the corresponding batch process.

For the reaction of dimethylamine with acrylonitrile, there is a distinct difference between using undiluted + pressurized and diluted aqueous dimethylamine (Fig. 6.23) [38].

The reaction times for the four high-yield additions were considerably decreased to 1–8 min instead of the corresponding 17–25 h in batch operations [38]. In this way, a high-p,T operation was established for the Michael addition. The

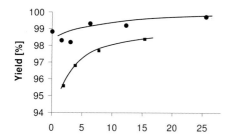

Fig. 6.23 Yield as a function of residence time for the reaction of dimethylamine (■ pure, 20 °C, molar ratio 1.1, 30 bar; ● 40 mass% aqueous solution, 80 °C, molar ratio 2.0, 3–4 bar) with acrylonitrile in the microreactor. (Courtesy of the American Chemical Society [38].)

Table 6.3 Reaction enthalpies of the Michael additions studied.

Reaction[a]	A + D	B + D	C + D	A + E	B + E	C + E
Enthalpy (kJ mol^{-1})	–6.8	+0.5	–9.0	–8.7	+0.7	–6.4

a) See Scheme 6.8 for the structures of compounds A and D.

space–time yield for the microstructured reactors was, in the best case, higher by a factor of about 650 than for batch processing. Thus, high productivities, e.g., 100-g piperidino-propionitrile in 13 min ≈ 9 L day^{-1} were possible also with the laboratory rig. The activity of amines and α,β-unsaturated compounds were determined as piperidine > dimethyl amine > diethyl amine/acrylonitrile > acrylic acid ester.

In addition to the yield investigations, the reaction enthalpies were determined, as many Michael additions exhibit considerable heat release when carried out rapidly [38]. Indeed, four of the reactions were found to be highly exothermic, while two are mildly endothermic (Table 6.3).

The data were gathered by a rough analysis, measuring the temperature shift in a more or less insulated reaction volume [38]. It was found that the microstructured reactors, unlike conventional processing, do not show any temperature-induced shift of conversion, even when approaching high temperatures (90 °C). This is indicative for efficient heat removal. Thus, classical limitations due to hot-spots, setting off reverse reactions, seem not to exist.

6.7.2
Kolbe–Schmitt Synthesis (IMM)

6.7.2.1 Process Development Issues

The Kolbe–Schmitt synthesis using resorcinol with water as solvent to give 2,4-dihydroxybenzoic acid (Scheme 6.9) is a very reactive example of such reactions, since resorcinol is electron rich as compared to other phenols [39, 40]. Although this allows one to circumvent the high-pressure autoclave process typically used, and to establish liquid-phase operation under ambient pressure, reaction times of about 2 h are still needed at 100 °C (reflux conditions).

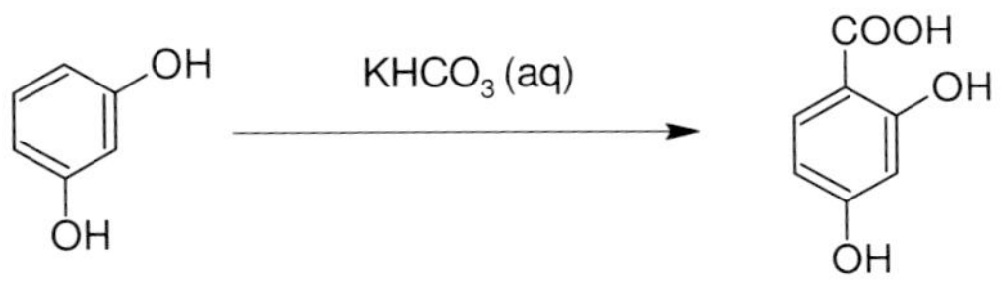

Scheme 6.9

Microreactors allow, by simple means, the operational range to be broadened of liquid-phase reactions that are typically limited from cryogenic temperatures

(> −70 °C) to about 150 °C, and in rare cases to 200 °C or more [39, 40]. Even at the latter highest temperature, organic reactions may remain much too slow for an efficient use in microstructured reactors, which are favorably operated from milliseconds to seconds. Using a microreactor setup, a facile high-p,T operation (p being pressure, T being temperature) beyond 200 °C and 50 bar allows one to speed up reactions by orders of magnitude. The question then is to what extent selectivity may be decreased, since typical side reactions profit if more energy is available.

6.7.2.2 Microreactor Plant and Processing Solution

In a first-hour rig, only the very basic functions concerning heat exchange, pressure hold-up and reaction time provision were included, to enable the initial studies on the high-p,T concept and since problems due to fouling were expected [39, 40]. Thus, initially simply a tube reactor (1–2 mm ID), immersed in an oil bath, with back-pressure regulation was used (Fig. 6.24). Drawbacks were the need for use of pre-mixed solutions (as no mixer was employed) and the potential for their oxidation and pre-reaction (which, however, was assumed to be very small). Also, the temperature rise and decrease before and after reaction were slow, which limited the investigations to low flow rates (< 500 mL h^{-1}) and, thereby, long reaction times. Furthermore, a considerable part of the tube reactor will be used for heating and the reaction will be carried out under quite irregular conditions.

Subsequently, an electrically heated micro heat exchanger was added in front of the tube reactor. Processing at high flow rates (1–4 L h^{-1}), and thus short residence times, was then feasible, while rapidly reaching the set temperature [39, 40]. Another cross-flow heat exchanger following the tube reactor served to quench the hot reaction solution to ambient temperature. In a still further improved version, the laboratory rig was equipped with a micromixer to feed the two reactant solutions separately into the tube reactor (Fig. 6.25). A high-pressure slit-type interdigital micromixer was used that can be operated up to at least 600 bar.

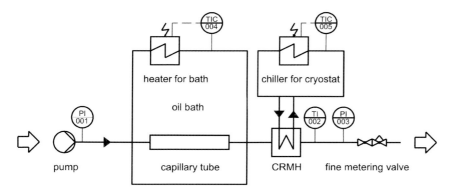

Fig. 6.24 Flow scheme of a simplified rig for performing continuously the aqueous Kolbe–Schmitt synthesis with resorcinol in a microstructured reactor. (Courtesy of the American Chemical Society [39, 40].)

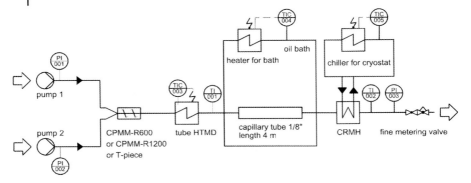

Fig. 6.25 Flow scheme of the upgraded rig for performing continuously the aqueous Kolbe–Schmitt synthesis with resorcinol in a microstructured reactor. (Courtesy of the American Chemical Society [39, 40].)

Process Development Results

The operational range of the Kolbe–Schmitt synthesis using resorcinol with water was extended by about 120 °C to 220 °C, as compared with a standard batch protocol under reflux conditions (100 °C) [39, 40]. The first-hour rig already allowed for facile high-p,T operation, but encountered also many problems with fouling that impacted the quality of product analysis. The yields were at best close to 40% (160 °C; 40 bar; 500 mL h^{-1}; 56 s) at full conversion, which approaches good practice in a laboratory-scale flask under reflux (100 °C). The expansion in the processing window by microreactor operation up to 220 °C afforded a 2000-fold decrease in reaction time and a 440-fold increase in space–time yield. The use of still higher temperatures, however, is limited by the increasing decarboxylation of the electrophilic substitution product (Scheme 6.10), as shown by the yield–temperature relationship (Fig. 6.26).

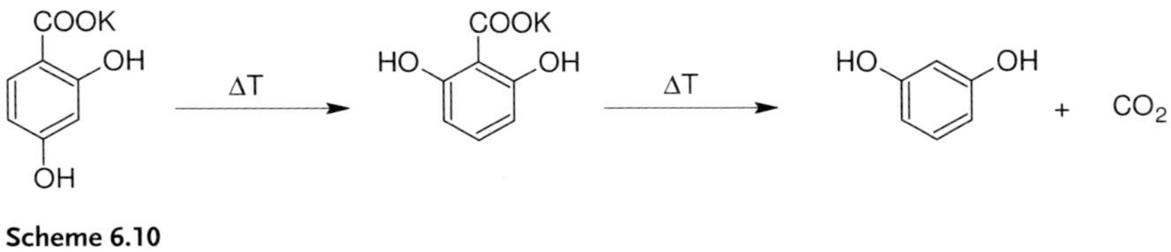

Scheme 6.10

In less than one minute, half of the 2,4-dihydroxybenzoic acid is decomposed already at 160 °C in the microreactor setup [39, 40]. This demonstrates the need to find optimal process parameters for temperature and residence time to achieve efficient high-p,T operation, which is achieved by detailed parameter variation (e.g., as statistical analysis, Design of Experiments, DoE). In this way, the best operation point was found (200 °C; 40 bar; 2000 mL h^{-1}; 16 s). There is also the question of substrate selectivity, e.g., the isomeric product 2,6-dihydroxybenzoic acid exhibits lower decarboxylation rates.

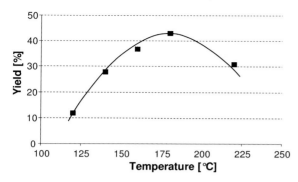

Fig. 6.26 Increase of yield with temperature for the aqueous Kolbe–Schmitt synthesis by higher reaction rates at given residence time, followed by a decrease of yield due to thermal decarboxylation.
(Courtesy of the American Chemical Society [39, 40].)

As a second reaction example, 1,4-dihydroxybenzene (hydroquinone) and 1,3,5-trihydroxybenzene (phloroglucinol) were also converted into their respective carboxylic acids via Kolbe–Schmitt syntheses (Scheme 6.11) [40]. Here, substitution isomerism and electron-richness of the aromatic are changed.

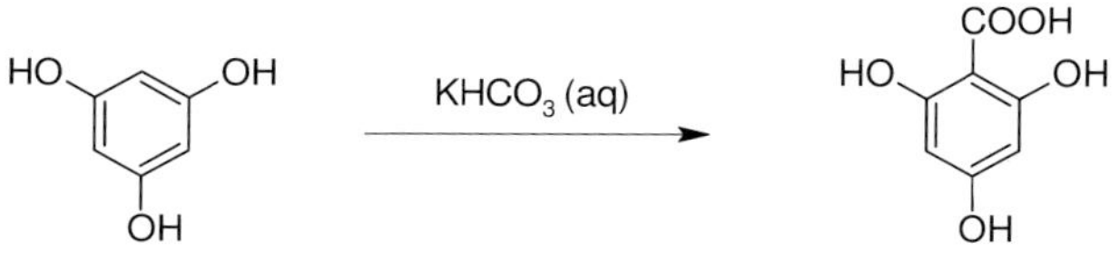

Scheme 6.11

For hydroquinone hardly any product formation was observed, owing to oxidation to benzoquinone, which was identified as a side product formed at large conversion (~50%) [40]. Phloroglucinol is a more electron-rich aromatic than resorcinol. Thus, the corresponding Kolbe–Schmitt reaction proceeded more rapidly and, also, the yields were notably higher (approaching 60%) but, probably for the same reason, decarboxylation was also favored. Whereas for resorcinol this notably set in above 200 °C, it occurred already at 160 °C for phloroglucinol. A more stable operation was achieved here that resulted in many parametric investigations using a more reliable and accurate product analysis.

6.7.3
Hydrogenation of Nitrobenzene (UCL + IMM)

6.7.3.1 Process Development Issues
Hydrogenations of nitro aromatics to the respective anilines (Scheme 6.12), which are valuable intermediates, e.g., for pharmaceuticals, have high intrinsic reaction rates. This reactive potential, however, cannot be exploited by conventional reactors

as they are unable to cope with the large heat releases owing to the large reaction enthalpies (500–550 kJ mol^{-1}) [41, 42]. Decomposition of the nitro aromatics or of partially hydrogenated intermediates would otherwise be the consequence [41]. For this reason, the hydrogen supply is restricted, thereby controlling reaction rate. The hydrogenation of nitrobenzene over supported noble metal catalysts was also among the first realized gas–liquid–solid processes in microchannels so that there is some generic experience with such processing, e.g., concerning the influence of the wall-coated catalyst on the falling film hydrodynamics, catalyst preparation and reactivation, and mass transfer in a three-phase system.

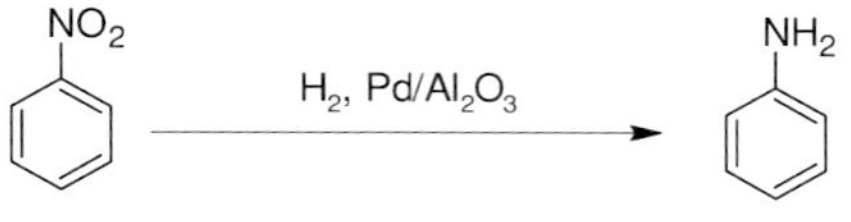

$$\text{PhNO}_2 \xrightarrow{H_2,\ Pd/Al_2O_3} \text{PhNH}_2$$

Scheme 6.12

6.7.3.2 Microreactor Plant and Processing Solution

Investigations were performed in a falling film microreactor (Fig. 6.27) [41, 43]. The microchannels of the falling film plates were wall coated with catalyst support by a slurry route, onto which then the active species was transferred by an impregnation or incipient-wetness technique. Alternatively, sputtered and UV-decomposed catalysts were tested, yielding denser, thinner films onto the channels. The liquid phase was transported to the reaction plate by pumping and also sucked off after reaction in the same way. Fluid motion during the reaction was due to gravity force (falling film) so that the two pumping actions had to be adjusted. Gas supply came from a gas bomb.

The processing done was rather conventional [41]. A normal falling film operation was set and the pressure was ambient. Temperatures were chosen

Fig. 6.27 Falling film microreactor. (Source: IMM.)

according to batch protocols. However, the catalyst reactivation route was tailored for the microreactor process, involving exposure of the reaction plate to cleaning solutions. This example shows that catalyst know-how, partly along with new preparation techniques, is a crucial step for such syntheses.

Process Development Results

First, the overall mass transfer coefficient $k_L a$ of the microreactor was estimated to be 3–8 s^{-1} [43]. For intensified gas liquid contactors, $k_L a$ can reach 3 s^{-1}, while bubble columns and agitated tanks do not exceed 0.2 s^{-1}. Reducing the flow rate and, accordingly, the liquid film thickness is a means of further increasing $k_L a$, which is limited, however, by liquid dry-out at very thin films. Despite such large mass transfer coefficients, gas–liquid microreactors such as the falling film device may still operate between mass transfer and kinetic control regimes, as fundamental simulation studies on the carbon dioxide absorption have demonstrated [44]. Distinct concentration profiles in the liquid, and even gas, phase are predicted.

Four preparation procedures of distinctively different nature were tested for the palladium catalyst for the hydrogenation of nitrobenzene [41]. A sputtered palladium catalyst gave only low conversion and low selectivity. Even then, deactivation was pronounced. An oxidation/reduction cycle led to slightly improved performance. After a steep initial deactivation with initial selectivity of nearly 100%, a stable operation at 2–4% conversion and about 60% selectivity followed. All mechanistically expected side products were found with the exception of phenylhydroxylamine. The conversion of a UV-decomposed palladium catalyst was slightly higher than for the sputtered one. A similar spectrum of side products as for the sputtered catalyst was found. For an impregnated palladium catalyst, complete conversion was achieved and maintained for six hours. Selectivity

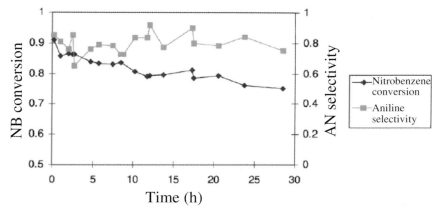

Fig. 6.28 Comparison of nitrobenzene (NB) conversion and aniline (AN) selectivity as a function of reaction time for the incipient-wetness catalyst. (Courtesy of the Royal Society of Chemistry [41].)

decreased with time, but remained still at a high level. The best performance of all catalysts investigated was found for an incipient-wetness palladium catalyst. Having initially over 90% conversion, a 75% conversion with selectivity of 80% was reached for long times on stream (Fig. 6.28).

The life-time of the four types of catalysts depends on the catalyst loading, which is related to the preparation route [41]. The larger the loading, the longer the catalysts could be used before reactivation. The four catalysts had the following sequence of life-time and activity: wet-impregnation > incipient wetness > UV-decomposition of precursors > sputtering.

Several reactivation routes of the used catalyst were tested, such as dissolution of organic residues by dichloromethane or their burning by heating in air. In this way, initial activity was recovered, thus regaining complete conversion [41].

6.7.4
Brominations of *m*-Nitrotoluene (IMM)

6.7.4.1 Process Development Issues
Without the use of radiation and catalyst the brominations of aromatics and alkylaromatics are often sluggish [45, 46]. Even with such aid, it may take hours for acceptable conversion. During this time, the whole reaction volume is exposed to the light and heated up considerably. Both photo-induced and thermal side reactions may result. The shallow penetration of light into big vessels wastes valuable energy. Furthermore, large amounts of bromine, especially under such conditions, pose a safety problem. This also hinders a solvent-free route using simply pure bromine and the undiluted aromatic. By transformation from batch into continuous processing, and thus to low inventory, a safe handling of bromine even at over 170 °C is provided [44]. This allows for a decrease of reaction time and an increase in space–time yields.

The side-chain bromination of *meta*-nitrotoluene was selected (Scheme 6.13) as this aromatic reacts very poorly without aid of light (even worse than toluene).

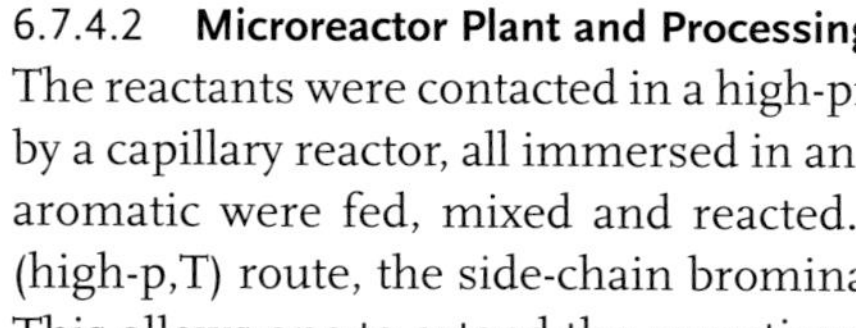

Scheme 6.13

6.7.4.2 Microreactor Plant and Processing Solution
The reactants were contacted in a high-pressure interdigital micromixer followed by a capillary reactor, all immersed in an oil bath [45, 46]. Pure bromine and pure aromatic were fed, mixed and reacted. By a high-temperature, high-pressure (high-p,T) route, the side-chain bromination of *meta*-nitrotoluene was achieved. This allows one to extend the operational range much beyond the boiling point, which is the technical limit of many reactions carried out batchwise.

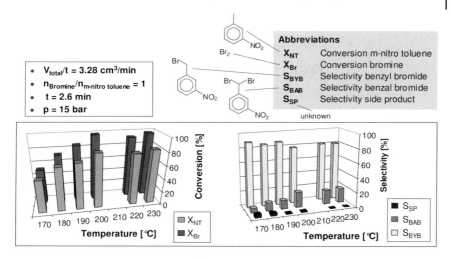

Fig. 6.29 High-p,T operation for the radical side-chain bromination of m-nitrotoluene (NT) in a micromixer-reactor setup. The large increase in operational temperature increases conversion at good selectivities, which tend to decline slightly with temperature. The two-fold substituted product, m-nitrobenzal bromide, is formed in larger amounts above 200 °C.
(By courtesy of Elsevier [47].)

Process Development Results
At temperatures of about 200 °C nearly complete conversion is achieved (Fig. 6.29) (see Ref. [46]). Selectivity to the target product benzyl bromide is reasonably high with 85%. Above 200 °C, selectivity slightly decreases to 80%. The main side product is the nitro-substituted benzal bromide, i.e., the two-fold brominated side-chain product.

6.7.5
Brominations of Thiophene (Fresenius + IMM)

6.7.5.1 Process Development Issues
The main issue here was to have a solvent-free continuous bromination process in a microreactor rig using pure bromine and aromatic liquid feeds [45, 46]. Favorable room temperature processing should be established, without the need for cooling this fast reaction. A further simplification of the process was to work without any catalyst. Aromatic compounds are typically brominated by the aid of Lewis catalysts, to generate an attacking electrophile. Thiophene was chosen since the corresponding 2,5-substituted derivative is an intermediate for making OLED materials by means of polyaddition (Scheme 6.14). This demands control over selectivity of the various mono- to tetra-brominated species, which are formed by consecutive reactions, e.g., by setting the residence time.

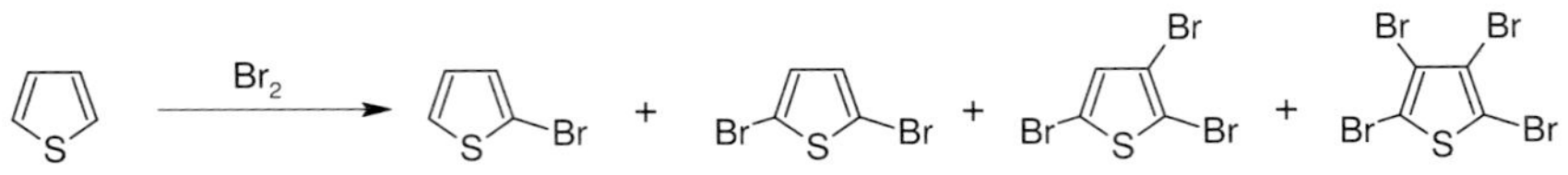

Scheme 6.14

6.7.5.2 Microreactor Plant and Processing Solution

A triangular interdigital or a caterpillar micromixer followed by a tube were used (Fig. 6.30) [45]. The micromixer-tube reactor was submersed into a thermostat bath for temperature setting. The temperature was set to −10 °C to room temperature. Piston and HPLC pumps fed the bromine and aromatic flows, respectively. The use of bromine demands special materials that are stable under such harsh conditions. Fluorinated reactor materials like PVDF or glass turned out to be suitable.

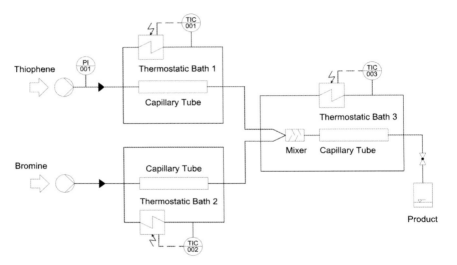

Fig. 6.30 Flow scheme of the rig for performing the bromination of thiophene continuously in a microstructured reactor. (Courtesy of the American Chemical Society [45].)

Process Development Results

Microreactor operation led to yields of up to 86%, at nearly complete conversion, which is better than for home-made (77% yield) and literature (50% yield) batch processing [45, 46]. Using the pure feeds and higher temperature, the reaction time was decreased from about 2 hours (for batch) to < 1 second (for micromixer-reactor). Correspondingly, the space–time yields were an order of magnitude higher for the continuous microreactor process. Owing to the ease with which reactant ratios and temperatures can be changed in the microreactor rig, a fast parametric study was performed to find the optimal operating conditions. In this way, the degree of the thiophene bromination could be controlled to optimize the share of the target product (Fig. 6.31).

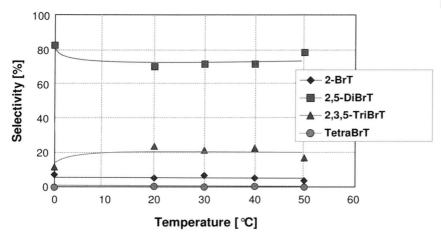

Fig. 6.31 Product distribution at a bromine-to-thiophene molar ratio of 2 : 1 for various temperatures (interdigital mixer; 93.4 mL h^{-1}). (Courtesy of the American Chemical Society [45].)

6.8
Free-radical Polymerizations (Uni Strasbourg, IMM)

6.8.1
Process Development Issues

Polymerizations as part of liquid-phase organic reactions are also influenced by mass and heat transfer and residence time distribution [37, 48]. This was first shown with largely heat-releasing radical polymerizations such as for butyl acrylate (evident already at dilute concentration) [49]. Here, a clear influence of microreactor operation on the polydispersity index was determined. Issues of mass transfer and residence time distribution in particular come into play when the solution becomes much more viscous during the reaction. Polymerizations change viscosities by orders of magnitude when carried out at high concentration or even in the bulk. The heat released is then even more of an issue, since tremendous hot spots may arise locally and lead to thermal runaway, known in polymer science as the Norrish–Tromsdorff effect.

Further extension of such work (albeit not reported here) aims to achieve a living radical polymerization by modifying the active species or decreasing its concentration, as, for example, given for the nitroxide mediated (NMP) radical polymerization. Here, the distribution between active and dormant chains affects the polydispersity index. Living polymerizations have the potential to result in very uniform polymer weight distributions. Also, uniform block copolymers may be formed in this way when adding a second monomer to a living polymer chain.

6.8.2 Microreactor Plant and Processing Solution

An interdigital slit-type high-pressure micromixer followed by a tube (0.9 mm inner diameter; 2.6 m long) was used [48]. The micromixer-tube reactor was submersed into a thermostat bath for temperature setting. HPLC pumps fed the monomer and initiator flows. A back-pressure cartridge (70 bar) and a pressure sensor served for pressure control. The temperature was set to 105 °C. The processing protocol was carried out conventionally, using the specs of similar batch processes.

6.8.2.1 Process Development Issues

A numerical study of the free-radical polymerization of styrene (Scheme 6.15) compared the behavior of an interdigital micromixer with a T-junction and a straight tube [37, 48]. The diffusion coefficient of the reactive species was varied to "simulate" the viscosity increase during a polymerization. The performance of the polymerization turned out to be largely dependent on the radial Peclet number. This dimensionless number is defined as the ratio of the characteristic time of diffusion in the direction perpendicular to the main flow to the characteristic time of convection in the flow direction (i.e., the mean residence time) and, therefore, is directly proportional to the characteristic length of the reactor.

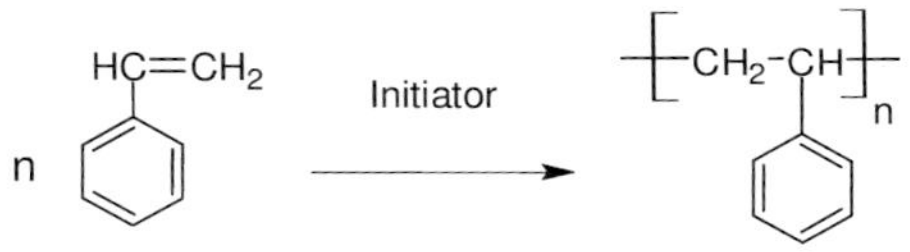

Scheme 6.15

For values of this Peclet number well below 1, as encountered in microreactors, a narrow molecular weight distribution can be achieved, while higher values, like those encountered in macroscale reactors, induce a drastic increase in the polydispersity index (Fig. 6.32) [37, 48]. Therefore, microreactors can lead to better control over bulk or semi-dilute polymerization processes.

Such theoretical predictions were confirmed experimentally for two solutions of different monomer concentration [49]. The microreactor polymerization results were compared with those of flask or T-junction reactors. The conversions are all similar, the flask showing slightly the highest value. The number average molecular weight was higher for the two continuous-flow processes – in the microreactor and the T-junction – than for the batch processing. The most remarkable effect was given for the polydispersity index. The microreactor approached the ideal value of 1.5 for radical polymerizations, similar to the T-junction, while the flask-type polymerization yielded less uniform polymers (Table 6.4).

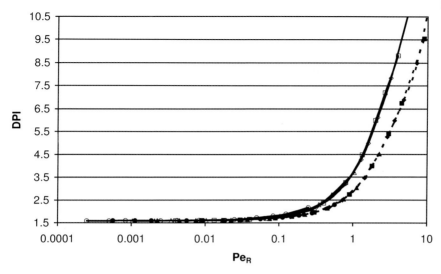

Fig. 6.32 Comparison of the polydispersity index (DPI) obtained in a multilamination micromixer (open symbols) and in a tube reactor (filled symbols) as a function of the radial Peclet number. (Courtesy of the Royal Society of Chemistry [48].)

Table 6.4 Performance data for the free-radical polymerization of styrene at two monomer concentrations. The behavior of three reactors is compared – flask, T-junction (TJ), and high-pressure interdigital micromixer (HPIMM) by experiment and by modeling [49].

Content of solvent (%)		20 (exp.)	20 (model.)	30 (exp.)	30 (model.)
Conversion (%)	Flask	56	54	59	55
	TJ	50		55	
	HPIMM	53		56	
M_n (Number average molecular weight)	Flask	28 000	46 000	28 000	40 000
	TJ	49 000		34 000	
	HPIMM	32 000		43 500	
I_p (Polydispersity index)	Flask	1.98	1.56	1.92	1.56
	TJ	1.52		1.60	
	HPIMM	1.56		1.52	

6.9
Future Directions – Establishing a Novel Chemistry by Enabling Function

6.9.1
Traditional Chemistry – Processes Follow Limitations of Reactors

The potential of microreactors for organic reactions will only be fully revealed when future chemical processes are tailored to the specific specs of the microreactors. Most important, the reaction has to be completed within a short residence time. Among the current chemical processes only a few percent fit this request. The question is whether this is intrinsic (i.e., kinetically limited) or – now that the new technology is available – if there is a chance to change the chemistry itself. Since chemistry has been done over decades mainly by batch processing, the organic chemist has voluntarily chosen to restrict his syntheses to slow reactions or, if being fast, to artificially slow them down – which is some sort of *"subduing chemistry"* [14]. Microreactors can overcome such limitations, paving the way to a *novel chemistry* – some corresponding generic paths are explained in the following.

6.9.2
Novel Chemistry – Tailoring Protocols to Increase Microreactor Performance

6.9.2.1 High-p,T Processing
Low pressure operations under reflux conditions are typically favored for laboratory flasks and agitated tanks. Accordingly, the maximum temperature of many organic routes is often simply defined by the solvent boiling point [14]. Microreactor rigs, however, allow a simple operation of liquid phases under high pressures and high temperatures. For instance, a system pressure of 50 bar is enough to maintain single-phase operation (i.e., no gas content and no boiling) even at temperatures up to 100 °C higher than boiling points of typical solvents. This has been termed high-p,T processing. The faster operation at higher temperatures typically is paid for by more side and consecutive reactions. Thus, efficient mixing and shortening of the residence time to the kinetic limit become important drivers for process optimization.

6.9.2.2 Solvent-free Processes – Contacting "All-at-once"
In batch processing aggressive reactants are, typically, diluted to prevent thermal overshooting and runaway. Even then they often are added drop-wise, to allow heat transfer to be adjusted to heat release. In some cases, this may take over half an hour or so. This unnecessarily prolongs processing time and, also, the reaction then is carried out for a considerable part under totally changing reactant concentrations (from zero to full-load content). Conversely, microreactors with their efficient heat and mass transfer have the potential to contact the full reactant load "all at once". In addition, microreactors can cope with concentrated solutions or even pure liquid reactants. Several examples are known for which such "all at once" or solvent-free procedures are feasible in microreactors with reasonable selectivity, whereas the

same contacting otherwise led to vigorous reactions and even explosions (when done under special safety precautions with miniature volumes) [14].

6.9.2.3 Use of Hazardous Elements and Building Blocks – Direct Routes
One way of process simplification is to make molecular complex compounds out of much simpler building blocks (e.g., by multi-component one-pot syntheses like the Ugi reaction), at best directly out of the elements. Especially in the latter case, this is often quoted as a "dream reaction" [14]. Typically, such routes have been realized so far with hazardous elements, easily undergoing reaction, but lacking selectivity. One example is direct fluorination starting with elemental fluorine, which has been performed both with aromatics and aliphatics. Since the heat release cannot be controlled with conventional reactors, the process is deliberately slowed down. While, for this reason, direct fluorination needs hours in a laboratory bubble column it is completed within seconds or even milliseconds when using a miniature bubble column operating close to the kinetic limit. Also, conversions with the volatile and explosive diazomethane, commonly used for methylation, have been conducted safely with microreactors in a continuous mode [14].

6.9.2.4 Routes in the Explosive Regime
Several examples have been reported for conducting routes in the explosive regime [14]. Among them, and most prominent, was the detonating-gas reaction using pure hydrogen and oxygen mixtures. This can also be considered as a direct route from the elements. With special catalysts, hydrogen peroxide, and not water, is obtained as a valuable product, avoiding the circuitous Anthraquinone process used at the industrial scale. Calculations of explosion limits clearly demonstrate that there is a considerable shift, when explosive reactions are carried out in microchannels. Safety is not only related to avoiding thermal runaway, but relates to mechanistic reasons of breaking the radical chain by enhanced wall collision in the small channels with their large specific interfaces.

6.9.2.5 Simplified Protocols
Some examples of simplified protocols have already been given above – solvent-free and direct-element processes. A further simplification can achieved for processes involving the presence of another species besides the reactants only, such as catalysts, moderators, promoters, etc. In this context, a microreactor route has been reported where the addition of a base to activate a homogeneous catalyst is omitted [14]. Basic functions can be provided in microchannels by surface groups such as deprotonated hydroxyls, which, for example, are created by electroosmotic flows in glass microchannels. Since the specific surface is large and diffusion distances are short, the surface may be part of the reaction in a way not observed for conventional reactors. In this way, a base-free Suzuki–Miyaura coupling has used tetrakis(triphenylphosphine)palladium(0) as catalyst. Similarly, an acid-free esterification of pyrenyl-alkyl acids has been established using only acidic functions of the microchannel surface. Thus, surface-chemistry routes may replace bulk-chemistry routes in small microchannels.

6.10
Summary

In summary, chemical microprocess engineering, i.e., the use of microstructured reactors for chemistry and chemical engineering, offers promising perspectives for tomorrow's processing. This will not really change the large world-scale plants on a mid-term view, but certainly will affect industrial scale-up of pilot and laboratory plants to an extent. In addition, laboratory plants for research synthesis and analytics will change. Actually, the two latter processes have started already, as the examples given in this chapter have tried to demonstrate [35]. To reach this level requires a combined interdisciplinary development. Microsystems technology skills, most pronounced at the laboratory scale, need to be supplemented by competence in plant and process engineering.

List of Company Abbreviations

Clariant	Clariant GmbH, Industriepark Höchst, Frankfurt am Main, Germany
DSM	DSM Fine Chemicals Austria Nfg GmbH & Co KG, Linz, Austria
Fresenius	Europa Fachhochschule Fresenius, Idstein, Germany
FZK	Forschungszentrum Karlsruhe, Karlsruhe, Germany
Idemitsu Kosan	Idemitsu Kosan Co. Ltd., Chiba next to Tokyo, Japan
IMM	Institut für Mikrotechnik Mainz GmbH, Mainz, Germany
LG Chem	LG Chem, Daejeon, Korea
SK Corporation	SK Corporation, Daejeon, Korea
Trust Chem	Trust Chem, Hangzhou, China
UCL	University College London
UOP	Universal Oil Products, Des Plaines in Chicago, USA
Uni Strasbourg	University Louis Pasteur Strasbourg (ULP), School of Chemistry, Polymers and Materials (ECPM); Engineering Laboratory for High-Technology Polymers (LIPHT-cnrs)

References

1 Hessel, V., Löwe, H., Müller, A., Kolb, G., *Chemical Micro Process Engineering – Processing and Plants*, Wiley-VCH, Weinheim (**2005**).
2 Hessel, V., Hardt, S., Löwe, H., *Chemical Micro Process Engineering – Fundamentals, Modelling and Reactions*, Wiley-VCH, Weinheim (**2004**).
3 Fletcher, P. D. I., Haswell, S. J., Pombo-Villar, E., Warrington, B. H., Watts, P., Wong, S. Y. F., Zhang, X., Micro reactors: principle and applications in organic synthesis, *Tetrahedron* 58(24) (**2002**) 4735–4757.
4 Haswell, S. T., Watts, P., Green chemistry: synthesis in micro reactors, *Green Chem.* 5, (**2003**) 240–249.
5 Gavriilidis, A., Angeli, P., Cao, E., Yeong, K. K., Wan, Y. S. S., Technology and application of microengineered reactors, *TransIChemE* 80/A(1) (**2002**) 3–30.
6 Jensen, K. F., Microreaction engineering – is small better? *Chem. Eng. Sci.* 56, (**2001**) 293–303.
7 Jensen, K. F., Smaller, faster chemistry, *Nature* 393(6) (**1998**) 735–736.
8 Jähnisch, K., Hessel, V., Löwe, H., Baerns, M., Chemistry in microstructured reactors, *Angew. Chem. Int. Ed.* 43(4) (**2004**) 406–446.
9 Hessel, V., Löwe, H., Micro mixers – a review on passive and active mixing priciples, *Chem. Eng. Sci.* 60(8–9), (**2005**) 2479–2501.
10 Hessel, V., Angeli, P., Gavriilidis, A., Löwe, H., Gas/liquid and gas/liquid/solid microstructured reactors – contacting principles and applications, *Ind. Eng. Chem. Res.* 44(25) (**2005**) 9750–9769.
11 Pennemann, H., Watts, P., Haswell, S., Hessel, V., Löwe, H., Benchmarking of microreactor applications, *Org. Proc. Res. Dev.* 8(3) (**2004**) 422–439.
12 Pennemann, P., Hessel, V., Löwe, H., Chemical micro process technology – from laboratory scale to production, *Chem. Eng. Sci.* 59(22–23) (**2004**) 4789–4794.
13 Hessel, V., Löwe, H., Organic synthesis with microstructured reactors, *Chem. Eng. Technol.* 28(3) (**2005**) 267–284.
14 Hessel, V., Löb, P., Löwe, H., Development of reactors to enable chemistry rather than subduing chemistry around the reactor – potentials of microstructured reactors for organic synthesis, *Curr. Org. Chem.* 9(8) (**2005**) 765–787.
15 Ehrfeld, W., Hessel, V., Löwe, H., *Microreactors*, Wiley-VCH, Weinheim (**2000**).
16 Müller, A., Cominos, V., Horn, B., Ziogas, A., Jähnisch, K., Grosser, V., Hillmann, V., Jam, K. A., Bazzanella, A., Rinke, G., Kraut, M., Fluidic bus system for chemical micro process engineering in the laboratory and for small-scale production, *Chem. Eng. J.* 107(1–3) (**2005**) 205–214.
17 Keoschkerjan, R., Richter, M., Boskovic, D., Schnürer, F., Loebbecke, S., Novel multifunctional microreaction unit for chemical engineering, *Chem. Eng. J.* 101(1–3) (**2004**) 469–475.
18 Hessel, V., Löwe, H., Chemical micro process engineering – current trends and issues to be resolved, *Microreactor Technology and Process Intensification*, ACS Symposium Series, Vol. 914 (eds. Wang, J., Holladay, J. D.), American Chemical Society, Washington D. C. (**2005**) pp. 23–46.
19 Ferstl, W., Loebbecke, S., Antes, J., Krause, H., Haeberl, M., Schmalz, D., Muntermann, H., Grund, M., Steckenborn, A., Lohf, A., Hassel, S., Bayer, T., Leipprand, I., Development of an automated microreaction system with integrated sensorics for process screening and production, *Chem. Eng. J.* 101(1–3) (**2004**) 431–438.
20 Löbbecke, S., Ferstl, W., Panic, S., Türcke, T., Concepts for modularization and automation of microreaction technology, *Chem. Eng. Technol.* 28(4) (**2005**) 484–493.
21 Werner, B., Hessel, V., Löb, P., Mixers with microstructured foils for chemical production purposes, *Chem. Eng. Technol.* 28(4) (**2004**) 401–407.
22 Schirrmeister, S., Markowz, G., Bulk chemicals in microreactors: development and scale-up of a new synthesis

route into the pilot plant scale, in Proceedings of the 27th International Exhibition-Congress on Chemical Engineering, Environmental Protection and Biotechnology, ACHEMA, p. 3, (19–24 May **2003**), DECHEMA, Frankfurt.
23 Bayer, T., Jenck, J., Matlosz, M., IMPULSE – a new approach to process design, *Chem. Eng. Technol.* 28(4) (**2005**) 431–438.
24 Hardt, S., Pennemann, H., Schönfeld, F., Theoretical and experimental characterization of a low-Reynolds number split-and-recombine mixer, *Microfluidics Nanofluidics* 2(3) (**2006**) 237–248.
25 Schönfeld, F., Hessel, V., Hofmann, C., An optimised split-and-recombine micro mixer with uniform 'chaotic' mixing, *Lab Chip* 4, 1, (**2004**) 65–69.
26 Hessel, V., Löwe, H., Micro mixers – a review on passive and active mixing principles, *Microreactor Technology and Process Intensification*, ACS Symposium Series, Vol. 914 (eds. Wang, J., Holladay, J. D.), American Chemical Society, Washington D. C. (**2005**) 334–359.
27 IMM results, unpublished.
28 Fournier, M.-C., Falk, L., Villermaux, J., A new parallel competing reaction system for assessing micromixing efficiency – experimental approach, *Chem. Eng. Sci.* 51(22) (**1996**) 5033–5064.
29 Werner, B., Hessel, V., Löb, P., Mischer mit mikrostrukturierten Folien für chemische Produktionsaufgaben, *Chem.-Ing. Tech.* 76(5) (**2004**) 567–574.
30 Hessel, V., Hofmann, C., Löwe, H., Meudt, A., Scherer, S., Schönfeld, F., Werner, B., Selectivity gains and energy savings for the industrial phenyl boronic acid process using micromixer/tubular reactors, *Org. Proc. Res. Dev.* 8(3) (**2004**) 511–523.
31 Pennemann, H., Hessel, V., Löwe, H., Forster, S., Kinkel, J., Improvement of dye properties of the azo pigment Yellow 12 using a micromixer-based process, *Org. Proc. Res. Dev.* 9(2) (**2005**) 188–192.
32 Veser, G., Experimental and theoretical investigation of H_2 oxidation in a high-temperature catalytic microreactor, *Chem. Eng. Sci.* 56(4) (**2001**) 1265–1273.

33 Kim, J., Park, J.-K., Kwak, B.-S., Development of pharmaceutical fine chemicals using continuous microreactor technology (MRT), in Proceedings of the 4th Asia-Pacific Chemical Reaction Engineering Symposium, APCRE05, (12–15 June **2005**), Gyeongju, Korea, pp. 441–442.
34 Choe, J., Song, K.-H., Kwon, Y., Microreaction technology in practice, in Proceedings of the 4th Asia-Pacific Chemical Reaction Engineering Symposium, APCRE05, (12–15 June **2005**), Gyeongju, Korea, pp. 435–436.
35 Thayer, A. M., Harnessing microreactors, *Chem. Eng. News* 83(22) (**2005**) 43–52.
36 Iwasaki, T., Yoshida, J.-I., Free radical polymerization in microreactors. Significant improvement in molecular weight distribution control, *Macromolecules* 38(4) (**2005**) 1159–1163.
37 Hessel, V., Serra, C., Löwe, H., Hadziioannou, G., Polymerisationen in mikrostrukturierten reaktoren: Ein Überblick, *Chem.-Ing. Tech.* 77(11) (**2005**) 39–59.
38 Löwe, H., Hessel, V., Löb, P., Hubbard, S., Addition of secondary amines to α,β-unsaturated carbonyl compounds and nitriles by using microstructured reactors, *Org. Proc. Res. Dev.* 10(6) (**2006**) 1144–1152.
39 Hessel, V., Hofmann, C., Löb, P., Löhndorf, J., Löwe, H., Ziogas, A., Aqueous Kolbe–Schmitt synthesis using resorcinol in a microreactor laboratory rig under high-p,T conditions, *Org. Proc. Res. Dev.* 9(4) (**2005**) 479–489.
40 Hessel, V., Hofmann, C., Löb, P., Löwe, H., Parals, M., Micro-reactor processing for the aqueous Kolbe-Schmitt synthesis of hydroquinone and phloroglucinol, *Chem. Eng. Technol.* 30(3) (**2007**), in print.
41 Yeong, K. K., Gavriilidis, A., Zapf, R., Hessel, V., Catalyst preparation and deactivation issues for nitrobenzene hydrogenation in a microstructured falling film reactor, *Catal. Today* 81(4) (**2003**) 641–651.
42 Losey, M. W., Schmidt, M. A., Jensen, K. F., A micro packed-bed reactor for chemical synthesis, in Ehrfeld, W.

(ed.) *Microreaction Technology: 3rd International Conference on Microreaction Technology, Proc. of IMRET 3*, Springer-Verlag, Berlin **(2000)** pp. 277–286.

43 Yeong, K. K., Gavriilidis, A., Zapf, R., Hessel, V., Experimental studies of nitrobenzene hydrogenation in a microstructured falling film reactor, *Chem. Eng. Sci.* 59(16) **(2004)** 3491–3494.

44 Zanfir, M., Gavriilidis, A., Zapf, R., Hessel, V., Carbon dioxide absorption in a falling film microstructured reactor: Experiments and modeling, *Ind. Eng. Chem. Res.* 44(6) **(2005)** 1742–1751.

45 Löb, P., Hessel, V., Klefenz, H., Löwe, H., Mazanek, K., Bromination of thiophene in micro reactors, *Lett. Org. Chem.* 2(8) **(2005)** 767–779.

46 Löb, P., Löwe, H., Hessel, V., Fluorinations, chlorinations and brominations of organic compounds in micro structured reactors, *J. Fluorine Chem.* 125(11) **(2004)** 1677–1694.

47 Hessel, V., Löb, P., Löwe, H., Chemical micro process engineering towards production – current trends and tasks to be solved, *Stud. Surf. Sci. Catal.* 159 New developments and application in chemical reaction engineering (eds.: Hyun-Ku, R., In-Sik, N., Park, J. M.), Elsevier, Amsterdam **(2006)**, 35–46.

48 Serra, C., Sary, N., Schlatter, G., Hessel, V., Numerical simulation of polymerization in interdigital multi-lamination micromixers, *Lab Chip* 5(9) **(2005)** 966–973.

49 Results Dr Serra, LIPHT-ECPM, Uni Strasbourg, unpublished.

7
Non-reactor Micro-component Development

Daniel R. Palo, Victoria S. Stenkamp, Jamie D. Holladay, Paul H. Humble, Robert A. Dagle, and Kriston P. Brooks

7.1
Overview

This chapter reviews research and development activities related to non-reactive applications of microchannel process technology. Non-reactive unit operations covered include heat transfer, mixing, emulsification, phase separation, phase transfer, biological processes, and body force applications. The ever-expanding research in non-reactive microchannel applications is striking, yet the fundamental advantages usually are found in the precise control made possible by small channel dimensions, rapid heat and/or mass transfer, and dominant interfacial phenomena. This chapter describes the advantages of microchannel process technology in each application space and summarizes the fundamental and applied research being conducted.

7.2
Introduction

In the early days of microchannel development, the key selling point involved non-reactive applications, in particular heat transfer [1–12]. Work in this area quickly demonstrated that microchannel components could provide rapid heat transfer while maintaining low pressure drop, despite the use of very narrow process channels [1, 2]. From there, microchannel research expanded to a range of non-reactive- and reactive-applications, where the major advantages of this technology not only included rapid heat and mass transfer but also the precise control of process conditions, low inventory of hazardous reactants, use of surface phenomena to significant advantage, and overall process intensification [13].

In one sense, non-reactive systems are simpler, since they do not incorporate chemical reactions, catalysts, and the associated complexities of such systems. However, some non-reactive systems can be even more complex, depending on the application, such as systems where multiple phases are present, or where surface phenomena are dominant. Complicating issues, such as surface forces, multiple phases, combined heat and mass transfer, are being investigated as part of application evaluations. A good illustration of progress in addressing these issues is technology that takes advantage of capillary forces, which can dominate to the point that gravity-independent operation can be achieved. The wicking phase separator developed by Pacific Northwest National Laboratory (PNNL), for example, uses surface forces, without gravity, to separate gas and liquid phases [14].

This chapter overviews ongoing activities in microchannel process technology development, from single-channel laboratory experiments to industrially-driven, multi-channel and multi-unit development at or near the prototype and pilot level. The non-reactive unit operations covered include heat transfer, mixing, emulsification, phase separation, phase transfer, biological processes, and body force applications.

Pertinent references are given throughout, and the reader is referred to the primary works for more details. Where applicable, previous reviews of the subject matter are cited. As with reactive microchannel systems, the centers of development for the technology and applications discussed here tend to be in Europe (mainly Germany) [15], the United States [16], and Asia (mainly Japan) [17].

7.3
Heat Exchange

Research on compact microchannel heat exchange spawned from early work on microchannel development [1–8]. Advances in device fabrication and analysis of heat and mass flow characteristics have led to more efficient designs, which in turn have enabled microchannel heat exchange technology to evolve. Some examples of several commercial or near-commercial applications emerging from this research include cooling for electronic devices, CO_2 refrigeration for automobile air-conditioning, and use in chemical processing (Table 7.1). Several companies such as Modine, Hydro Aluminum, Exergy, Heatric, and Ceramatec have developed microscale heat exchange technology, much of which is already available commercially for HVAC equipment, petrochemicals, and pharmaceuticals [18–21].

The function of many types of heat exchangers is to transfer heat as effectively as possible in a compact system, while often seeking to minimize pressure drop. One way to increase the heat transfer effectiveness is to increase the surface area-to-volume ratio. The resulting increased thermal intimacy is particularly important for heat exchange involving gases, since thermal resistance at the gas–solid interface usually dominates the overall heat transfer [22]. However, with conventional heat exchange technology, shrinking the system diameter to enhance thermal contacting would create a higher pressure drop due to higher flow velocities. Such designs would require relatively high length to hydraulic diameter ratios, L/D_H, resulting in undesirable pumping constraints. With micro

Table 7.1 Examples of commercial microchannel heat exchanger applications [16].

Company	Materials	Application areas
Modine	Metals	Vehicles, HVAC equipment, industrial equipment, refrigeration systems, fuel cells and electronics
Hydro Aluminum	Aluminum	HVAC equipment, refrigeration systems
Exergy	Stainless steel	Aerospace, pharmaceutical, semiconductor, petrochemical, and fuel cell
Ceramatec	Ceramics	Recuperators for high-temperature applications
Heatric	Metals	Heat exchangers for offshore platforms

heat exchangers, flow channel dimensions ranging from microns to millimeters significantly reduce the distances for heat and mass transfer, producing very high transfer coefficients and resulting in small thermal processing devices [16]. With proper design, multiple parallel heat exchange channels can provide the desired thermal fluxes while maintaining acceptable pressure drops.

One important potential application for micro heat exchange involves cooling of electronic devices. The demand for high-performance microprocessing devices has increased the levels of power density and required heat dissipation in microelectronic components [22]. As the reliability and lifespan of a microelectronic component is dependent on operating temperature, removing the generated heat has become a crucial issue for computer system designers. Over two decades ago, Tuckerman and Pease investigated micro heat sinks for electronic cooling [23]. A water-cooled heat sink with micro flow channels was demonstrated at high-power density with a heat flux as high as 790 W cm^{-2}. Since then more research has been carried out with micro heat sinks, mostly involving a fin approach [24]. In these heat sinks, heat is typically removed by conduction through solid and then dissipated away by interfacial convection. Conventional thermal control systems, such as air cooling with fans, thermoelectric cooling, heat pipes, and vapor chambers, are already reaching their practical application limits [22]. Heat dissipation values of 20–50 W are estimated to increase to 100 W or to even 200 W in coming years. As such, microelectromechanical systems (MEMS)-based microchannel cooling systems are being investigated. Such systems combine microchannels and microheat pipes to form a new type of heat sink.

Several research groups have carried out studies involving MEMS-based systems. Studies have involved numerical analysis of heat transfer characteristics as a function of channel design and fluid flow [24–29]. Findings include potential cooling capacities in excess of 200 W [26], optimized geometries [27], and an understanding of how combined conduction and convection effects vary with Reynolds numbers and design [24, 25, 28, 29]. Several experimental studies have researched similar effects, investigating channel geometry tradeoffs [30], varied flow velocities [31–34], and fractal network-type designs [35, 36]. Studies of experimental heat transfer correlations combined with analytical solutions have also been reported [37–40]. Novel, elaborate microchannel designs also include direct spot-cooling of microelectronic components [22] and utilizing microchannels narrow enough to generate capillary forces, making coolant circulation occur naturally, with no need for pumping [41].

Compact heat exchange has also been investigated for refrigeration systems, where an increased heat exchange effectiveness can be a trade-off for less-efficient refrigerants, such as CO_2, for applications like automotive cooling. Using microchannel technology, such less-efficient refrigerants can be used without adding size or weight to the system [42]. High heat fluxes and low pressure drops have been obtained using simulation models [43] and experimental approaches [44, 45]. Models have been developed to look at different refrigeration designs [46, 47]. A comprehensive review of evaporator and condenser designs for microchannel air-conditioning systems is given by Kim et al. [48]. Refrigeration systems outside of automobile applications have also been explored. For example, Mathias et al. have described a process for cooling a product in a microchannel heat exchanger, with direct application to the liquefying of natural gas [49].

Advances in microfabrication techniques have been instrumental in furthering microchannel heat exchanger development. Swift et al. have reported heat exchangers composed of stacks of thin metal sheets brazed together, using copper, forming a plurality of parallel slots for use in a Stirling engine [11]. These narrowly placed thin sheets provide many fluid flow channels, yielding a high surface area to volume ratio and resulting in improved heat transfer. Other early microchannels were developed by stacking foils of copper or steel with grooves and compressing them together [10]. Diffusion bonding techniques using patterned metal sheets were developed initially for heat exchange applications [9]. Further developments resulted in devices that were gasket-less, robust, leak-free, and capable of high-pressure operation. Such micro-devices could be made modularly, offering incremental scale-up and making viable potential economic advantages when compared with conventional singular component costs [50].

Microcomponent sheet architecture advances have created a more efficient microscale medium for several chemical operations to occur while allowing scale-up to a process level. Tonkovich et al. describe devices coupling microchannel heat exchange with chemical reactions [51]. Exothermic reactions supplying heat to other endothermic reactions, creating a more thermally efficient device, were suggested. These devices with minimized heat transfer resistances enable the exploitation of catalytic materials operating at or near their intrinsic kinetics. At PNNL further investigations into microchannel fuel processors and other microreactor related devices have been developed [51]. At the same time, several other groups were investigating microchannel reactors, including the Institut für Mikrotechnik Mainz GmbH (IMM), BASF, DuPont, Massachusetts Institute of Technology (MIT), University of Newcastle, University College London, University of Frankfurt, Bayer, the University of Chemnitz, and Forschungszentrum Karlsruhe GmbH (Karlsruhe) [50].

Microchannel heat exchange technology has advanced greatly in the past decade, from miniature, small-scale, shell-in-tube designs through advances in fabrication techniques that led to the development of highly robust, scalable, sheet architectures. From such advances, new possibilities emerged, recognizing microchannel devices capable of not only excellent heat transfer, but of thermal conversions coupled with mass transfer. The utility of such systems can be applied to several areas, from power conversion to chemical processes [52, 53], as illustrated in the following sections.

7.4 Mixing

Most chemical reactors require fluids mixing, and, consequently, much research has focused on developing micromixers to improve the process while maintaining throughput and decreasing the mixer volume [54–105]. Levenspiel and others have shown that for fast reactions with competitive pathways the degree of mixing will strongly impact the product distribution [54, 55]. Micromixers have great advantages as well as significant challenges. They allow for optimization of fast reactions [55, 56]; they also have small internal volumes, which are very useful when working with hazardous [55] or biological materials [57, 58], when working in explosive regimes [59], when the materials are rare or expensive, and for micro analyzers such as Micro Total Analysis Systems (μTAS) [60, 61]. Where ordered or controlled concentration or periodic concentrations needed, micromixers are ideal [62]. Further, since the microchannels used in microreactors and micromixers produce laminar flow profiles, micromixers are ideal when laminar mixing is desired (e.g., for viscous fluids). They can also be used for generating emulsions [63] and foams, and for mixing immiscible fluids [64]. Some challenges include lack of research in real-world environments, potential problems with fouling, and fabrication expense.

Micromixers have been developed for flows ranging from < 1 mL h^{-1} up to 10 000 L h^{-1}. At the low flow range, only single elements of the micromixer are required. These are often used in integrated systems for credit card sized fluidic chips [65]. For flows > 1 L h^{-1}, microstructured mixers are used for process intensification in pilot- and conventional-scale applications [65].

Since the flow in micromixers is laminar, the mixing is accomplished primarily by molecular interdiffusion [65, 66]. Laminar flow also has the advantage of being simpler to model than turbulent flow, enabling faster development and improved designs. Hessel et al. have noted that the diffusive flux is the product of the diffusion coefficient, the interfacial surface area, and species concentration gradient ($D \times A \times \nabla c$) [65]. Therefore, a micromixer strives to increase the interfacial surface area and the concentration gradient. This is most commonly achieved by improving the convective diffusion.

Mixer performance can be characterized using dimensionless units like the Reynolds number ($Re = U \times d/\nu$), the Fourier number ($Fo = D \times t/d^2 = Tr/Tm$),

Table 7.2 Examples of active and passive micromixers.

Active	Passive
Periodic flow switching [67, 68]	Y- or T-type [64, 69–72]
Acoustic bubble shaking [58, 73]	Multi-laminating [65, 74–78]
Electric Field Mixing [57, 79–83]	Split and recombination [60, 63, 84, 85]
Ultrasound/piezoelectric membranes [56, 86, 87]	Structured packing [88, 89]
Microimpellers [65]	Chaotic flow [90–98]
	Recirculation [61, 99–101]
	Colliding jet [102]
	Moving droplet [103, 104]

the Peclet number ($Pe = U \times d/D$), and, in the case of active mixers, the Strouhal number ($St = f \times d/U$) [65, 66]. For these dimensionless units, the U, d, t, v, and f are defined as average velocity, hydraulic diameter, time, kinematic viscosity, and frequency of vortices shed in the flow pattern (vortex street), respectively. Tr is defined as $Tr = L/U$, where L = longitudinal length, and $Tm = d^2/D$. The Pe is used to find the mixing length, Lm, in laminar uniaxial flows (Lm is proportional to Pe times channel width) [65, 66]. It is desirable in micromixers to minimize the Lm, which is accomplished by increasing the convective diffusion.

Various designs have been proposed to increase convective diffusion, and each has its strengths and weaknesses. The means to generate the mixing are categorized here as *active* (where energy is introduced to the fluids forcing mixing) and *passive* (which harnesses the flow energy of the fluid to expedite mixing) [65, 66]. Examples of each type are listed in Table 7.2 and described separately below.

7.4.1
Active Micromixers

7.4.1.1 Periodic Flow Switching

Periodic flow switching is achieved by pulsing the feed flow, thus creating axial mixing that normally does not occur in laminar flows [65, 67, 68]. Obviously, the number of pulsing streams, periodicity, and flow rate substantially affect mixing effectiveness. Either the pump can be pulsed or the valves can be constricted and opened for most fluids. Periodic switching in electroosmotic flows can also be performed by placing electrodes beneath the solid–liquid interface, causing non-uniform potentials along the microchannel walls [68]. To evaluate the effectiveness of this approach, Qian and Bau [68] used a simple two-electrode system. They found that chaotic mixing could be induced using long periodicity. They also surmised that as more electrodes are added the flow fields induced become more complex [68].

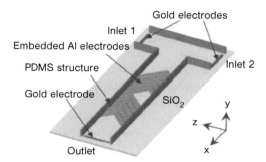

Fig. 7.1 T-shaped micromixer for electrokinetic instability induced mixing [79].

7.4.1.2 Electric Field Mixing

Electric fields are used to induce mixing in two different scenarios: electrowetting [65, 83] and the more common electrokinetic instability [57, 79–82]. Electrowetting uses an array of electrodes to move and merge droplets by electrowetting and then shaking them to expedite the mixing [65, 83]. The droplets are typically sized to be larger than a single electrode, causing it to overlap on a second one. The mixer can be designed to utilize electrowetting to only move/merge the droplet(s) or to shake the droplet(s), as well as move/merge them [65]. However, without the shaking, the mixing is not very efficient [65].

Electrokinetic instability mixers take advantage of the fact that most microchannel surfaces develop a small charge, by either chemical reaction, ionization, or ion adsorption, when in contact with an electrolyte solution [79–82]. This surface charge attracts opposite-charged ions from the liquid to the wall, creating an electrical double layer (EDL). The EDL ions can be moved, dragging surrounding fluid along with them when an external electric field is applied. The resultant bulk fluid movement is referred to as electric osmosis [79–82]. By varying the electric field potential, effective mixing is achieved. Wu et al. have designed and demonstrated a T-type mixer (cross section was 200×50 μm) with a mean velocity of 10 μm s^{-1} that achieved over 90% mixing efficiency in a 5-mm-long microchannel with an alternating voltage between +100 and –50 V at 0.5 Hz [79]. There were ten electrodes in a herringbone configuration (Fig. 7.1). Lastochkin et al. were able to combine an electroosmotic pump with an electrokinetic mixer [80]. The device was a 2-cm array with 16 T-type electrode elements able to generate up to 1 mm s^{-1} velocity. The power consumption at 20 V was 0.1 mW.

7.4.1.3 Ultrasound/Piezoelectric Membranes

Ultrasound is used for effective mixing in conventional systems and microfluidics, where a piezoelectric membrane is used [56, 65, 86, 87]. For example, Yaralioglu et al. have employed ZnO piezoelectric membranes oriented perpendicular to the fluid flow [87]. Using a dye technique they were able to observe rapid mixing at flow rates from 1 to 100 μm min^{-1} with the transducers driven at 1.2 V$_{rms}$ sinusoidal voltages and 450 MHz (Fig. 7.2). Challenges often associated with ultrasonic micromixers include bubble formation and heating.

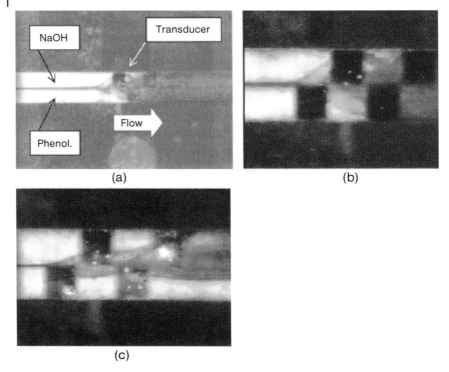

Fig. 7.2 Ultrasound micromixer.
(a) Laminar flow before the transducer and mixing after the transducer at 10 μL min^{-1} with a transducer power of 30 mW. The image is 1.5 × 2 mm.
(b) Magnified transducer view so the image is 650 × 860 μm, same flow rate.
(c) Increased flow rate of 60 μL min^{-1}. On all the images, lighting conditions were kept constant, but the brightness levels were auto balanced to improve the appearance of the images [87].

7.4.1.4 Microimpellers

Magnetically driven microimpellers have been proposed as micromixers [65]. This concept is based on the most common way of conventional mixing. A proposed mixer system based on this technology was assembled from a cap, hub, and two rotary blades composed of Permalloy, a ferromagnetic material. The system was fabricated using electroplating in a mixing chamber 2.5 mm long and 40 μm deep. Complete mixing of a 0.17 μL min^{-1} volume was achieved in approximately 1 min with the stirrer operating at 600 rpm. Proposed advantages of this type of system include the ability to turn the mixer on and off, the potential to make the impeller diameter equal to the mixing chamber, and the ability to perform large area mixing [65].

7.4.1.5 Acoustic Bubble Shaking

Acoustic bubble shaking is the process of using sound waves to vibrate (shake) a gas bubble surface, which acts like a vibrating membrane in liquid, causing mixing to occur [58, 65, 73]. The process is also referred to as cavitation microstreaming or acoustic microstreaming and is dependent on the wave type (square or sinusoidal) used as well as amplitude. The use of acoustics without gas bubbles does not result in significant mixing. One study of this technique compared a single bubble with multiple bubbles to better understand the mixing, using dye dilution in a cylindrical chamber 15 mm in diameter and 300 μm deep [58, 73]. The time it took to fill the chamber with dye for one or four bubbles was 110 s and 45 s respectively. This experiment illustrated that the number of bubbles decreases the mixing time, though not linearly.

7.4.2 Passive Micromixers

7.4.2.1 T and Y Type Micromixers

A simple micromixer design is the T or Y mixer. This design has been used for single phase (gas or liquid) [65, 69–71], as well as for dual phase flows [64, 72]. Hessel et al. estimated the mixing length as $Lm = Pe \times$ channel width, assuming laminar flow [65]. However, the mixing efficiency is highly dependent on the details of the structure. For example, Gobby et al. studied the effect of orientation on the mixing length of a Y-mixer using gaseous oxygen and methanol at an inlet velocity of 0.3 m s^{-1}; $Pe = 8.08$, channel width = 0.5 mm (Fig. 7.3) [71]. They found that varying the Y angle would decrease the Lm from 2.12 mm at (–45° orientation of inlets) to 2.03 mm (at +45°). However, the smallest foot print was the –45° orientation. The shortest mixing length of 0.5 mm was achieved by integrating a Venturi in the Y mixer [65, 71]. In another design, Wong et al. [69] examined a T-type micromixer with a hydraulic diameter of 67 μm and found that increasing the pressure caused turbulent conditions at the confluence point, making complete mixing possible within less than a millisecond.

Fig. 7.3 Y-type micromixers:
–45° (left), +45° (middle), and +45° with venturi (right).
Methanol mass fraction contours for mixing gaseous methanol
and oxygen (diameter = 160 μm, 0.3 m s^{-1}, $Pe = 8.08$).
(Source Gobby et al. [71].)

The application of conventional pressure drop correlations for single and dual phase flow has been examined by Yue et al. [64]. They reported that the single fluid behavior still obeys classical theory in microchannels with diameters of several hundred micrometers. For dual phase flow, they proposed a modified correlation, but recognized that further research would be necessary [64].

7.4.2.2 Multi-laminating Mixers

Multi-laminating mixers can be generalized into four configurations: bifurcation, focused flow, cyclone, and star. Bifurcation designs feed alternate components in microchannels into a main chamber and then into an inverse bifurcation structure, often followed by a delay loop where mixing takes place [65]. The liquid lamellae from the feed are designed to be < 100 μm to aid in the diffusive mixing. The focused flow design is extremely similar to the bifurcation structure, except that the liquid lamellae are larger (> 100 μm), and, therefore, the inlets are directed (focused) using geometric constraints (Fig. 7.4) [65, 76]. Hessel et al. [65] observed that the benefit of the focusing can be represented numerically with the Fourier number, which for a straight rectangular channel of two fluid lamellae is given by $Fo = Tr/Tm = 4 \, (h \, L/w) \, (D/\Phi)$, where h, L, w, and Φ are channel height, length, width, and volumetric flow rate respectively.

Cyclone mixers can be made by focused injection of the fluids to be mixed in radial and tangential streams. Hardt et al. [77] have produced a CFD model of this type of mixer and optically verified its performance (Fig. 7.5).

Star laminators are high-capacity multi-lamination micromixers. Unlike the other mixers discussed here, these devices often exhibit a turbulent flow regime caused by their large flows and internal volume. They are made by stacking plates with characteristic star-like openings. Mixers using this configuration have been reported to effectively mix 3.5 m^3 h^{-1} [65].

Fig. 7.4 Focus design for multi-lamination mixer.
This design utilizes 138 feed channels (flow = 350 L h^{-1} at 3.5 bar) [76].

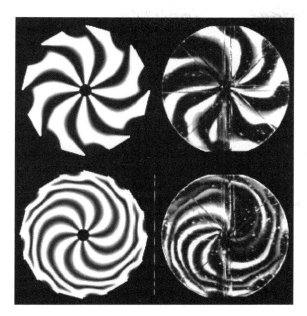

Fig. 7.5 Cyclone micromixer. CFD predictions (left) and experimental patterns using a diffusion experiment [77].

7.4.2.3 Split and Recombination

While the multi-lamination micromixers use a parallel approach, the split and recombination units use a linear approach by splitting the flow, recombining it and by rearranging it, almost always with some recirculation flow [60, 63, 65, 66, 84, 85]. Designs used to achieve this flow include, but are not limited to, fork-like, ramp-like, cross-like, and curved. Mae et al. reported a two-phase mixer designed to create a water-oil emulsion that could process up to 5 L h^{-1} (YM-1) and 20 L h^{-1} (YM-2) [63]. Figure 7.6 shows the YM-1 structure.

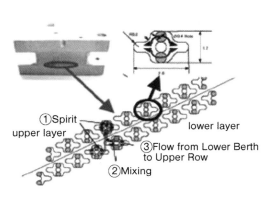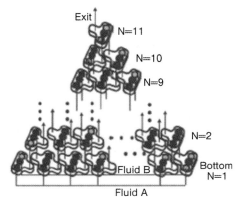

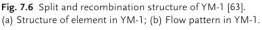

Fig. 7.6 Split and recombination structure of YM-1 [63].
(a) Structure of element in YM-1; (b) Flow pattern in YM-1.

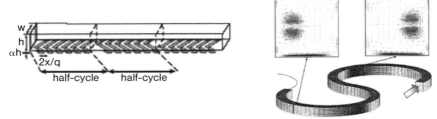

Fig. 7.7 Geometry of a staggered herringbone micromixer (left) and a curved channel to induce Dean Vortexes with corresponding secondary flow patterns (right). Arrow indicates flow [92].

7.4.2.4 Chaotic Flow

Chaotic flow mixing occurs when vortex(es) periodically change position between two or more locations. Much research has been performed in this area [90–98]. The mixing in chaotic flow micromixers often has spots of fast mixing and others that are essentially dormant. The first chaotic flow micromixers had microstructures only on one side of the microchannels [65, 92, 95, 96]. CFD was used to examine the mixing effectiveness of this design, finding that small and large vortexes occurred in a double helical flow [95, 96]. This design had very effective mixing and short mixing lengths. Another design used planar curved channels to induce helical flows without the need of a 3D structure [92]. In the laminar regime the Dean number, K [$K = Re\,(d/R)^{0.5}$], can be used to determine when vortexes will form. In this equation, d is the hydraulic diameter and R is the channel's mean radius of curvature. Four counter-rotating vortexes occur at $K > 140$–150 (Fig. 7.7) [92]. Jiang et al. [92] have used an imaging technique developed by Hessel et al. [105], called iron rhodanide reactive imaging, to determine that the two curves were required to achieve complete mixing.

7.4.2.5 Recirculation Mixers

Recirculation mixers use a zigzag pattern [61, 99–101]. The upper and lower teeth of the zigzag pattern cause eddies, which border the main flow at sufficiently high Re numbers ($Re > 267$) [61, 65, 99–101]. However, at lower Re and constant Pe the mixing is very poor [65]. The mixing efficiency is also related to the zigzag period, with too little or too much resulting in a loss of efficiency [100].

7.4.2.6 Structured Packing

Structured packing mixers are similar to packed beds, but have the packing material designed, controlled, and engineered to provide the desired mixing with minimized pressure drop [88, 89]. The concept is similar to split and recombination, but the structures are smaller and the time "split" is considerably less.

7.4.2.7 Colliding Jet Mixers

In conventional mixers, one way to obtain good mixing is to merge high-velocity jet streams. Even at lower velocities, this concept can be useful in certain cases, such as particle production. Nagasawa et al. have used this technique with the Villermaux/Dushman reaction, forming polymer particles [102]. Their results indicated that rapid mixing was achieved and that high flow rates could be used [102].

7.4.2.8 Moving Droplet Mixers

Modern microfluidic systems often utilize diminutive-scale chemistry. One such system involves the injection of aqueous droplets into a hydrophobic liquid [103, 104]. The droplets act as isolated chemical reactors and reagent transport systems. For this type of system, diffusion alone is not sufficient to achieve suitable mixing. Therefore, it was proposed to inject the droplets into a pressure driven flow in a serpentine channel, as shown in Fig. 7.8 [103, 104]. Flows internal to the drop were caused by changing the geometry of the channels, and resulted in intra-drop mixing.

Fig. 7.8 Moving droplet mixer. Two aqueous solutions were injected into immiscible oil in a serpentine channel 28 μm wide and 45 μm deep. Flow is from left to right [104].

7.5
Microchannel Emulsification

Emulsions are defined as disperse multi-phase systems of immiscible liquids. In simplest terms, they are either oil-in-water (O/W) or water-in-oil (W/O), where the dispersed phase is listed first and exists as discrete droplets within the continuous phase. In some cases, multiple emulsions can be formed, such as water-in-oil-in-water (W/O/W). Both natural and manufactured emulsions have been studied and utilized extensively, especially within the past 200 years, and Lissant has given an historical overview of the subject [106].

Emulsions are used extensively in the food, pharmaceutical, and cosmetics industries, where their most important properties include stability (both physical and qualitative), rheology, reactivity, shelf life, texture, appearance, and flavor. All of these properties are affected by droplet size and/or size distribution [107, 108], which in turn are functions of the method of production. Because emulsions are thermodynamically unstable, they require energy input for production and usually rely on a stabilizing agent (emulsifier, surfactant) to remain stable over long periods [109].

Traditional methods of emulsification include high-pressure homogenizers, rotor-stator systems, and ultrasound homogenizers [110, 111]. The first two examples employ high mechanical shear rates to produce small droplets of the dispersed phase, while ultrasound techniques rely on cavitation [112, 113]. All three methods, however, require a pre-mixed oil/water solution be fed to the device. These traditional mechanical methods are advantageous in that they are continuous processes with high capacity. This is most useful in industries where high volume products are being processed and narrow droplet size distribution is not as critical. However, since these methods rely on high mechanical shear rates, the processed ingredients often cannot be shear-sensitive compounds such as proteins or pharmaceuticals.

More recently, membrane and microchannel emulsification techniques have been developed and reported. Nakashima and Shimizu first proposed the membrane technique in the 1980s [114, 115], while Nakajima and coworkers first reported the microchannel technique in 1997 [12], apparently as an outgrowth of visualization techniques developed around microfiltration [116] and blood rheology [117, 118]. Both of these emulsification methods can produce emulsions without pre-mixing the solutions. In these cases, referred to collectively as microporous systems, the dispersed phase is forced through the microporous structure (membrane or microchannel array) into the continuous phase, forming μm-scale droplets one at a time from each pore or channel [119, 120]. These methods are advantageous in that they produce droplets with very narrow size distribution using very low shear. However, these same techniques suffer from low throughput, making them applicable only to systems that require low shear, have low volume output requirements, benefit from narrow size distribution, or are highly value-added to make up for the low production rate.

Given the two microporous processing options, their comparison can be summarized as in Fig. 7.9 [109]. Generally speaking, membrane emulsification provides higher flux and smaller droplet sizes than the microchannel process. Microchannels, however, are less susceptible to fouling, require little or no

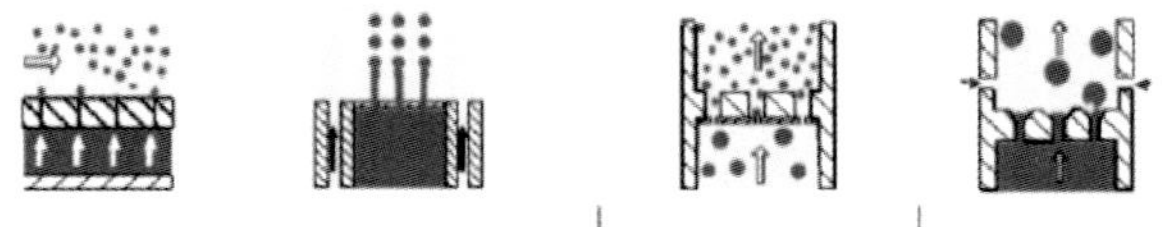

process:	membrane emulsification		premix-membrane-emulsification	microchannel-emulsification
	single droplets	liquid jets		
droplet detachment (formation) by:	wall shear stress	Rayleigh instabilities	flux	instability
droplet size / μm	0.1 - 10	> 0.2 ?	0.1 - 10	> 3
droplet size distribution:	narrow	narrow (?)	narrow	monodisperse
flux / m³/(m²·h)	J_d < 0.4 (0.2)	J_d > 1 (?)	1 < J_{em} < 10 (20)	J_d<0.01
danger of membrane fouling:	medium	low	high	low

Fig. 7.9 Comparison of membrane and microchannel emulsification [109].

shear [121, 122], and have narrower droplet size distribution [122]. Membrane emulsification has been thoroughly reviewed elsewhere [109, 120] and will not be further discussed here.

Several groups have been active in this field [63, 104, 123–135], and some of this work [123, 124] predates that of Nakajima. However, Nakajima et al. (also collaborating with Tragardh and others in Sweden) are recognized as pioneers in microchannel emulsification [12], and have progressed through several major improvements of the technique, including channel layout [136], geometry [137–141], and surface treatment [142, 143]. Additionally, they have examined the driving forces of the process [122, 142–146], focusing on the effect of surfactants, surface tension, and channel surface modification. In other studies, they have modeled droplet formation phenomena [146–149], and examined non-silicon materials of construction for the microchannel devices, including stainless steel [150] and polymers [151–153]. Both of these materials would offer advantages in cost and manufacturability. Stainless steel was found to have severe limitations in utility, while polymer materials have shown great promise, with investigations continuing.

Figure 7.10 shows the geometry of the system employed by Nakajima and coworkers, where the dispersed phase is forced from a reservoir (left-hand side)

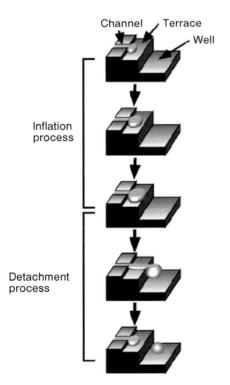

Fig. 7.10 Geometry of the microchannel systems employed by Nakajima and coworkers [137].

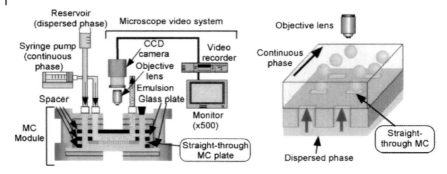

Fig. 7.11 Schematic of the "straight through" microchannel emulsification system [138].

through microchannels into the continuous phase (right-hand side), which is usually flowing past the exit of the microchannel array. The geometry of the exit portion of the microchannel array is important in droplet formation, requiring a "terrace" and a "well" at the end of the microchannel array, and much work has investigated this factor [136–141]. More recent developments in seeking high-throughput systems have led to designs such as that shown in Fig. 7.11, known as "straight through" microchannel emulsification.

Nakajima and coworkers have also examined the potential size range of microemulsion droplets, from the very small (~4 μm) [154], to mid-range (16–50 μm) [136, 140, 142, 155], to large (100–200 μm) [150, 156, 157]. Of great importance has been the very narrow size distribution of the droplets formed, generally having coefficients of variation less than 5% [138, 140, 144, 146, 156].

The microchannel emulsion technique has been extended to the formation of multiple emulsions [158–163], encapsulation [123, 158, 164–166], polymer bead formation [123, 125, 167–169], demulsification [116, 158, 170], and even microbubble formation [171]. New methods of stabilizing emulsions have also been investigated in this realm, including particle-stabilized [172] and protein-stabilized emulsions [173], with some work in emulsification without surfactants [135, 146]. In the case of multiple emulsions, microchannel architecture can enable the formation of W/O/W emulsions in which two water droplets of different compositions can be encased in the same oil droplet [163].

Other groups have taken different approaches to the emulsification challenge, employing T-junction [104, 127–129, 132, 133, 160, 167], Y-junction [134], and co-current flow [125, 126, 130] architectures. Flow focusing has also been investigated as a means of producing droplets that have smaller diameter than the channel from which they are created [126]. Low throughput has been a shortcoming of the microchannel emulsification process, but recent work has focused on scale-up of the microchannel process [140, 141] using "straight through" microchannel emulsification.

In addition to the dozens of publications in this area in less than 10 years, several patents have been granted or are in process. The patents granted have captured applications in food [174], imaging [175, 176], pharmaceuticals [177, 178], and general emulsion production [179–184], with more applications to follow as this area continues to develop.

As with many other applications of microchannel process technology, the advantage in emulsification relates primarily to the precise control that is possible in such systems. As discussed above, droplet size can be tailored from < 10 μm up to ~200 μm, with excellent monodispersity. Using a different approach, researchers at Velocys, Inc. (Velocys) have employed microchannel architecture in emulsification demonstration (Fig. 7.12). Here, the microchannels are not used as the "pores" from which the droplets are formed, but rather are used as the architecture for flow of the continuous and dispersed phases. This use enables rapid heat transfer and precise control of continuous phase flow patterns [135] in a membrane-based emulsification system. Velocys has demonstrated a moisturizing lotion emulsification process using such an architecture. This system was shown to have several advantages over conventional methods, including rapid heating or cooling before or after emulsification, tight control of the shear rate (which controls droplet size), and optimization of product formulations.

New opportunities and future directions in the area of microchannel emulsification are most likely in the areas of scale-up [140, 141], encapsulation/polymerization [123, 125, 158, 164–169], rapid quenching of droplets [135], and the use of emulsions as templates for uniform macroporous particle structure formation [172]. Microchannel emulsification is also likely to open up new opportunities with systems that are highly shear-sensitive [120, 135, 173]. The ability to scale up the process will spur new markets that require high production rates and the production of monodisperse capsules and polymer particles. Such developments will be useful in drug delivery applications and will contribute to the further quantification of micro-particle properties. Additionally, the use of monodisperse emulsions as particle templates is likely to enhance the utility of highly functional nanoparticles in need of a deployment mechanism [172].

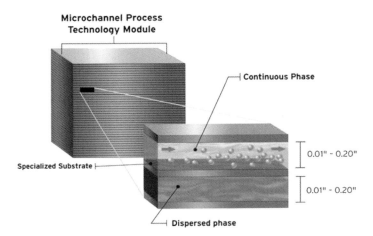

Fig. 7.12 Microchannel architecture for membrane emulsification. (Courtesy of Velocys, Inc.)

7.6
Phase Separation

In phase separation two immiscible fluids are physically separated. Microchannels offer the ability to separate phases in an orientation-independent manner, since capillary and surface tension forces are more dominant in these high-surface-area devices. Various microchannel phase separators have been developed to separate organic and aqueous phases for use in unit processes such as solvent extraction or reactions conducted at an aqueous organic interface [185–188]. The approach is to hydrophobize half of the channel with a non-polar agent so that the organic phase is constrained to the hydrophobic half and the aqueous phase to the hydrophilic half. Phase separation is simply then a matter of splitting the flow at the hydrophobic–hydrophilic junction of the flow.

Wicking technology has been developed to separate two immiscible fluids [14]. In this technology, the two mixed fluids are separated through the use of capillary forces that cause the liquid to be preferentially sorbed into a porous structure. Both single channel and multichannel devices have successfully separated non-condensable gas from water in both normal and microgravity environments. In both environments and in various orientation in normal gravity, the multichannel system developed by PNNL was able to meet the design requirements for a 5-kW PEM fuel cell system, removing 40 mL min^{-1} of water from 2.7 to 54.3 SLPM of gas at 3 to 5 bar and 80 °C (Fig. 7.13). The technology has also been extended to

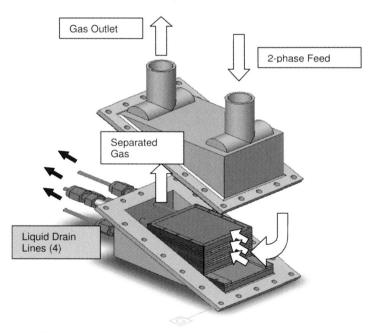

Fig. 7.13 Schematic of a wick-based phase separator sized for a 5-kW PEM fuel cell and demonstrated in both normal and microgravity environments [14].

include partial condensation with phase separation, with overall heat transfer coefficients ranging between 500 and 2000 $W\ m^{-2}\ K^{-1}$ [14].

7.7 Phase Transfer Processes

7.7.1 Adsorption

The adsorption process involves the reversible physical or chemical fixation of vapor or liquid to a porous solid. These processes are caused by van der Waals forces or electrostatic interactions. Generally, adsorption is used either for purification where the product has little value, as a separation technique as in chromatography [189–191], or as a collection technique if the product has sufficient value [194–198]. When the product is to be collected, such techniques as temperature swing adsorption (TSA), pressure swing adsorption (PSA), or displacement are commonly used. Ion exchange has also been included in this section as it has many parallel uses and can be considered as chemical adsorption [192, 193].

Adsorption processes can benefit from microchannel devices, because the apparent adsorption kinetics in conventional reactors are often much slower than intrinsic adsorption kinetics as a result of thermal and diffusional resistances. Microchannels provide short transfer distances for both mass and energy.

7.7.1.1 Microchannel Chromatography

Gas, liquid, and ion chromatography have been studied in microchannel configurations, leading to reduction in analysis time, system size, sample consumption, waste production, and cost. Multiple chromatographic separations can also be performed in a single device. For instance, a 2-m long column has been fabricated on a silicon glass chip only 20 by 25 mm, where the stationary phase consisted of a silicon organic created by chemical vapor deposition [189]. The channels were etched 27×70 μm in silicon with a 0.5-μm thick stationary phase. The column successfully separated methane and ethane. In another application, a GC column has been fabricated with single-crystal silicon channel walls supported on glass (Fig. 7.14). The 3-m long column wall had a poly(dimethylsiloxane) non-polar stationary phase. This column resulted in 6000 theoretical plates, a number that is ~35% of ideal [190].

In liquid phase chromatography, Henderson et al. have developed a liquid chromatograph on a chip in which the separation column and electrodes are integrated into the device [191]. The patent authors claim that good resolution is obtained using this approach.

Kang et al. have developed a multiple parallel channel ion exchange device in which each channel contained a stationary phase and its own dedicated conductivity detection electrode [192]. By reducing the channel width, these researchers

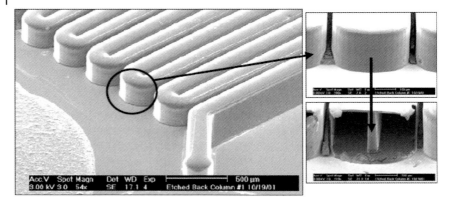

Fig. 7.14 Images of a microchannel gas chromatograph developed by the Engineering Research Center for Wireless Integrated Microsystems at the University of Michigan [190].

predict an improvement of nearly two orders of magnitude in separation impedance (combined separation efficiency and energy cost) as compared with conventional chromatographic separators. Kuban et al. have used ion exchange materials either dissolved in an organic phase or blended as a solid in an aqueous phase [193]. While initial results of this ion exchange device were encouraging, they were not competitive with macroscale ion exchange. This could be improved by using smaller microchannels to reduce the diffusion lengths.

7.7.1.2 Microchannel Adsorption for Component Collection

In addition to electrophoresis-based collection of DNA in microchannels (see Section 7.7.3), adsorbents have been used to collect and purify DNA. In these cases, microchannels are generally used because of the small quantity of material requiring collection. Using solid phase extraction, sorbent particles of nano- and micro-silica and micro-sized octadecylsilica were immobilized using sol–gel chemistry to fill the microchannels of the microfluidic device [194]. DNA as well as several organic compounds were evaluated for adsorption and desorption. They showed excellent adsorption, but poor recovery because they were difficult to extract.

Adsorbents tend to have a higher capacity for the target adsorbate species when cool than when hot. Figure 7.15 illustrates the adsorption and desorption processes used in TSA [195]. Because of this characteristic, TSA is a method of collecting a target analyte. Microchannel devices, due to their reduced thermal mass and intimate contact between the cooling/heating microchannel and the adsorbent microchannel, have also been used for TSA. These systems have higher throughput and reduced energy requirements than conventional TSA systems.

A single-channel TSA system was developed for NASA CO_2 collection and compression from the Martian atmosphere [196]. Accordingly, there was a need for a lightweight, highly efficient compressor. Based on the results of single-

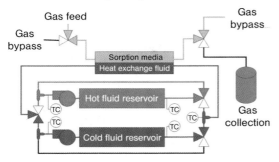

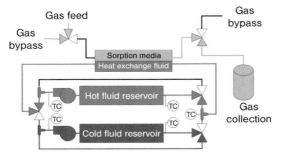

Fig. 7.15 Adsorption and desorption processes for a temperature swing adsorption (TSA) system. By using microchannel devices, the thermal mass is low and heat transfer is rapid, resulting in a small device [195].

channel testing, the Zeolite 13X adsorbent loaded inside of a microchannel can be cycled from 0 to 100 °C in approximately 2 min and produce greater than 10× compression ratio. Furthermore, an eight-cell microchannel device was designed in which alternating cells are adsorbing and desorbing, resulting in near-constant flow of pressurized product. This device is considerably smaller than mechanical compressors and similar conventional TSA systems.

Microchannel TSA devices using rapid swing have also been patented for purification of hydrogen using palladium as the adsorbent material [197]. With only a very narrow temperature swing, surprisingly high hydrogen throughputs are possible. Other gas separations have also been considered with this technique [198].

7.7.2
Extraction

Solvent extraction is a means of transferring a soluble analyte between two immiscible liquids. It is used to separate, concentrate, and purify, based on the principle that an analyte will not partition equally between two different phases. In many cases, the analyte is extracted from an aqueous stream into an organic stream and then back-extracted into a separate aqueous stream for analysis or further processing. Solvent extraction is generally performed in a well-stirred contactor where the phases are well mixed to increase interfacial surface area. The mixture is then collected in settlers where the phases re-coalesce. Solvent extraction is improved if either the diffusion of the key analyte or separation of the phases can be improved.

Microchannel devices, when applied to solvent extraction, allow intimate contacting of two immiscible fluids as they flow through very thin channels that are smaller than the normal mass transfer boundary layer. By reducing the mass transfer resistances in each phase, overall mass transfer rates can be significantly enhanced and either the overall size of the equipment can be reduced [199] or the extraction can be performed more rapidly [200]. Furthermore, microchannel extraction does not require stirring, mixing, or shaking. In one study, extraction time was 60 s in a microchannel as compared with 20 min shaking time for mechanical extractor with very similar extraction efficiency [201].

Most work in microchannel extraction focuses on improving the extraction efficiency or the phase separation or system development through adding multiple unit operations on a single chip or by scaling up. Because laminar flow exists in the microchannel devices, the intimate mixing of turbulent flow in traditional contactors is not present. Most studies have shown that the dissimilar phases flow parallel to each other with movement of solute molecules caused by molecular diffusion only. Thus, extraction is governed by contact time between phases [202].

To improve the extraction efficiency, several approaches have been taken. In one study intermittent partition walls stabilized the two-phase flow and allowed good phase separation, but also induced a slight turbulence and improved the extraction efficiency two- to three-fold at a contact time of 0.12 to 0.24 s despite the reduced interfacial area. Yttrium ions were successfully extracted from an aqueous phase into a n-heptane phase [203]. Ueno et al. have used zigzag side-walled channels to improve extraction [204]. The non-symmetric zigzags produced a sinusoidal oil/water interface that resulted in improved extraction efficiency.

A study conducted by Zhao et al. created intricate patterns of hydrophilic and hydrophobic regions down the microchannel to improve extraction [205]. This was achieved by laying down photocleavable SAM (self-assembled monolayer) and exposing with UV light to create exposed hydrophilic regions. The unexposed regions remained hydrophobic. If pressure is controlled properly, once patterned, the aqueous phase will remain separated from the organic phase.

A microchannel contactor has been developed and tested with water and cyclohexane streams extracting cyclohexanol [199]. Using this device, the relative importance of mass transfer resistance in the flow channels versus the contactor plate was explored. Both micromachined contactor plates and commercial polymeric membranes were configured with various channel heights both on the feed and solvent sides. Data indicate that contactor plate mass transfer becomes limiting as the channel height decreases below 300 μm for current contactor plates. However, as the contactor plates become more efficient, the need is anticipated for operating with substantially smaller channel heights. Wang et al. [206], using supported liquid membrane extraction, also found that shallower channels improved the enrichment factor, which was as high as 65.

Phase separation improvements are based on either surface modification, fluid property control, or physical separation. Studies have shown that organic liquid membranes can be developed in a microchannel device using surface modification [207, 208]. An organic liquid membrane consists of an organic phase with aqueous phases on either side. An analyte can be extracted from the aqueous phase, into the organic phase and then back-extracted into the second aqueous phase. These three phases can flow stably within a single microchannel, but better separation of the three phases is possible with surface modification of the organic phase channel (Fig. 7.16).

Countercurrent microchannel extraction has been performed by chemical modification of the microchannel wall with octadecylsilane groups on the bottom of the glass plate. Additionally, the two phases flowed in separate microchannels and crossed each other at an angle. If the correct flow rate and pressure differences were maintained, phase separation would occur [209].

Reddy and Zahn have performed extraction of non-DNA material from an aqueous phase into an organic phase [210]. The ability to contain the two immiscible liquids separated was found to be a function of the viscosity of the two phases and the interfacial tension. High viscosity and low interfacial tension facilitate a stable flow. The reduction in interfacial tension was performed using surfactants. Tests were performed to determine the amount of surfactant required while still maintaining stable flow.

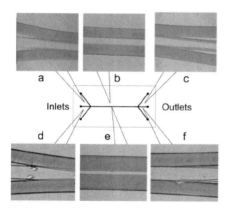

Fig. 7.16 Photographs of three-phase flow in a microchannel:
(a) near the inlet junction, (b) center, and (c) near exit junction
of the microchannel without surface modification; and
(d) near inlet junction, (e) center, and (f) exit junction of the
microchannel with octadecylsilane group surface modification [207].

A microchannel device has been developed for coalescence of dispersed droplets after solvent extraction [211]. Results of the study showed that the microchannel depth must be smaller than the diameter of the oil droplets for coalescence to occur. By using rectangular plates and controlling the contact angle of the liquid at the surface of the wall, the speed of the dispersed phase can be slowed with respect to the continuous phase, allowing coalescence [212].

Liquid–liquid extraction has been performed with a separation membrane to allow immiscible fluids to come in contact. In this case, stainless steel membranes were produced by photochemical etching of 130-μm holes. Multichannel extractors were constructed by diffusion bonding. Polymeric membranes were also studied with 15-μm micromachined holes [213].

Shaw et al. focused their work on scale-up of two different extraction processes [214]. Using microchannels etched in glass containing aqueous and organic phases joined by anodic bonding, they developed microchannel-based extraction systems both for parallel organic extractants and a single high-throughput system. The parallel processing system incorporated ten separate organic fluids and a single aqueous feed. These microchannel devices were reconfigured such that 50 contactors in parallel, each with 120 channels, yielded an output of 250 mL h^{-1}.

Solvent extraction has been combined with chelation and purification in a single microchannel device for Co(II) analysis. Putting all of these unit operations onto a single chip resulted in faster analysis time and reduced complexity [215]. This device included a chelation reaction occurring within the same microchannels as an aqueous organic extraction. This step was followed by an organic washing step that included three-phase acid/organic/base extraction.

7.7.3
Microchannel Electrophoresis

Electrophoresis refers to the movement of electrically charged molecules in the presence of an electric field. The major application of electrophoresis has been as an analytical separation technique for charged molecules, especially large biomolecules such as DNA, peptides, and proteins. Microchannels have been used extensively in electrophoretic separations. Capillary electrophoresis (CE) is a high-efficiency separation technique that employs narrow-bore (10–100 μm ID) capillaries generally made from fused silica. CE is a mature separation technique with numerous reviews and books, and several hundreds of papers having been published on the topic [216–219]. More recently, there has been a great interest in developing microfluidic chips that utilize electrophoresis and related electrokinetic separation techniques [220–223]. These developments have been pursued on the assumption that a microchip format offers advantages over traditional capillaries in terms of separation speed, the ability to integrate electrophoretic separations with other chemical systems or separation techniques, and the ability to integrate multiple separation channels in a single device.

One of the more visible applications of capillary electrophoresis was in the Human Genome Project. The development of capillary array electrophoresis

(CAE) greatly reduced the time required to complete the project compared with sequencing using the slab gel technology that was available when the project began [224–226].

Capillary electrophoresis is conducted in capillaries filled with an electrolyte solution. Buffered electrolytes are generally used, since biomolecule mobilities and electroosmotic flow (EOF) are sensitive to pH. The ends of the filled capillaries are placed in electrolyte reservoirs that contain electrodes, and the electrodes are positioned so that electrolysis products do not enter the capillary. A small plug of solution containing the analytes to be separated is pressure- or electrokinetically-injected into one end of the capillary, and a voltage difference is applied to the electrodes such that the analytes of interest migrate toward the other end of the capillary, where they are detected. Analytes with different electrophoretic mobilities migrate at different speeds and become separated as they transverse the capillary.

In the absence of EOF, positively charged analytes will migrate toward the cathode and negatively charged analytes will migrate toward the anode, which enables either cationic or anionic species to be separated in a given run. However, the direction of migration can be different when significant EOF is present. EOF refers to the bulk flow of fluid in the presence of an electric field, and is a result of surface charges on the capillary or microchannel surface. For example, at high pH the surface of a fused silica capillary has a negative charge. To maintain charge balance, a layer of cations builds up near this surface, creating an electric double layer.

The resulting potential difference across the double layer is known as the zeta potential. When an electric field is applied to the ends of the capillary, those ions in the diffuse part of the double layer migrate along the length of the capillary and drag the bulk solution with them.

Depending on the sign of the zeta potential, EOF can be towards either the anode or cathode, and the apparent mobility of an analyte is the sum of its electrophoretic mobility and EOF. Electroosmotic forces act near the capillary wall, which results in a flat flow profile that is less dispersive than the parabolic flow profile associated with pressure-driven flows. EOF can be beneficial since it can enable the analysis of both anions and cations in a single run. Electroosmosis has been investigated extensively as a means for producing fluid flow for various processes in microfluidic devices [227–232].

In capillary and microchip electrophoresis, the separation efficiency is limited by longitudinal dispersion, dispersion-generated temperature gradients in the capillary, electroosmotic flow, and solute-wall interactions. When only longitudinal dispersion is considered, the variance (σ_D^2) of the analyte band is given by Eq. (1), where D is the diffusivity of the analyte and t is the separation time. The time (t) required for an analyte to migrate through the separation channel is given by Eq. (2), where μ is the analyte mobility, L is the length of the separation channel and V is the applied voltage:

$$\sigma_D^2 = 2\,D\,t \tag{1}$$

$$t = \frac{L^2}{\mu V} \tag{2}$$

The number of theoretical plates (N) is a measure of the efficiency of an analytical separation technique, as defined by Eq. (3):

$$N = \frac{L^2}{\sigma^2} \tag{3}$$

Combining the three equations, it is found that N is independent of the length, but is proportional to the voltage applied across the separation channel as shown in Eq. (4). This contrasts with other separation mechanisms, such as chromatography, where column efficiency increases with increased column length.

$$N = \frac{\mu V}{2 D} \tag{4}$$

Since separation time is inversely proportional and efficiency is directly proportional to applied voltage, higher applied voltages can result in fast and efficient separations. However, the voltage that can be applied across the separation channel is governed by the amount of Joule heating that can be tolerated. Microchannels offer enhanced heat transfer characteristics, and their small cross section reduces the current and heat generated. This is the major premise behind capillary electrophoresis, and is a motivation to use increasingly smaller channels for electrophoretic separations.

Several electrophoresis modes are used frequently in capillaries and other microfluidic devices. Capillary zone electrophoresis (CZE) is the simplest mode, using a capillary filled with a buffered electrolyte. In CZE, analytes migrate through the capillary and separate according to their electrophoretic mobility. Capillary gel electrophoresis (CGE) is the technique generally used for DNA analysis. In CGE the capillary is filled with a gelled electrolyte that acts as a sieving matrix, and produces a separation that is based on size. Micellar electrokinetic chromatography (MEC) can be considered an electrokinetically driven liquid/liquid chromatography. MEC uses a buffered electrolyte containing micelles. The micelles move in the electric field and the solutes are separated by partitioning between the micelles and electrolyte. This technique allows for the separation of neutral analytes. Capillary isoelectric focusing (CIF) separates ampholyte analytes, such as proteins, based on their isoelectric points (pI). The analytes are mixed with a solution containing a series of ampholytes that cover the pH range of interest. When voltage is applied, a pH gradient is formed along the length of the capillary, and analytes migrate and focus at the point where their pI matches the pH. Capillary isotachophoresis (CITP) uses a leading electrolyte, which contains ions with higher electrophoretic mobilities than any of the analytes of interest, and a trailing electrolyte containing ions with lower mobilities than those of the analytes. When voltage is applied, the analytes migrate at the same velocity, but separate into zones according to mobility.

As previously noted, one of the great successes of microchannel electrophoresis has been in the analysis of DNA using CGE. There is ongoing interest in developing new devices and instruments that increase the speed of separation, reduce the amount of sample required, and integrate sample preparation. Much of this work has been directed towards the development of microfluidic chips. DNA sequencing has been performed on microfluidic chips and has given data comparable to that obtained with CAE instruments [233–235]. The separation times for these devices was less than half of those for commercial CAE instruments. Lagally et al. have combined microfabricated heaters, temperature sensors, and membrane valves into a polymerase chain reaction (PCR) amplification and DNA capillary electrophoretic cell for pathogen detection. Combining the two components resulted in an analysis time less than 30 min [236]. Chiesl et al. have developed a microfluidic electrophoresis device that also adsorbed proteins and allowed the purification of DNA. Separations generally use centrifugation and precipitation of DNA prior to electrophoresis. Combining this process into a single microchannel device thus simplifies it [237].

Microchannel electrophoretic devices are being developed to meet the need for better methods for protein separation and analysis. Interactions between analytes and the wall of the separation channel can reduce the efficiency of microchannel electrophoretic separations. These analyte–wall interactions are especially problematic for protein separations. Consequently, a large volume of the literature is focused on preventing protein adsorption on the walls of the microchannel by varying the microchannel material [238], by employing surfactants [239], or by modifying the microchannel surfaces [240].

One of the challenges associated with analyzing biological fluids is the wide range of analyte concentrations in these fluids. The analytes of interest often must be concentrated to be analyzed. Several preconcentration methods have been developed for use in capillary and microchip electrophoresis. Porous membrane techniques, for example, have been incorporated into microchannels. A porous bonding layer between two sets of microchannels allowed ionic current to pass but not DNA or protein molecules, thus concentrating the sample before electrophoresis by 100-fold or more [241]. A promising microfluidic preconcentration device, developed by Wang et al. [242], uses a nanofluidic (40 nm deep) channel in place of a membrane. Owing to a negative surface charge on the silica surfaces in this device, the nanofluidic channel functioned as a cation exchange membrane. When an electric potential was applied to the device, an ion depletion zone formed in one of the microfluidic channels adjacent to the nanofluidic side channel, and created an electric field gradient with an opposed electroosmotic flow that was able to trap and concentrate charged analytes, such as proteins. The most promising feature of this device is its ability to achieve high (10^7-fold) concentration factors in relatively short (1 hour) times.

The nanofluidic preconcentration device can be considered an equilibrium gradient device using a steep electric field gradient and an opposing bulk flow to focus analytes [243]. Other equilibrium gradient techniques have also been implemented in microfluidic channels. These techniques include temperature gradient focusing [244] and other forms of electric field gradient focusing [245–249].

7.7.4
Absorption

In many cases where mass transfer occurs between a liquid and a vapor, microtechnology offers the advantage of process intensification through control of the liquid phase thickness. This intensification often occurs because the diffusivity is typically orders of magnitude lower in the liquid phase than in the gas phase; hence, control of the liquid thickness decreases the primary impedance to mass transfer. This control benefits operations such as absorption and desorption, and processes based on these phenomena.

Absorption in microchannels has been studied intensively, for both physical absorption and absorption accompanied by reaction. For physical absorption, Sato and Goto [250] have shown that absorption of carbon dioxide diluted by nitrogen into water did not occur in 500-μm and 1000-μm channels but that significant absorption of the both gases occurred in channels under 250 μm, with complete absorption occurring for some conditions in 100-μm channels. Garimella has developed a falling film absorber with upward vapor flow and microchannel tubes to transfer heat on the coolant size for use in residential and lightweight commercial heat pump systems [251–254]. With ammonia–water systems, Garimella achieved overall heat transfer coefficients of 133 to 403 W m^{-2} K^{-1}, with flow distribution improvements resulting in overall heat transfer coefficients of 638 to 1648 W m^{-2} K^{-1}. Wicking absorbers have also been developed for heat pumps, where the benefits of microchannels can be utilized on the mass transfer side as well as the heat transfer side. The microchannels are air cooled with cross flow finned heat exchangers, while wicks are used to control the liquid film thickness [255, 256]. For absorption with reaction, Zanfir et al. have shown that, in some cases, such as CO_2 absorption into sodium hydroxide, the reaction is so fast that depletion of the CO_2 occurs within 25% of the film thickness [257]. Hence, increased efficiency occurs through controlling the liquid film thickness to reduce the consumption of the absorbent solution. Variation of the gas film thickness between 2.5 and 5.5 mm had little effect, since the shortened residence time in the thin channel was compensated for by the decreased gas diffusion length. Ono et al. have demonstrated the use of microchannel absorption through a porous glass plate for the detection of nitrogen dioxide [258].

7.7.5
Desorption

Desorption technology has been developed primarily for absorption-cycle heat pumps. The counter flow falling film absorber developed by Garimella has been tested by Determan as a desorber, requiring only modification of the flow distribution and enclosure [259]. The desorber (18 × 18 × 51 cm) could transfer

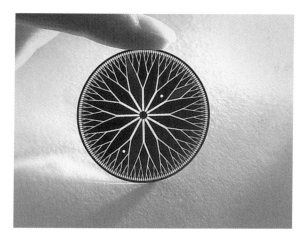

Fig. 7.17 Fractal-based droplet desorber developed by Oregon State University [261].

from 5.4 to 17.5 kW of heat with overall heat transfer coefficients from 388 to 617 W m^{-2} K^{-1}. Significant rectification occurs in the counter flow desorber, allowing the downstream rectifier to be much smaller. Two co-flow droplet desorbers have been developed by Oregon State University, also in conjunction with development of absorption heat pumps. One droplet desorber uses stainless microtubes to convey the ammonia–water solution, which is heated on the shell side [260]. A second droplet desorber uses fractal branching of the channels to send the ammonia–water solution radially from its center (Fig. 7.17). The fluid is heated in the fractals before it is discharged. Heat fluxes of 3 to 8 W cm^{-2} have been achieved [261]. Heat pump systems are described in more detail in Section 7.7.8.

7.7.6
Pervaporation

Pervaporation is a membrane process in which a liquid is maintained on the feed side of a membrane and permeate is removed as a vapor on the downstream side of the membrane. Pervaporation is used, because of its low energy consumption and low cost, to separate dissolved organics from water, purify waste water or volatile chemicals, and break azeotropes. Pervaporation plants range from processing a few grams per hour up to thousands of tons per year. For waste water treatment flow of less than 76 L min^{-1}, pervaporation is more cost-effective than other treatment options, such as chemical oxidation, ultraviolet destruction, air stripping followed by carbon adsorption, steam stripping, or distillation/incineration [262].

Pervaporation can be used in microfluidic devices for sample concentration. Microchannels 500 nm high, 4 mm long, and 2 to 30 µm thick have been fabricated with polyimide using thin film deposition [263]. Pervaporation occurs through the polyimide polymer membrane (Fig. 7.18). In this study, water was

Fig. 7.18 Branched polyimide membrane channels used for pervaporation and osmosis experiments [263].

pervaporated from a 0.1 M KNO_3 stream inside the microchannels into a 30% RH air stream outside the channels, concentrating the stream by 7× in 50 min. A similar microchannel concentrator device has been studied with a hydrophobic vapor permeable membrane for solution concentration with a countercurrent nitrogen flow [264].

Pervaporation can also be used as a micropump. Isopropanol was placed inside the polyimide membrane microchannels described above, and water was deposited on top of the permeation area. The large selectivity for water transport over alcohol resulted in water permeation at a rate of 70 μm s^{-1} [263]. In a similar study, a micropump was developed using pervaporation to transport a volume of 300 μL of Ringer's solution out of a membrane over a 6-day period. Capillary forces then induced additional flow into the membrane device to produce a very constant flow of 35 nL min^{-1} [265]. Although this device did not utilize microchannel architectures, the low fabrication costs and high reliability of such a system make pervaporation an attractive approach to pumping small flow rates for microfluidic devices.

7.7.7
Distillation

Distillation represents an ideal case for process intensification, since it accounts for 95% of the total separation energy used in the refining and chemical process industry and 6% of all domestic energy used in the United States [266]. Unlike absorption, however, distillation research is not as advanced, partially due to the more stringent mass transfer requirement to reach near equilibrium conditions in multiple stages. In one of the earliest embodiments, micron-sized holes were used on column sieve plates instead of the typically large perforations [267]. The small holes allowed surface tension to keep the liquid from weeping through the plate, which in turn allowed a decrease in the vapor flow rate and an accompanying increase in mass transfer.

Process intensification through the use of microchannels as opposed micron-sized holes has been manifested in a few different distillation approaches, including the use of nanostructures by the University of Tokyo [268], wicks by PNNL [269], and rotating surfaces by the University of Sheffield [270]. Secondary flow patterns in the rotating system enhance mass transfer, with the relative magnitude of dispersion Dc being orders of magnitude greater than the molecular diffusivity D and dependent on the angular velocity Ω. Figure 7.19 shows a schematic of this device, and Fig. 7.20 illustrates the resulting secondary flow patterns. In the wicking microchannels, early values of 2 cm for the height equivalent of a theoretical plate (HETP) have been achieved, which compares favorably with HETPs of typical advanced packings of 30 to 45 cm [271]. Application of the wicking

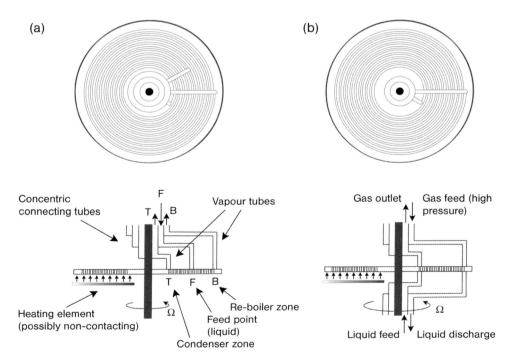

Fig. 7.19 Schematic design for spinning micro-disk gas–liquid contactor: (a) distillation and (b) absorption [270].

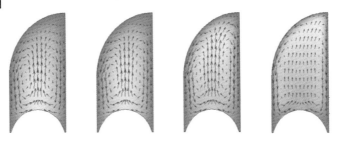

Fig. 7.20 Calculated secondary flow patterns in the liquid. Conditions are: $W = 100$ µm, $R = 0.02$ m, properties of standard water and, from left to right, $\Omega = 300, 1000, 3000$, and $10\,000$ rad s^{-1} [270].

technology is anticipated to reduce energy consumption by 20% for separations of ethane from ethylene [272], and Velocys has reported HETPs of less than 1 cm for this C_2 splitting. Promise has also been shown for this technology in the catalytic distillation of C_4 hydrocarbons [269].

7.7.8
Heat Pump Systems

According to the US Department of Energy [273]:

> The equipment and systems used to provide thermal comfort and adequate indoor air quality for residential and commercial buildings consume 39% of the total energy used in buildings. Energy consumption in buildings is a major cause of acid rain, smog, and greenhouse gas emission in the US, representing 35% of carbon dioxide emissions, 48% of sulfur dioxide emissions, and 21% of nitrogen oxide emissions.

Microchannel technology offers many methods of decreasing this environmental impact by increasing efficiency and reducing chemical inventory, when applied to heat pump systems [274].

In conventionally air coupled vapor compression heat pumps, size reductions of 50 to 75% have resulted through the use of microchannel, multi-louvered fin heat exchangers. For R-290 refrigerants, heat fluxes of up to 135 kW m^{-2} and overall heat transfer coefficients of 10 kW m^{-2} K^{-1} have been achieved in a microchannel condenser [275]. The size reduction can be increased to an order of magnitude through the use of hydronic loops incorporating microchannels, which in turn results in an order of magnitude reduction of refrigerant consumption. The refrigerant is confined to a small area and the system affords zoned comfort levels, because the fluid from the hydronic loop is used to convey heat to the building. The loss in efficiency caused by indirect heating of the hydronic loop is compensated for by the use of microchannels.

Conventional chlorofluorocarbon (CFC) and hydrofluorocarbon (HFC) consumption can be completely eliminated by using CO_2 as a refrigerant. The potentially lower coefficient of performance (COP) is compensated for through the use of microchannels, whose small characteristic dimensions offer the additional advantage of easily handling the high pressures on the compression side of the cycle. Innovative near-counter flow gas coolers offer the advantage of higher efficiency associated with counter flow on the CO_2 side, while still affording low pressure drops associated with cross flow on the air side [276]. In an alternative approach, a separator is used to enable only liquid to be fed to the evaporator to enhance distribution of the refrigerant into the microchannels and to raise the heat transfer coefficients to 12 kW m^{-2} K^{-1} [277].

Absorption heat pumps offer additional benefits over compression heat pumps in that the use of a liquid pump to drive the refrigerant to high pressures, instead of a compressor, reduces the electricity consumption by orders of magnitude (Fig. 7.21). In addition, these systems are heat activated as opposed to mechanically driven so they can use waste heat and reduce the use of electricity in areas where it is in short supply. The absorber is the largest component of the system, and its size results in reduced viability of absorption cycles for small capacity systems. However, commercial viability is increased by high heat and mass transfer coefficients, low pressure drops, and large transfer areas [278], all of which can be addressed through the use of microchannel technology.

As discussed in the section on absorbers (Section 7.7.4), falling films used in microchannels have shown potential for ammonia–water systems [279]. The military has shown interest in these absorption systems due to their inherently low electrical requirements and low noise. However, military application requires orientation independence, even more restrictive size requirements, and a heat rejection temperature of 50 °C (120 °F). To address this need, wicking and fractal microchannel technology has been developed that allows orientation independence and control of the liquid film thickness, which in turn controls the majority of mass

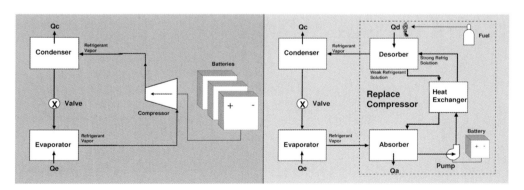

Fig. 7.21 Conventional vapor compression cycle heat pump (left) compared with absorption cycle heat pump (right).
In absorption cycle heat pumps, the compressor is replaced by a smaller pump and an absorption/desorption loop, greatly reducing the necessary power for moving the refrigerant through the system [256].

transfer resistance [255, 256, 261]. A breadboard system has been developed [255] showing a COP of 0.28, based on fuel energy content, with a cooling duty of 190 W. Subsequent results yielded over 250 W. The size of a currently available commercial system is $1.05 \times 1.12 \times 0.76$ m ($41 \times 44 \times 30$ in.) and 268 kg (590 lb), in addition to a 82 kg (180 lb) condenser that is $0.94 \times 0.97 \times 0.46$ m ($37 \times 38 \times 18$ in.). Second-generation targets for the microchannel technology include a 132-kg (290 lb) system that is $0.51 \times 0.76 \times 0.74$ m ($20 \times 30 \times 29$ in.) with the same COP [280].

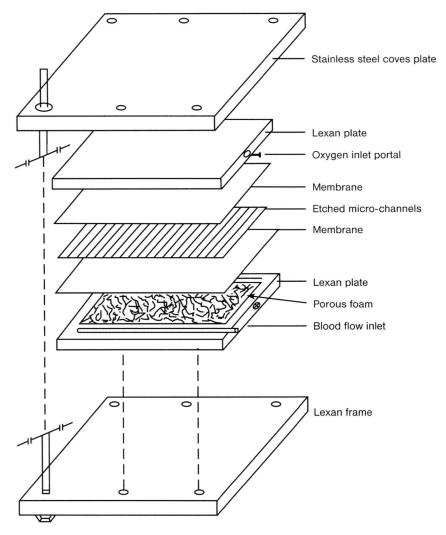

Fig. 7.22 Schematic of a microchannel membrane oxygenator for *in vitro* testing [281].

7.8
Biological Processes

7.8.1
Cell Testing with Microchannel Systems

Biological applications of microchannels are driven by the need for cost-effective handling of very small sample volumes along with parallel processing. The "Lab-on-a-Chip" concept could meet this need. One example involves techniques for cell testing or processing.

In conjunction with membrane-based blood oxygenators for example, microchannels offer improved gas exchange efficiency and reduce the volume of blood required for initial priming. A device developed by Hung et al. [281] involves a stack of 16 plate units incorporating 110 μm deep by 230 μm wide microchannels sandwiched with oxygen permeable membranes (Fig. 7.22). Experimental results not only show improved gas exchange efficiency over both macrochannels and theoretical predictions, but also a reduction in apparent blood viscosity.

Oxygenation of bio-artificial liver cells has been studied in a microchannel reactor using 140-μm high channels both with and without a membrane [282]. Microchannels were used to provide sufficient heat exchange so that the effects of the membrane could be seen. Theoretical work showed the advantages of a membrane in a microchannel oxygenator versus no membrane. In the case where a membrane is used, better control of oxygen gradients is possible.

Even studies involving cell shape and function [283] and cell membrane potential [284] are best performed in a microchannel device. The microchannels allow the cells to elongate and be imaged without high shear flow. The microchannel devices for measuring flow are used to minimize the amount of reagent and control its mixing.

7.8.2
Hemodialysis

Researchers at Oregon State University have demonstrated the advantages of microchannel architecture in improving the hemodialysis process. Using microchannel architecture, they were able to show 70–80% reductions in the necessary transfer area relative to commercial hollow fiber systems for the clearance of creatinin (Fig. 7.23) and urea (Fig. 7.24) from a simulated blood stream [285]. The microchannel advantage, as has been seen in other applications, comes in the form of well-defined and narrow channels that facilitate rapid mass transfer into and out of the fluid media. This approach is expected to change the current paradigm in hemodialysis from clinical treatment to at-home use, and may allow for the creation of a wearable hemodialyzer [286].

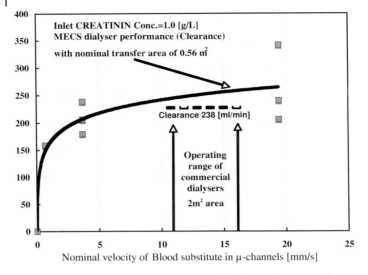

Fig. 7.23 Creatinin clearance in a microchannel dialyzer with a nominal area of 0.56 m^2 (solid line and data points), devised by Oregon State University, compared with the creatinin clearance in a typical hollow fiber commercial dialyzer (dashed line) [285].

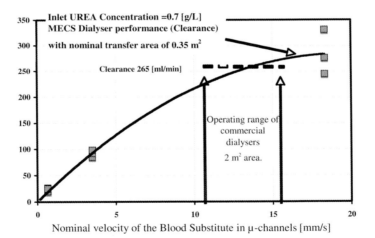

Fig. 7.24 Urea clearance in a microchannel dialyzer with a nominal area of 0.35 m^2 (solid line and data points), devised by Oregon State University, compared with urea clearance in a typical hollow fiber commercial dialyzer (dashed line) [285].

7.9
Body Force Driven Processes

In addition to the typical thermal and pressure forces exerted on fluids in microchannels, acoustic/ultrasonic and electric field forces have also been applied to manipulate fluids as a body – referred to as body force driven processes.

7.9.1
Electric Fields

Several uses have been developed for electric fields in microchannel devices:

- electroosmotic flow
- electric field actuated valves
- electromagnets
- mixers
- electrophoresis (concentration and separations)
- piezoelectric and other pumps with moving parts.

The first three uses are summarized below. Mixers and electrophoresis were covered in Sections 7.4 and 7.7.3, respectively. Piezoelectric and other pumps with moving parts are beyond the scope of this chapter.

7.9.1.1 Electroosmotic Flow (EOF)

Electroosmotic flow uses an applied electric field at the surface of the channel to stimulate fluid flow [287, 288]. Specifically, EOF takes advantage of small charges that often develop on the microchannel surface walls with most fluids. The surface charges are caused by chemical reactions, ionization, or ion adsorption when an electrolyte is in contact with the microchannels. Since opposite charges attract, opposite charges in the liquid move toward the microchannel walls, creating an electric double layer (EDL). Using external electrodes, an electric field is generated that moves the EDL ions, which in turn move the surrounding fluid [289].

This pumping technique uses no moving parts, however, and does not require sophisticated fabrication techniques. It is relatively easy to implement and operate in microchannels and other miniaturized systems [287]. The disadvantage of EOF is its dependency on fluid physiochemical properties (ionic strength, pH, organic content, etc.) to develop the surface charge, so its effectiveness will change for each fluid used in the pump [290].

EOF has been applied to electrophoresis (Section 7.7.3), electrospraying [291, 292], and electrochromatography [293]. Pressures of over 5 bar have been generated using EOF techniques [290]. Utilizing high pressures, mists or aerosols can be generated that are often sent to analyzer equipment such as electrophoresis, mass spectrometry, and electrospray ionization.

Electrochromatography uses EOF to drive a mobile phase through a packed microchannel or capillary (50–150-μm diameter) [293]. The Helmholtz–Smoluchowski equation is used to calculate the linear flow through the packed bed: $u = \varepsilon_o \varepsilon_r E \zeta/\eta$ where u is the linear flow, ε_o is the vacuum permeability of the packing, ε_r is the relative permeability of the packing, E is the applied electric field, ζ is the zeta-potential, and η is the solvent viscosity. Analysis of this equation reveals that there is, in theory, no dependence on particle size. This has been shown to be the case, as long as there is no overlap in the double layers [294, 295]. By using smaller packing particles, it was shown that electrochromatography was more efficient than conventional pressure-driven alternatives [294, 295]. Smith and Carter-Finch have published an in-depth review of electrochromatography [293].

A final application of EOF is the ability to direct and control the stream profile. Besselink et al. have developed a system that is designed to accurately position the middle of three streams in a laminar flow chamber [287]. They proposed that the system can be used for sending the middle fluid to selected parts of a chip, which contains a multi-sample array. The system is designed such that three parallel microchannels feed into the flow chamber (a wider channel). The two outer streams sandwich the middle stream to be analyzed. The guiding action of the two outer streams causes the center stream to flow across the chamber. By increasing the flow of one outer stream and decreasing the flow of the other, the middle stream is pushed to the side, as shown schematically in Fig. 7.25 and pictorially in Fig. 7.26. Control of the middle stream's width is adjusted by the flow rate ratio of the guiding streams and the sample stream [287].

Fig. 7.25 Flow guiding illustration. The position of the middle stream is controlled by adjusting the outer two flow streams [287].

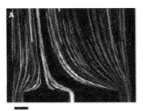  

Fig. 7.26 Three micrographs showing the stream profile of suspended beads in the flow chamber. (A and C) Guiding flows at unequal voltages (flows) with the higher voltage (flow) at the right reservoir (A) and the left reservoir (C). (B) Guiding flows operating at the same voltage, resulting in a central flow [287].

7.9.1.2 Electric Field Actuated Valves

Several research groups have proposed using electric fields for controlling flow in microfluidic systems. The two techniques discussed here include electroosmotic valves [290] and electrocapillary force actuation [296]. Electroosmotic valving uses critical dimensions so that the valve is "open" to EOF flows and closed to pressure-driven flows. The electric field is turned off in channels where flow is not desired, and the capillary forces prevent the fluids from progressing. Electric field is applied to channels where flow is desired. This systems appears to be a fairly simple method for valving in microfluidics.

Electrocapillary force actuation is somewhat more complex. It entails the use of a droplet that can be moved using electroosmotic forces, but will not disperse or mix with the working fluid. Thus, the droplet can be positioned to plug a channel to prevent flow, and then moved out of the way, allowing the fluid to flow. Le Berre et al. have demonstrated this valving technique using a mercury droplet in nitric acid [296].

7.9.1.3 Electromagnets

Electromagnets have been used for filtration, concentration, and separation in biochemical systems since the early 1970s [297]. This technique has been applied to microchannel devices using polymer encapsulated superparamagnetic nanoparticles (magnetic beads) [298, 299]. The polymer surface of the magnetic beads is functionalized to enable it to capture the desired materials. The beds are mixed with the sample fluid where they capture the desired materials. A magnetic field is then applied to filter the magnetic beads out of the sample, and the captured material can be concentrated and analyzed. For example, Smistrup et al. have used three microelectromagnets (copper coil semi-encapsulated in a dielectric layer with a soft magnetic yoke of nickel on top) [298]. The electromagnets were fabricated using standard microfabrication techniques. Three electromagnets were used, enabling a capture efficiency of ~89% [298]. The advantage of this type of system is the concentration of otherwise dilute material, which allows a fast analysis. The disadvantage is the loss of magnetic beads.

7.9.2
Acoustic/Ultrasonic Forces

Acoustic and ultrasonic forces have been used in microchannel devices primarily for mixing (Section 7.4) and separation purposes. Using acoustic forces to manipulate biomaterials has several benefits: there are no moving parts; the design is simple and compact; the same device can be used to mix or separate, depending on the frequency; and in situ operation enables contamination-free handling [300]. Gröschl in a three-part series describes the fundamentals [301], design and operation of separation devices [302], and the application in biotechnology [303] of these processes. Essentially, high-frequency acoustic/ultrasonic standing waves can separate and concentrate materials that have different characteristics [301, 304]. The standing waves are primarily used to agglomerate or detain particles, or move them in the flow. Other applications being investigated include solid particle removal and emulsion splitting [301]. The volume required to capture particles depends on the acoustic characteristics; however, in general for biological systems, mammalian cells (diameters of the order of 20 µm) require an active volume (chamber) of 75 mL [305], whereas bacteria (diameter of 1 to 2 µm) require only 2–8 mL [306]. Indeed, bacteria are difficult to manipulate in the larger volumes. As the channel gets smaller, the frequency must increase in order to obtain the manipulation desired. Even though the frequency increases, the power does not necessarily need to be increased, as the larger channels generally require relatively more power input to achieve the equivalent separation achieved in smaller channels. This is because in larger channels there is more material that needs to be moved and the material must be moved a greater distance to achieve separation. Importantly, since energy is being added to the fluid via the ultrasonics, the fluid temperature does increase, which may cause sample damage, boiling, intercellular component leakage, or other problems [300–304].

Many models for simple ultrasonic separator cells (single inlet rectangular cell, with several outlets at the opposite face) have been proposed, and some examples are discussed here. Hill et al. have developed a model based on a multi-layered resonant structure [307, 308]. They used this model to design a device that allows

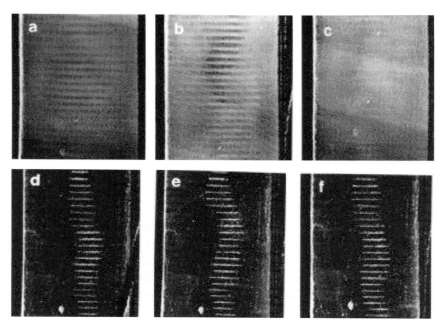

Fig. 7.27 Observed banding patterns as the gravity changes from 0 to 1.8 g [312]. Two suspensions of latex polystyrene are shown: 1.3-µm particles (a–c) and 12-µm particles (d–f). The time after entering 0 g are 1 s (a and d) and 20 s (b and e). Patterns in (c) and (f) are for 1.8 g at approximately 8 s after leaving 0 g.

a sample of particle-free fluid to be extracted from a flow (fluid clarifier). Harris et al. [309] have also used this model to develop a device to concentrate particles into the middle of a chamber. Gröschl [301–303] and Higashitani et al. [310] have developed models that relate the acoustic energy and frequency for a particle–fluid mixture to the flow velocity. Yasuda and Kamakura [311] have developed models based on acoustic radiation. CFD modeling can be used to study the effect of Reynolds number, chamber size, and geometry [307].

An example of ultrasonics to manipulate particles is the work reported by Hawkes et al. [312]. They tested a device in microgravity during a KC-135 parabolic flight. The chamber was a thin-walled cellulose acetate acoustic chamber (50 mm high × 11.7 mm ID) mounted on a 1-MHz PZT 26 transducer. It had a square glass plate reflector (1.2 mm thick × 12 mm × 12 mm). The transducer disk was 40 mm in diameter with electrodes (8 mm in diameter) overlap at the center at front and back transducer faces. Using this device, yeast cells and latex particles (1.3, 12, and 20 µm) in aqueous suspension were manipulated in 0 g, 1 g, and 1.8 g. They found that at 1 g the 1.3-µm particles formed bands, but dispersed after a few seconds. However the same particles formed stable bands at 0 g (Fig. 7.27a, b). These bands were lost after leaving 0 g (Fig. 7.27c). Yeast cells tested at 1 g formed more stable bands, but these bands broke up during transition from 0 g to 1.8 g. The larger particles (12 and 20 µm) were stable over the entire range, including the transitions (see for example the 12 µm particles in Fig. 7.27d–f showing the bands for entry into 0 g and 8 seconds after leaving 0 g when accevelation was 1.8 g). Analysis revealed that thermal convective forces caused thermal streaming, resulting in breakdown of the bands for the 1.2-µm particles. Since thermal streaming requires a gravitational field, thermal streaming did not occur during the 0 g tests, thus preserving the bands [312].

7.10
Summary and Future Directions

Microchannel process technology has broad applications, even when limited to non-reactive systems. The research and development activities reviewed here clearly show that each technological area is at a different point in its development, whether fundamental science studies and analytical development, or highly applied industrial or pre-industrial applications. The primary references cited in this chapter provide more details on the specific technologies. Continued work in these areas will no doubt lead to additional breakthroughs and continued interest among industry, civilian and military government agencies, investors, and consumers.

References

1 T. S. Ravigururajan, J. Cuta, C. E. McDonald, M. K. Drost, *Proceedings of the 31st National Heat Transfer Conference*, **1996**, 7, 157–166.
2 A. L. Y. Tonkovich, C. J. Call, D. M. Jimenez, R. S. Wegeng, M. K. Drost, *Proceedings of the 31st National Heat Transfer Conference*, ASME, New York, **1996**, 7, pp. 167–178.
3 A. L. Y. Tonkovich, C. J. Call, D. M. Jimenez, R. S. Wegeng, M. K. Drost, *AIChE Symposium Series*, **1996**, 310, 119–125.
4 C. J. Call, M. K. Drost, R. S. Wegeng, *AIChE Symposium Series*, **1996**, vol. 310.
5 T. S. Ravigururajan, J. Cuta, C. E. McDonald, M. K. Drost, *Proceedings of the 31st National Heat Transfer Conference*, ASME, New York, **1996**, 7, 167–178.
6 M. K. Drost, M. Friedrich, *Proceedings of the Intersociety Energy Conversion Engineering Conference*, AIChE, New York, **1997**, 32nd 1271–1274.
7 W. E. TeGrotenhuis, R. Cameron, M. G. Butcher, P. M. Martin, R. S. Wegeng, *10th Symposium on Separation Science and Technology for Energy Applications*, Gatlinburg, TN, **1997**.
8 D. W. Matson, P. M. Martin, A. L. Y. Tonkovich, G. L. Roberts, *Proc. SPIE- Int. Soc. Opt. Eng.* (**1998**), 3514 (Micromachined Devices and Components IV), pp. 386–392.
9 R. Wegeng, K. Drost, C. McDonald, *U.S. Patent*, US-1997-5611214, **1997**.
10 W. Bier, W. Keller, G. Linder, D. Seidel, *Chem. Eng. Process.*, **1993**, 32, 33–43.
11 G. W. Swift, A. Migliori, J. C. Wheatley, *U.S. Patent*, US-1985-4516632, **1985**.
12 T. Kawakatsu, Y. Kikuchi, M. Nakajima, *J. Am. Oil Chem. Soc.*, **1997**, 74, 317–321.
13 The reader is referred to the 1st through 8th *Proceedings of the International Conference on Microreactor Technology (IMRET)*, 1997–2005 (excluding 2004) for a broad overview of microchannel development activities since **1997**.
14 W. Tegrotenhuis, S. Stenkamp, A. Twitchell, in *Microreactor Technology and Process Intensification*, Y. Wang, J. Holladay (Eds.), American Chemical Society, Washington, D.C., **2005**.
15 T. Bayer, M. Kinzl, in *Advanced Micro and Nano Systems – Micro Process Engineering – Fundamentals, Devices, Fabrication, and Applications*, ed. N. Kockmann, Wiley-VCH, Weinheim, **2006**, pp. 415–438.
16 D. R. Palo, S. S. Stenkamp, R. A. Dagle, G. N. Jovanovic, in *Advanced Micro and Nano Systems – Micro Process Engineering – Fundamentals, Devices, Fabrication, and Applications*, N. Kockmann, Ed., Wiley-VCH, Weinheim, **2006**, pp. 387–414.
17 J. Yoshida, H. Okamoto, in *Advanced Micro and Nano Systems – Micro Process Engineering – Fundamentals, Devices, Fabrication, and Applications*, N. Kockmann, Ed., Wiley-VCH, Weinheim, **2006**, pp. 439–462.
18 R. Babyak, in *Appliance Design*, June 6, **2000**.
19 Modine Manufacturing Company, *ASHRAE Winter Meeting*, Denver, CO, Feb. 5–9, **2005**.
20 M. A. Wilson, K. P. Recknagle, K. P. Brooks, *Proceedings of the AIChE Spring Meeting*, Atlanta, GA, April 10–14, **2005**.
21 Y. Zhao, M. M. Ohadi, R. Radermacher, *Microchannel Heat Exchangers with Carbon Dioxide*, final Report for the Air Conditioning and Refrigeration Technology Institute, ARTI-21CR/10020-01.
22 A. H. Monfarad, *U.S. Patent* US-2004-6741469, **2004**.
23 D. B. Tuckerman, R. F. W. Pease, *IEEE Electron Dev. Lett.* **1981**, 2, 126–129.
24 C. Y. Zhao, T. J. Lu, *Int. J. Heat Mass Transfer*, **2002**, 45, 4857.
25 A. Fedorov, R. Viskanta, *Int. J. Heat Mass Transfer*, **2000**, 43(3), 399.
26 R. Chein, G. Huang, *Appl. Thermal Eng.*, **2004**, 14–15, 2207.
27 C. P. Tso, S. P. Mahulikar. *Int. J. Heat Mass Transfer*, **2000**, 43(6), 1007.
28 S. Belhardj, S. Mimouni, A. Saidane, M. Benzohra, *Microelectronics J.*, **2003**, 34(4), 247.
29 J. Li, G. P. Peterson, P. Cheng, *Int. J. Heat Mass Transfer*, **2004**, 47, 4215–4231.

30 C. Gillot, A. Bricard, C. Schaeffer, *Int. J. Thermal Sci.*, **2000**, 39(8), 826.
31 S. P. Jang, S. J. Kim, K. W. Paik, *Sens. Actuators*, **2003**, 105, 211.
32 G. Hetsroni, A. Mosyak, Z. Segal, G. Ziskind, *Int. J. Heat Mass Transfer*, **2002**, 45(16), 3275.
33 M. J. Kohl, S. I. Abdel-Khalik, S. M. Jeter, D. L. Sadowski, *Int. J. Heat Mass Transfer*, **2005**, 48, 1518–1533.
34 T. M. Harms, M. J. Kazmierczak, F. M. Gerner, *Int. J. Heat Mass Transfer*, **1999**, 20, 149–157.
35 Y. Chen, P. Cheng, *Int. J. Heat Mass Transfer*, **2002**, 45(13), 2643–2648.
36 L. Ghodoossi, *Energy Conversion Management*, **2005**, 46(5), 771–788.
37 R. Chein, C. Yehong, *Int. J. Refrigeration*, **2005**, 28(6), 828.
38 P. S. Lee, S. V. Garimella, D. Liu, *Int. J. Heat Mass Transfer*, **2004**, 48, 1688–1704.
39 C. W. Liu, C. Gau, C. G. Liu, C. S. Yang, *Sens. Actuators A*, **2005**, 122, 177–183.
40 S. H. Chong, K. T. Ooi, T. N. Wong, *Appl. Therm. Eng.*, **2002**, 22(14), 1569–1585.
41 J. Go, T. Kim, K. Cho, *U.S. Patent*, US-2004-6812563, **2004**.
42 J. Pettersen, A. Hafner, G. Skaugen, *Int. J. Refrigeration*, **1998**, 21(3), 180.
43 P. Asinari, L. Cecchinator, E. Fornasieri, *Int. J. Refrigeration*, **2004**, 577–586.
44 J. Pettersen, *Exper. Therm. Fluid Sci.*, **2004**, 28, 111–121.
45 R. Yun, Y. Kim, M. S. Kim, *Int. J. Heat Mass Transfer*, **2005**, 48, 235–242.
46 L. Y. Xiong, G. Kaiser, A. Binneberg, *Cryogenics*, **2004**, 44, 801–807.
47 T. Kulkarni, C. W. Bullard, K. Cho, *Appl. Therm. Eng.*, **2004**, 24, 759–776.
48 M. H. Kim, S. Y. Lee, S. S. Mehendale, R. L. Webb, *Adv. Heat Transfer*, **2003**, 37, 297–429.
49 J. Mathias, R. Arora, W. Simmons, J. McDaniel, A. L. Tonkovich, W. Krause, L. Silva, *U.S. Patent*, US-2003-6622519, **2003**.
50 A. L. Y. Tonkovich, Y. Wang, in *Microreactor Technology and Process Intesification*, ed. Y. Wang, J. D. Holladay, ACS Symposium Series, American Chemical Society, Washington, D.C., **2005**, 914, 47–65.
51 A. L. Y. Tonkovich, Y. Wang, R. Wegeng, Y. Gao, *U.S. Patent*, US-2003-6540975, **2003**.
52 R. Wegeng, K. Drost, C. McDonald, *U.S. Patent*, US-1998-5811062, **1998**.
53 R. Arana, A. Franz, K. Jensen, S. Schaevitz, M. Schmidt, *U.S. Patent*, US-2005-6939632, **2005**.
54 O. Levenspiel, *Chemical Reaction Engineering*, 2nd ed., Wiley, New York, **1972**.
55 V. Hessel, S. Hardt, H. Löwe, *Chemical Micro Process Engineering – Fundamentals, Modeling, and Reactions*. Wiley-VCH, Weinheim, **2004**.
56 L. Zeng, J. Palmer, in *Microreactor Technology and Process Intesification*, ed. Y. Wang, J. D. Holladay, ACS Symposium Series, American Chemical Society, Washington, D.C., **2005**, vol. 914, pp. 322–333.
57 L. M. Fu, R. J. Yang, C. H. Lin, Y. S. Chien, *Electrophoresis*, **2005**, 26(12), 1814–1824.
58 R. H. Liu, R. Lenigk, P. Grodzinski, *J. Microlithogr. Microfabrication Microsyst.*, **2003**, 2(3), 178–184.
59 K. Haas-Santo, O. Görke, K. Schubert, J. Fiedler, H. Funke, in *Microreaction Technology*, ed. M. Matlosz, W. Ehrfeld, J. P. Baselt, Springer, Berlin, **2001**, pp. 313–321.
60 D. S. Kim, S. H. Lee, T. H. Kwon, C. H. Ahn, *Lab Chip*, **2005**, 5(7), 739–746.
61 C. I. Huang, K. C. Wang, C. K. Chyou, *JSME Int. J. Ser. B*, **2005**, 48(1), 17–24.
62 R. Schenk, M. Donnet, V. Hessel, Ch. Hofmann, N. Jongen, H. Löwe, in *Microreaction Technology*, ed. M. Matlosz, W. Ehrfeld, J. P. Baselt, Springer, Berlin, **2001**, pp. 489–498.
63 K. Mae, T. Maki, I. Hasegawa, U. Eto, Y. Mizutani, N. Honda, *Chem. Eng. J.* **2004**, 101(1–3), 31–38.
64 J. Yue, G. W. Chen, Q. Yuan, *Chem. Eng. J.*, **2004**, 102 (1), 11–24.
65 V. Hessel, H. Löwe, F. Schönfeld, *Chem. Eng. Sci.*, **2005**, 60, 2479–2501.
66 N. T. Nguyen, Z. G. Wu, *J. Micromech. Microeng.*, **2005**, 15(2), R1–R16.
67 X. Niu, Y-K. Lee, *J. Micromech. Microeng.*, **2003**, 13, 454–462.
68 S. Qian, H. H. Bau, *Anal. Chem.*, **2002**, 74(15), 3616–3625.

69 S. H. Wong, M. C. L. Ward, C. W. Wharton, *Sens. Actuators B*, **2004**, 100(3), 359–379.
70 P. Svasek, E. Svasek, B. Lendl, M. Vellekoop, *Sens. Actuators A.*, **2004**, 115(2–3), 591–599.
71 D. Gobby, P. Angeli, A. Gavriilidis, *J. Micromech. Microeng.*, **2001**, 11, 126–132.
72 C.-H. Chang, S.-H. Liu, Y. Tennico, J. T. Rundel, V. T. Remcho, E. Blackwell, T. Tseng, B. K. Paul, *AIChE Spring National Meeting, Conference Proceedings*, Atlanta, GA, April 10–14, **2005**.
73 R. H. Liu, R. Lenigk, R. L. Druyor-Sanchez, J. Yang, P. Grodzinski, *Anal. Chem*, **2003**, 75, 1911–1917.
74 Y. Yamaguchi, K. Ogino, K. Yamashita, H. Maeda, *J. Chem. Eng. Jpn.*, **2004**, 37(10), 1265–1270.
75 F. G. Bessoth, A. J. deMello, A. Manz, *Anal. Commun.*, **1999**, 36, 213–215.
76 P. Löb, K. S. Dress, V. Hessel, S. Hardt, C. Hofmann, H. Löwe, R. Schenk, F. Schönfeld, B. Werner, *Chem. Eng. Technol.*, **2004**, 27(3), 340–345.
77 S. Hardt, T. Dietrich, A. Freitag, V. Hessel, H. Löwe, C. Hoffman, A. Oroskar, F. Schönfeld, K. VandenBussche, in *Sixth International Conference on Microreaction Technology, IMRET 6*, ed. I. Rinard, B. Hoch, New Orleans, AIChE Publications No 164. **2002**, pp. 329–344.
78 B. Werner, V. Hessel, P. Löb, *Chem. Eng. Technol.*, **2005**, 28(4), 401–407.
79 H. Y. Wu, C. H. Liu, *Sens. Actuators A*, **2005**, 118(1), 107–115.
80 D. Lastochkin, R. H. Zhou, P. Wang, Y. X. Ben, H. C. Chang, *J. Appl. Phys.*, **2004**, 96(3), 1730–1733.
81 E. Biddiss, D. Erickson, D. Q. Li, *Anal. Chem.*, **2004**, 76(11), 3208–3213.
82 Y. Wang, Q. Lin, T. Mukherjee, *Lab Chip*, **2005**, 5(8) 877–887.
83 P. Paik, V. K. Pamela, R. B. Fair, *Lab Chip*, **2003**, 3, 253–259.
84 P. B. Howell, D. R. Mott, S. Fertig, C. R. Kaplan, J. P. Golden, E. S. Oran, F. S. Ligler, *Lab Chip*, **2005**, 5(5), 524–530.
85 V. Hessel, C. Hofmann, H. Lowe, A. Meudt, S. Scherer, F. Schonfeld, B. Werner, *Org. Proc. Res. Dev.*, **2004**, 8(3), 511–523.
86 C. Y. Lee, G. B. Lee, J. L. Lin, F. C. Huang, C. S. Liao, *J. Micromech. Microeng.* **2005**, 15(6), 1215–1223.
87 G. G. Yaralioglu, I. O. Wygant, T. C. Marentis, B. T. Kuri-Yakub, *Anal. Chem.*, **2004**, 76(13), 3694–3698.
88 J. Melin, G. Gimenez, N. Roxhed, W. van der Wijngaart, G. Stemme, *Lab Chip*, **2004**, 4(3), 214–219.
89 B. He, B. J. Burke, X. Zhang, R. Zhang, F. E. Regnier, *Anal. Chem.*, **2001**, 73(9), 1942–1947.
90 P. B. Howell, D. R. Mott, J. P. Golden, F. S. Ligler, *Lab Chip*, **2004**, 4(6), 663–669.
91 D. S. Kim, I. H. Lee, T. H. Kwon, D. W. Cho, *J. Micromech. Microeng.*, **2004**, 14(10), 1294–1301.
92 F. Jiang, K. S. Drese, S. Hardt, M. Kupper, F. Schonfeld, *AICHE J.*, **2004**, 50(9), 2297–2305.
93 D. S. Kim, S. W. Lee, T. H. Kwon, S. S. Lee, *J. Micromech. Microeng.*, **2004**, 14(6), 798–805.
94 C. H. Lin, C. H. Tsai, L. M. Fu, *J. Micromech. Microeng.*, **2005**, 15(5), 935–943.
95 J. Aubin, D. F. Fletcher, C. Xuereb, *Chem. Eng. Sci.*, **2005**, 60(8–9), 2503–2516.
96 F. Schönfeld, S. Hardt, *AIChE J.*, **2004**, 50, 771.
97 A. D. Stroock, S. K. W. Dertinger, A. Ajdari, I. Mezic, H. A. Stone, G. M. Whitesides, *Science*, **2002**, 295, 647–651.
98 T. G. Kang, T. H. Kwon, *J. Micromech. Microeng.*, **2004**, 14(7), 891–899.
99 M. K. Jeon, J. H. Kim, J. Noh, H. G. Park, S. I. Woo, *J. Micromech. Microeng.*, **2005**, 15(2), 346–350.
100 V. Mengeaud, J. Josserand, H. H. Girault, *Anal. Chem.*, **2002**, 74, 4279–4286.
101 D. J. Kim, H. J. Oh, T. H. Park, J. B. Choo, S. H. Lee, *Analyst*, **2005**, 130(3), 293–298.
102 H. Nagasawa, N. Aoki, K. Mae, *Chem. Eng. Technol.*, **2005**, 28(3), 324–330.
103 Z. B. Stone, H. A. Stone, *Phys. Fluids*, **2005**, 17(6), AR063103.

104 H. Song, J. D. Tice, R. F. Ismagilov, *Angew. Chem. Int. Ed.*, **2003**, 42(7), 767–772.

105 V. Hessel, S. Hardt, H. Löwe, F. Schönfeld, *AIChE J.*, **2003**, 49, 566.

106 K. J. Lissant, *Colloids Surf.*, **1988**, 29, 1–5.

107 T. G. Mason, A. H. Krall, H. Gang, J. Bibette, D. A. Weitz, in *Encyclopedia of Emulsion Technology*, ed. P. Becher, Marcel Dekker, New York, **1996**, Vol. 4, Chapter 6.

108 D. J. McClements, *Food Emulsions: Principles, Practice, and Techniques*; CRC Press, Boca Raton, FL, **1999**.

109 U. Lambrich, H. Schubert, *J. Membr. Sci.*, **2005**, 257, 76–84.

110 G. T. Vladisavljevic, H. Schubert, *Desalination*, **2002**, 144, 167–172.

111 Y. F. Maa, C. C. Hsu, *Pharmaceut. Dev. Tech.*, **1999**, 4(2), 233–240.

112 O. Behrend, K. Ax, H. Schubert, *Ultrason. Sonochem.*, **2000**, 7, 77–85.

113 O. Behrend, H. Schubert, *Ultrason. Sonochem.*, **2001**, 8, 271–276.

114 T. Nakashima, M. Shimizu, *Proceedings of the Autumn Conference of the Society of Chemical Engineers, Japan*, **1988**, p. SB214.

115 T. Nakashima, M. Shimizu, *Ceram. Jpn.*, **1986**, 21, 408–412.

116 T. Kawakatsu, Y. Kikuchi, M. Nakajima, *J. Chem. Eng. Jpn.*, **1996**, 29(2), 399–401.

117 Y. Kikuchi, K. Sato, H. Ohki, T. Kaneko, *Microvasc. Res.*, **1992**, 44, 226–240.

118 Y. Kikuchi, K. Sato, Y. Mizuguchi, *Microvasc. Res.*, **1994**, 47, 126–139.

119 G. T. Vladisavljevic, U. Lambrich, M. Nakajima, H. Schubert, *Colloids Surf. A*, **2004**, 232, 199–207.

120 C. Charcosset, H. Fessi, *Rev. Chem. Eng.*, **2005**, 21(1), 1–32.

121 S. Sugiura, M. Nakajima, S. Iwamoto, M. Seki, *Langmuir*, **2001**, 17, 5562–5566.

122 S. Sugiura, M. Nakajima, N. Kumazawa, S. Iwamoto, M. Seki, *J. Phys. Chem. B*, **2002**, 106, 9405–9409.

123 H. Korstvedt, R. Bates, J. King, A. Siciliano, *Drug Cosmet. Ind.*, **1984**, 36–37, 40.

124 N. Schwesinger, T. Frank, H. Wurmus, *J. Micromech. Microeng.*, **1996**, 6(1), 99–102.

125 W. J. Jeong, J. Y. Kim, J. Choo, E. K. Lee, C. S. Han, D. J. Beebe, G. H. Seong, S. H. Lee, *Langmuir*, **2005**, 21, 3738–3741.

126 S. L. Anna, N. Bontoux, H. A. Stone, *Appl. Phys. Lett.*, **2003**, 82(3), 364–366.

127 D. R. Link, S. L. Anna, D. A. Weitz, H. A. Stone, *Phys. Rev. Lett.*, **2004**, 92(5), 054503(4).

128 J. D. Tice, H. Song, A. D. Lyon, R. F. Ismagilov, *Langmuir*, **2003**, 19, 9127–9133.

129 T. Thorsen, R. W. Roberts, F. H. Arnold, S. R. Quake, *Phys. Rev. Lett.*, **2001**, 86(18), 4163–4166.

130 P. B. Umbanhowar, V. Prasad, D. A. Weitz, *Langmuir*, **2000**, 16, 347–351.

131 L. Silva, A. L. Tonkovich, D. Qiu, K. Pagnotto, P. Neagle, S. Perry, R. Lochhead, *Polym. Preprints*, **2005**, 46(2), 510–511.

132 J. Collins, Y. C. Tan, A. P. Lee, *Am. Soc. Mech. Eng. FED Publ.*, **2003**, 259, 579–582.

133 D. Dendukuri, P. S. Doyle, T. A. Hatton, *Abstr. Papers ACS Natl. Mtg. 229th*, **2005**, COLL-507.

134 A. Kawai, T. Futami, H. Kiriya, K. Katayama, K. Nishizawa, *Proceedings of 6th micro-TAS 2002 Symposium*, **2002**, 1, 368–370.

135 L. Silva, A. L. Tonkovich, D. Qiu, P. Neagle, K. Pagnatto, S. Perry, *Cosmetics Toiletries*, **2005**, 120(8), 41–46.

136 T. Kawakatsu, H. Komori, M. Nakajima, Y. Kikuchi, T. Yonemoto, *J. Chem. Eng. Jpn.*, **1999**, 32(2), 241–244.

137 S. Sugiura, M. Nakajima, M. Seki, *Langmuir*, **2002**, 18, 5708–5712.

138 I. Kobayashi, M. Nakajima, K. Chun, Y. Kikuchi, H. Fujita, *AIChE J.*, **2002**, 48(8), 1639–1644.

139 I. Kobayashi, S. Mukataka, M. Nakajima, *J. Colloid Interface Sci.*, **2004**, 279, 277–280.

140 I. Kobayashi, S. Mukataka, M. Nakajima, *Langmuir*, **2005**, 21, 7629–7632.

141 I. Kobayashi, S. Mukataka, M. Nakajima, *Ind. Eng. Chem. Res.*, **2005**, 44, 5852–5856.

142 T. Kawakatsu, G. Tragardh, C. Tragardh, M. Nakajima, N. Oda, T. Yonemoto, *Colloids Surf. A*, **2001**, 179, 29–37.

143 S. Sugiura, M. Nakajima, H. Ushijima, K. Yamamoto, M. Seki, *J. Chem. Eng. Jpn.*, **2001**, 34(6), 757–765.

144 I. Kobayashi, M. Nakajima, S. Mukataka, *Colloids Surf. A*, **2003**, 229, 33–41.

145 S. Sugiura, M. Nakajima, T. Oda, M. Satake, M. Seki, *J. Colloid Interface Sci.*, **2004**, 269, 178–185.

146 I. Kobayashi, S. Mukataka, M. Nakajima, *Langmuir*, **2005**, 21, 5722–5730.

147 S. Sugiura, M. Nakajima, M. Seki, *Langmuir*, **2002**, 18, 3854–3859.

148 S. Sugiura, M. Nakajima, M. Seki, *Ind. Eng. Chem. Res.*, **2004**, 43, 8233–8238.

149 I. Kobayashi, S. Mukataka, M. Nakajima, *Langmuir*, **2004**, 20, 9868–9877.

150 J. Tong, M. Nakajima, H. Nabetani, Y. Kikuchi, Y. Maruta, *J. Colloid Interface Sci.*, **2001**, 237, 239–248.

151 H. Liu, M. Nakajima, T. Nishi, T. Kimura, *ASAE Annual Meeting*, **2003**.

152 H. Liu, M. Nakajima, T. Nishi, T. Kimura, *Jpn. J. Food Eng.*, **2004**, 5, 259.

153 H. Liu, M. Nakajima, T. Kimura, *J. Am. Oil Chem. Soc.*, **2004**, 81(7), 705–711.

154 I. Kobayashi, M. Nakajima, H. Nabetani, Y. Kikuchi, A. Shohno, K. Satoh, *J. Am. Oil Chem. Soc.*, **2001**, 78(8), 797–802.

155 S. Sugiura, M. Nakajima, J. Tong, H. Nabetani, M. Seki, *J. Colloid Interface Sci.*, **2000**, 227, 95–103.

156 S. Sugiura, M. Nakajima, M. Seki, *J. Am. Oil Chem. Soc.*, **2002**, 79(5), 515–519.

157 Y. Izumida, S. Sugiura, T. Oda, M. Satake, M. Nakajima, *J. Am. Oil Chem. Soc.*, **2005**, 82(1), 73–78.

158 T. Kawakatsu, G. Tragardh, C. Tragardh, *Colloids Surf. A*, **2001**, 189, 257–264.

159 S. Sugiura, M. Nakajima, K. Yamamoto, S. Iwamoto, T. Oda, M. Satake, M. Seki, *J. Colloid Interface Sci.*, **2004**, 270, 221–228.

160 S. Okushima, T. Nisisako, T. Torii, T. Higuchi, *Langmuir*, **2004**, 20, 9905–9908.

161 T. Nisisako, S. Okushima, T. Torii, *Soft Matter*, **2005**, 1(1), 23–27.

162 I. Kobayashi, X. Liu, S. Mukataka, M. Nakajima, *J. Am. Oil Chem. Soc.*, **2005**, 82(1), 65–71.

163 T. Nisisako, S. Okushima, T. Torii, *J. Mater. Chem.*, **2005**, 15(22), 23–27.

164 S. Iwamoto, K. Nakagawa, S. Sugiura, M. Nakajima, *AAPS PharmSciTech*, **2002**, 3(3), article 25.

165 K. Nakagawa, S. Iwamoto, M. Nakajima, A. Shono, K. Satoh, *J. Colloid Interface Sci.*, **2004**, 278, 198–205.

166 F. Ikkai, S. Iwamoto, E. Adachi, M. Nakajima, *Colloid Polym. Sci.*, **2005**, 283, 1149–1153.

167 T. Nisisako, T. Torii, T. Higuchi, *Chem. Eng. J.*, **2004**, 101, 23–29.

168 S. Sugiura, M. Nakajima, M. Seki, *Ind. Eng. Chem. Res.*, **2002**, 41, 4043–4047.

169 S. Sugiura, M. Nakajima, H. Itou, M. Seki, *Macromol. Rapid Commun.*, **2001**, 22(10), 773–778.

170 T. Kawakatsu, R. M. Boom, H. Nabetani, Y. Kikuchi, M. Nakajima, *AIChE J.*, **1999**, 45(5), 967–975.

171 M. Yasuno, S. Sugiura, S. Iwamoto, M. Nakajima, *AIChE J.*, **2004**, 50(12), 3227–3233.

172 Q. Y. Xu, M. Nakajima, B. P. Binks, *Colloids Surf. A.*, **2005**, 262, 94–100.

173 M. Saito, L. J. Yin, I. Kobayashi, M. Nakajima, *Food Hydrocolloids*, **2005**, 19, 745–751.

174 T. Shimada, M. Nakajima, *Jpn. Patent*, JP-2005-118619, **2005**.

175 Y. Ichikawa, E. Nagasawa, F. Shiraishi, *Jpn. Pat.*, JP-2003-215741, **2003**.

176 M. Nakajima, H. Nabetani, H. Ito, K. Mukai, *Jpn. Pat.*, JP-2001-181309, **2001**.

177 M. Nakajima, T. Oda, S. Sugiura, *PCT Int. Appl.*, WO-2004-026457, **2004**.

178 T. Higuchi, T. Torii, T. Nishisako, T. Taniguchi, *PCT Int. Appl.*, WO-2002-068104, **2002**.

179 D. Qiu, A. L. Tonkovich, L. Silva, R. Q. Long, B. L. Yang, K. M. Trenkamp, *U.S. Pat. Appl.*, US-2004-0234566, **2004**.

180 D. Qiu, A. L. Tonkovich, L. Silva, R. Q. Long, B. L. Yang, K. M. Trenkamp, *U.S. Pat. Appl.*, US-2004-0228882, **2004**.

181 M. Hohmann, M. Schmelz, G. Brenner, H. Wurziger, N. Schwesinger, *PCT Int. Appl.*, WO-2001-089693, **2001**.

182 M. Nakajima, Y. Kikuchi, T. Kawakatsu, H. Komori, T. Komemoto, *Jpn. Pat.*, JP-2000-015070, **2000**.

183 K. Schubert, W. Bier, G. Linder, E. Herrmann, *World Pat.*, WO-9833582, **1998**.

184 W. Ehrfeld, V. Hessel, *World Pat.*, WO-2001-043857, **2001**.
185 A. Hibara, M. Nanoaka, H. Hisamoto, K. Uchiyama, Y. Kikutani, M. Tokeshi, T. Kitamori, in *Micro Total Analysis Systems*, ed. J. M. Ramsey A. Vanden Berg, Kluwer Academic Publishers, Netherlands, **2001**, pp. 549–550.
186 M. Tokashi, T. Minagawa, K. Uchiyama, A. Hibara, K. Sato, H. Hisamoto, T. Kitamori, *Anal. Chem.*, **2002**, 74, 1565–1571.
187 A. Hibara, M. Nonaka, H. Hisamoto, K. Uchiyama, Y. Kikutani, M. Tokeshi, T. Kitamori, *Anal. Chem.*, **2002**, 74, 1724–1728.
188 T. Maruyama, J. Uchida, T. Ohkawa, T. Futami, K. Katayama, K. Nishizawa, K. Sotowa, F. Kubota, N. Kamiya, M. Goto, *Lab Chip*, **2003**, 3, 308–312.
189 U. Lehmann, O. Krusemark, J. Miller, in *Micro Total Analysis Systems 2000*, ed. A. Vanden Berg, Kluwer Academic Publishers, Netherlands, **2000**, pp. 167–170.
190 K. D. Wise, *Wireless Integrated Microsystems: Coming Revolution in the Gathering of Information*, University of Michigan Center for Wireless Integrated MicroSystems, Stanford University, January 13, **2004**.
191 H. Henderson, N. deGouvea-Pinto, *U.S. Pat.* 6,258,263, **2001**.
192 Q. Kang, N. Golubovic, N. Pinto, H. Henderson, *Chem. Eng. Sci.*, **2001**, 56, 3409–3420.
193 P. Kuban, P. Dasgupta, K. Morris, *Anal. Chem.*, **2002**, 74, 5667–5675.
194 M. Karwa, D. Hahn, S. Mitra, *Anal. Chem. Acta*, **2005**, 546, 22–29.
195 S. Rassat, Pacific Northwest National Laboratory, personal communication.
196 K. Brooks, S. Rassat, R. Wegeng, V. Stenkamp, W. Tegrotenhuis, D. Caldwell, *AIChE 2002 Spring National Meeting*, **2002**.
197 B. Monzyk, A. Tonkavich, Y. Wang, D. VanderWiel, S. Perry, S. Fitzgerald, W. Simmons, J. McDaniel, A. Weller, C. Cucksey, *U.S. Pat.* 6,824,592, **2004**.
198 A. Tonkavich, B. Monzyk, Y. Wang, D. VanderWiel, S. Perry, S. Fitzgerald, W. Simmons, J. McDaniel, A. Weller, *U.S. Pat.* 6,508,862, **2003**.

199 W. Tegrotenhuis, R. Cameron, M. Butcher, P. Martin, R. Wegeng, *Sep. Sci. Technol.*, **1999**, 34(6–7), 951–974.
200 F. Kubota, J. Uchida, M. Goto, *Solv. Extr. Res. Dev.-Jpn.*, **2003**, 10, 93–102.
201 M. Tokeshi, T. Minagawa, T. Kitamori, *Anal. Chem.*, **2000**, 72, 1711–1714.
202 H. Kim, K. Ueno, M. Chiba, O. Kogi, N. Kitamura, *Anal. Sci.*, **2000**, 16, 871–876.
203 T. Maruyama, T. Kaji, T. Ohkawa, K. Sotowa, H. Matsushita, F. Kubota, N. Kamiya, K. Kusakabe, M. Goto, *Analyst*, **2004**, 129, 1008–1013.
204 K. Ueno, H. Kim, N. Kitamura, *Anal. Sci.*, **2003**, 19(3), 391–394.
205 B. Zhao, J. Moore, D. Beebe, *Science*, **2001**, 291(5506), 1023–1026.
206 X. Wang, C. Saridara, S. Mitra, *Anal. Chem. Acta*, **2005**, 543, 92–98.
207 T. Maruyama, H. Matsushita, J. Uchida, F. Kubota, N. Kamiya, M. Goto, *Anal. Chem.* **2004**, 76, 4495–4500.
208 M. Surmeian, M. Slyadnev, H. Hisamoto, A. Hibara, K. Uchiyama, T. Kitamori, *Anal. Chem.*, **2002**, 74, 2014–2020.
209 A. Hibara, M. Nonaka, H. Hisamoto, K. Uchiyama, Y. Kikutani, M. Tokeshi, T. Kitamori, *Anal. Chem.*, **2002**, 74, 1724–1728.
210 V. Reddy, J. Zahn, *J. Colloid Interface Sci.*, **2005**, 286, 158–165.
211 J. Xu, G. Luo, G. Chen, B. Tan, *J. Membr. Sci.*, **2005**, 249, 75–81.
212 Y. Okubo, M. Toma, H. Ueda, T. Maki, K. Mae, *Chem. Eng. J.*, **2004**, 101, 39–48.
213 P. Martin, D. Matson, W. Bennett, *Chem. Eng. Commun.*, **1999**, 173, 245–254.
214 J. Shaw, R. Nudd, B. Naik, C. Turner, D. Rudge, M. Benson, A. Garman, in *Micro Total Analysis Systems 2000*, **2000**, ed. A. Vanden Berg, Kluwer Academic Publishers, Netherlands, pp. 371–374.
215 M. Tokeshi, T. Minagawa, K. Uchiyama, A. Kibara, K. Sato, H. Hisamoto, T. Kitamori, *Anal. Chem.*, **2002**, 74, 1565–1571.
216 S. Beale, *Anal. Chem.*, **1998**, 70, 279R–300R.
217 R. Kennedy, I. German, J. Thompson, S. Witowski, *Chem. Rev.*, **1999**, 99, 3081–3131.

218 R. St. Claire, *Anal. Chem.*, **1996**, 68, 569R–586R.
219 S. Krylov, N. Dovichi, *Anal. Chem.*, **2000**, 72, 111R–128R.
220 S. Liu, A. Guttman, *Trends Anal. Chem.*, **2004**, 23(6), 422–431.
221 A. Gawron, R. Martin, S. Lunte, *Eur. J. Pharmaceut. Sci.*, **2001**, 14(1), 1–12.
222 J. Lian, J. Ferrance, J. Landers, *BioTechniques*, **2001**, 31(6), 1332–1353.
223 R. Kelly, A. Woolley, *Anal. Chem.*, **2005**, March 1, 99A–102A.
224 J. Prober, G. Trainor, R. Dam, F. Hobbs, C. Robertson, R. Zagursky, A. Cocuzza, M. Jensen, K. Baumeister, *Science*, **1987**, 238(4825), 336–341.
225 I. Kheterpal, R. Mathies, *Anal. Chem.*, **1999**, 71, 31A–37A.
226 E. Zubrisky, *Anal. Chem.*, **2002**, 74, 23A–26A.
227 L. Chen, J. Ma, F. Tan, Y. Guan, L. Chen, *Sens. Actuators B*, **2003**, 88(3), 260–265.
228 T. Johnson, L. Locascio, *Lab Chip*, **2002**, 2(3), 135–140.
229 Y. Takamura, H. Onoda, H. Inokuchi, S. Adachi, A. Oki, Y. Horiike, *Electrophoresis*, **2003**, 24(1), 185–192.
230 L. Szekely, R. Freitag, *Electrophoresis*, **2005**, 26(10), 1928–1939.
231 R. Schasfoort, S. Schlautmann, J. Hendrikse, A. Vanden Berg, *Science*, **1999**, 286(5411), 942–945.
232 M. Peterman, J. Noolandi, M. Blumenkranz, H. Fishman, *Anal. Chem.* **2004**, 76(7), 1850–1856.
233 D. Schmalzing, A. Adourian, L. Koutny, L. Ziaugra, P. Matsudaira, D. Ehrlich, *Anal. Chem.*, **1998**, 70, 2303–2310.
234 S. Liu, Y. Shi, W. W. Ja, R. A. Mathies, *Anal. Chem.*, **1999**, 71, 566–573.
235 O. Salas-Solano, D. Schmalzing, L. Koutny, S. Buonocore, A. Adourian, P. Matsudaira, D. Ehrlich, *Anal. Chem.* **2000**, 72, 3129–3137.
236 E. T. Lagally, J. R. Scherer, R. G. Blazej, N. M. Toriello, B. A. Diep, M. Ramchandani, G. F. Sensabaugh, L. W. Riley, R. A. Mathies, *Anal. Chem.* **2004**, 76, 3162–3170.
237 T. Chiesl, W. Shi, A. Barron, *Anal. Chem.*, **2005**, 77, 772–779.
238 M. Kim, S. Cho, K. Lee, Y. Kim, *Sens. Actuators B*, **2004**, 107, 818–824.
239 L. Roach, H. Song, R. Ismagilov, *Anal. Chem.*, **2005**, 77, 785–796.
240 H. Makamba, J. Kim, K. Lim, N. Park, J. Hahn, *Electrophoresis*, **2003**, 24, 3607–3619.
241 R. Foote, J. Khandurina, S. Jacobson, J. Ramsey, *Anal. Chem.* **2005**, 77, 57–63.
242 Y. Wang, A. Stevens, J. Han, *Anal. Chem.*, **2005**, 77, 4293–4299.
243 H. Tolley, Q. Wang, D. LeFebre, M. Lee, *Anal. Chem.*, **2002**, 74, 4456–4463.
244 D. Ross. L. Locascio, *Anal. Chem.*, **2002**, 74, 2556–2564.
245 Q. Wang, H. Tolley, D. LeFebre, M. Lee, *Anal. Bioanal. Chem.*, **2002**, 373, 125–135.
246 P. Myers, K. Bartle, *J. Chromatogr. A*, **2004**, 1044, 253–258.
247 Q. Wang, S. Lin, K. Warnick, H. Tolley, M. Lee, *J. Chromatogr. A*, **2003**, 985, 455–462.
248 P. Humble, R. Kelly, A. Woolley, H. Tolley, M. Lee, *Anal. Chem.*, **2004**, 76, 5641–5648.
249 D. Petsev, G. Lopez, C. Ivory, S. Sibbett, *Lab Chip*, **2005**, 5, 587–597.
250 M. Sato, M. Goto, *Sep. Sci. Technol.*, **2004**, 39, 3163–3167.
251 S. Garimella, *International Sorption Heat Pump Conference*, **1999**, Munich.
252 J. M. Meacham, S. Garimella, *International Sorption Heat Pump Conference*, **2002**, Shanghai, China, pp. 270–276.
253 J. M. Meacham, S. Garimella, *Winter Meeting of the ASHRAE*, Jan 26–29, **2003**, Chicago, 412–422.
254 J. M. Meacham, S. Garimella, *ASHRAE Trans.*, **2004**, 110, 525–532.
255 W. E. TeGrotenhuis, V. S. Stenkamp, B. Q. Roberts, J. M. Davis, C. M. Fischer, K. Drost, D. Pence, R. Cullion, G. Mouchka, *International Sorption Heat Pump Conference*, ISHPC-042-2005, June 22–24, **2005**, Denver, CO USA.
256 V. S. Stenkamp, W. E. TeGrotenhuis, B. Q. Roberts, M. Flake, *Proceedings of the Eighth International Conference on Microreaction Technology*, April 10–14, **2005**, Atlanta, GA, AIChE, New York.
257 M. Zanfir, A. Gavriilidis, Ch. Wille, V. Hessel, *Ind. Eng. Chem. Res.*, **2005**, 44, 1742–1751.
258 Y. Ono, H. Yamashita, K. Uchiyama, T. Korenaga, in *Micro Total Analysis*

systems, ed. Y. Baba et al., Kluwer Academic Publishers, Netherlands, **2002**, 1, pp. 530–532.
259 M. D. Determan, *Experimental and Analytical Investigation of Ammonia-Water Desorption in Microchannel Geometries*, Master's Thesis, Georgia Institute of Technology, August, **2005**.
260 M. K. Drost, V. Naragyanan, D. V. Pence, *U.S. Pat. Appl.* US2005/0126211
261 D. V. Pence, J. N. Reyees, N. Phillips, Q. Wu, J. T. Groome, *US Pat.*, 6,688,381.
262 G. Cox, R. W. Baker, Pervaporation, in *Industrial Wastewater*, **1998**, 6, 35–38.
263 J. C. T. Eijkel, J. G. Bomer, A. VandenBerg, *Appl. Phys. Lett.*, **2005**, 87, 114103.
264 B. H. Timmer, K. M. van Delft, W. Olthuis, P. Bergveld, A. VandenBerg, *Sens. Actuators B*, **2003**, 91, 342–346.
265 C. S. Effenhauser, H. Harttig, P. Krämer, *Biomed. Microdev.*, **2002**, 4(1), 27–32.
266 R. B. Eldridge, A. F. Seibert, S. Robinson, *Hybrid Separations/Distillation Technology Research Opportunities for Energy and Emissions Reduction*, Industrial Technologies Program, United States Department of Energy, Energy Efficiency and Renewable Energy.
267 B. C. Almaula, *US Pat.* 4,597,947, July 1, **1986**.
268 A. Hibara, K. Toshin, T. Taukahara, K. Mawatari, T. Kitamori, *Ninth International Conference on Miniaturized Systems for Chemistry and Life Sciences*, Boston, MA, October 9–13, **2005**.
269 W. E. TeGrotenhuis, R. A. Dagle, V. S. Stenkamp, *AIChE Spring National Meeting*, Atlanta, GA, April 10–14, **2005**.
270 J. M. MacInnes, R. W. K. Allen, *7th World Congress of Chemical Engineering*, Glasgow, July 10–14, **2005**.
271 J. L. Humphrey, G. E. Keller, II, *Separation Process Technology*, McGraw-Hill, San Francisco, **1997**.
272 A. L. Tonkovich, L. Silva, R. Arora, T. Hickey, *AIChE Spring National Meeting, Distillation Topical Conference*, Atlanta, GA, April 10–14, **2005**.
273 *Energy Efficient Building Equipment and Envelop Technologies, Round III*, Morgantown, WV, US Department of Energy, **2000**.
274 S. Garimella, *Energy*, **2003**, 28, 1593–1614.
275 S. T. Munkejord, H. S. Maehlum, G. R. Zakeri, P. Neksa, J. Pettersen, *20th International Congress of Refrigeration*, IIR/IIF/Sydney, **1999**.
276 S. Garimella, *ASHRAE Trans.*, **2002**, 108(1) 492–499.
277 A. Hafner, *22nd International Congress of Refrigeration*, IIF-IIR, Guangzhou, China, **2002**.
278 A. Beutler, F. Ziegler, G. Alefeld, *International Absorption Heat Pump Conference*, Montreal, Canada, **1996**, 303–309.
279 S. Garimella, *ASHRAE Trans.*, **2000**, 106(1), 453–462.
280 J. Thorud, Army/Ft. Belvoir program objectives and technology needs, *Oregon Nanoscience and Microtechnologies Institute Leadership Retreat*, September 15–16, **2005**.
281 T. Hung, H. Borovetz, M. Weissman, *Ann. Biomed. Eng.*, **1977**, 5, 343–361.
282 P. Roy, H. Barkaran, A. Tilles, M. Yarmush, M. Toner, *Ann. Biomed. Eng.*, **2001**, 29, 947–955.
283 B. Gray, D. Lieu, S. Collins, R. Smith, A. Barakat, *Biomed. Microdev.*, **2002**, 4(1), 9–16.
284 J. Farinas, A. Chow, H. Wada, *Anal. Biochem.*, **2001**, 295, 138–142.
285 G. Jovanovic, Chemical Engineering Department, Oregon State University, personal communication.
286 G. Kleiner, *Med. News Today*, February 3, **2004**.
287 G. A. J. Besselink, P. Vulto, R. G. H. Lammertink, S. Schlautmann, A. Vanden Berg, W. Olthuis, G. H. M. Engbers, R. B. M. Schasfoort, *Electrophoresis*, **2004**, 25, 3705–3711.
288 C. S. Lee, in *Handbook of Capillary Electrophoresis* 2nd ed., CRC Press, New York, **1997**, pp. 717–739.
289 V. Pretorious, B. J. Hopkins, J. D. Schieke, *J. Chromatogr.*, **1974**, 99, 23–30.
290 I. M. Lazar, B. L. Karger, *Anal. Chem.*, **2002**, 74, 6259–6268.
291 T. Wachs, J. Henion, *Anal. Chem.*, **2001**, 73, 632–638.
292 J. S. Kim, D. R. Knapp, *J. Am. Soc. Mass Spectrom.*, **2001**, 12, 463–469.

293 N. W. Smith, A. S. Carter-Finch, *J. Chromatogr. A*, **2000**, 892, 219–255.
294 J. H. Knox, I. H. Grant, *Chromatographia*, **1987**, 24, 135–143.
295 J. H. Knox, I. H. Grant, *Chromatographia*, **1991**, 32, 317–328.
296 M. Le Berre, Y. Chen, C. Crozatier, Z. L. Zhang, *Microelect. Eng.*, **2005**, 78–79, 93–99.
297 J. H. P. Watson, *J. Appl. Phys.*, **1973**, 44, 4209–4213.
298 K. Smistrup, O. Hansen, H. Buus, M. F. Hansen, *J. Magn. Magn. Mater.*, **2005**, 293, 597–604.
299 T. Deng, M. Prentiss, G. M. Whitesides, *Appl. Phys. Lett.*, **2002**, 80, 461–463.
300 K. Yasuda, *Sens. Actuators B*, **2000**, 64, 128–135.
301 M. Gröschl, *Acta Acustica*, **1998**, 84, 432–447.
302 M. Gröschl, *Acta Acustica*, **1998**, 84, 632–642.
303 M. Gröschl, *Acta Acustica* **1998**, 84, 815–822.
304 J. J. Hawkes, D. Barrow, W. T. Coakley, *Ultrasonics*, **1998**, 36, 925–931.
305 O. Doblhoff-Dier, Th. Gaida, H. Katinger, W. Burger, M. Groschl, E. Benes, *Biotechnol. Prog.*, **1994**, 10, 428–432.
306 M. S. Limaye, J. J. Hawkes, W. T. Coakley, *J. Microbiol. Meth.*, **1996**, 27, 211–220.
307 M. Hill, R. J. K. Wood, *Ultrasonics*, **2000**, 38, 662–665.
308 M. Hill, Y. Shen, J. J. Hawkes, *Ultrasonics*, **2002**, 40, 385–392.
309 N. Harris, M. Hill, Y. Shen, T. J. Townsend, S. Beeby, N. White, *Ultrasonics*, **2004**, 42, 139–144.
310 K. Higashitani, M. Fukushima, Y. Matsuno, *Chem. Eng. Sci.*, **1981**, 36, 1877–1882.
311 K. Yasuda, T. Kamakura, *Appl. Phys. Lett.*, **1997**, 13, 1771–1773.
312 J. J. Hawkes, J. J. Cefai, D. A. Barrow, W. T. Coakley, L. G. Briarty, *J. Phys. D*, **1998**, 31, 1673–1680.

8
Microcomponent Flow Characterization

Bruce A. Finlayson, Pawel W. Drapala, Matt Gebhardt, Michael D. Harrison, Bryan Johnson, Marlina Lukman, Suwimol Kunaridtipol, Trevor Plaisted, Zachary Tyree, Jeremy VanBuren, and Albert Witarsa

8.1
Introduction

Microfluidic devices involve fluid flow and transport of energy and species. When designing such devices it is useful for engineers to have correlations that allow predictions before the device is built, so that the design can be optimized. Unfortunately, most correlations in the literature are for high-speed, or turbulent, flow. Thus, it is useful to develop correlations for slower, laminar flow. While most three-dimensional flow situations can be modeled in laminar flow using computational fluid dynamics (CFD), the calculations are extensive and time consuming. Correlations are especially useful when the CFD capability is not present. Owing to the multi-dimensional character of microfluidics, it is important to understand both the power and the limitations of concepts based on simple one- or two-dimensional cases. In this chapter we outline correlations and pertinent theoretical concepts for slow speed flow and diffusion: pressure drop, entry length, and mixing due to convection and diffusion.

Most of the work is conducted using CFD with the program Comsol Multiphysics (formerly FEMLAB). This program is easy enough to use, such that undergraduates can perform useful work with little guidance. Undergraduates in the chemical engineering program at the University of Washington first solved the problems described here over the past five years. The cases have been recomputed using more elements to improve accuracy using state-of-the-art computers and programs.

8.2
Pressure Drop

8.2.1
Friction Factor for Slow Flows

First, consider the flow of an incompressible fluid in a straight pipe. The pressure drop is a function of the density and viscosity of the fluid, the average velocity of the fluid, and the diameter and length of the pipe. The relationship is usually expressed by means of a friction factor defined as in Eq. (1):

$$f \equiv \frac{1}{4}\frac{d}{L}\frac{\Delta p}{\frac{1}{2}\rho \langle v \rangle^2} \tag{1}$$

The Reynolds number is:

$$Re \equiv \frac{\rho \langle v \rangle d}{\eta} \tag{2}$$

The usual curve of friction factor is given in Fig. 8.1(a) in a log–log plot.

For any given Reynolds number, one can read the value of friction factor, and then use Eq. (1) to determine the pressure drop. In the turbulent region, for $Re > 2200$, one correlation is the Blasius formula [Eq. (3)]:

$$f = 0.0791/Re^{0.25} \tag{3}$$

In the laminar region, for $Re < 2200$, the correlation is given by Eq. (4):

$$f = 16/Re \tag{4}$$

When the definitions are inserted into Eq. (4) for laminar flow, one obtains for the pressure drop:

$$\Delta p = 32 \frac{L}{d}\frac{\eta \langle v \rangle}{d} \tag{5}$$

Thus, in laminar flow in a straight channel, the density of the fluid does not affect the pressure drop. To use the correlation for friction factor, or Fig. 8.1(a), one must know the density. This problem is avoided if one plots the product of friction factor and Reynolds number versus Reynolds number, since from their definition:

$$f\,Re = \frac{\Delta p\, d^2}{2\,L\,\eta \langle v \rangle} \tag{6}$$

This is shown in Fig. 8.1(b). Notice that for $Re < 1$ the only thing needed is the numerical constant, here 16. For fully developed pipe flow, the quantity $f\,Re$

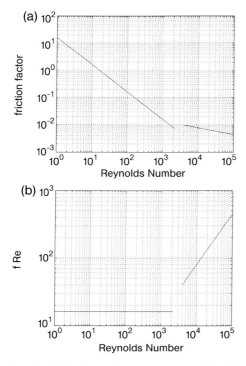

Fig. 8.1 Friction losses (pressure decreases) for fully-developed pipe flow.
(a) Standard presentation of friction factor versus Reynolds number.
(b) Low Reynolds number presentation of *f Re*.

does not change as the Reynolds number increases to 2200, but it does in more complicated cases. This covers the range that most microfluidic devices operate in. To predict the pressure drop, all that is needed is the single number for the geometry and flow situation in question. The way these numbers are determined is best illustrated by looking at the mechanical energy balance for a flowing system.

8.2.2
Mechanical Energy Balance for Turbulent Flow

Consider the control volume shown in Fig. 8.2, with inlet at point 1 and outlet at point 2. In the following, Δ-variable means the variable at point 2 (outlet) minus the variable at point 1 (inlet).

The mechanical energy balance is given by Eq. (7) [1]:

$$\Delta \left(\frac{1}{2} \frac{\langle v^3 \rangle}{\langle v \rangle} \right) + g \, \Delta h + \int_1^2 \frac{dp}{\rho} = \hat{W}_m - \hat{E}_v \tag{7}$$

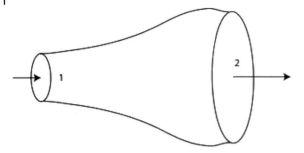

Fig. 8.2 Geometry and notation for mechanical energy balance.

For no height change $g \, \Delta h = 0$; for no pump in the control volume $\hat{W}_m = 0$; for an incompressible fluid with constant density:

$$\int_1^2 \frac{dp}{\rho} = \frac{1}{\rho}(p_2 - p_1) = \frac{1}{\rho}\Delta p$$

Thus, under all these assumptions we have Eq. (8):

$$\Delta\left(\frac{1}{2}\frac{\langle v^3 \rangle}{\langle v \rangle}\right) + \frac{1}{\rho}\Delta p = -\hat{E}_v \tag{8}$$

By continuity the mass flow rate is the same, $w_1 = w_2$. Viscous losses are defined by:

$$E_v = -\int_V \tau : \nabla \mathbf{v} \, dV = \int_V \eta \, [\nabla \mathbf{v} + (\nabla \mathbf{v})^T] : \nabla \mathbf{v} \, dV \geq 0 \quad \text{and} \quad \hat{E}_v = \frac{E_v}{w}$$

Rewriting Eq. (8) in the form of Eq. (9):

$$\Delta\left(\frac{1}{2}\frac{\langle v^3 \rangle}{\langle v \rangle}\right) + \hat{E}_v = -\frac{1}{\rho}\Delta p = \frac{p_1 - p_2}{\rho} \tag{9}$$

8.3
Dimensionless Mechanical Energy Balance

We make all variables non-dimensional by dividing them by a dimensional standard (which has to be identified in each case) and denoting the standard with a subscript s. For the velocity standard, we use an average velocity somewhere in the system:

$$v' = \frac{v}{v_s}, \quad v_s \equiv \langle v \rangle$$

8.3 Dimensionless Mechanical Energy Balance

The pressure is made non-dimensional by dividing it by twice the kinetic energy:

$$p'' \equiv \frac{p}{\rho v_s^2}$$

The units of viscous dissipation are

$$E_v [=] \eta \frac{v_s^2 x_s^3}{x_s^2} = \eta v_s^2 x_s \quad \text{and} \quad \hat{E}_v = \frac{E_v}{w} [=] \frac{\eta v_s^2 x_s}{\rho v_s x_s^2} = \frac{\eta v_s}{\rho x_s}$$

Thus, we define:

$$\hat{E}'_v = \frac{\hat{E}_v \, \rho \, x_s}{\eta \, v_s}$$

The non-dimensional form of the mechanical energy balance is then:

$$v_s^2 \, \Delta \left(\frac{1}{2} \frac{\langle v'^3 \rangle}{\langle v' \rangle} \right) + \frac{\eta \, v_s}{\rho \, x_s} \hat{E}'_v = v_s^2 \, (p''_1 - p''_2)$$

Dividing by v_s^2 gives:

$$\Delta \left(\frac{1}{2} \frac{\langle v'^3 \rangle}{\langle v' \rangle} \right) + \frac{\eta}{\rho \, x_s \, v_s} \hat{E}'_v = (p''_1 - p''_2) \quad \text{or} \tag{10}$$

$$\Delta \left(\frac{1}{2} \frac{\langle v'^3 \rangle}{\langle v' \rangle} \right) + \frac{1}{Re} \hat{E}'_v = (p''_1 - p''_2)$$

where $Re = \rho \, v_s \, x_s / \eta$. This is the approach one uses for high-speed flow, including all turbulent cases.

For fully developed flow through a tube with a constant diameter, d, the kinetic energy does not change from one end to the other. We already have a formula for the pressure drop in terms of the friction factor, so that:

$$p''_1 - p''_2 = \frac{\Delta p}{\rho \langle v \rangle^2} = 2 f \frac{L}{d}$$

Thus the viscous dissipation term in Eq. (10) is also given by the friction factor for flow in a tube:

$$\frac{1}{Re} \hat{E}'_v = 2 f \frac{L}{d}$$

8.3.1
Mechanical Energy Balance for Laminar Flow

When the flow is slow and laminar, we use a different non-dimensionalization for the pressure. Divide the pressure by the viscosity times the velocity divided by a characteristic length:

$$p' \equiv \frac{p}{p_s}; \quad \text{with} \quad p_s \equiv \frac{\eta \, v_s}{x_s}, \quad p' = \frac{p \, x_s}{\eta \, v_s}$$

Then the mechanical energy balance is:

$$v_s^2 \, \Delta \left(\frac{1}{2} \frac{\langle v'^3 \rangle}{\langle v' \rangle} \right) + \frac{\eta \, v_s}{\rho \, x_s} \hat{E}'_v = \frac{\eta \, v_s}{\rho \, x_s} (p'_1 - p'_2)$$

We then multiply by $\rho \, x_s / \eta \, v_s$ to get another non-dimensional form of the mechanical energy balance:

$$\frac{\rho \, v_s \, x_s}{\eta} \Delta \left(\frac{1}{2} \frac{\langle v'^3 \rangle}{\langle v' \rangle} \right) + \hat{E}'_v = (p'_1 - p'_2) \quad \text{or} \tag{11}$$

$$Re \, \Delta \left(\frac{1}{2} \frac{\langle v'^3 \rangle}{\langle v' \rangle} \right) + \hat{E}'_v = (p'_1 - p'_2)$$

As noted above, for a fully developed flow through a tube with a constant diameter, d, the kinetic energy does not change from one end to the other. As we, again, already have a formula for the pressure drop in terms of the friction factor we find:

$$p'_1 - p'_2 = \frac{\Delta p \, d}{\eta \, \langle v \rangle} = 32 \frac{L}{d}$$

Thus, the viscous dissipation term in Eq. (11) for laminar flow in a straight channel is:

$$\hat{E}'_v = 32 \frac{L}{d}$$

Both Eqs. (10) and (11) express the same physics. Equation (10) is more useful at high velocity in turbulent flow when f is a slowly varying function of Reynolds number, and Eq. (11) is more useful at low velocity in laminar flow. The non-dimensional pressures in these two equations are different, but they are related:

$$p' = Re \times p''$$

8.3.2
Pressure Drop for Flow Disturbances

Equation (11) is the form of the mechanical energy balance that is useful in microfluidics. Note that a Reynolds number multiplies the term representing the change of kinetic energy, which is small in microfluidics. Consider the case when the Reynolds number is extremely small. Then, any flow situation is governed by Stokes equation:

$$\eta \nabla^2 \mathbf{v} = -\nabla p$$

In non-dimensional form this equation is:

$$\eta \frac{v_s}{x_s^2} \nabla'^2 \mathbf{v}' = -\frac{p_s}{x_s} \nabla' p' \quad \text{or} \quad \nabla'^2 \mathbf{v}' = -\frac{p_s \, x_s}{\eta \, v_s} \nabla' p'$$

But with the definition of $p_s = \eta \, v_s/x_s$:

$$\nabla'^2 \mathbf{v}' = -\nabla' p'$$

Thus, the non-dimensional flow does not depend upon the Reynolds number. This means that both:

$$\Delta \left(\frac{1}{2} \frac{\langle v'^3 \rangle}{\langle v' \rangle} \right) \quad \text{and} \quad \hat{E}'_v$$

are constants for the particular geometry being studied, since they can be calculated knowing the velocity field distribution. For turbulent flow, a reasonable approximation is that the velocity is constant with respect to radial position in the inlet and outlet ducts, and then $\langle v'^3 \rangle / \langle v' \rangle = \langle v'^2 \rangle$. This is not true for laminar flow, although both terms can be calculated for fully developed flow at the inlet and outlet. Furthermore, the kinetic energy term is negligible at small Reynolds number. To correlate the pressure drop, the main task is to correlate the viscous dissipation term, $\hat{E}'_v$.

A convenient way to correlate data for excess pressure drop when there are contractions, expansions, bends and turns, etc., when the flow is turbulent is to write them as:

$$\Delta p_{\text{excess}} = K \frac{1}{2} \rho \langle v \rangle^2$$

Then, to calculate the pressure drop for any given fluid and flow rate, one only needs to know the values of K, a few of which are tabulated in Table 8.1.

As shown above, for slow flow the pressure drop is linear in the velocity rather than quadratic, and a different correlation is preferred:

$$\Delta p_{\text{excess}} = K_{\text{L}} \frac{\eta \langle v \rangle}{d}$$

//188 | 8 Microcomponent Flow Characterization

Table 8.1 Frictional loss coefficient for turbulent flow.[a]

	$K (= \Delta p_{excess}/\tfrac{1}{2} \rho \langle v \rangle^2)$
90° "ell", standard	0.75
90° "ell", long radius	0.35
180° bend, close return	1.5
Tee, branching flow	1.0

a) From Table 6-4, pp. 6–18, *Perry's Chemical Engineers' Handbook* [2].

Table 8.2 gives values of K_L for different geometries, as determined by finite element calculations.

Table 8.2 Coefficient K_L for negligibly small *Re*. The value in parentheses that found if the entire length through the centerline is used to calculate the fully developed pressure drop. The excess pressure drop beyond this is negligible.

Description	Picture	K_L
Sharp bend, 2D		9.1
Smooth bend, 90 degrees, short radius, 2D (centerline radius = gap size)		19 (0.5)
Smooth bend, 90 degrees, long radius, 2D (centerline radius = 1.5 × gap size)		29 (0.36)
Bend, 445 degrees, sharp change, 2D		5.2 (0.2)
Bend, 45 degrees, long radius, 2D (centerline radius = 1.5 × gap size)		14.3 (0.2)
Square corner, 3D		4.2 (−2.1)
Round pipe, 3D		75.2 (−0.2)

v_s = average velocity, x_s = thickness or diameter.

8.3.3
Pressure Drop for Contractions and Expansions

Consider a large circular tube emptying into a smaller circular tube (Fig. 8.3). The pressure drop in this device is due to viscous dissipation in the fully developed regions of the large and small channels, plus the extra viscous dissipation due to the contraction, plus the kinetic energy change. When the Reynolds number is small, the kinetic energy change is negligible.

For the flow illustrated in Fig. 8.3, we write the excess pressure drop for the contraction as the total pressure drop minus the pressure drop for fully developed flow in the large and small channels. The pressure drop for fully developed laminar flow is known analytically, of course [Eq. (5)]:

$$\Delta p_{\text{excess}} = \Delta p_{\text{total}} - \Delta p_{\text{large channel}} - \Delta p_{\text{small channel}}$$

The excess pressure drop is correlated using $\Delta p_{\text{excess}} = K_L \eta \langle v \rangle / d$, and K_L can be determined from finite element calculations. Note that one must specify which average velocity and distance are used in this correlation, and the usual choices are the average velocity and the diameter or total thickness between parallel plates, all at the narrow end. The value of K_L depends upon the contraction ratio, too. Values are given in Table 8.3, as determined from finite element calculations. Table 8.3 indicates that K_L is approximately 8–15, depending upon the geometry.

When the Reynolds number is not vanishingly small, K_L depends on the Reynolds number, too, and the dependence is easily determined using CFD. For example, Fig. 8.4 shows the results for a 3 : 1 contraction in a pipe. Note that K_L is approximately constant for $Re < 1$. The departure from a constant is due partially to extra viscous dissipation at higher Reynolds number, but mostly due to kinetic energy changes.

For an axisymmetric pipe or channel, in a contracting flow, the kinetic energy change is derived as follows. Let $\beta = d_2^2 \, d_1^2$ be the area ratio ($d_2 < d_1$ for contraction), and use $v_i = v_{0i} [1 - (r/R_i)^2]$ for the inlet and outlet. Then:

$$\int_0^{R_1} v^3(r) \, r \, dr \, / \int_0^{R_1} v(r) \, r \, dr = \frac{1}{2} v_{01}^2 = 2 \langle v_1 \rangle^2$$

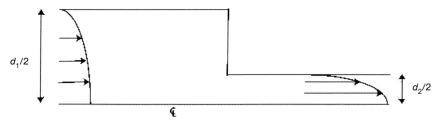

Fig. 8.3 Contraction flow geometry.

Table 8.3 Coefficient K_L for contractions and expansions for negligibly small Re.

Description	Picture	K_L
2 : 1 pipe/planar		7.3/3.1
3 : 1 pipe/planar		8.6/4.1
4 : 1 pipe/planar		9.0/4.5
45 degrees tapered, planar, 3 : 1		4.9
28.07 degrees tapered, planar, 3 : 1		10.8
3 : 1 square (quarter of the geometry)		8.1

v_s = average velocity, x_s = thickness or diameter, both in the small section.

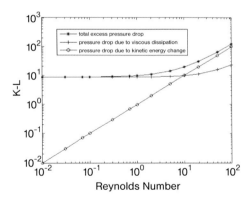

Fig. 8.4 Laminar flow excess pressure drop for a 3 : 1 contraction in a circular channel. K-L is the total pressure minus the fully developed pressure drop in the two regions, expressed as. $K_L = \Delta p_{excess}\, d_2/\eta\, \langle v_2 \rangle$.

By continuity:

$$d_1^2 v_{01} = d_2^2 v_{02}, \quad \text{or} \quad v_{01} = \beta v_{02}$$

The kinetic energy term is then:

$$\frac{1}{2}[2\langle v_{02}\rangle^2 - 2\langle v_{01}\rangle^2] = \langle v_{02}\rangle^2 (1 - \beta^2)$$

The total balance, Eq. (11), is then:

$$p_1' - p_2' = Re \langle v_2'\rangle^2 (1 - \beta^2) + \hat{E}_v' \quad \beta = d_2^2/d_1^2, \quad \beta < 1 \tag{12}$$

If the velocity standard is the average velocity in the narrow tube, $\langle v_2' \rangle = 1$. The viscous dissipation is $\hat{E}_v' = K_L$. In dimensional terms Eq. (12) is:

$$\frac{(p_1 - p_2) d_2}{\eta \langle v_2 \rangle} = \frac{\rho \langle v_2 \rangle d_2}{\eta} (1 - \beta^2) + K_L$$

or

$$p_1 - p_2 = \rho \langle v_2 \rangle^2 (1 - \beta^2) + \frac{\eta \langle v_2 \rangle}{d_2} K_L$$

When performing calculations or an experiment, it is the total pressure drop that is calculated or measured. When the analytic expressions for pressure drop in both tubes are subtracted, what remains is the excess pressure drop due to the contraction. The change of kinetic energy can be calculated so that the remaining term, viscous dissipation, is readily available. As seen in Fig. 8.4, the viscous term predominates at low Reynolds numbers (i.e., in microfluidics) whereas the kinetic energy term predominates at high Reynolds number (> 100). The equivalent expressions for other geometries are:

Flat plates, contraction, $x_s = H_2 =$ thickness between plates, $v_s = \langle v_2 \rangle$ so that $\langle v' \rangle = 1$:

$$p_1' - p_2' = Re \langle v_2'\rangle^2 \frac{27}{35}(1 - \beta^2) + \hat{E}_v', \quad \beta = H_2/H_1, \quad \beta < 1 \tag{13}$$

or

$$p_1 - p_2 = \rho \langle v_2 \rangle^2 \frac{27}{35}(1 - \beta^2) + K_L \frac{\eta \langle v_2 \rangle}{H_2}$$

Flat plate with symmetry conditions upstream, contraction:

$$p_1' - p_2' = Re \langle v_2'\rangle^2 \left(\frac{27}{35} - \frac{1}{2}\beta^2\right) + \hat{E}_v', \quad \beta = H_2/H_1, \quad \beta < 1 \tag{14}$$

$$p_1 - p_2 = \rho \langle v_2 \rangle^2 \left(\frac{27}{35} - \frac{1}{2}\beta^2 \right) + K_L \frac{\eta \langle v_2 \rangle}{H_2}$$

Expansions work in a similar fashion, and the viscous dissipation at negligible Reynolds number is the same in a contraction and an expansion. For expansion, the average velocity in the small tube is $\langle v_1 \rangle$ and we take $v_s = \langle v_1 \rangle$ so that $\langle v' \rangle = 1$.

With circular tubes, expansion gives Eq. (15):

$$p_1' - p_2' = Re \langle v_1' \rangle^2 \left(\frac{1}{\beta^2} - 1 \right) + \hat{E}_v', \quad \beta = \frac{d_2^2}{d_1^2} > 1 \tag{15}$$

or

$$p_1 - p_2 = \rho \langle v_1 \rangle^2 \left(\frac{1}{\beta^2} - 1 \right) + K_L \frac{\eta \langle v_1 \rangle}{d_1}$$

With flat plates, expansion gives Eq. (16):

$$p_1' - p_2' = Re \langle v_1' \rangle^2 \frac{27}{35} \left(\frac{1}{\beta^2} - 1 \right) + \hat{E}_v', \quad \beta = \frac{H_2}{H_1} > 1 \tag{16}$$

A flat plate with symmetry conditions upstream, expansion gives Eq. (17):

$$p_1' - p_2' = Re \langle v_1' \rangle^2 \left(\frac{27}{35} \frac{1}{\beta^2} - \frac{1}{2} \right) + \hat{E}_v', \quad \beta = \frac{H_2}{H_1} > 1 \tag{17}$$

However, in an expansion the kinetic energy change is opposite in sign to that in a contraction, causing the pressure to increase due to the kinetic energy and decrease due to the viscous dissipation.

8.3.4
Manifolds

When a flowing stream approaches a manifold with many holes, other effects must be considered. Figure 8.5 shows a typical situation. Calculations can be carried out for one cell of the manifold simply by using a symmetry condition upstream of the manifold, as illustrated. The case considered here is for a planar manifold, but 3D manifolds could be analyzed in the same way provided there was symmetry – otherwise the entire device must be modeled.

Kays has analyzed this situation for turbulent flow [3], but he gives curves for laminar flow, too. Here we show that those curves are over-simplified for laminar flow.

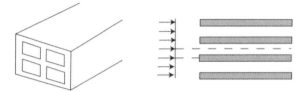

Fig. 8.5 Manifold with square holes (left) or flat plates (right).

Kays [3] presents results for a loss coefficient, K_c, defined as:

$$\frac{\Delta p}{\frac{1}{2}\rho \langle v \rangle^2} = K_c + 1 - \sigma^2 \tag{18}$$

where the loss coefficient depends upon σ, the ratio of free area in the channels to the free area upstream (see Figures 3, 6 in Ref. [3]). However, there is only one curve for laminar flow, although there are multiple curves for different Reynolds numbers in turbulent flow. The term $1 - \sigma^2$ represents the kinetic energy change assuming a flat velocity profile, which is not true for laminar flow between flat plates. Since the flow will be considerably different at low Reynolds numbers (when the pressure changes are linear in velocity) from the flow at higher, but still laminar, Reynolds numbers (when the pressure changes are quadratic in velocity), multiple curves are expected for different laminar Reynolds numbers, too. Finite element calculations have been performed for various contraction ratios and the K_c is calculated using Eq. (18) along with the computed excess pressure drop. Figure 8.6 shows that there are different curves for different Reynolds numbers, as expected, but that the curves for $Re = 100$ are close to those given by Kays. However, microfluidic devices seldom operate in this regime.

The same calculated results are plotted in terms of K_L defined using Eq. (14) and the computed excess pressure drop (Fig. 8.7). Now the values approach an asymptote at small Reynolds number, which is more useful for microfluidic flows.

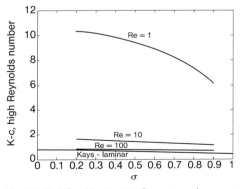

Fig. 8.6 K_c defined by Eq. (18) for various planar contractions for a manifold.

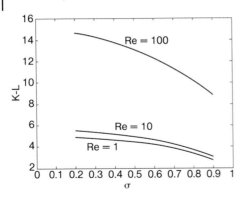

Fig. 8.7 K_L defined by Eq. (14) for various planar contractions for a manifold; the curve for $Re = 0$ is indistinguishable from that for $Re = 1$.

8.4
Entry Lengths

When the diameter or size of a flow channel changes, the velocity profile must change to adapt to the new size. It takes a certain length in the new size for the flow to become fully developed (if that is possible). In a microfluidic device, it is of interest to know how long that region is, so that the effect of a developing velocity field can be ignored if it is unimportant. The developing velocity field affects the pressure drop and it also complicates the diffusion of species in the device. This section explains the approach length and entry length for laminar flows.

8.4.1
Contraction Flows

Consider flow into a circular channel or pipe with the velocity a constant at the entrance. As the fluid enters the channel, the velocity soon develops a parabolic profile. The velocity at the centerline increases from $\langle v \rangle$ (at the entrance) to $2\langle v \rangle$ (far downstream) while the velocity at the wall decreases suddenly from $\langle v \rangle$ to zero. The entry length is defined as the length of channel it takes before the centerline velocity is 99% of its ultimate value, $2\langle v \rangle$. Sometimes the entry length is defined using a 98% or 95% criterion. Atkinson et al. [2, 4] have solved this problem using the finite element method and correlated the results with the equation:

$$\frac{L_e}{d} = 0.59 + 0.056\,Re, \quad 1\% \text{ criterion}$$

This equation is based on the Stokes flow solution ($Re = 0$, giving the 0.59) plus a boundary layer solution (giving the 0.056 Re). It is not actually a curve fit of the results for small Reynolds numbers. The calculation for $Re = 0$ in Ref. [4] and the experimental data in Ref. [5] give the following correlation:

$$\frac{L_e}{d} = 0.594 + 0.0024\ Re + 0.0036\ Re^2, \quad Re \leq 6.34, \quad r^2 = 1.0, \quad 1\%\ \text{criterion}$$

For a 5% criterion, the correlation of Atkinson's data is:

$$\frac{L_e}{d} = 0.43 + 0.0037\ Re + 0.0028\ Re^2, \quad Re \leq 6.34, \quad r^2 = 0.9993, \quad 5\%\ \text{criterion}$$

It is impossible to realize a constant velocity at the inlet to the circular channel unless the wall is "paper thin" and the flow can go past the outer wall, too. A more realistic case is where a larger channel decreases to a smaller channel (Fig. 8.3). Then a fully developed velocity profile can be used far upstream in the larger channel. The flow rearrangement occurs before the contraction (Fig. 8.8). This is important in laminar flow because vorticity is generated at the corner and diffuses upstream. As the velocity increases, this region gets pushed against the inlet and becomes smaller. Thus, for turbulent flow the effect is much shorter.

The approach length is presented in Fig. 8.9 as a function of Reynolds number. Inside the pipe the entry length for a 4 : 1 contraction in a circular channel or pipe is derived from finite element calculations including the upstream segment:

$$\frac{L_e}{d} = 0.1264 + 0.0113\ Re + 0.0002\ Re^2, \quad Re \leq 10, \quad r^2 = 1.0, \quad 5\%\ \text{criterion}$$

The effect of Reynolds number is clearly displayed in Fig. 8.10. When the upstream region is included the entry length is much shorter ($L_e/d = 0.1264$ rather than 0.43) since the velocity has already rearranged somewhat upstream.

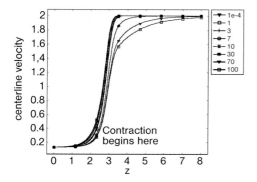

Fig. 8.8 Velocity rearrangement both before and after a 4 : 1 contraction (cylindrical geometry) for different Reynolds numbers based on the downstream section.

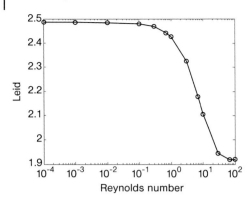

Fig. 8.9 Approach length as a function of Reynolds number; 5% criterion, 4 : 1 contraction, cylindrical geometry.

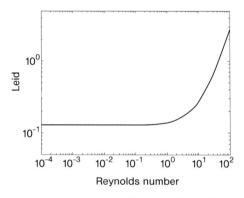

Fig. 8.10 Entry length as a function of Reynolds number; 5% criterion, 4 : 1 contraction, cylindrical geometry.

8.5
Diffusion

Many microfluidic devices involve diffusion of concentration as well as fluid flow. The velocity crossing a plane normal to the flow is seldom uniform, so the diffusion must be examined in the midst of non-uniform velocity profiles. The simplest illustration of this effect is with Taylor dispersion.

For simplicity, consider flow between two flat plates as shown in Fig. 8.11. The flow is laminar and the velocity profile is quadratic. Now suppose the entering fluid has a second chemical with the concentration shown, 1.0 in the top half and 0 in the bottom half. As this fluid moves downstream (to the right), the velocity is highest in the center, so the second chemical moves fast there. However, it can also diffuse sideways, and the chemical that does so first then moves more slowly down the flow channel. At a point not on the centerline, the concentration

Fig. 8.11 Diffusion in a fully developed planar flow.

is determined by this slower velocity in the flow direction and diffusion sideways because the concentration on the centerline is always higher than that in adjacent fluid paths. This is called Taylor dispersion [6]. Taylor [7] and Aris [8] showed how to model the average concentration as a function of length using a simpler equation (dispersion in the flow direction only) and an effective dispersion coefficient given by Eq. (19):

$$K = D + \frac{\langle v \rangle^2 d^2}{192\, D} \quad \text{or} \quad \frac{K}{D} = 1 + \frac{\langle v \rangle^2 d^2}{192\, D^2} = 1 + \frac{Pe^2}{192} \tag{19}$$

Next, consider fully developed flow in a narrow channel, which is typical of a microfluidic device. Figure 8.12 shows the normal view, with flow going into the paper. The velocity varies in x and y according to the equation:

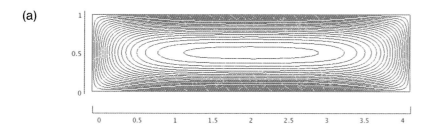

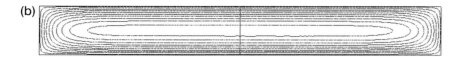

Fig. 8.12 Fully developed velocity profile in
(a) a 1 × 4 rectangular channel,
(b) a 1 × 8 rectangular channel, $\langle v \rangle = 1$, $v(0,0) = 1.628$, and
(c) a 1 × 8 rectangular channel with slip on the side walls, $\langle v \rangle = 1$, $v(0,0) = 1.500$.

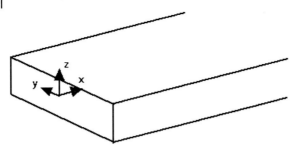

Fig. 8.13 Three-dimensional rectangular channel.

$$\eta\left(\frac{\partial^2 v}{\partial x^2} + \frac{\partial^2 v}{\partial y^2}\right) = -\frac{\Delta p}{L}$$

Typical solutions are shown in Fig. 8.12(a, b). If one uses a model ignoring the edge effects, one is essentially using the velocity profile shown in Fig. 8.12(c).

When diffusion is added into the problem, these two flow situations (b and c) are different. Take the three-dimensional flow situation shown in Fig. 8.13, with the velocity profile being the same for all x. Let the material on one side have a dilute concentration of some species (taken as $c = 1$), and on the other side there is no concentration of that species ($c = 0$). What happens at different planes downstream? Figures 8.14 and 8.15 below show typical concentration profiles; a quantitative description is also given below.

8.5.1
Characterization of Mixing

To evaluate mixing it is necessary to have a quantitative measure. The variance from the mean is appropriate, but in a flowing system some fluid elements have a higher velocity than others. The concept in chemical engineering that is used to account for this is the mixing cup concentration. This is the concentration of fluid if the flow emptied into a cup that was well stirred. In mathematical terms, the mixing cup concentration is:

$$c_{\text{mixing cup}} = \frac{\int_A c(x,y,z)\, v(x,y)\, dx\, dy}{\int_A v(x,y)\, dx\, dy} \tag{20}$$

Note that the average velocity multiplied by the mixing cup concentration gives the total mole/mass flux through an area at position z. It is then natural to define the variance of the concentration from the mixing cup concentration:

$$c_{\text{variance}} = \frac{\int_A [c(x,y,z) - c_{\text{mixing cup}}]^2\, v(x,y)\, dx\, dy}{\int_A v(x,y)\, dx\, dy} \tag{21}$$

8.5.2
Average Concentration along an Optical Path

Optical measurements are frequently made through the thin layer of a microfluidic device. The average concentration is then determined by the integral along the path, such as by Eq. (22):

$$c_{\text{optical}} = \int_0^L c(x,y,z)\, dy \bigg/ \int_0^L dy \tag{22}$$

When the velocity is variable in the y-direction some regions of the fluid are moving faster than others. Thus, the optical measurement, or integration along the optical path, may not be comparable to the mixing cup concentration. Indeed, it is entirely possible that several different flow and concentration distributions can give the same average along an optical path. In that case, computational fluid dynamics can be used to interpret the optical measurements.

8.5.3
Peclet Number

The convective diffusion equation is:

$$\frac{\partial c}{\partial t} + \mathbf{v} \cdot \nabla c = D \nabla^2 c$$

$$\frac{\partial c}{\partial t} + \mathbf{v} \cdot \nabla c = D \nabla^2 c$$

If this equation is made non-dimensional using:

$$c' = \frac{c}{c_s}, \quad \mathbf{v}' = \frac{\mathbf{v}}{v_s}, \quad \nabla' = x_s \nabla, \quad t' = \frac{t\, v_s}{x_s},$$

it is:

$$Pe\, \frac{\partial c'}{\partial t'} + Pe\, \mathbf{v}' \cdot \nabla' c' = \nabla'^2 c' \quad \text{or} \quad \frac{\partial c'}{\partial t'} + \mathbf{v}' \cdot \nabla' c' = \frac{1}{Pe} \nabla'^2 c' \tag{23}$$

where Pe is given by Eq. (24):

$$Pe \equiv \frac{v_s\, x_s}{D} \tag{24}$$

The Peclet number is a ratio of the time for diffusion, $x_s^2\, D$, to the time for convection, x_s/v_s. Typically, Pe is very large, and this presents numerical problems.

The second form of Eq. (22) shows that the coefficient of the term in the differential equation with highest derivative goes to zero; this makes the problem singular. For large but finite *Pe* the problem is still difficult to solve. For a simple one-dimensional problem it can be shown [9] that the solution will oscillate from node to node unrealistically unless:

$$\frac{Pe \, \Delta x'}{2} \leq 1, \quad \text{or} \quad \frac{v_s \, x_s}{2D} \frac{\Delta x}{x_s} \leq 1$$

This means that as *Pe* increases, the mesh size must decrease. Since the mesh size decreases, it takes more elements or grid points to solve the problem, and the problem may become too big. One way to avoid this is to introduce some numerical diffusion, which essentially lowers the Peclet number. If this extra diffusion is introduced in the flow direction only, the solution may still be acceptable. Various techniques include upstream weighting (finite difference [10]) and Petrov-Galerkin (finite element [11]). Basically, if a numerical solution shows unphysical oscillations, either the mesh must be refined, or some extra diffusion must be added. Since it is the *relative* convection and diffusion that matter, the Peclet number should always be calculated even if the problem is solved in dimensional units. The value of *Pe* will alert the chemist, chemical engineer, or bioengineer whether this difficulty would arise or not. Typically, v_s is an average velocity, x_s is a diameter or height, and the exact choice must be identified for each case.

Sometimes an approximation is used to neglect axial diffusion since it is so small compared with axial convection. If the flow is fully developed in a channel then one solves (for steady problems):

$$w(x,y) \frac{\partial c}{\partial z} = D \left(\frac{\partial^2 c}{\partial x^2} + \frac{\partial^2 c}{\partial y^2} \right) \quad (25)$$

This is a much simpler problem since the *D* can be absorbed into the length, *z*,

$$w(x,y) \frac{\partial c}{\partial (zD)} = \left(\frac{\partial^2 c}{\partial x^2} + \frac{\partial^2 c}{\partial y^2} \right)$$

In non-dimensional terms this is given by Eq. (26):

$$w'(x',y') \frac{\partial c'}{\partial (z'/Pe)} = \left(\frac{\partial^2 c'}{\partial x'^2} + \frac{\partial^2 c'}{\partial y'^2} \right) \quad (26)$$

Then all results should depend upon z'/Pe not z' and *Pe* individually. The velocity is not a constant; it depends upon *x* and *y*. Thus, the problem is different from the following formulation [Eq. (27)]:

$$\langle w'\rangle \frac{\partial c'}{\partial(z'/Pe)} = \left(\frac{\partial^2 c'}{\partial x'^2} + \frac{\partial^2 c'}{\partial y'^2}\right) \quad \text{or} \quad \frac{\partial c'}{\partial t'} = \left(\frac{\partial^2 c'}{\partial x'^2} + \frac{\partial^2 c'}{\partial y'^2}\right), \quad (27)$$

where $t' = \dfrac{z'}{\langle w\rangle' > Pe}$

where

$$\langle w'\rangle = \int_A w'(x',y')\,dx'\,dy' \Big/ \int_A dx'\,dy'$$

8.5.4
Diffusion in a Rectangular Channel

The quantitative measures are calculated for fully developed flow in rectangular channels. The concentration profile is shown in Fig. 8.14. Figure 8.15(a) shows contours of the concentration on the exit plane. The material near the upper and lower surface has a slower velocity (see Fig. 8.12b); thus it has a longer residence time in the channel and has more time to diffuse sideways. In fact, the concentration profile resembles a butterfly, and this effect was called the butterfly effect by Kamholz et al. [12]. If the velocity is changed to that shown in Fig. 8.12(c), but with the same average velocity, the concentration profiles are altered slightly but the butterfly effect still exists [Fig. 8.15(b)]. The peak velocities are different in Fig. 12(b) and (c), though, even when the average velocities are the same, so that the butterfly effect would not be calculated correctly with a velocity allowing slip on the side walls (Fig. 8.12c). When Eq. (27) is used with a uniform velocity, the butterfly effect disappears (Fig. 8.15c) since every element of fluid has the same velocity, and hence the same residence time.

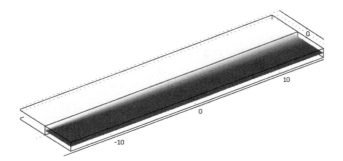

Fig. 8.14 Concentration profile in channel; $Pe = 100$, width = 8, height = 1, length = 30, $\langle v'\rangle = 1$. Fully developed velocity profile determined with 28 160 elements, 56 897 degrees of freedom. Concentration determined with 46 426 elements, 68 845 degrees of freedom.

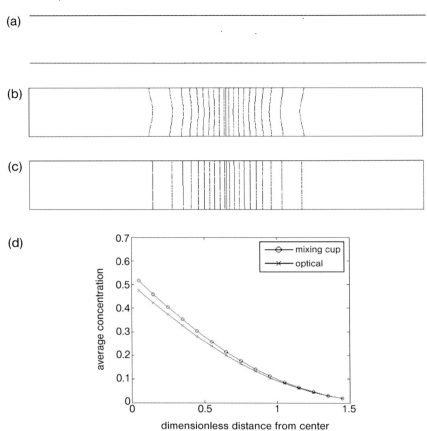

Fig. 8.15 Butterfly effect; concentration at exit at $z' = 30$.
(a) Concentration at exit with fully developed velocity profile; variance = 0.193.
(b) Concentration at exit with velocity profile allowing slip on the side walls, for the same average velocity as (a); variance = 0.195.
(c) Concentration at exit with uniform velocity, for the same average velocity as (a); variance = 0.195.
(d) Comparison of mixing cup and optical path-length concentrations.

Figure 8.15(d) illustrates the difference between the mixing cup concentration and the optical path concentration. The differences are not large (maximum of 8% difference for this case).

8.5.5
Diffusion in a T-Sensor [13, 14]

Consider next a two-dimensional case with flow coming in top and bottom on the left (see Fig. 8.16). The fluid is mainly water, but a small concentration of a second species is in the upper flow. Along the top take $c = 1$, and along the bottom take $c = 0$. How will they mix? Figure 8.17 shows the variance as a function of length/Peclet number. Points are computed when the problem is solved with different Peclet numbers and the curves are close to each other, justifying Eq. (26). If this problem is solved with a large Peclet number, oscillations appear (Ref. [15] p. 217) and the mesh must be refined or some stabilization (like Petrov-Galerkin or upstream weighting) must be applied. The stabilization, of course, smoothes the solution and adds unphysical diffusion. It is up to the analyst to decide if that effect can be tolerated.

The effect of placing a knife-edge in the device is very small; it is negligible because the amount of diffusion across the line where the knife-edge would be is a very small portion of the total mass transfer, and the flow field is affected only slightly.

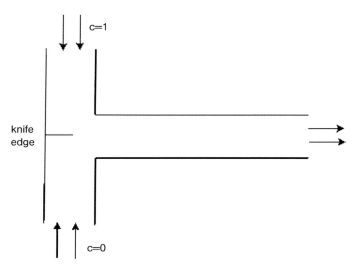

Fig. 8.16 T-sensor; the knife edge can be added as shown.

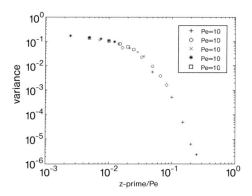

Fig. 8.17 Variance for mixing in T-sensor.

8.5.6
Serpentine Mixer

A serpentine mixer (Fig. 8.18) can be used to mix two chemicals in a shorter length than happens in a straight channel. Figure 8.19 shows a typical solution for a Reynolds number of 1.0 and Peclet number of 1000. Comparing these data with those for a T-sensor (Fig. 8.17) shows that the total length/width necessary has been reduced from 250 to 19.6 when the variance is 2.6×10^{-4}.

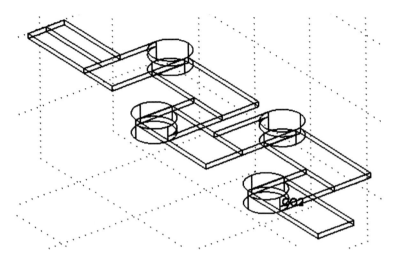

Fig. 8.18 Serpentine mixer; width = 1, height = 1/8.

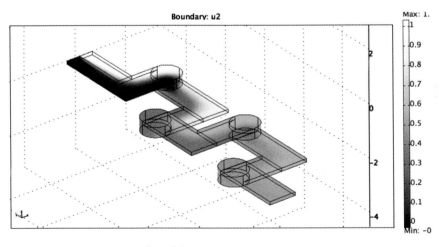

Fig. 8.19 Concentration on surface of the serpentine mixer; $Re = 1$, $Pe = 1000$, 28 380 elements, 145 924 degrees of freedom for flow, 46 272 degrees of freedom for concentration; variance at exit is 2.6×10^{-4}.

Table 8.4 Variance at different points in a serpentine mixer.

Re	Pe	Variance	Equivalent length[a]
0.25	250	5.31×10^{-8}	
0.5	500	1.56×10^{-5}	225
1.0	1000	2.60×10^{-4}	250
2	2000	1.12×10^{-3}	400
4	4000	2.10×10^{-3}	700
8	8000	3.03×10^{-2}	600

a) The equivalent length is the length of a straight channel of the same dimensions that gives the same variance; the lengths are determined from Fig. 8.17. All calculations have a dimensionless pressure drop of $\Delta p' = \Delta p \, H/\eta \, \langle v \rangle = 2331$, which means that the dimensional pressure drop increases linearly with Reynolds number. $Sc = 1000$. The case with $Pe = 8000$ is solved with 88 986 elements and 136 299 degrees of freedom for the concentration.

Another aspect of the serpentine mixer is to define how much mixing occurs for a given pressure drop. The flow solution gives the pressure drop and the variance gives the mixing. Table 8.4 shows results at different velocities for the serpentine mixer.

8.5.7
Reactor System

The importance of diffusion for mixing (or the lack of it) is illustrated by an example of a reactor system developed at Dow Chemical Company. Figure 8.20 shows the geometry of a small part of the device.

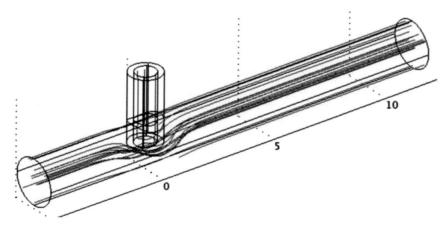

Fig. 8.20 Reactor with two inlets for two reagents and a catalyst.

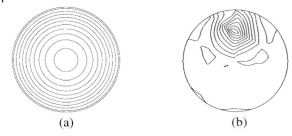

Fig. 8.21 Outlet conditions: (a) velocity; (b) concentration of product.

One chemical comes in at the left, and another chemical plus a catalyst enters at the top inlet. The concentration distributions are found by solving the convection-diffusion-reaction equations for the two chemicals (A and B) plus the catalyst (C). The system is assumed to be dilute so that the total concentration does not change appreciably. The reaction is taken as:

$$A + 2B \rightarrow D \text{ in the presence of } C$$

Naturally, the reaction only occurs where both chemicals and the catalyst are all in the same place. Since little mixing occurs in laminar flow, and diffusion is slow, the product concentration coming out is not uniform in space (Fig. 8.21b) even though the velocity is fully developed at the exit (Fig. 8.21a). In this complicated three-dimensional case only CFD can give details of the mixing.

8.6
Conclusion

Engineering correlations have been presented for pressure drop and entry length in common geometrical flow channels with laminar flow. The correlations differ from their counterparts with turbulent flow (available in handbooks, depending upon density and velocity) since in slow flow the pressure drop is proportional to the velocity and the fluid viscosity. The entry length in laminar flow is also longer than is the case for turbulent flow, which means that some section of flow in microfluidics is in the entry region.

Mixing is more difficult to achieve in laminar flow than in turbulent flow. The mixing cup variance is introduced as a quantitative measure of mixing when there are flow and concentration variations. Diffusion is the mechanism in the T-sensor, and the amount of mixing depends upon the Peclet number. In the serpentine mixer, however, mixing is due primarily to the intertwining of flow streams, with diffusion across a short distance. A reaction system illustrated the importance of characterizing the flow and diffusion using computational fluid dynamics.

Nomenclature

c	concentration (kmol m^{-3})
d	diameter of tube (m)
D	diffusivity (m^2 s^{-1})
E_v	viscous dissipation (Pa m^3 s^{-1})
$\hat{E}_v$	viscous dissipation per mass flow rate, $= E_v/w$ (Pa m^3 kg^{-1})
f	friction factor, defined by Eq. (1) (–)
g	acceleration of gravity (m s^{-2})
h	height above a datum (m)
H	distance between two flat plates (m)
K	friction loss coefficient, $= \Delta p_{excess}/½ \rho \langle v \rangle^2$ (–)
K_L	friction loss coefficient for laminar flow, $= \Delta p_{excess} x_s/\eta\, v_s$ (–)
K_c	friction loss coefficient for a manifold, defined by Eq. (18) (–)
L	length of channel (m)
L_e	entry length for velocity to come within a percentage of the fully-developed value (m)
p	pressure (Pa)
Pe	Peclet number, $= v_s\, x_x/D$ (–)
r	radial position (m)
R	radius of tube or correlation coefficient (m)
Re	Reynolds number, $= \rho\, v_s\, x_x/\eta$ (–)
Sc	Schmidt number, $= \eta/\rho\, D$ (–)
t	time (s)
u	velocity in the x-direction (m s^{-1})
v	velocity, and velocity in the y-direction (m s^{-1})
V	volume (m^3)
$\mathbf{v}$	vector velocity (m s^{-1})
w	mass flow rate (kg s^{-1}) or velocity in z-direction (m s^{-1})
$\hat{W}_m$	work of pump per mass flow rate (Pa m^3 kg^{-1})
x	coordinate (m)
y	coordinate (m)
z	coordinate (m)

Greek Letters

β	(–) ratio of diameters or heights, defined by Eqs. (12–17)
η	viscosity (Pa s)
ρ	density (kg m^{-3})
σ	ratio of free area in manifold to area upstream (–)
τ	shear stress (Pa)

Special Symbols

$'$	non-dimensional pressure, using $p_s = \eta\, v_s/x_s$
$''$	non-dimensional pressure, using $p_s = \eta\, v_s^2$
$\langle \rangle$	average over an area

∇ gradient operator
∇^2 Laplacian operator

Subscripts
mixing cup   concentration defined by Eq. (20)
optical   concentration defined by Eq. (22)
variance   variance defined by Eq. (21)
s   standard quantity
1   inlet value
2   outlet value
0   centerline value

References

1 R. B. Bird, W. E. Stewart, E. N. Lightfoot, *Transport Phenomena*, 2nd ed., Wiley, **2002**.
2 *Perry's Chemical Engineers' Handbook*, 7th ed., ed. Perry, R. H., Green, D. W., McGraw-Hill, **1997**.
3 W. M. Kays, *Trans. ASME* **1950**, 1067–1074.
4 B. Atkinson, M. P. Brocklebank, C. C. H. Card, J. M. Smith, *AIChE J.* **1969**, 15, 548–553.
5 B. Atkinson, Z. Kemblowski, J. M. Smith, *AIChE J.* **1967**, 13, 17.
6 W. M. Deen, *Analysis of Transport Phenomena*, Oxford, **1998**.
7 G. Taylor, *Proc. Roy. Soc. London A* **1953**, 219, 186–203.
8 R. Aris, *Proc. Roy. Soc. London A* **1956**, 235, 67–77.
9 B. A. Finlayson, *Numerical Methods for Problems with Moving Fronts*, Ravenna Park, **1992**.
10 R. Peyret, T. D. Taylor, *Computational Methods for Fluid Flow*, Springer Verlag, **1983**.
11 T. Hughes, A. Brooks, in *Finite Element Methods for Convection Dominated Flow*, ed. T. Hughes, ASME, **1979**, pp. 19–35.
12 A. E. Kamholz, B. H. Weigl, B. A. Finlayson, P. Yager, *Anal. Chem.* **1999**, 71, 5340–5347.
13 B. H. Weigl, P. Yager, *Science* **1999**, 283, 346–347.
14 A. Hatch, A. E. Kamholz, K. R. Hawkins, M. S. Munson, E. A. Schilling, B. H. Weigl, P. Yager, *Nat. Biotechnol.* **2001**, 19, 461–465.
15 B. A. Finlayson, *Introduction to Chemical Engineering Computing*, Wiley, **2006**.

9
Selected Developments in Micro-analytical Technology

9.1
Introduction

Melvin V. Koch

As the ability to generate more samples per unit time increases with improvements in the capabilities of combinatorial and microreactor systems, there is a growing need for a faster analytical response. To keep the "Development Cycle" (react, analyze, handle data, and design next experiment) functioning at a productive rate there is need to characterize samples and convert the corresponding analytical data into valuable information. This information will then feed the "Design of Experiment" part of the cycle. Significant effort has been put into approaches to speed up the ability to analyze samples and this has often been aided by miniaturization of the analytical equipment.

This chapter describes various selected technologies that exemplify improvements in analytical science for achieving a faster analytical response, often involving miniaturized components. Each selection covers a wide range of technologies and types of characterization. They were chosen from technologies being developed and studied at CPAC that were presented at the CPAC Summer Institute in 2004 and 2005. These developments aim to address a range of needs indicated by CPAC industrial members, including:

- mixing efficiency
- cleaning validation
- particle sizing
- chemical composition
- coating characterization
- vapor characterization
- viscosity/rheometrics
- moisture
- bio-assay.

Micro Instrumentation for High Throughput Experimentation and Process Intensification – a Tool for PAT
Edited by M. V. Koch, K. M. VandenBussche and R. W. Chrisman
Copyright © 2007 WILEY-VCH Verlag GmbH & Co. KGaA, Weinheim
ISBN: 978-3-527-31425-6

It is envisioned that many of these technologies will ultimately be compatible with the NeSSI (New Sampling and Sensor Initiative) platform and, thus, will be added to the growing toolbox of characterization technologies that could be used in an array format for the generation of meaningful data. The technologies described include:

- surface plasmon resonance sensors
- fringe field dielectric spectroscopy
- Raman spectroscopy
- grating light reflection spectroscopy
- mini liquid chromatography
- ultra micro gas analyzers
- mini nuclear magnetic resonance.

9.2
Application of On-line Raman Spectroscopy to Characterize and Optimize a Continuous Microreactor

Brian Marquardt

9.2.1
Introduction

Raman spectroscopy is a well-established vibrational spectroscopy technique for determining both qualitative and quantitative molecular information from almost any type of sample (e.g., solid, liquid or gas). A Raman spectrum is obtained by exciting a sample with a laser and measuring the inelastic scattering of photons from the sample. Raman scattering is a weak phenomenon and, therefore, not inherently sensitive. This is because approximately 1 in 10^7 photons is scattered at an optical frequency (Raman effect) different from that of the source radiation [1, 2]. Raman spectroscopy is a versatile and useful tool for determining and monitoring both organic and inorganic molecular species. Much of the versatility of the technique is due to the lack of sample preparation needed to perform an analysis. Raman analyses can be performed on very small samples with or without the aid of a microscope. Since water is a weak scatterer, Raman is also a very useful technique for determining the molecular composition of aqueous solutions.

Raman spectroscopy is also attractive for on-line analysis due to short analysis times (1–30 s on average) and ease of optical sampling. A line of sight to the sample is all that is required to perform a measurement. The use of optical fibers for both delivery of the excitation laser and collection of the scattered signal also increases the flexibility of performing Raman measurements *in situ*. The sampling geometry (epi-illumination) in Raman spectroscopy also facilitates the use of various optical sampling approaches, such as long working distance telescopes, immersion probes and non-contact lens systems. Recently, Raman has been used to monitor a diverse variety of industrial processes to improve product quality and process understanding. These applications include monitoring polymerization reactions and polymer film characterization in the polymer industry [3, 4], an on-line monitor for the analysis of pharmaceutical tablet quality [5, 6], a process measurement of the concentration of glucose and fructose in aqueous sugar mixtures [7], and, most recently, Raman has been applied to monitor various stages of distillation processes, curing of paints and coatings and various other chemical processes [8–18].

Owing to the information rich data (qualitative and quantitative), the ease of optical sampling and simplicity of application Raman spectroscopy is a very viable analytical technique for characterizing and monitoring chemical reaction in microreactors. One of the challenges of using microreactors for developing novel chemistries is gaining a fundamental understanding of the reactor itself. Microreactors are unique chemical systems that allow for improved heat transfer, effective mixing and well understood flow dynamics, which can all be leveraged to improve overall chemical reaction efficiencies. Microreactors have become an interesting aspect of performing continuous chemical production over the last 5–8 years. However, little information is found in the literature describing the application of analytical sensors or instrumentation to microreactors to improve reaction optimization or control. The present chapter demonstrates the use of Raman spectroscopy as a tool for analyzing and understanding a chemical reaction performed in a continuous microreactor.

9.2.2
Optical Sampling

A key element to monitoring any chemical process on-line is to sample the system effectively and reproducibly. Towards this goal an immersion probe has been developed at the Center for Process Analytical Chemistry for the analysis of single or mixed phase chemical reactions that meets these requirements. The immersion probe (Fig. 9.2.1a and b), incorporating a spherical lens, was designed and optimized for the purpose of performing Raman measurements in both laboratory and industrial reactor environments [18–21]. The spherical lens probe is an efficient sampling interface for the analysis of heterogeneous multiphase samples including solids, liquids and gases. The probe design is compact, durable and straightforward with no moving or easily fouled components. In comparison to other commercially available immersion probes, the spherical lens probe demonstrates greatly improved measurement precision in various samples. The design of the probe has also been shown to be effective with other types of spectroscopy, including UV/Vis, NIR and fluorescence. The improved measurement precision of the probe is due to the novel design whereby a spherical lens is used as both the focusing element and the optical sample interface. This optical design provides a constant focal length for the excitation source, which leads to increased measurement precision. Further precision enhancement is gained by the focal point being within 100 μm of the surface of the spherical lens. The fact that the focal volume of the probe is a constant ensures near perfect optical focus with any type of sample. Therefore, the only sampling requirement is that the spherical lens is in contact with the sample to perform an optimal optical measurement. The probe also exhibits several attributes not previously available in a commercial probe: (a) precise focus on any surface or material, (b) no need for sample alignment, (c) ease of sampling – simply place probe into sample, (d) use in flowing or static sampling systems, (e) not affected by directional flows or variable contact points, (f) not affected by differential light scattering or particle distribution of solid particles, (g) probe element is fully sealed (> 600 psi He, > 3000 psi H_2O), highly durable and has been proven effective in harsh process environments. The immersion probe has been effectively demonstrated for the analysis of various solids, powders, slurries, suspensions, particles, and liquids with very good analytical performance. The simple and robust optical design of this probe has enabled effective Raman analysis of both batch and continuous reactors.

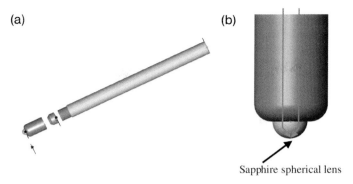

Fig. 9.2.1 Sketch of (A) a Raman immersion probe and (B) an immersion probe tip and spherical lens.

9.2.3
Continuous Reaction Monitoring

To illustrate the concept of combining analytics to improve process understanding an example chemical reaction was run using a Cellular Process Chemistry Systems (CPC) continuous feed microreactor. This microreactor is configured to operate as a small-scale chemical production plant. It has two reactant input lines and two solvent/wash lines. The thermally controlled microreactor block of the continuous feed reactor has a total internal volume of 50 µL. The reactor system contains active control for both temperature and feed rate of the two reactants. The system flows product from the microreactor block to a residence time module (12 mL volume) and then out of the reactor for product collection and work-up.

The initial goal of this microreactor project was to effectively couple an inline Raman probe to the continuous reactor system to evaluate the reactor performance and to apply Raman spectroscopy as a process control tool for the screening and optimization of reaction parameters. Towards this goal the reactor was initially fitted with a custom designed low-volume flow cell incorporating a Raman ballprobe at the outlet port of the residence time module (Fig. 9.2.2). Owing to earlier problems with bubble formation in the sample line and phase separation of the product, a back pressure regulating valve was installed downstream of the Raman probe to improve the sampling precision. This system worked quite well for the work presented in this chapter; recently, however, a small NeSSI (New Sampling Sensor Initiative) sampling block incorporating a Raman probe adapter and a back pressure regulator has been installed at the output of the reactor to further improve the sampling precision. The following sections describe the reaction optimization studies for the esterification of methanol in the continuous microreactor, utilizing on-line Raman spectroscopy. Figure 9.2.3 shows the reaction describing the esterification of methanol with acetic acid.

The primary objective of this study was to develop a general chemistry test bed for evaluating and optimizing various chemistries and on-line analytical sensors.

Fig. 9.2.2 CPC continuous feed reactor with on-line Raman detection utilizing a commercial Raman probe equipped with a Raman ballprobe at the outlet port of the reactor.

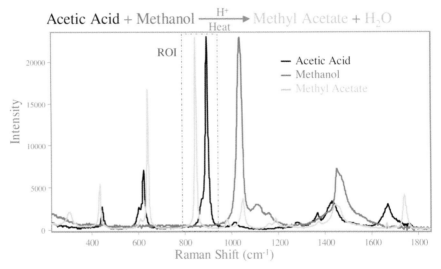

Fig. 9.2.3 Methyl acetate reaction data, including standard Raman spectra of acetic acid, methanol and methyl acetate. The acid catalyst used in this reaction was conc. sulfuric acid.

This was initiated by first choosing a simple test bed chemical reaction to evaluate and understand the functionality, flexibility and limitations of the microreactor platform. The reaction of acetic acid and methanol to form methyl ester was selected because the reaction was temperature sensitive and of minimal toxicity. This chemistry has been extensively studied in the author's laboratory previously by Raman spectroscopy in a typical batch reactor. The batch reactor results were a very useful foundation when trying to understand the reaction processes in the microreactor. The microreactor experiments were structured to study reaction response to reactor parameter changes (temperature and flow rate) using Raman spectroscopy.

The formation of methyl acetate was monitored using a modified flow cell equipped with a Raman ballprobe and a back pressure regulating valve to reduce perturbations in the flow. The Raman system was set to acquire the average of two 7.5-second integrations for each spectrum (15 s total acquisition time). The reaction was run at 50 °C for a total of 70 min. The flow rate through the reactor varied from less than 1 mL min^{-1} to greater than 10 mL min^{-1}. The reactor temperature was crudely controlled using the reactor operating software.

The Raman results of the first reaction were promising and the raw data from that reaction are shown in Fig. 9.2.4(a) graphically, illustrating the changing intensities of the Raman peaks of the reactants methanol (1040 cm^{-1} peak) and acetic acid (890 cm^{-1} peak) compared with the peaks for the product methyl acetate (850 and 1050 cm^{-1}). Figure 9.2.4(b) is a principle components analysis (PCA)

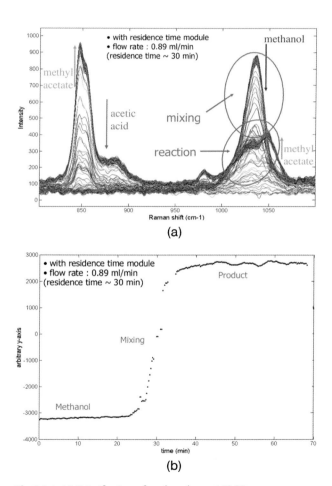

Fig. 9.2.4 (a) Esterification of methanol run at 50 °C.
(b) Scores plot describing the mixing and reaction profile of the esterification in (a).

scores plot describing the reaction profile at 50 °C. The reaction first began with the reactor charged with methanol and then at approximately 26 min the acetic acid began to appear and mix into the reactant stream. Then, at approximately 40 min, the reactor reached a steady state output concentration of product at the prescribed reactor conditions. The scores plot represents the relative intensity of the methyl acetate band as a function of reaction time. The reaction kinetics of interest in this continuous reaction were overwhelmed by the large sources of variance described in the PCA model during the mixing stage of this reactor. This was a difficulty encountered in working with a continuous microreactor for performing kinetic studies with a single point stationary analyzer. When observing only one point in a flowing stream (this single point measurement can be related to a measurement at a constant time in a batch reaction) the reactor mimics a very small volume batch reactor.

Therefore, it was determined that the reactor had to be dynamically perturbed after it reached continuous product formation to measure any kinetic response. To better describe and illustrate the reaction kinetics and the change in reaction yield, a dynamic temperature step was performed while continuously monitoring the esterification reaction. The results of that experiment are shown in Fig. 9.2.5. The Raman spectra in Fig. 9.2.5(a) illustrate the change in reactant and product ratios as a function of temperature change. The reaction profile data is more fully explained in the first scores plot of the PCA analysis in Fig. 9.2.5(b). Note that the data shown in Fig. 9.2.5(b) does not include the highly variable mixing regime shown in Fig. 9.2.4(b) to improve clarity. The data in Fig. 9.2.5(b) describe the initial production of methyl ester at 25 °C between 15 and 22 min and then the temperature was stepped to 40 °C at approximately 23–25 min. The reaction then reached its new equilibrium product output concentration at 40 °C after approximately 35 min. The temperature curves for both the microreactor capillary plate and the residence time module are shown in the figure for clarity. This experiment began to clarify that we could change and control the microreactor parameters in real-time and monitor the chemical response to that control variable change with Raman spectroscopy. This success led to an experiment involving several temperature steps during a reaction in the continuous microreactor.

To further understand the chemical response of both the reaction and the reactor to temperature an experiment was conducted that incorporated several temperature steps. The experiment encompassed stepping the reactor from a starting temperature of 30 °C to a final temperature setting of 60 °C by three separate 10 °C increments. The Raman spectral data from this experiment are plotted in Fig. 9.2.6(a). Note that the ratio between the reactants and products that was so evident in the one temperature step in Fig. 9.2.5(a) is more obscured due to the multiple temperature steps. However, the first scores plot from the PCA analysis, Fig. 9.2.6(b), describes product yield as a function of time and temperature and is full of relevant kinetic and chemical information about the reaction. The PCA data effectively relate the reaction yield efficiency to an increase in reactor temperature. The initial reactor temperature was 30 °C and then the temperature was then systematically stepped to 40, 50 and 60 °C as shown in

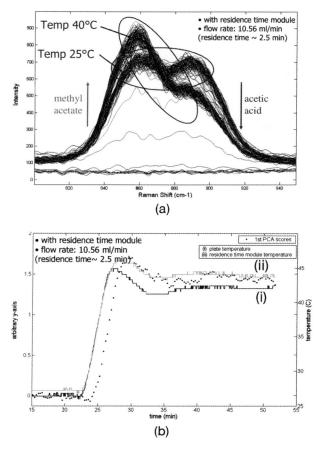

Fig. 9.2.5 (a) Esterification of methanol run with a temperature step of 25–40 °C. (b) Scores plot describing the mixing and reaction profile of the esterification. The Raman response is correlated in time to both the plate and residence time temperature responses.

the figure. Note the overshooting of the temperature at every step. This is due mainly to slow feedback control of the temperature regulating circuitry in the microreactor control software.

Successes from the initial microreactor experiments using Raman spectroscopy included effectively coupling the Raman ballprobe to the microreactor to monitor reactions in the microreactor with good optical sampling precision. Using the on-line Raman instrumentation we were able to monitor changes in reaction yield as a response to temperature and determine kinetic profiles. Some of the limitations encountered were (a) the need to operate far from reaction equilibrium to see significant changes in reaction kinetics, (b) the initial mixing zone in the reaction dominates any multivariate analysis and needs to be removed before

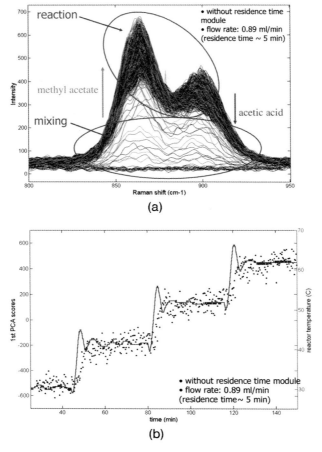

Fig. 9.2.6 (a) Esterification of methanol run with multiple temperature steps of 30–40, 40–50, and 50–60 °C. (b) Scores plot describing the mixing and reaction profile of the esterification. The Raman response is correlated in time to reactor temperature response.

analysis, (c) the current CPC microreactor temperature control software is not very efficient and (d) improvements could be made to ensure good optical sampling in more difficult matrices. Further experiments were performed after this initial reactor profiling work to react butanol and acetic acid to form butyl acetate and water as insoluble products. Immediately, problems were encountered with the original optical sampling apparatus due to multi-phase products from this reaction system. This optical sampling problem was solved by using a simple sampling system built using modular sample handling components. The NeSSI sampling system designed for use with the continuous microreactor is described in the next section.

9.2.4
Improved Reactor Sampling Using NeSSI Components

A new suite of process sampling hardware components have become available over the last few years that offer flexibility in design and implementation of complicated flow systems for process sampling and analysis. These general components are better known as NeSSI components and they have been successfully adopted by numerous industries as a cost-effective way of sampling both gas and liquid streams for process analysis. During a recent experiment expanding on the above-described esterification reaction, in which butyl alcohol was used in place of methanol, an optical sampling problem was encountered with our custom-made Raman sampling apparatus seen at the far right-hand side of Fig. 9.2.2. Butyl alcohol and acetic acid react to form butyl acetate and water. The problems encountered with our original sampling apparatus were two-fold. First, butyl acetate is a more viscous product than methyl acetate. This increased viscosity led to a larger pressure drop when the product exited the reactor and led to the formation of air bubbles in the Raman sampling apparatus tubing. The bubbles were problematic because they greatly affected the measurement to measurement precision of the Raman analysis. The second problem with the butyl acetate chemistry was that the products were phase disassociating between exiting the heated reactor block and reaching the Raman ballprobe tip in the sampling apparatus.

Several modifications were made of the original sampling system and small gains were made with bubble formation by increasing the back pressure of the system at the point of measurement. The problem of product dissociation could be reduced but not eliminated by reducing the length of tubing between the reactor and the sample system. This was determined to be due to a loss of heat in the products in the short distance between the reactor and the Raman flow cell. After the problems were identified and several modifications made to the original sample system to improve our optical sampling of the reactor it was concluded

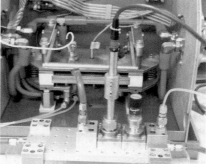

Fig. 9.2.7 NeSSI sampling system designed to reduce optical sampling variables for performing on-line Raman analysis. The arrow indicates the position of the Raman ballprobe.

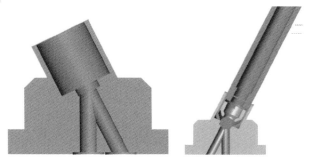

Fig. 9.2.8 NeSSI top mount component for use with a Raman ballprobe. The Raman ballprobe inserted into the component, indicating flow channels to probe optic.

that a better system was needed to completely eliminate the sampling problems with the butyl acetate reaction. The solution to the sampling problems was a simple NeSSI sampling system purchased from Parker-Hannifin (Fig. 9.2.7). The Raman ballprobe is the second component from the left with the black fiber-optic cable attached to the top.

The sample system consisted of a substrate heated by plumbing our reactor thermostat to both the reactor and the sampling system to ensure a constant reaction temperature at the Raman probe (red tubing into reactor plate and sampling system substrate are plumbed to heating unit). Heating the sampling system eliminated the multiphase sampling problems that occurred in the original system and allowed for reproducible continuous analysis of the butyl acetate and water products. The rest of the sampling system consisted of only two modular face mount components to effectively monitor the continuous microreactor. A 0.5-inch internal diameter pressure fitting was used to mount the Raman ballprobe into the sampling system. Figure 9.2.8 illustrates the probe mount component and the reaction flow path. After the probe coupling component, a surface mount needle valve component was installed. The needle valve was used to increase the backpressure of the system at the Raman ballprobe interface in order to greatly reduce the formation of air bubbles in the sampling system. This simple commercially available sampling system allowed for increased control of the sample stream at the Raman probe interface and, therefore, vastly improved the reproducibility and accuracy of the Raman analysis.

References

1 L. A. Woodward, *Raman Spectroscopy*, ed. H. A. Szymanski, Plenum Press, New York, **1967**, ch 1, pp. 4–6.
2 B. J. Bulkin, *Analytical Raman Spectroscopy*, ed. J. G. Graselli, B. J. Bulkin, John Wiley & Sons, New York, **1991**, ch 1, pp. 3–5.
3 N. Everall, K. Davis, H. Owen, M. J. Pelletier, J. Slater, *Appl. Spectrosc.* **1996**, 50, 388.

4 N. Everall, B. King, *Macromol. Symp.* **1999**, 141, 103.
5 D. S. Hausman, R. T. Cambron, A. Sakr, *Int. J. Pharm.* **2005**, 299, 19.
6 D. S. Hausman, R. T. Cambron, A. Sakr, *Int. J. Pharm.* **2005**, 298, 80.
7 J. J. Freeman, D. O. Fisher, G. J. Gervasio, *Appl. Spectrosc.* **1993**, 47, 1115.
8 G. J. Gervasio, M. J. Pelletier, *J. Process Anal. Chem.* **1997**, 3(1,2), 7.
9 F. Adar, R. Geiger, J. Noonan, *InTech.* **1997**, 44(7), 57.
10 S. Higuchi, T. Hamada, Y. Gohshi, *Appl. Spectrosc.* **1997** 51(8), 1218.
11 M. J. Pelletier, K. L. Davis, R. A. Carpio, *Proceedings of the Symposium on Process Control*, (Electrochem. Soc., Reno). **1993**, pp. 282–293.
12 M. Z. Martin, A. A. Garrison, M. J. Roberts, P. D. Hall, C. F., *Proc. Control Quality*, **1993**, 5(2–3), 187.
13 S. Farquharson, S. F. Simpson, *Proc. SPIE* **1993**, Vol. 1796, 272.
14 P. E. Flecher, W. T. Welch, S. Albin, J. B. Cooper, *Spectrochim. Acta.* **1997**, 53A(2), 199.
15 E. D. Lipp, R. L. Grosse, *Appl. Spectrosc.* **1998**, 52, 42.
16 I. M. Clegg, N. J. Everall, B. King, H. Melvin, C. Norton, *Appl. Spectrosc.* **2001**, 55, 1138.
17 S. Romero-Torres, J. D. Perez-Ramos, K. R. Morris, E. R. Grant, *J. Pharm. Biomed. Anal.* **2005**, 38, 270.
18 M. Skrifvars, P. Niemela, R. Koskinen, O. Hormi, *J. Appl. Polym. Sci.* **2004**, 93, 1285.
19 B. J. Marquardt, T. Le, L. W. Burgess, *SPIE Proc.* **2001**, vol. 4469.
20 J. P. Wold, B. J. Marquardt, B. K. Dable, D. Robb, B. Hatlen, *Appl. Spectrosc.* **2004**, 58(4), 395.
21 B. J. Marquardt, L. W. Burgess, Optical immersion probe incorporating a spherical lens, *US Pat.* # 6,831,745 B2, issued December **2004**.

9.3
Developments in Ultra Micro Gas Analyzers

Ulrich Bonne, Clark T. Nguyen, and Dennis E. Polla

9.3.1
Overview

In addition to the need to process large quantities of samples that technology produces more and more prolifically, the demand for faster analytical response is also borne from applications where analysis time is inherently critical, as with first-responders and field soldiers sampling the air to check for the presence of chemical warfare agents (CWA). DARPA (Defense Advanced Research Projects Agency), which commissions advanced research for the Department of Defense, is funding the development of micro gas analyzers (MGA) suitable for portable and virtually instant CWA detection.

By striving towards order-of-magnitude reductions in size, energy consumption, and total analysis time, these micro-analyzers will enable a host of new strategic applications and deployment scenarios, including handheld or wearable sensors for first-responders or soldiers, projectile-delivered sensors for remote detection, and unattended ground sensors for advance warning and perimeter protection [1]. (The small size has the added benefit of resilience against high g forces, which, coupled with miniature wireless technologies targeted by other DARPA programs, enables projectile and remote monitoring scenarios.) In the civilian sector, applications could include environmental monitoring for toxic industrial compounds (TICs), gas analyzers for industrial process controls and equipment health monitoring, explosives-related compound (ERC) detection, onboard bleed-air and cabin air quality monitors for commercial aviation, medical testing for gastro-intestinal and cardio-pulmonary anomalies, and air quality control in space-conditioned environments.

The military already employs portable CWA detectors, but this research is being funded in large part because of the unacceptably high incidence of false positives, or false alarm rates (FAR), exhibited by current technologies. So, accuracy and reliability are actually a stronger driving force than size, energy consumption, and speed – perhaps the more intuitive requisites for portable sensors. Notably, this refers to accuracy in field operating conditions, i.e., without the use of extensive laboratory equipment to control pressure, moisture, carrier gas, or temperature, and to an unprecedented sensitivity for such a device.

Since false positives are a central issue of this R&D effort, it was important to *quantify* false positive results, particularly for the purpose of guiding and evaluating sensor design at the theoretical stage, to avoid exhaustive empirical testing (and the amount of time involved in such "trial-and-error" testing). To that end, a simple concept evolved based on the number of orthogonal or "linearly independent" measurements, n, also referred to as orthogonal channel capacity

(OCC), that the sensor system can apply to an unknown sample of one or more known analytes. The larger the value of n, the smaller (and more desirable) the probability of false positives (P_{fp}) would be. Furthermore, to qualitatively find a relationship between n, FAR and P_{fp}, we made assumptions and imposed constraints, in line with reasonable expectations about the behavior of P_{fp}, as discussed in Section 9.3.5.4.

9.3.2
MGA Performance Goals

Micro-miniaturization offers precisely the coveted combination of reduced size, time, and energy consumption mentioned above. DARPA's specific design goals, therefore, include a total unit size of less than 2 cm^{-3}, energy consumption per sample analysis of less than 1 joule, total analysis time less than 4 s, and a level of detection less than 1 part per trillion toxic compound. Achieving the combined goals of sensitivity, energy consumption and low FAR pose the largest hurdle for current technologies.

The sensitivity and selectivity of gas chromatography make it a strong fundamental technology for addressing DARPA's needs in the area of CWA detection. No other "spectrometric" analyzer offers to separate the components of a mixture and then make small amounts of them available for further identification and quantification. However, current GC systems are unacceptably large and slow for the envisioned applications.

9.3.3
PHASED Air-based Analyzer

All the research groups developing such MGAs (Micro Gas Analyzers) are concentrating on gas chromatography technology, either using air [2] or H_2 [3] as a carrier gas. Using the air carrier, Honeywell's PHASED (Phased Heater Array Structure for Enhanced Detection) MGA includes sample collection and transport functions, preconcentration, separation, and detection in a single, on-chip integrated unit, as depicted in the conceptual model (Fig. 9.3.1) and in the layout of a functional chip (Fig. 9.3.2) of approximately 1×1 cm. This micro GC system is based on channels etched into silicon (Fig. 9.3.2b) in which GC stationary phases are deposited onto thermally isolated, low-mass, unsupported, heatable, thin-film membranes with thermal response times of less than 1 ms. Column-channels, GC-detectors, flow sensors and ultimately control/driver electronics will be integrated on a single MEMS (Micro Electro-Mechanical Systems) platform of bonded silicon wafers. It will be compatible with battery and wireless operation.

The high sensitivity of this MGA is due to the large number of stages (20–40) and short response time of the preconcentrator (left-hand side of Fig. 9.3.2a), consisting of an array of rapidly heatable micro-elements that can adsorb gas components and later rapidly and precisely desorb them via a thermal pulse into the sample stream,

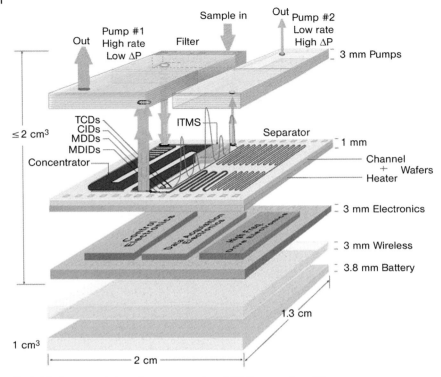

Fig. 9.3.1 Conceptual view of a compact micro GC/MS gas analyzer [2].

enhancing the concentration of the analyte by a factor of 100–1000×, depending on the partition function of individual analytes. This means of preconcentration has already enabled precise, controlled injection of the focused and concentrated analyte sample without the need for bulky mechanical valving. To date, the flow sensor and differential thermal conductivity sensor (Fig. 9.3.2c) have also been integrated on-chip with the preconcentrator and separator (right-hand side of Fig. 9.3.2a). One of the novel MGA features under study is the use of carbon nanotubes (CNTs) developed at Lawrence Livermore National Laboratory [4] for any of the three MGA functions: preconcentration, separation and detection.

Grown *in situ* on the thermally isolated heaters before covering them with the flow channels of the top wafer, the nanotubes use iron as a growth initiation site. A 5-nm thick layer of iron is deposited into the channel in the desired pattern using chemical vapor deposition, where the pattern of metal deposition controls the geometry and density of the arrays of nanotubes. Bakajin et al. are optimizing nanotube geometries for both preconcentration beds and separation columns [4, 7]. Carbon nanotubes yield a surface-to-volume ratio 2 to 3 orders of magnitude greater than traditional bead packed or surface coated columns, thus offering that much faster, and experimentally found greater, adsorption and desorption.

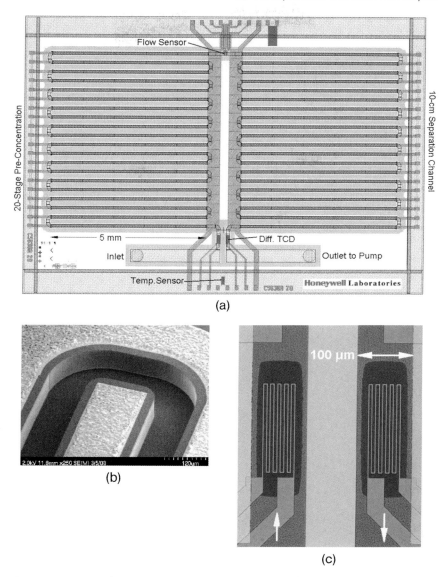

Fig. 9.3.2 (a) Layout of PHASED MGA gen.1 chip, with integrated 20-stage preconcentrator, flow and temperature sensors and differential TCD [11]. (b) SEM photograph of one Si-microchannel loop of the PHASED MGA, before wafer–wafer bonding [11]. (c) Microphoto of the differential TCD [11].

The LLNL team fabricated single wall nanotube (SWNT) stationary phases on one wall of a 100 × 100 µm (cross section) channel 50 cm long. Figure 9.3.3 shows an electron microscope image of recently grown SWNTs, which are unique for their short length [7] – near 1 µm and of a uniformity further to be improved.

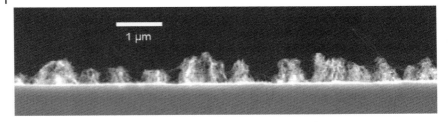

Fig. 9.3.3 Cross sectional view of grown single wall nanotubes (SWNTs) on Si, with a Fe seed layer [7].

Other stationary phase materials are also being investigated, such as coatings of dimethylpolysiloxane (PDMS) (with or without 5% the diphenyl doping), carbosilanes and Nanoglass, provided by Ohio Valley, SeacoastSciences, Dow Chemical and Honeywell SM (Nanoglass), respectively, which may or may not feature nanopores.

After thermal injection (no moving parts) and separation of the preconcentrated sample, Honeywell's MGA detects eluting compound peaks using orthogonal detectors based on thermal conductivity (integrated), optical emission spectra (to be integrated) from micro-discharge devices (MDDs), and micro ion-trap mass spectrometry (ITMS) – the latter being developed at Oak Ridge National Laboratory [5, 6]; Fig. 9.3.4(a) shows a general layout with an array of ion traps of only 40 μm in diameter.

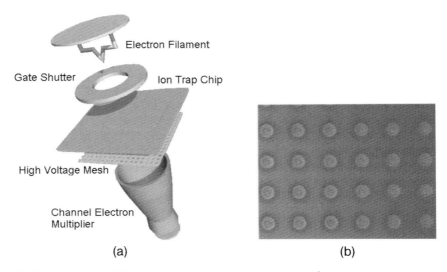

Fig. 9.3.4 (a) Micro ITMS (ion-trap mass spectrometry) experimental set-up for a 40-μm ID ion trap array of 256 elements [6].
(b) Microphoto of a first fabrication of a 1000 × 1000 array of 1-μm ID ion traps [5].

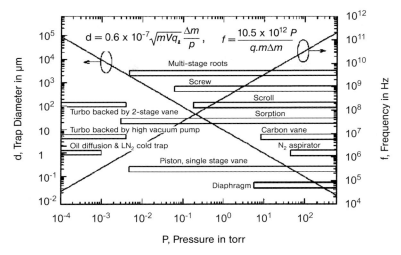

Fig. 9.3.5 Relationships between ITMS operating pressure, field frequency, trap radius and pump capabilities [5, 6].

The operation of mass spectrometers (MSs) has traditionally required a very low-pressure environment (typically $< 1 \times 10^{-4}$ torr), produced by bulky vacuum pumps, in order to yield meaningful results. Drawing on their unique experience in ITMS to develop submillimeter ion traps, the ORNL team has achieved a world-first in terms of demonstrating 1-mm-ID trap functionality above 1 torr, thus minimizing pumping requirements. Figure 9.3.4(b) shows the first fabrication of a 1000 × 1000 trap array of only 1 μm in diameter intended for operation above 100 torr. Their goal is to increase the MS working pressure to such levels as to make MS operation possible with simpler, low-power and more compact pumps (Fig. 9.3.5). This reduction to one-tenth normal atmospheric pressure can be produced with MEMS technology, enabling the integration of MS into a microscale gas analyzer. However, microfabricating ion trap arrays with trap diameters of < 1 μm will be a challenge.

Another detector selected for its compactness and its "orthogonality" with respect to GC and MS, is spectro-chemical or optical emission as generated, e.g., by micro discharge detectors (MDD). Figure 9.3.6 presents photos of such emission for both in CO_2 and in He (at left) and a cross section on an envisioned MDD structure, coupled into a linear-filter-based micro-spectrometer, which is compact but not efficient.

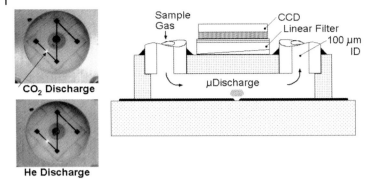

Fig. 9.3.6 Photo of a micro-discharge device (MDD) in operation with visible glow emission, compliments of Caviton, Inc. [9], and envisioned MDD assembly with spectrometer (right-hand side).

9.3.4
Sandia's Hydrogen-based Analyzer

Using hydrogen as a carrier gas after the air-sample collection and preconcentration, Sandia National Laboratory also uses GC as the fundamental separation technology, but is exploring construction of the GC channels using metal, in addition to silicon etching, and is developing element-specific detection technologies.

Simonson et al. have taken advantage of three properties of H_2 as carrier gas: distinctiveness, combustibility, and viscosity. Its distinctly low mass and high thermal conductivity enhance the signal-to-noise ratio of many detectors. The released energy from its combustion fuels the detectors, and heat from catalytic oxidation enables thermal ramping of the micro separation column without consuming additional battery power. Its low viscosity (~2× lower than air) enables high flow rates and allows the use of smaller dimensions in the design of the microfluidic channels of the analyzer. The challenge, however, is micro-generating the hydrogen in a controlled, reliable way.

Sandia's design thermally actuates an exothermic reaction of metal hydride with water to generate the hydrogen on board the microsystem. The pressure created by this reaction and the low viscosity of the hydrogen enable the high flow rate of the mobile phase, which, in turn, enables the desired 2–4-second chromatographic separation.

Simonson et al. are also pursuing an integrated, monolithic unit (Fig. 9.3.7), where the critical analytical components such as the injector and detector, along with the thermally isolated membranes, are embedded on the column microstructure to reduce dead volume and cold spots. Their design retains modularity (for easier customization to different compounds) in the preconcentrators and in the materials used for the detectors.

Like Honeywell, Sandia's team is fabricating flow channels using deep reactive ion etching (DRIE) in silicon but, unlike Honeywell, they are also pursuing

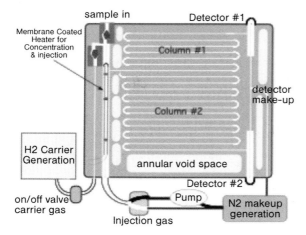

Fig. 9.3.7 MGA (micro gas analyzer) designed for sampling air and GC analysis after switch-over to H_2 [3].

the micro-manufacture of rectangular-cross section, high-aspect-ratio nickel channels using LIGA (Lithographie und Galvanoformung, or lithography and electroforming) technology. Specifically, they are integrating Louisiana State University's demonstrated fabrication of 50×500 µm, 2-m long nickel columns. They have developed new techniques for depositing stationary phases on both silicon and nickel, and have achieved over 20 000 theoretical plates (TP) per meter for high-resolution CG separations.

They have already designed adsorbing materials that efficiently collect chemical warfare agents (CWA) while excluding interferents, and the new hurdle will be to develop preconcentrators on a microscale that are efficient enough for ppt sensitivity. They have recently made an important stride towards that goal by demonstrating a gain in concentration of over 100× that of the original sample, making downstream detection more reliable and sensitive.

Chosen for their sensitivity and selectivity, thermionic ionization detectors (TIDs), like the nitrogen phosphorus detector (NPD), are the heart of Sandia's detection technology. The NPD is commonly used to sense organic compounds that contain nitrogen or phosphorus at a ratio of 10 000 : 1 selectivity over carbon. This makes it very selective to CWAs. The NPD contains a small rubidium (Rb) silicate bead that is heated to 600–800 °C by a fine platinum wire. This bead is the basis for the thermo-chemical operation of the detector. A sensitive electrometer circuit counts the negative ions produced by the bead. A typical circuit can detect 0.4 pg of nitrogen and 0.1 pg of phosphorus (0.1 pg in a 100-nL sensor dead-volume corresponds to 0.833 ppm in air or 12 ppm in H_2). Simonson et al. plan to microminiaturize this sensor to maximize ionization efficiency and mass sensitivity. The challenge will be to develop new catalytic materials for modular micro-detectors with element specificity, such as for organosulfur compounds, to further enhance the selectivity of the system.

9.3.5
Preliminary Results

Both MEMS gas analyzer systems described above have recently demonstrated preconcentration gains of over 100 after a sampling time of ≤ 4 s, and separation of eight analytes in less than 4 s. We will now describe some details of the measurements obtained with the PHASED MGA system.

9.3.5.1 Preconcentration

Figure 9.3.8 presents results obtained with a 40-stage PHASED chip, in which each stage measures just 5 mm in length along the 100 × 100 μm channel etched into a silicon wafer. Drawing a test mixture of hexane-in-air, the sampling time lasted 4 s, although the mass spectrometer (see experimental set-up in the insert) trace of mass 57 (one of the hexane fragments, represented by the black curve) shows that less than 200 ms were needed to saturate the 100-nm thick film of Nanoglass, while the desorption and concentration time was under 100 ms [peak marked C6]. So, in less than 300 ms the hexane concentration was boosted ~77-fold and generated an injection pulse of 3 ms in half-width ($W_{1/2}$), which is also shown at the bottom of Fig. 9.3.8, after correcting for the thermal conductivity detector (TCD) signals caused by the individual desorption gas flow pulses. The small but observable non-simultaneous desorption of water and hexane could be improved by raising the temperature of the desorption pulse. The achievable preconcentration gain increases with adsorption enthalpy and desorption temperature as computed and plotted in Fig. 9.3.9. Such gain also increases with film thickness, number of stages and decreasing adsorption temperature, although the plot only represents the performance for 600 nm,

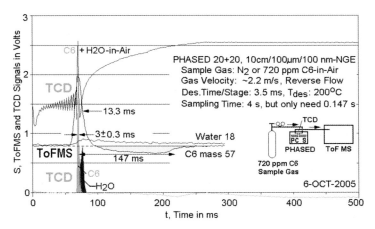

Fig. 9.3.8 PHASED MGA sample collection, preconcentration (PC) and separation, demonstrated with hexane and water vapors. Time to saturate PC: ~ 147 ms. PC gain: 1.70/(0.022 ± 50%) = 77 [13].

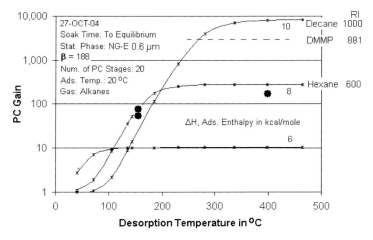

Fig. 9.3.9 Preconcentration of C_6–C_{10} alkanes on the 50 cm/100 × 100 μm SWNT microchannel of Fig. 9.3.3 [10].

20 stages and 20 °C, respectively. In summary, even these preliminary results show the unprecedented gain, speed and injection pulse width achievable with such MEMS preconcentrators.

9.3.5.2 Separation

In collaborating with LLNL, R. Synovec's team at the University of Washington demonstrated very repeatable separations of alkanes and other analytes with the SWNTs grown in LLNL's 50-cm channels, under both isothermal and temperature ramped (as high as 60 °C s^{-1}) conditions, with H_2 as carrier gas and ~3 ms injection pulses [7]. The four-compound test mixtures were separated within 1–4 s (Fig. 9.3.10 shows an example). The derived peak or channel capacities for separations under isothermal conditions of 90 and 140 °C, and ramped from 90 to 150 °C (Fig. 9.3.10), were 8.5, 11 and 16.5, respectively, when one counts from hexane (as a surrogate for the zero retention time, t_o) to an elution time of t_r = 4 s, using the measured t_r and $W_{1/2}$-values from the chromatograms. The exhibited peak tailing is thought to be caused by less-than-perfect operation of the 3.5-ms injection system, and some remaining SWNT non-uniformities, both of which are being refined. The computed GC separation in Fig. 9.3.11 illustrates what an ideal chromatogram would look like, without peak tailing.

For speed and resolution comparison, the nearest GC separation on carbon nanotubes known to the authors is the one by Saridara and Mitra at the N.J. Inst. Tech. [13], which took 60× longer, or 4 min, to achieve a separation of C_2–C_6 alkenes and a comparable peak capacity of 16. They used a 300-cm capillary column with 500-μm ID, temperature ramping of 50 °C min^{-1}, and a carrier gas velocity of ~85 cm s^{-1}. We will return later to the importance of peak capacity and its influence on FAR in the section on FAR (Section 9.3.5.4).

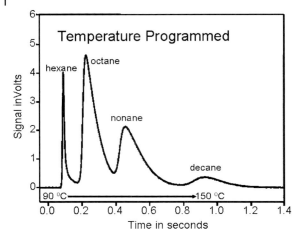

Fig. 9.3.10 Experimental separation of C_6–C_{10} alkanes on the 50 cm/100 × 100 μm microchannel of Fig. 9.3.2, coated on one side with SWNTs ~1 μm long. Achieved peak capacity: ~16 between t_o and 4 s [7].

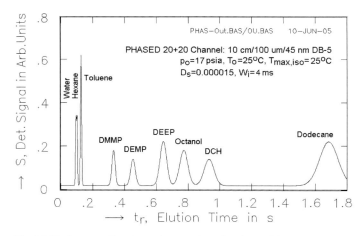

Fig. 9.3.11 Computed isothermal separation of analytes with a microchannel 10 cm/100 × 100 μm and 45 nm of PDMS (DB-5), with an air flow of 120 cm s^{-1}, achieving a peak capacity of > 20 between t_o and 1.7 s [13].

9.3.5.3 Detection

Use of the classical FID (hydrogen Flame Ionization Detector) in GC analysis would be quite inhibiting in a portable, compact device, with a reasonable operational life of 100–1000 h before its battery [14] and/or hydrogen supply [15] would need recharging or replacement. Therefore, alternate approaches are under development for the PHASED MGA, with preliminary results as follows:

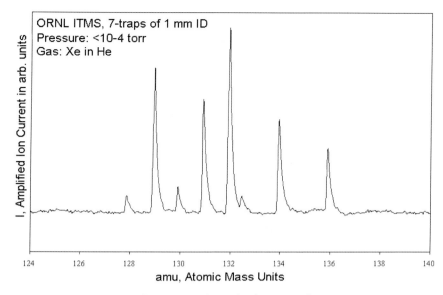

Fig. 9.3.12 Mass spectrum of Xe isotopes obtained with an array of seven 1-mm-traps of an ITMS below 10^{-4} torr. (Courtesy of W. B. Whitten [5].)

- TCD, thermal conductivity detector (Fig. 9.3.2c): Its response time is less than 1 ms (Fig. 9.3.8) and compatible with MGA requirements. However, it only provides one channel of information.
- ITMS, micro Ion Trap Mass Spectrometer (Fig. 9.3.4): To achieve a combination of high sensitivity (i.e., high S/N) and millisecond-level response time, the use of arrays has provided promising results. To date, mass spectra with near 1 amu mass resolution were obtained with 1-mm ID arrays (7 traps) and with 40-µm arrays (256 traps) [6]. Figure 9.3.12 shows the individual separation and detection of Xe isotopes. The development goal is to achieve such results at pressures near atmospheric, i.e., 6–7 orders of magnitude above the traditional level of 10^{-4} torr, and good progress is being made in that direction [5, 6]. Such a "detector," with a typical mass resolution of ≤ 1 amu, adds a new dimension to the traditional GC peak capacity, with its own large channel capacity of 200–300, corresponding to the mass range commonly used in environmental applications.
- MDD, micro discharge detector (Fig. 9.3.6): For air with and without "contamination" by trichloroethane (TCE), Fig. 9.3.13(a) shows exemplary emission spectra, with bands corresponding to nitrogen, and atomic and molecular chlorine, which results from the fragmentation of TCE in the discharge plasma. Visual isolation of the emission due to chlorine was achieved by forming the difference and ratio of spectrometer outputs between pure air and the sample with TCE, resulting in the spectral outputs at the middle and bottom of Fig. 9.3.13(a), respectively. Similar differentiation is achievable for other contaminants containing elements such as phosphorous, carbon, sulfur, halogens, metals [Fig. 9.3.13(b)]. Tell-tale

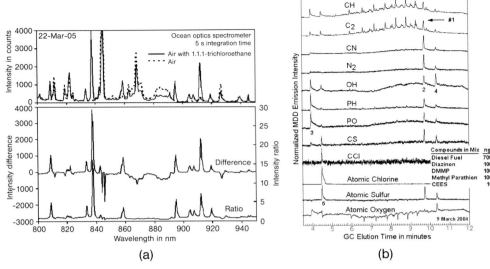

Fig. 9.3.13 (a) Recorded micro-discharge device (MDD) emission spectra in air, with and without the presence of ~1% of trichloroethane, showing Cl emission. (Courtesy of C. Herring et al. [11].)
(b) MDD outputs for a GC analysis of diesel fuel laced with pesticide simulants, for 12 wavelength bands. HP-GC, capillary of 1000 cm/100 μm ID/400 nm RTX1; OceanOptics spectrometer set at 300 ms point^{-1} and $\Delta\lambda \sim 1$ nm. (Courtesy of C. Herring et al. [11].)

detection and intensity of molecular "radical" emission, such as that of CH, CN and C2, are indicative of the presence and concentration of hydrocarbons, OH for water, PO, PH and OH for phosphonates, and CCl and CS for organic Cl and S compounds, so that, together with the total discharge emission and current or voltage, over 8–10 orthogonal measurement channels (OCC) may be obtained with MDDs.

Other detectors such as PID, CID and ECDs may successfully compete with TCD, but do not provide the additional dimensionality to boost MGA OCC as spectrometric "detectors" as much as MDD and ITMS, to help reduce FAR.

9.3.5.4 False Alarm Rate Metrics

The identification of analytes detected during environmental, industrial or security monitoring, or as part of medical breath analysis has been a traditional challenge to spectrometry, whether using optical, mass or gas-chromatographic type of analyses. Errors in such identification lead to undesirable false alarms, or to failure in detecting/identifying actually present noxious analytes, in other words, false positives and false negatives and their repercussions. The search continues for a method that can combine accuracy with practicality for gas analyzers, despite

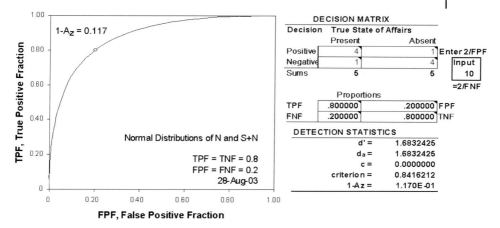

Fig. 9.3.14 Left: Receiver operating characteristic (ROC) for FPF = 1/5.
Right: Decision matrix and detection statistics for this ROC curve.

the routine use of Receiver Operating Characteristic (ROC) curves, first for radar-identification of aircraft type and since then to quantify the certainty of medical diagnostics and explosives detection. The discussion below represents an update of work in progress.

To provide a method for the evaluation, or at least comparison between two analyzers' probabilities of false alarms, a model for the prediction of FAR was devised, which is based on relating the total number of the analyzer's orthogonal measurement channels and the analyzer's signal-to-noise ratio (R_{NS}) ratio to the probability of analysis errors obtained under *specific test conditions*, which also corresponds to errors predicted by ROC curves [8], as illustrated by the example in Fig. 9.3.14. The area above the ROC curve, 1-Az, represents the total instrumental error of the involved analyzer, and may be plotted by simply using the result of tests yielding values of FPF (False Positive Fraction) and FNF [16]. One set of specific test conditions assumed in early model computations consisted of a simple symmetry for FPF and FNF, leading to the symmetrical Decision Matrix and Proportions shown in Fig. 9.3.14.

Starting with a simple analyzer composed of an array of polymer film sensors, it is agreed among some researchers that, despite the inclusion of 10–20 different polymer films in the array, its maximum OCC, or n, is 5 to 6. Next, a microGC might be able to achieve a peak capacity of $n = 50$, if the time and energy constraints in this MGA Program were relaxed.

Similar arguments can be made for the remaining analyzers in Table 9.3.1, taken individually ($Y = 1$), i.e., without the benefit of networking. Networking just 10 MGAs may boost the OCC by a factor of 100 because of the intelligence derived above mere redundancy, such as associated with wind direction and speed. Such considerations generated the three sets of curves of Fig. 9.3.15.

The signal/noise ratio, R_{SN}, was set to $R_{SN} = 10$ for all the listed functions:

Table 9.3.1 Estimated, relative false positive fraction (FPF) and area above the ROC curve for analyzer examples. (Test assumption: Target analytes are actually present 50% of the time, that TPF ~ TNF; FPF = P_{fp}.)

Analyzer Example	Y	$n^{a)}$	$1/P_{fp}^{b)}$	TPF	d'	1-Az (ROC)
Array of different polymers, CID (e.g. nanocant.)	1	5.5	5	0.800000	1.6832	1.17×10^{-1}
MGC with non-selective detector (TCD or FID)	1	50	22	0.954545	3.3812	8.40×10^{-3}
NDIR spectral correlation (e.g. Polychromator)	1	100	37	0.972973	3.8528	3.22×10^{-3}
Std. GC with non-selective detector (TCD or FID)	1	200	63	0.984127	4.2952	1.19×10^{-3}
MGC with one multi-channel det. (MDD), 50×10	1	500	130	0.992308	4.8464	3.05×10^{-4}
MGC with 2 multi-channel det. (MDD and FD), $50 \times 10 \times 4$		2000	392	0.997451	5.6015	3.74×10^{-5}
MGC-MGC-MDD combination, $50 \times 10 \times 10$	1	5000	815	0.998774	6.0583	9.19×10^{-6}
Std. GC with 2 multi-channel det. (MDD and FD), 200×40		8000	1187	0.999158	6.2818	4.46×10^{-6}
MGC-MS combination, 50×300	1	15 000	1962	0.999490	6.5705	1.69×10^{-6}
Std. GC-MS "Gold Std." combination, 200×300	1	60 000	5946	0.999832	7.1712	1.98×10^{-7}
MGC-MGC-CID-PID-MDID-MDD, $50 \times 10 \times 6 \times 4 \times 10$		120 000	10 352	0.999903	7.4552	6.77×10^{-8}
MGC-MGC-CID-MDID-ITMS, $50 \times 6 \times 2 \times 00$	1	180 000	14 318	0.999930	7.6182	3.59×10^{-8}
MGC-MGC-CID-MDID-ITMS, $50 \times 6 \times 2 \times 300$	100	180 000	18 000 000	1.000000	~11.7	~3.0×10^{-14}
MGC-MGC-MDD combination, $50 \times 10 \times 10$	100	$5000^{b)}$	500 000	0.999998	9.2387	3.24×10^{-11}

a) Explanation of OCCs for:
n (TCD, FID, MDID) = 1, n (FD) = 4, n (CID) = 6, n (MDD, 2nd-GC) = 10, n (MGC) = 50, n (GC) = 200, n (ITMS, MS) = 300;
TCD = Thermal Conductivity D., MDID = MicroDischarge Ionization D., FD = Fluorescence D., CID = Chemi-Impedance D. (polymer film), MDOD = MicroDischarge Optical D., ITMS = Ion Trap MS, MS = Mass Spectrometer.
b) See equation for P_{fp1} in Fig. 9.3.15.

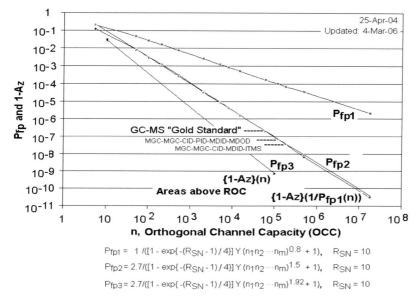

Fig. 9.3.15 Probability of false positives and ROC output vs. orthogonal channel capacity (OCC).

- The top curve for $P_{fp1}(n)$ represents the 3rd and 4th columns of Table 9.3.1 [8].
- The middle two (closely overlapping) curves represent $P_{fp2}(n)$ and 1-Az, where the former was made to closely match 1-Az when using $1/P_{fp1}(n)$ as input to the ROC curve calculation.
- The bottom curve $P_{fp2}(n)$ is matched to 1-Az when using n as input to the ROC curve calculation.

Note that only two parameters of the P_{fp} curves needed adjustment to achieve the match with the 1-Az curves obtained under the symmetrical test conditions mentioned above.

The implications for a development project would be that, to meet exemplary $\leq 1/200\,000$ and $\leq 1/10\,000\,000$ FAR goals, assuming that FAR = 1-Az, we use either of:

- FAR = 1-Az = P_{fp2} and strive to develop MGAs with n = OCC = 1/FPF $\geq$ 7400 or 93 000, respectively.
- FAR = 1-Az = P_{fp3} and strive to develop MGAs with n = OCC = 1/FPF $\geq$ 1120 or 8500, respectively, using the Decision Matrix of Fig. 9.3.14.

The above method of determining FAR, if confirmed and validated experimentally, would save 20 000–200 000 tests for each analyzer but, initially, generated significant skepticism. Since then, related work by NRC [17] came to light, thanks to inputs by J. Adams [19], in which the NRC committee tries to relate information

Table 9.3.2 Comparison of the informing power of IMS and MS analytical techniques.

Technique	P_{inf} informing power bits	Increase in P_{inf} factor[a]
IMS	1×10^3	1
QMS	1.2×10^4	12
Capillary GC-QMS	6.6×10^6	6 600
QMS/QMS	1.2×10^7	12 000
Capillary GC-QMS/QMS	6.6×10^9	6 600 000

a) Informing power of IMS estimated by the committee [17].
   Remaining values in columns 2 and 3 are from Yost and Fetterolf, 1984 [18].

content ("Informing Power," P_{inf}) of IMS (ion mobility spectroscopy) and MS analyzers to Transportation Security. Table 9.3.2 shows some of their results [17]. But comparing P_{inf} to the n = OCC or $1/P_{fp}$ values in Table 9.3.1 shows the values to be about 100× higher. The NRC committee assigned a P_{inf} of 1000 to IMS, based on an "information theory" method by Yost and Fetterolf [18], which includes resolvable amplitude levels. Without such levels, channel capacities of IMS and DMS (differential mobility spectroscopy) analyzers were determined to be in the range 14–20 [20], which would be a factor of 50–70 lower than 1000 and would bring P_{inf} much closer to the n values in Table 9.3.1 for the GC/MS analyzer. One may conclude that the idea of relating OCC to FAR may not be so far fetched after all.

However, several issues still need to be resolved, such as:

- The extent to which the orthogonal channel capacity (OCC) is available for each analysis, as the full capacity cannot always be harnessed, e.g., with an array of polymer film sensors where some polymers may not have any interaction with an analyte, or with some mass values of an analyte mixture not being represented in a mass spectrum. Appropriate OCC "discounts" may need to be derived.
- Influence of R_{SN} on 1-Az, other than the heuristic approach discussed above.
- Determine which OCC–ROC curve and FAR relation would best be supported by statistics or information theory.

9.3.6
Conclusions

Although conventional GC/MS systems may not compete with the speed and sensitivity of these micro gas analyzers, the former's strength lies in their ability to resolve smaller analyte differences, which would be highly desirable for FAR reduction. The challenge of this research is to achieve a similar resolution using fast, more compact and multidimensional analysis approaches, such as µGC-µITMS and µGC-µITMS-MDD systems.

The MGA energy per analysis requirements may threaten the compactness of a battery-operated unit, even if the ambitious goal of 1 J per analysis is met, because a lithium battery pack capable of 24 h of analyses repeated every 4 s would occupy a volume of 50.7 cm^3 equivalent to a 3.7-cm cube [14], but would still be acceptable for hand-held analyzers. A volumetric reduction by factor of 4–6 may be achievable by the use of metal hydride sources of H_2 for fuel cells [15].

In any case, the size constraints for this DARPA commissioned research far exceed those of NeSSI's (New Sampling and Sensing Initiative) modular and standardized micro-analytic elements, so that these sensors will easily fit on a NeSSI substrate for even further application possibilities. The drastic reduction in size, analysis time, cost, and energy consumption, ease of implementation and maintenance, and compliance with NeSSI standards make this technology highly marketable to the civilian sector.

Acknowledgments

The work described above was funded by DARPA-MTO, DOE-ITP and Honeywell Labs, and led to the publications listed under Refs. [1–12], from which the material was excerpted, with permission from the authors. For their friendship and collaboration, and for the able help provided by tech writer Marc G. Bonne, the authors are very grateful indeed.

References

1 D. Polla, C. Nguyen, Vision statement, DARPA-MTO webpage, **2004**, at http://www.darpa.mil/mto/mga/vision/index.html

2 U.Bonne, R. E. Higashi et al., PHASED Micro Gas Analyzer, DARPA-MTO webpage, **2004**, at http://www.darpa.mil/mto/mga/summaries/2004_summaries/honeywell.html

3 J. Simonson et al., Micro-scale gas separation analyzer, DARPA-MTO webpage, **2004**, at http://www.darpa.mil/mto/mga/summaries/2004_summaries/snl.html

4 O. Bakajin, Micro-separation and pre-concentration columns based on carbon nanotube arrays, DARPA-MTO webpage, **2004**, at http://www.darpa.mil/mto/mga/summaries/2004_summaries/llnl.html

5 W. B. Whitten, Micro ion trap mass spectrometer, DARPA-MTO, **2004** webpage at http://www.darpa.mil/mto/mga/summaries/2004_summaries/oakridge.html

6 S. Pau, Y. L. Low, C. S. Pai, J. Moxom, P. T. A. Reily, W. B. Whitten, J. M. Ramsey, Microfabricated quadrupole ion trap for mass spectrometer applications, *Phys. Rev. Lett.* **2006**, 96, 120801.

7 M. Stadermann, A. D. McBrady, B. Dick, V. Reid, A. Noy, R. E. Synovec, O. Bakajin, Fast gas chromatography on single-wall carbon nanotube stationary phases in microfabricated channels, submitted for publication to *Anal.Chem.* **2006**, 78, 5639.

8 U. Bonne (Honeywell Labs), G. Eden (U.Illinois), G. Frye-Mason (Nomadics), R. Sacks (U.Michigan), R. Synovec (U.Washington), Micro gas chromatography tradeoff study. Final report, to DARPA/AFRC/UTC No.: AFRL-PR-WP-TR-2004-2060; Contract No. 03-S530-0013-01-C1, Plymouth, MN,

1 Dec. **2003**, http://www.stormingmedia.us/89/8975/A897524.html
9 U. Bonne, R. Higashi, T. Marta, F. Nusseibeh, T. Rezachek (Honeywell Labs), and C. Herring, D. Kellner, K. Kunze and M. Castelein (Caviton, Inc), MicroGas analyzer for NeSSI and DHS: Measurements and simulations, PittCon 2006, Orlando, FL, 13–16 March **2006**, Paper # 2020-6.
10 U. Bonne, R. Higashi, K. Johnson, N. Iwamoto, R. Sacks, R. Synovec, Stationary phase films for micro-analytical measurements, PittCon 2005, Orlando, FL, 27 Feb.–4 Mar. **2005**, Paper Number: 420-1
11 U. Bonne, K. Johnson, R. Higashi, F. Nusseibeh, K. Newstrom-Peitso, T. Marta, E. Cabuz, and T.-Y. Wang, J. Mosher, N. Iwamoto, N. Lytle (Honeywell International); R. Synovec, V. Reid, B. Prazen, G. Gross, and D. Veltkamp (Chemistry Department/CPAC, University of Washington, Seattle, WA); C. Herring, K. Kunze and D. Kellner (Caviton, Inc., Champaign, IL), PHASED*: Development of a µGC-on-a-chip with interface to NeSSI, Sensors Expo, Chicago, IL, 6–9 June **2005**, Session on 2005 ITP Sensors & Automation Annual Portfolio Review, http://www.eere.energy.gov/industry/sensors_automation/pdfs/meetings/0605/bonne_0605.pdf
12 N. Iwamoto, U. Bonne, Molecular modeling of analyte adsorption on MEMS GC stationary phases, Eurosime, Como, Italy, 24–26 April **2006**; paper #293.
13 C. Saridara, S. Mitra, Chromatography of self-assembled carbon nanotubes, *Anal. Chem.*, **2005**, 77, 7094; achieved separation of C2–C6 alkenes in under 4 min at ~85 cm s^{-1} in a 300-cm column of 500-µm ID, temp ramping at 50 °C min^{-1}.
14 PowerStream, Lir2450 Lithium-Ion Battery, 400 Wh L^{-1}, 1166 J, 3.6 V, 90 mAh, for 1.3 hours of analyses at 1 J each every 4 s. Vol.: 2.74 cm^3, 24.5 mm OD × 5.4 mm height, $1.36/ea (for 100–1000 units); http://www.powerstream.com/LiCoin.htm
15 A. Wood, S. Eickhoff, Active micro power generator "AMPGen", DARPA-MTO webpage, **2004**, http://www.darpa.mil/mto/mpg/summaries/2004_summaries/honeywell.html
16 D. E. Berger, M. Healy (Claremont Graduate University, Claremont, CA), private communications; http://wise.cgu.edu/sdt/sdt.html computes and displays ROC curves on line [8].
17 Committee on Assessment of Security Technologies for Transportation, *Opportunities to Improve Airport Passenger Screening with Mass Spectrometry*, National Research Council, Report, 56 pp (**2004**), http://www.nap.edu/catalog/10996.html, as part of *Assessment of Technologies Deployed to Improve Aviation Security: 2nd Report, Progress Towards Objectives*, NRC, Washington, DC, **2002**.
18 R. A. Yost, D. D. Fetterolf, Added resolution elements for greater informing power in tandem mass spectrometry, *Int. J. Mass Spectrom. Ion Processes*, **1984**, 62, 33–49.
19 J. Adams, Nevada Nanotech Systems, Inc., Reno, NV, jdadams@nevadanano.com, private communication, 530-368-2870, http://www.nevadanano.com, April **2006**.
20 E. G. A. Eiceman, E. V. Krylov, N. S. Krylova (Dept. Chem. & Biochem., NM State U), E. G. Nazarov, R. A. Miller (Sionex, Inc.), Separation of ions from explosives in differential mobility spectrometry by vapor-modified drift gas, *Anal. Chem.* **2004**, 76, 4937.

9.4
Nuclear Magnetic Resonance Spectroscopy

Michael McCarthy

Nuclear magnetic resonance (NMR) spectroscopy and magnetic resonance imaging (MRI) have many advantageous properties for use in process analysis, process control and high throughput systems. The measurement is noncontacting and noninvasive. The magnetic resonance signal is directly proportional to the number of nuclei in the sampling volume and the response is linear from the detection limits (~100 ppt) to 100% [1]. The signal can be made to be very specific by obtaining information from only one type of nucleus, e.g., ^{13}C, ^{31}P, ^{23}Na, ^{19}F or ^{1}H, or a combination of nuclei. The specific information available includes concentration, self-diffusion coefficients, droplet size distribution, shear viscosity or co-polymer blend ratios. These properties enable NMR to be applied to a wide range of materials from polymers to consumer products to foods for wide range of measurements, including reaction rates, molecular weight distribution and phase distribution. Despite the advantages of NMR, application in process analysis and control has been very limited in comparison with other analytical methodologies, primarily a result of difficulties in implementing NMR in an in-line or on-line mode.

Barriers to more widespread use of NMR and MRI include cost, complexity of equipment, equipment size, and difficulty incorporating the equipment into a high throughput system. The major components of an NMR/MRI system are a computer for control/data analysis, a spectrometer, a radiofrequency coil, a magnet and a shim/gradient system. Except for the computer system the cost of each of these components decreases as the size of the sample volume is decreased. This suggests a significant opportunity to use micro-machining and micro-fabrication techniques to achieve a relatively low cost spectrometer for application to combinatorial and microreactor systems as well as process control applications. Figure 9.4.1 shows a micro-fabricated radiofrequency coil. The coil consists of five turns with an average diameter of 3.5 mm, substrate thickness is 3.5 mm, trace width is 120 µm, pitch is 160 µm, and the trace thickness is 15 µm. Shown passing through the coil is Teflon tubing with an inner diameter of 1.02×10^{-3} m. The tubing is used for sample delivery to the coil. This radiofrequency coil is intended for both concentration measurements and viscosity measurements. The coil is approximately one order of magnitude smaller than standard spectroscopy/imaging coils.

The performance of the spectrometer is impacted by changes in the dimensions of the radiofrequency coil. The signal-to-noise ratio (S/N) of an NMR experiment is increased by decreasing the size of the radiofrequency coil. Decreasing the size of the receiver coil results in increasing the S/N ratio since this ratio is proportional to the inverse of the coil diameter (for saddle and solenoid geometries). This functionality holds until the diameter of the coil is approximately 100 µm and

Fig. 9.4.1 A five-turn copper microcoil mounted on a borosilicate glass with sample handling tubing inserted below the glass.

then changes to an inverse square root dependence on the coil diameter [2]. These predicted enhancements to sensitivity have been demonstrated by Schlotterbeck et al. [3] using aqueous sucrose solutions at 600 MHz proton frequency. The mass sensitivity was five times larger for a 1 mm diameter coil than for a 5 mm diameter coil [3]. When comparing the sensitivity of different coils it is useful to recall both mass sensitivity and concentration sensitivity. When the total mass of a sample is limited and high concentrations are possible, decreasing the coil diameter is helpful. However, if only low concentrations are possible and total mass of the solute is not limited, larger coils yield higher S/N ratios. At intermediate solubility and mass availability the optimum size of the coil will be determined by considering both of these factors [4].

Reviews of microcoil applications to spectroscopy and imaging have recently been published by Webb and by van Bentum et al. [4, 5]. Applications center around two different approaches of detecting the NMR signal, induction detection and force detection [6]. Force detection works well for pure fluids and has theoretical signal-to-noise advantages over induction detected NMR as the sample size decreases below 500 µm in diameter. This advantage in signal-to-noise has yet to be demonstrated experimentally. This technique is applicable to pure liquid streams (or streams with very small particle sizes) since the dimensions of the channels would be on the order tens of µm. Conventional induction detection techniques have proven very successful for micro-NMR systems. Boero et al. [7, 8] have constructed a partially integrated magnetometer based on continuous-wave proton NMR. This magnetometer consists of a sensor using planar coils of diameter down to 2 mm. This coil is deposited on glass and is integrated with a CMOS circuit on silicon. The CMOS circuit is used for signal detection and amplification. This device was demonstrated using proton detection. More practical examples have been demonstrated, including measurement of reaction kinetics, mixing and rheological characterization.

Reaction kinetics have been measured by monitoring imine formation from benzaldehyde and aniline [9]. A microreactor was constructed from Borofloat with a channel depth of 450 μm by 500 μm wide. The channel cross-section had the shape of a rounded letter V. The microcoil used to detect the proton spectra of the reactants and products was patterned/deposited on top of the reactor. The microcoil was formed from 24 windings with 20 μm separation. The inner diameter of the microcoil was 200 μm. The reactor-coil system was placed in a traditional laboratory based permanent magnet (1.4 Tesla). Good signal-to-noise was achieved by using high reactant concentrations, 4.95 M for both aniline and benzaldehyde in deuterated nitromethane. Proton spectra were well resolved and rate constants readily calculated from the spectra. A rate constant of 6.6×10^{-2} M^{-1} min^{-1} was calculated by fitting a second-order rate expression to the data.

MRI has also been applied to study micro-scale system to characterize both flow [10] and fluid rheology [11]. The flow and mixing in a commercial interdigital micromixer were characterized by building a homemade radiofrequency coil around the device. Non-uniformities in the flow and clogging in flow channels were observed in the device. Imaging of the velocity profiles in small channels may also be used to quantify the shear viscosity of the fluid as a function of shear rate. Goloshevsky et al. [11, 12] have demonstrated flow imaging and viscosity measurements using radiofrequency microcoils and micro-gradient coils. Figure 9.4.2 shows a combined radiofrequency and micro-gradient set. The set uses only two gradients, the Y and the Z, since only unidirectional flow was measured.

The MRI-based viscosity measurement relies on local calculations of the velocity gradient based on using MRI velocity profiles to provide a wide range of shear viscosity–shear rate data. Two observations are used to characterize fully developed and steady laminar flow in a MRI-based viscometer: the velocity profile and the pressure drop per unit tube length measurements. Viscosity measurements were acquired on strawberry milk samples using HPLC tubing for the sample channel.

Fig. 9.4.2 Microfabricated radiofrequency and gradient set.

The tubing inserted into the middle of a six-turn microfabricated Helmholtz spiral coil [11] was used as both a transmitter and receiver. The radiofrequency coil was fabricated on borosilicate glass substrate using copper metallization for magnetic susceptibility matching. The average diameter of the coil was 7 mm with a 3.5 mm separation between the opposing coils of the Helmholtz to optimize the field homogeneity over the sample volume. Figure 9.4.3 shows a flow image of the strawberry milk. The profile is well resolved and the fluid is in laminar flow. The derived rheogram for this fluid is also shown in Fig. 9.4.3. The range of shear rates covered by the miniaturized radiofrequency coil is narrow compared with laboratory rheological instruments as a result of the limitations in fluid residence time.

The volume of the sample in the coil is small and hence the residence time of the fluid becomes important to the range of shear rates that can be measured. As the volumetric flow-rate increases for the fluid the time before the fluid begins exiting the active region of the radiofrequency coil decreases. Thus, the residence time of the fluid must be less than or equal to the time required to measure one phase encoding step. The time to complete the measurement in one phase encode step is the time from the beginning of the first radiofrequency pulse to the end of data acquisition for that step. The reduced shear rate range for the MRI based measurements is a direct result of the small volume in the system. For the measurements reported by Goloshevsky et al. [11, 12] the time to complete one phase encode step is approximately 200 ms. Since the active volume in the coil is approximately 3.5×10^{-3} m in length the maximum velocity that can be

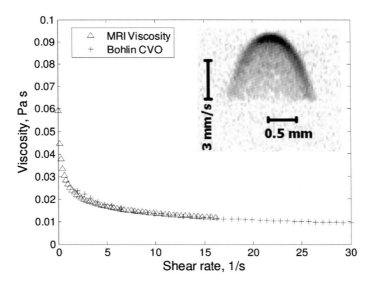

Fig. 9.4.3 Viscosity versus shear rate from both MRI data and a laboratory research grade Bohlin CVO rheometer for a strawberry milk sample. Insert: MRI velocity profile used for the viscosity calculations.

measured is 0.0175 m s^{-1} (3.5×10^{-3} m/0.2 s). This maximum velocity then limits the maximum shear rate observable in the system.

In summary, miniaturized NMR spectrometers have been demonstrated to be useful for concentration, flow and viscosity measurements on materials in microchannels and should be useful detectors in high throughput systems.

References

1. T. W. Skloss, A. J. Kim, J. F. Haw, High resolution NMR process analyzer for oxygenates in gasoline, *Anal. Chem.* **1994**, 66(4), 536–542.
2. T. L. Peck, R. L. Magin, P. C. Lauterbur, *J. Magn. Reson. Ser. B* **1995**, 108, 114–124.
3. G. Schlotterbeck, A. Ross, R. Hochstrasser, H. Senn, T. Kuhn, D. Marek, O. Schett, *Anal. Chem.* **2002**, 74, 4464–4471.
4. A. G. Webb, Microcoil nuclear magnetic resonance spectroscopy, *J. Pharm. Med. Anal.* **2005**, 38, 892–903.
5. P. J. M. van Bentum, J. W. G. Janssen, A. P. M. Kentgens, Towards nuclear magnetic resonance µ-spectroscopy and µ-imaging, *The Analyst* **2004**, 129, 793–803.
6. L. A. Madsen, M. Leskowitz, D. P. Weitekamp, Observation of force-detected nuclear magnetic resonance in a homogeneous field, *Proc. Natl. Acad. Sci. U.S. A.* **2004**, 101(35), 12804–12808.
7. G. Boero, C. de Raad Iseli, P. A. Besse, R. S. Popovic, An NMR magnetometer with planar microcoils and integrated electronics for signal detection and amplification, *Sens. Actuators A* **1998**, 67, 18–23.
8. G. Boero, P. A. Besse, R. S. Popovic, Hall detection of magnetic resonance, *Appl. Phys. Lett.* **2001**, 79, 1498–1500.
9. H. Wensink, F. Benito-Lopez, D. C. Hermes, W. Verboom, H. J. G. E. Gardeniers, D. N. Reinhoudt, A. van den Berg, Measuring reaction kinetics in lab-on-a-chip by microcoil NMR, *Lab Chip* **2005**, 5, 280–284.
10. S. Ahola, F. Casanova, J. Perlo, K. Munnemann, B. Blumich, S. Stapf, Monitoring of fluid motion in a micromixer by dynamic NMR microscopy, *Lab on a Chip* **2006**, 6(1), 90–95.
11. A. G. Goloshevsky, J. H. Walton, M. V. Shutov, J. S. De Ropp, S. D. Collins, M. J. McCarthy, Nuclear magnetic resonance imaging for viscosity measurements of non-Newtonian fluids using a miniaturized RF coil, *Meas. Sci. Technol.* **2005**, 16, 513–518.
12. A. G. Goloshevsky, J. H. Walton, M. V. Shutov, J. S. De Ropp, S. D. Collins, M. J. McCarthy, Integration of biaxial planar gradient coils and an RF microcoil for NMR flow imaging, *Meas. Sci. Technol.* **2005**, 16, 505–512.

9.5
Surface Plasmon Resonance (SPR) Sensors

Clement E. Furlong, Timothy Chinowsky, and Scott Soelberg

9.5.1
Introduction

Recent advances in the miniaturization and manufacturing of integrated SPR "chips" have provided the capability for developing miniaturized sensors capable of real-time monitoring of many different product streams [1]. In the simplest implementation, the chips can be used to provide very sensitive monitoring of refractive index (RI) in real-time. Attachment of receptors, antibodies or analytes to the sensor surfaces provides the capability of monitoring presence of product or contaminants in production streams either in or near real-time [2]. Several examples of these applications are provided below, along with a description of the current University of Washington sensor system capable of analyzing up to 24 separate streams simultaneously or up to 24 different analytes in a single stream simultaneously in real-time. These examples are not an exhaustive list of the different capabilities of the technology, but provide some examples that may be useful to the reader.

9.5.2
Monitoring Refractive Index, Fundamentals of SPR

Figure 9.5.1 shows the fundamental principle of SPR sensing based on the Kretschmann design [3]. The intensity of transverse magnetic polarized light reflected off a thin layer of gold on the surface of a prism exhibits a dependence on the angle of incidence and wavelength of the incident light. Plotting the intensity of reflection against the angle of reflection produces SPR curves or profiles (middle section of Fig. 9.5.1). Minima in the SPR curves are observed when the frequency and momentum of the incident light are matched with that of the surface plasmons (free electrons on the metal surface that undergo resonant oscillation with the illuminating light at angles above the critical angle). The angle at which the

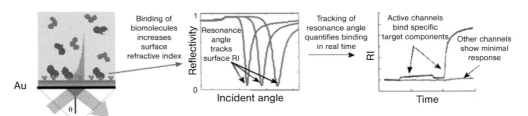

Fig. 9.5.1 Basic principle of SPR sensing.

minimum of reflection occurs depends on the RI of the medium on the outside of the gold layer. Because biological substances typically have a much higher RI than water, the presence of tiny quantities of such substances on the surface can be detected and quantified in real-time. To make the sensor specific for a particular biological analyte, the surface is treated with chemistry such that the sensor surface will tend to bind the selected analyte and resist binding of other substances.

9.5.3
Instrumentation

9.5.3.1 Spreeta Sensor Elements

Miniaturized and high-performance instruments for SPR-based measurements may be readily constructed using Spreeta sensing chips (Fig. 9.5.2) developed by Texas Instruments in collaboration with the University of Washington [2, 4–7]. These chips consist of a plastic prism molded onto a small printed circuit board (Fig. 9.5.2B). Light emitted by the sensor's light emitting diodes (LEDs) strikes the gold-coated SPR surface, reflects from a mirror on top of the sensor, and strikes a linear diode array. Because the incident angle varies with position along the diode array, reading the array gives a measurement of reflectivity as a function of angle. This reflectivity measurement may then be analyzed to obtain a measurement of biomolecular binding in real-time.

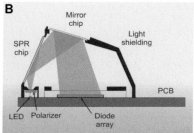

Fig. 9.5.2 Photograph and schematic of Texas Instruments' Spreeta sensing chip.

Spreeta devices exhibit very low noise when used in conjunction with optimized readout hardware and data analysis techniques [7]. Because the Spreeta is extremely compact, robust in construction, and commercially available in large quantities for low cost, the Spreeta sensors provide an ideal technology for the development of low-cost distributed instrumentation.

9.5.4
Automated Hand-held 24-Channel SPR Instrument with Modular Fluidics

Figure 9.5.3 shows a 24-channel Spreeta-based instrument constructed by the University of Washington. Instrument functions are divided between an electronics unit (in the base of the clamshell case) and a fluidics assembly located in the lid. The fluidics assembly is built around eight three-channel Spreeta devices fastened into a flowcell fabricated from cast silicone. Individual Spreeta sensors (Fig. 9.5.4) clip into this assembly, sealing the sensor surface onto the silicone flowcell. The flowcell is held in an aluminum backer that serves to equilibrate the sensors at a constant temperature, equilibrate the fluid stream to constant temperature, and mate with a thermoelectric heater/cooler attached to a heat sink extending from the lid of the instrument. Sensors are exposed to samples using a flow injection system coupled to the sensor flowcell. A sample injection port leads through a three-way valve into a sample loop backed with a pressurized buffer reservoir. The injection valve switches between the injection port and the flowcell input to control flow of sample over the sensors path. Additional valves allow for automated purging of the injection loop and flow of auxiliary reagents across the sensors. All fluidics components are mounted on a printed circuit board held in the lid of the sensor unit.

The base of the instrument contains electronics for (a) generation of signals necessary for driving the Spreeta devices, including stable powering of the sensors LEDs; (b) rapid readout and processing of data from the Spreeta devices; (c) control of fluidics and temperature stabilization; and (d) a touch screen user interface. The instrument reports RI values for each channel every second, with a noise level less than 10^{-6} RI units. The sensor surface for each of the 24 independently monitored channels may be functionalized as desired for specific applications of interest.

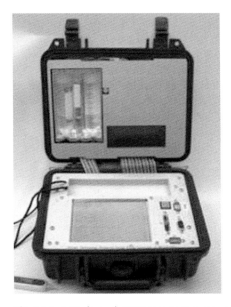

Fig. 9.5.3 A 24-channel SPR instrument constructed by the University of Washington.

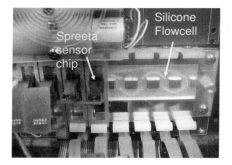

Fig. 9.5.4 Clip-in flowcell for the UW SPR instrument.

9.5.5
Examples of Specific Applications

9.5.5.1 Analysis of Small Organics

The direct interaction of small organic molecules with receptors immobilized on the sensor surface is difficult to detect. Two approaches may be used to detect or quantify the presence of small analytes in a process stream, displacement assays or competition assays.

Displacement assays can be designed in at least two different ways. (a) Analyte may be immobilized on the sensor surface and the sensor loaded with anti-analyte

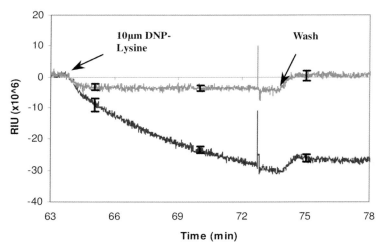

Fig. 9.5.5 Detection of the small molecule dinitrophenol-lysine (MW 349) with a displacement assay. In this type of assay a negative signal indicates a positive detection. The lower channel surface is the DNP–BSA sensor with anti-DNP antibodies bound; the upper channel is the reference sensor with BSA only. Each line represents an average of three channels with error bars every 5 min. The first arrow indicates when the sample was injected. Wash indicates a return to PBS-Tween buffer alone.

antibodies. These are displaced at rates proportional to analyte concentration. (b) An analyte haptenized carrier (e.g., large protein) can be bound to surface-immobilized antibodies. Rates of displacement of the haptenized carrier are easily observed and are proportional to analyte concentration. The displacement assay format is convenient in that special flow cells can be designed that separate the process stream from the sensing components with a semipermeable membrane (Fig. 9.5.5).

The most common approach for running competition assays is to inhibit the rate of binding of analyte-specific antibodies to surface immobilized analyte. The inhibition of antibody binding is proportional to the concentration of analyte present in the process stream. One main advantage of the competition assay is high sensitivity. If antibodies have a very high affinity for the analyte, very low levels of antibodies can be used in the assay. Binding, in turn, is inhibited by very low levels of analyte in the process stream. It is also possible to separate the process stream from the analytical system with the design of appropriate flow cells.

9.5.5.2 Analysis of Larger Analytes such as Production Proteins or Contaminants

Larger analytes are readily detected by direct interaction with antibodies immobilized on the sensor surface (Fig. 9.5.6). Sensitivities for this type of assay are typically in the high nanomolar to high picomolar range. The sensitivity and specificity can be increased simultaneously with the use of secondary antibodies that are specific for an epitope other than the one bound by the surface immobilized antibodies. If very high sensitivity is required, several stages of amplification can be carried out, allowing for detection of femtomolar levels of protein. For example, mouse monoclonal capture antibodies can be immobilized on the sensor surface.

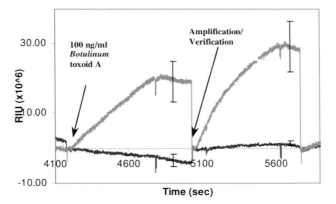

Fig. 9.5.6 Direct detection and amplification of the protein toxin *Clostridium botulinum* toxin A (toxoid). The upper line is an average of three channels with anti-*botulinum* antibody on the surface. The lower line (reference) is an average of three reference channels. Arrows indicate sample (100 ng mL^{-1} *botulinum* toxoid A) and amplifier (anti-*botulinum* antibodies) injection.

Binding of analyte allows for the binding of a number of rabbit polyclonal anti-analyte antibodies. Even higher levels of goat anti rabbit antibodies can be bound to the bound rabbit antibodies. A further step of amplification can be achieved with donkey anti-goat antibodies.

9.5.5.3 Monitoring of Larger Analytes such as Viruses, Whole Cells or Spores

Viruses, whole cells and spores can also be detected by direct interaction with antibodies immobilized on the sensor surface (Fig. 9.5.7). If a genus specific antibody is used as the capture antibody, species-specific antibodies can be used to speciate the captured analyte.

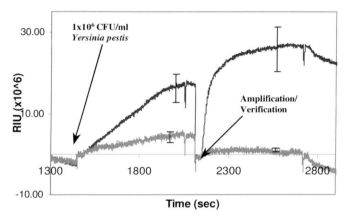

Fig. 9.5.7 Direct detection and amplification of the bacterium *Yersinia pestis* (non-pathogenic strain, irradiated) at 1×10^6 colony forming units (CFU) per mL. The upper line is an average of three sensors with an anti-*Y. pestis* antibody surface. The lower line is an average of three channels with a reference antibody on the surface. Arrows indicate when sample is injected and when the amplifier was injected into the 24-channel box.

References

1. J. Melendez, R. Carr, D. Bartholomew, H. Taneja, S. Yee, C. Jung, C. Furlong, Development of a surface plasmon resonance sensor for commercial applications, *Sens. Actuators, B* **1997**, 39, 375–379.
2. R. G. Woodbury, C. Wendin, J. Clendenning, J. Melendez, J. Elkind, D. Bartholomew, S. Brown, C. Furlong, Construction of surface plasmon resonance biosensors using a gold-binding polypeptide and a miniature integrated sensor, *Biosens. Bioelectron.*, **1998**, 13, 1117–1126.
3. E. Kretschmann, The determination of the optical constants of metals by excitation of surface plasmons, *Z. Phys.* **1971**, 241, 313–324.
4. www.spreeta.com http://www.ti.com/snc/products/sensors/spreeta.htm
5. J. Melendez, R. Carr, D. U. Bartholomew, K. Kukanskis, J. Elkind, S. Yee, C. Furlong, R. Woodbury, A commercial solution for surface plasmon sensing, *Sens. Actuators, B* **1996**, 35, 212–216.

6 A. N. Naimushin, S. D. Soelberg, D. Nguyen, L. Dunlap, D. Barthalomew, J. Elkind, J. Melendez, C. E. Furlong, Detection of Staphylococcus aureus enterotoxin B in femtomolar amounts by a miniature integrated two-channel SPR sensor, *Biosens. Bioelectron.* **2002**, 17, 573–584.

7 T. M. Chinowsky, J. G. Quinn, D. U. Bartholomew, R. Kaiser, J. L. Elkind, Performance of the Spreeta 2000 integrated surface plasmon resonance affinity sensor, *Sens. Actuators B* **2003**, 6954, 1–9.

9.6
Dielectric Spectroscopy: Choosing the Right Approach

Alexander Mamishev

9.6.1
Introduction

Dielectric spectroscopy has been an integral part of the analytical chemistry toolset for several decades. While the fundamentals remain unchanged, the laboratory aspects of using this technique rapidly evolve as the advances in materials science, microelectronics, and signal processing present new opportunities for more accurate and faster measurements, in-line feedback control loops, and imaging of complex geometries. In addition to the new opportunities for instrumentation design due to component technology advances, new application parameters are posed by rapidly developing pharmaceutical, chemical, food, and other industries. As the result, new versions of sensors and algorithms for interpretation of their output are introduced to the field every year. This chapter should help the reader make a more educated choice when evaluating the potential of using dielectric spectroscopy instrumentation in laboratory, field, or plant applications.

The use of fringing electric field (FEF) impedance spectroscopy sensors as a potential PAT tool for monitoring of multiple physical properties of pharmaceutical samples in real time and in-line is described. Major advantages of this measurement technique include one-sided access to material under test, profiling and imaging capabilities, and model-based signal analysis. The sensors use variable excitation frequency to obtain impedance spectrum of the material under test. Sensor array output signal can be converted into such physical variables as thickness of coating, drug signature, concentration of API, estimated dissolution dynamics, and tablet hardness. This chapter helps in understanding the strengths and weaknesses of dielectric spectroscopy, as compared with other tools.

9.6.2
Dielectric Spectroscopy in Comparison with Other Techniques

Process analytical offers a wide array of sensing and measurement technologies for laboratory practitioners as well as for manufacturing engineers and scientists. How does dielectric spectroscopy compare with other analytical tools? Generally speaking, dielectric spectroscopy techniques are great for evaluating bulk physical properties of materials. In contrast to GC chromatography or NIR spectroscopy, dielectric spectroscopy sensors are not likely to detect the presence of specific compounds through measurement of sharp spikes in the spectra associated with an analyte. Instead, spectra obtained from dielectric measurements usually consist of smoothly varying dependencies with few resonant peaks. Contributions of individual compounds overlap in the frequency domain. However, changes of physical properties, even without a change in chemical composition, are readily reflected in a dielectric response. For example, a change of temperature or porosity of material is very likely to lead to a clear signature in the sensor response. Multiple types of physical phenomena contribute to dielectric spectrum signatures. Authoritative references on the subject of dielectric spectroscopy in general include Refs. [1–9].

The following examples illustrate situations where dielectric spectroscopy may be a preferred analytical method. Each example provides a hypothetical inferior alternative sensing method.

- Measurements need to be conducted in non-contact mode. For example, a biologically active layer is deposited on moving sheet of fabric substrate. An acoustic sensor may be able to measure the uniformity of the moving material, but only if direct coupling is provided. A dielectric spectroscopy sensor can glide above the moving sheet, without touching any parts, and the measurements will be just as accurate as the contact measurements would have been.
- Measurements of analyte properties in a microreactor taken through a separating wall. The separator has to be electrically insulating, otherwise, the electric field lines will not go through to the other side. Alternatively, a chromatograph could be used to take measurements, but the sample would have to be removed from the microreactor to conduct the measurement.
- Profiling or imaging of a physical variable, e.g., monitoring of moisture diffusion process through an organic material is readily accomplished with low frequency dielectric sensors. Alternatively, a microwave probe can be used, but the sample penetration depth of the microwave probe would be limited, in many cases, to a few microns, whereas the penetration depth of low frequency dielectric measurements can be controlled by changing the separation between the sensor head electrodes.

9.6.3
Electrode Patterns

9.6.3.1 Control of Spatial Wavelength
The shape of the electrodes used for dielectric spectroscopy measurements varies widely. Classic shapes used for material characterization are parallel-plate electrodes (usually round or square patches) and coaxial cylinders. These two shapes have the advantage of having a simple linear relationship between the complex dielectric permittivity of the material under test and the complex impedance between the electrodes (assuming that the material is uniform). The disadvantage of such shapes is that imaging and profiling of the material under test is hardly feasible. For imaging applications, various forms of fringing electric field (FEF) electrodes are used. Figure 9.6.1 shows a schematic cross sectional view of an FEF sensor. Metal strips, shown as rectangles on the top of the substrate, can be excited with different patterns. The figure shows three possible excitation patterns. Variations in penetration depth obtained from the different excitations

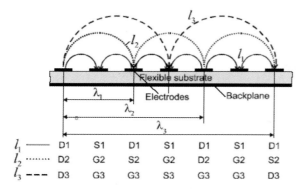

Fig. 9.6.1 Cross sectional view of a fringing electric field sensor (see text for details). Different connection schemes can be used to vary the penetration depth of the sensor.

provide the sensor with access to different layers of the material under test. The "D" in Fig. 9.6.1 represents the driving electrodes, "S" the sensing electrodes, and "G" the ground electrodes. An AC voltage signal is applied to the driving electrodes and the induced voltage or current is measured at the sensing electrodes. The ground electrodes are either connected to ground or kept at the same voltage potential as the sensing electrodes by using unity gain buffer amplifiers. The spatial wavelength of the sensor is increased $(\lambda_3 > \lambda_2 > \lambda_1)$ by using different connection schemes.

9.6.3.2 Control of Signal Strength

By changing the area of the sensor, the number of fingers, and the spacing between them, one can control the strength of the output signal. A trade-off between the signal-to-noise ratio and the minimum sensing area is selected based on the application requirements. While this aspect of sensor design is rarely important in stand-alone applications, it becomes critical for sensors integrated into microreactors, due to the limited space for sensor heads. Methods to maximize signal strength are extensively discussed in Ref. [10].

9.6.3.3 Imaging Capability

Either by moving sensor heads or by forming sensor arrays one can interrogate different regions of material under test. A simple interdigital structure can be moved up and down to measure the depth profile or float above the material surface to measure variation of changes at a specified depth. A combination of both potentially provides a 3D image of the material under test, but simple interdigital sensors are rarely used in this manner because it can take a long time to scan a sample.

9.6.4
Direct Sensing vs. Pre-concentrators

In this context, direct sensing means that the electric field generated by the sensor penetrates the material under test, and the changes in the material properties cause a change in the impedance between the electrodes. This method is used to measure property distributions in solid and liquid samples. However, in many cases the changes in material properties are rather minute, and the direct sensing approach does not have a sufficient resolution. In such cases, the measurement system needs an additional element, a pre-concentrator, that will attract species of interest to a specific volume. This is a generic case for many types of sensors, e.g., optical or acoustic. With dielectrometry sensors, typical pre-concentration techniques include deposition of hydrophilic layers on the surface of the electrodes (for moisture measurement), selectively absorbent and catalytic layers (for chemical measurements), and anti-bodies (for biological measurements). To give a feel for typical magnitude of the effect, consider polyimide used for humidity measurements. Polyimide is hydrophilic and can absorb water for up to 3% of its dry weight [11, 12]. Owing to the high dielectric permittivity of water ($\varepsilon_r \approx 80$) in comparison with that of dry polyimide ($\varepsilon_r \approx 3$), even relatively small changes in humidity may increase the dielectric permittivity of polyimide by as much as 30% [13, 14]. Whenever pre-concentrators are used, sensor response is subject to hysteresis, because the sensitive layer cannot reach immediate equilibrium with the environment.

9.6.5
Frequency of Excitation

Depending on the frequency of the excitation voltage and thus the electric fields generated by the sensor driving electrodes, different aspects of material properties can be measured by dielectric sensors. The excitation signal may be sinusoidal, step, triangle, or pulse waves. For non-sinusoidal signals, Fourier transform is often performed on the sensor response. The lowest boundary for frequency in existing dielectric spectroscopy is around 1 µHz, and the highest is in the THz range. It is rarely practical to go to such extremes; most practical industrial measurements are accomplished in the range 1 Hz to 100 MHz.

9.6.6
Conclusions

Dielectric spectroscopy techniques are promising for numerous applications that require non-invasive, non-destructive, non-contact, and real-time measurements. Non-invasive measurements with gas, liquid, solid, and mixed samples are possible on distance scales from nanometers to meters and a frequency of excitation from microhertz to terahertz. The main advantages of fringing electric field dielectric sensors include one-side access to material under test, convenience of application of sensitive coatings, a possibility of spectroscopy measurements through variation of frequency of electrical excitation, and material imaging capability.

References

1. P. Hedvig, *Dielectric Spectroscopy of Polymers*, Wiley, New York, **1977**.
2. C. J. F. Böttcher, P. Bordewijk, *Theory of Electric Polarization*, Elsevier Scientific Publishing Company, Amsterdam, Vols. I and II, **1978**.
3. R. Coelho, *Physics of Dielectrics for the Engineer*, Elsevier Scientific Publishing Company, Amsterdam, **1979**.
4. A. K. Jonscher, *Dielectric Relaxation in Solids*, Chelsea Dielectrics Press, London, **1983**.
5. J. R. MacDonald, *Impedance Spectroscopy: Emphasizing Solid Materials and Systems*, Wiley New York, **1987**.
6. A. K. Jonscher, *Universal Relaxation Law*, Chelsea Dielectrics Press, London, **1996**.
7. S. J. Havriliak, *Dielectric and Mechanical Relaxation in Materials*, Carl Hanser Verlag, Munich, **1997**.
8. J. P. Runt, J. J. Fitzgerald, *Dielectric Spectroscopy of Polymeric Materials: Fundamentals and Applications*, American Chemical Society, Washington, D.C., **1997**.
9. D. Q. M. Craig, *Dielectric Analysis of Pharmaceutical Systems*, Taylor & Francis Group, London, **1995**.
10. X. B. Li, S. D. Larson, A. S. Zyuzin, A. V. Mamishev, Design principles for multi-channel fringing electric field sensors, *IEEE Sens. J.*, Apr. **2006**, 6(2), 434-440.
11. P. A. von Guggenberg, Application of Interdigital Dielectrometry to Moisture and Double Layer Measurements in Transformer Insulation, Ph.D. Thesis, Department of Electrical Engineering and Computer Science, Massachusetts Institute of Technology, Cambridge, MA, **1993**.
12. J. Melcher, Y. Daben, G. Arlt, Dielectric effects of moisture in polyimide, *IEEE Trans. Electrical Insulation*, Feb. **1989**, 24(1), 31–38.
13. Kapton Polyimide Film: Summary of Properties, Du Pont Company, Wilmington, DE, **1998**.
14. D. D. Denton, D. R. Day, D. F. Priore, S. D. Senturia, E. S. Anolick, D. Scheider, Moisture diffusion in polyimide films in integrated circuits, *J. Electron. Mater.*, **1985** 14(2), 119–136.

9.7
The Future of Liquid-phase Microseparation Devices in Process Analytical Technology

Scott E. Gilbert

9.7.1
Introduction

Microreactor technology has made rapid strides in the last five to ten years. However, progress in the development of analytical technologies appropriate to microreactor and process monitoring has lagged behind. For microreactors to find increasing application in academic research and industrial production, appropriate on-line separation devices must be developed to provide adequate monitoring capabilities. Such devices need to be miniaturized versions of conventional instrumentation, and be capable of parallel analysis for microreactor arrays. Microfluidic (chip-based) analytical separation devices, such as liquid chromatography or capillary electrophoresis, are highly amenable to this application, considering the fact that microreactors themselves are miniaturized microfluidic devices. Their compatibility, small size, extremely small sample volumes, the possibility to configure systems with several or many microchips operating in parallel to monitor microreactor arrays, and, finally, speed of analysis make them highly ideal candidates for microreactor monitoring instrumentation.

Microseparation devices have proven to be faster, more efficient and less expensive (both in capital cost and maintenance) than their conventional counterparts. Despite these advantages, only four companies have developed commercial products, none of which have been intended for on-line process monitoring or for high throughput experimentation. Current commercial chip-based microseparation instruments have been developed for high throughput analysis in pharmaceutical and biomedical research markets, which are far larger than the process monitoring market. With regards to microreactors and PAT, micro-Total Analytical Systems (μ-TAS) have been exploring the concept of integration of separation columns, usually electrophoretic, with a microreactor for biochemical analysis. Polymerase chain reaction (PCR) microreactors for DNA analysis directly coupled to a micromachined capillary electrophoresis column on the same chip were among the first examples of this, but the concept has been extended to protein and cellular analysis [1]. As of this writing, no examples exist in the literature where such systems have been applied to small molecule chemistry. One of the main reasons is that comparatively little research has been devoted to continuous monitoring on-line sampling techniques for microfluidic devices. Although advances have been made in this area that will be discussed in this section, none have as yet been exploited for direct monitoring of microreactors. Hopefully, this connection will be established soon, motivated by the growing necessity for integrating on-line analytical capabilities with microreactors, especially in the arenas of high throughput experimentation for process development, and direct industrial production.

This section focuses on microchip technologies that have been developed for liquid phase analysis, namely liquid chromatography and capillary electrophoresis, and sampling methods that have been developed for on-line monitoring to date. Although there are no examples in the literature of truly integrated microreactor/microseparation devices, hybrid systems have been reported but will not be covered in this review. We cover current trends in lab-on-chip microfluidic separation devices for greater understanding of their capabilities and applicability as analytical devices for microreactors.

9.7.2
Historical Perspectives of Microseparation Device Technology

The field of microchip-based separations opened up as early as 1979 with the seminal publication by Terry et al., then at Stanford University, of a microfabricated gas chromatograph (GC) featuring an etched capillary column and integrated injector and thermal conductivity detector on a single silicon substrate [2]. Strangely, interest in follow-up developments along the same line lay dormant for a good 11 years, until the 1990 publication by Andreas Manz and coworkers [3], then at Hitachi Corporation in Japan, describing a microchip liquid chromatograph (LC) microfabricated in silicon. This device also featured an etched injection system, open tubular liquid chromatography (OTLC) column, and finally a detector, integrated together on the silicon substrate. The latter publication apparently was enough to drive further interest in microfabricated separation devices, probably spurred on also by several papers touting the merits of such devices in the early 1990s [4, 5]. These sparked great interest in the field, as a plethora of papers have appeared in the literature since – however these focused primarily on electrically-driven liquid-based separation devices such as capillary electrophoresis (CE) chips and capillary electrochromatography (CEC) chips, with little interest shown in the development of gas and liquid chromatographies on-chip.

Since the first reports of successful application of on-chip CE in the early 1990s [6, 7], over 540 articles on microchip electrophoretic or electrochromatographic devices have been published as of this writing. By contrast, only 20 some odd papers have been devoted to pressure-driven LC over the last 16 years. The reasoning behind this is two-fold; first, application-based funding supporting the research tended to favor devices that would eventually benefit biotechnology, hence the emphasis on electrophoretic separations. Secondly, electrically-driven liquid handling proved to be easier than hydrodynamic (pressure-driven) fluid manipulation that required pumps and valves. By taking advantage of electro-osmotic flow (EOF), a phenomenon of liquid (a dilute salt solution) propulsion in capillaries being generated by an applied electric field, no mechanical pumps or valves are needed, greatly simplifying the experimental details. EOF-based liquid propulsion and routing is actuated by strategic placement of electrodes within the etched microchannels on the chip, an easy task in microfabrication. Computer controlled voltage switching is all that is needed to control fluid movement on-chip, and complicated routing schemes have been devised to manipulate streams carrying macrobiomolecules to whole living cells within the microchannels. In addition, EOF has a flat, plug-flow profile in contrast to the parabolic profile of hydrodynamic flow. Thus, electrokinetic injection produces flat plugs that are not subject to Taylor dispersion, resulting in virtually no column-induced band-broadening, and high separation efficiencies. An elegant visualization of the two types of flow has been demonstrated by Paul et al. [8]. The flow profiles are made visible by using a UV laser to uncage a fluorescent dye, defining a sharply delimited tracer plug (Fig. 9.7.1).

Historically, the development of chip-based EOF separation devices has been crucial to establishing the credibility of claims of highly miniaturized and integrated chemical analysis systems. The first CE chip demonstrated the feasibility of using electroosmotic pumping to control the flow in a microchannel manifold network without the use of mechanical valves. This point is important, as it opened the way to using this approach for complex sample handling on-chip development of integrated flow injection analysis techniques for sample preparation. These developments opened the way for the realization of micro-Total Analysis Systems

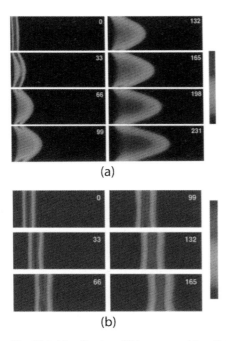

Fig. 9.7.1 Visualization of (a) pressure-driven Poiseuille flow (parabolic) in a 100 μm ID fused silica capillary and (b) electrokinetically-driven plug flow in a 75 μm ID capillary. Numbers denote time in milliseconds between micrographs in both sequences. See text for details.
(Reprinted with permission from Ref. [8]. ©1998 American Chemical Society.)

(μ-TAS) [4, 9, 10], which strive to develop true "lab on a chip", or fully integrated analytical systems involving sample introduction, preparation, analysis (may or may not include a separation) and result reporting on a single substrate. This field is still in its infancy, but has attracted many hundreds of researchers and now is the main topic of numerous international conferences and journals. Taken as a whole, development of chip-based separation capabilities helped establish the field of microfluidics, a branch of fluid mechanics devoted to fluid behavior in sub-millimeter diameter channels interconnected in simple and complex ways. Channels in this size range are commonly referred to as microchannels, whose cross sections are continually shrinking. A current "hot" research area deals with fluid behavior in nanochannels.

9.7.2.1 Development of Chip-based CE and CEC

The first reports of microchip-based capillary electrophoresis (CE) appeared in 1992 [6, 7]. These early devices were also the first to introduce an injection cross consisting of orthogonal intersecting etched channels in a glass substrate, one channel for sample and the other for running buffer, a geometry now ubiquitous in virtually all present-day CE chips. Until then, well-defined sample plugs were very difficult to produce in conventional CE systems. This first design used three reservoirs, where the waste reservoir was shared by the sample loading and separation channels. At the ends of the channels, wells formed from holes in the cover plate acted as sources and sinks for sample solution and running buffer. Platinum electrodes were positioned at the ends of channels as well. The injection procedure was to fill the sample channel using EOF by applying 3000 V at the beginning of the sample channel, then grounding the end of the separation channel. By automatically switching the voltages applied at the electrodes, injection was accomplished by momentarily imposing an electric field along the sample channel by applying 250 V between the sample reservoir and waste at the end of the separation channel for 30 s, allowing a plug of sample to flow into the separation channel. The voltages were then switched to re-establish an electric field down the separation channel by imposing high voltage at the buffer inlet reservoir, simultaneously floating the sample channel. Test separations of fluorescein and calcein were performed in a channel 13 cm long, 10 μm deep and 30 μm wide. The device achieved a maximum of 35 000 plates for calcein in about 6 min.

Subsequent important contributions to the field of microchip CE were made early on by Jed Harrison and Andreas Manz, who reported a spectacular enhancement of device performance by redesigning the channel layout and increasing the applied electric fields [11]. With an imposed field of 1 kV cm^{-1}, six fluorescence-tagged amino acids were separated with baseline resolution in 15 s, generating 40 000 to 75 000 theoretical plates. In the improved design, the separation channel length was reduced from 13 to 2.2 cm, utilizing the same cross section as in the previous work.

Michael Ramsey's group at Oakridge National Laboratories further improved device performance by introducing a 4-port crossed column geometry, giving independent control over sample and buffer flows [12]. This geometry has

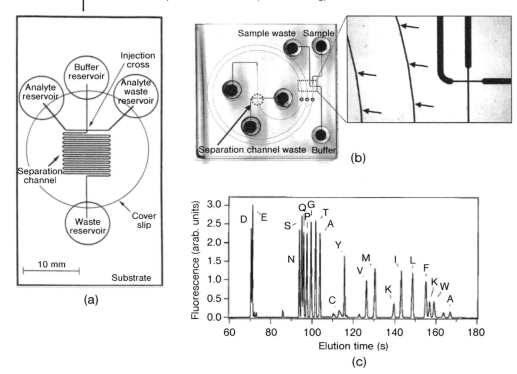

Fig. 9.7.2 (a) First electrophoresis chip with now standard 4-port cross-channel configuration developed by Michael Ramsey's group at Oakridge National Laboratories.
(b) Packing a long (25 cm) channel on a 5 × 5 cm chip format using spiral configuration to avoid dispersion due to small radii in the serpentine configuration.
(c) Electropherogram showing fast separation of 19 amino acids in 165 s, with average plate number of 280 000.
(Reprinted with permission from Refs. [12, 13]. ©1994 and 2000, American Chemical Society.)

become the standard configuration for all subsequent CE and CEC microchip layouts (Fig. 9.7.2a). Ramsey's group also showed that extremely fast separations, on the order of one second, were able to be done on a microchip format, with concomitantly high efficiencies (up to 10^6 theoretical plates) by taking into account radii of curvature of the serpentine separation channel (Fig. 9.7.2b), and in fact proposing a spiral-shaped channel [13]. In the same publication, the authors demonstrated a "real world" baseline separation with the same device of 19 amino acids in 165 s (Fig. 9.7.2c), an order of magnitude faster than the best conventional CE separation known at the time.

Capillary electrochromatography (CEC) is a method of chromatographically separating neutral analytes in an electroosmotically-generated flow. As with CE, the liquid matrix is propelled by an imposed electric field. Unlike capillary electrophoresis, however, solute molecules are not separated by differences in charge/mass ratio because they are neutral – thus a stationary phase is needed to separate these neutrals chromatographically. Electrochromatograms differ from pressure-driven or hydrodynamic chromatograms in the order of peaks in many cases because of the superposition of the electroosmotic flow (EOF) on the partitioning of solutes. CEC, like CE, is characterized by plug flow, since the radial velocity profile is flat. Again, in CEC like CE, the inherent resolution is much greater than for LC because no Taylor dispersion is present. These combined advantages have spurred on much progress in chip-based CEC technology.

The first report of on-chip open channel CEC was again made by Ramsey's group in a 1998 publication [14]. In a chip similar to that shown in Fig. 9.2(a), the channel walls were coated with octadecylsilane (ODS) as a reversed-phase chemically bonded coating. Isocratic and gradient elutions were tested and, under optimized conditions, the device performed a separation of four neutral coumarin dyes with base-line resolution within 20 s. Fred Regnier's group at Purdue University have created microfabricated arrays of micropillars within microchannels in quartz substrates to act as a packing with perfect geometry. Using reactive ion etching, arrays of diamond-shaped or cylindrical collocated monolith support structures (COMOSS) were easily defined in the bed of 10 μm deep microchannels etched on quartz substrates [15–18] and more recently on PDMS [poly(dimethylsiloxane)] substrates [19]. Because the packing array is defined photolithographically, the structure size and spacing can be adjusted at will to optimize the chromatographic behavior. A typical feature size is 5 μm, separated by interstitial spaces of 1 to 2 μm. These dimensions approach the size of interstitial spaces in conventional particle packings for similar-sized beads. Figure 9.7.3 shows a typical representation. The COMOSS column was estimated to be equivalent to a conventional 4 μm non-porous bead packing. In addition, the surfaces of the structures were functionalized with a silane to form a C18 reversed-phase monolayer. To obtain the nanoliter per minute flow rates required to operate the devices, the authors relied on electroosmotically-driven flow using an electric field of 770 V cm^{-1}, thus operating in CEC mode, since mechanical HPLC pumps available at the time were not capable of delivering such a small flow at the pressures required. However, the column efficiency obtained was remarkable, with some 35 000 plates (777 000 plate m^{-1}) in 110 s obtained for unretained Rhodamine 123 on a column 4.5 cm long, 10 μm deep and 150 μm wide [15]. These same workers also showed that the same C18-coated COMOSS device operating in isocratic CEC mode could resolve a peptide mixture derived from an ovalbumin digest in 10 min, almost as well as a conventional C18 packed HPLC column (4.6 mm × 2.5 m) using gradient elution could do in 50 min [17].

Although very amenable to miniaturization (no external pumps or valves needed) it is not clear that CE and CEC methodologies would be compatible with many small molecule organic chemical processes where non-polar solutes are involved. As an electrokinetic technique, it requires a conductive mobile phase and, as such, would be more suited to monitor fermentations and other bioprocesses. For very comprehensive treatises on EOF microchips covering their current applications, interested readers are referred to excellent reviews by Bruin [20], Lacher et al. [21], Verpoorte [22], Khandurina and Guttman [23] and Pumera [24].

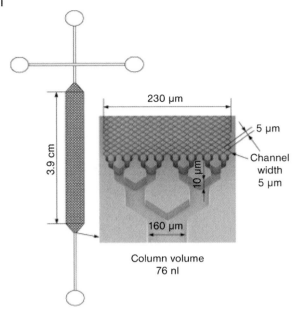

Fig. 9.7.3 (Left) Depiction of the COMOSS column configuration with injection cross and ports. (Right) Detail of a photolithographically-defined COMOSS stationary phase support structure.
(Reprinted from Ref. [19], ©2002, with permission from Elsevier.)

9.7.2.2 Development of Chip-based LC

Chip-based liquid chromatography has not been blessed with an extensive developmental history for reasons given above, but many reasons exist for justifying its emergence as a ubiquitous miniaturized process analyzer, eventually becoming a sensor-like instrument. First, for small molecule analysis, HPLC is still the separation method of choice used by many laboratories, especially for process monitoring applications in specialty chemical and pharmaceutical manufacture (off-line), and currently used HPLC methods can more readily be transferred to a microchip-based LC analyzer if one can be built than to a CEC or CE device. HPLC, however, is rarely used for on-line process analysis because it is slow and expensive to run as it consumes large quantities of solvent, and thus is normally run off-line. Despite these shortcomings, Dionex has recently marketed a true process-compatible HPLC [25].

Background

Despite the comparative paucity of literature on chip-based LC, developments in the field of chip-based pressure-driven LC to date have shown that rapid analyses are possible, with virtually no solvent consumed (flow rates in the nanoliter per minute range vs. milliliter or even microliter per minute ranges), demonstrating the feasibility of this technology for on-line process applications, and on-line parallel processing for microreactors. The technology is also amenable to sensorization, since electrokinetic nanoflow pumps (based on EOF) with no moving parts are now commercially available from Eksigent Technologies (Microfluidic Flow Control and EKPump™ technologies). Ultra-miniaturized valves have been fabricated *in situ* in microchannels [26–29] as well, with both developments favoring extensive miniaturization of the overall system. Details of these achievements are highlighted in the discussion below.

As mentioned above, purely hydrodynamic, or pressure-driven liquid chromatography on a micromachined substrate (chip) was first developed at Hitachi Corporation by Andreas Manz and coworkers in 1990 [3]. Their device featured a capillary column made by etching a 150 mm long separation channel in a silicon substrate, with a rectangular cross section of only 6×2 µm, including integrated electrodes for electrochemical amperometric detection. No chromatographic results were shown, but the advantages of moving to LC on a chip format were highlighted. The dimensions of the separation channel were chosen to be in the same range as particle diameters used in high performance packed columns, and are the ideal, but elusive, inner dimensions to achieve for capillary columns required for high performance OTLC. Micromachining techniques allow columns of any dimension to be fabricated at will. Use of anisotropic dry etching processes consisting of plasma, or reactive ion etching methods, normally produce channels of truly rectangular cross section in silicon substrates, with sharp corners. Isotropic dry etching is also possible for silicon, and in fact the column in the first LC chip was formed this way, producing a cylindrical cross section. Truly rectangular cross sections are also realized in polymer substrates, produced by laser cutting sheets and lamination of several layers of polycarbonate, PETE, or cyclic olefin copolymer (COC) sheets together to make the entire substrate. Rectangular microchannels are also very commonly molded in silicone rubber (PDMS) substrates using photolithographically defined masters. Glass and fused silica wafer substrates are wet-etched in hydrofluoric acid (HF) as the method of choice to date, because of historical precedent, and because of more general accessibility (it can be performed on any clean room wet bench or in any fume hood, but great care needs to be exercised). Only recently have rapid reactive ion etching methods for glass and fused-silica substrates been developed, allowing anisotropic etching in these materials by dry etch methods. Although only a few papers on this method have been published [30–34], it probably will replace wet etching as the technique catches on. Wet methods always etch isotropically, producing a cross section that only approximates a rectangle, as the cross sectional shape is really more trapezoidal with rounded corners. This is because lateral etching takes place at about the same rate as vertical etching, resulting in a progressive under-cutting at the mask borders, finally making the top of the channel wider than the bottom. At the top, a cover plate is bonded to the etched wafer, making a flat wall with corners forming acute angles. Figure 9.7.4 shows a typical cross section of a wet-etched channel in a Pyrex wafer [35].

Initially modeled as flow between infinite plates, Giddings et al. [36] showed that the performance of rectangular geometry for open channel LC columns

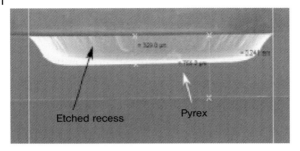

Fig. 9.7.4 Typical microchannel trapezoidal cross-section generated from wet etching in hydrofluoric (HF) acid baths. (Reprinted from Ref. [35], ©2004, with permission from Elsevier.)

was theoretically superior to that of circular cross-sections, principally because the cross sectional dimensions could be independently controlled. The smaller dimension (depth) can be tailored to allow for very short diffusion times for solute molecules to transfer to the top and bottom walls, while the width of the column can be made arbitrarily wide to reduce flow resistance, allowing for much greater flow rates in a rectangular geometry whose depth is approximately equal to the diameter of a circular cross section. However, the main drawback of the rectangular geometry is that in real channels the intrinsic band broadening is a factor of eight larger than in tubular geometries due to side wall effects, even in very high aspect ratio (width >> depth) channels. From experimental data in ribbon-like field-flow fractionation channels, Giddings et al. [36] had found that experimental plate heights were almost an order of magnitude larger than had been predicted theoretically, but could not explain this disparity. Later, Desmet and Baron [37], using a one-dimensional analytical model, explained this effect as a chromatographic-like mass transfer exchange between fast moving regions of mobile phase in the central part of the channel and stationary, stagnant areas along the sides, and analytically derived a factor of 7.95 for the increase in axial dispersion in rectangular channels (affecting the second term of the Golay equation for plate height). Poppe [38] reported in a simultaneous publication this same result obtained by a numerical two-dimensional model, essentially pointing to the stagnation of flow at the angles between the walls. Dutta and Leighton [39] have shown how modifications to the rectangular geometry in microchannels (deepening the channel at the sides) could reduce the enhanced dispersion.

Comparison between Open and Packed Channels
From a microfabrication point of view, OTLC has the appearance of being more easily implemented on a microchip than making packed channels. A major practical reason for choosing this approach over a packed column is the technological challenge in packing a microcolumn, and examples of this are discussed in the subsection below on "Column Preparation and Stationary Phases". Another very practical reason is the high permeability of OTLC capillary columns, requiring

very low pressures (< 5 bar) to operate the device, generating optimal linear mobile phase velocities in channels as short as 5–15 cm for reasonable separation efficiencies, even in microchannels 5 µm deep or less. This fact can greatly simplify fluid handling for such a device, and allow it to function at low pressures. However, there are certain practical issues with this approach, generally relating to thin stationary phases, possible clogging issues and detector technologies. For example, since very thin channels are required for efficient separations (< 5 µm), they can become easily blocked or clogged by small particles, rendering the chip unusable. Secondly, in such microchannels, very small injector and detector volumes become necessary, posing challenges to standard detection methods. In relation to the first consideration, the mass loadability of OTLC columns is more limited than for packed columns due to the smaller phase ratio (stationary phase volume/mobile phase volume), and this could have consequences for concentrated samples found, for instance, in reactors. During the 1990s, however, relatively thick polyacrylate based wall coatings were developed for capillary OTLC columns, and were shown to be successful with good mass loadabilities [40, 41].

In terms of column performance, on both a theoretical and practical note, when comparing OTLC with packed columns, OTLC outperforms the packed column version in terms of analysis speed only for very high efficiency separations involving plate numbers on the order of 500 000 to over 10^6 [42–44]. Put another way, OTLC would be the method of choice for difficult separations, or for separations involving hundreds of components. For instance, for a peak with $N = 10^6$, the retention time would be 55 h in a packed column operated at 100 bar, but only two hours in an 14-m long OTLC columns with a 10 µm bore, operated at the same pressure. The same OTLC will outperform a packed column when $N > 30 000$. The situation becomes less favorable for the OTLC when its bore is larger [43]. Thus, for more routine separations, OTLC may not be the best choice. More recent discussions on these comparisons have been brought forth by both Poppe [44] and Desmet and Baron [45] where a more analytical approach has been taken. Halász [46] has also made critical comparisons between OTLC and packed columns, showing only marginal gains in performance by the former. Modern monolithic phases have shown great promise to bridge the gap between OTLC and packed columns, as they exhibit very high permeabilities and excellent mass transfer characteristics. This results in short but highly efficient columns, producing low back pressures. More details on monoliths will be presented below.

On-chip Hydrodynamic Sampling/Injection Strategies
Among the several advantages of chip-based liquid chromatography, the possibility to integrate the injector, column (separation channel in the case of a microchip) and detection cell on a single chip substrate using microfabrication techniques stands out as one of the greatest. This is true because integration of these components on a single chip completely eliminates extra-column band-broadening normally associated with dead volumes from fittings and plumbing connections between parts in conventional HPLC. Sub-nanoliter injection volumes required for the extremely small columns are easily achieved by simple microchannel

configurations for injection plug definition. Conventional off-chip HPLC injection valves can not be fabricated small enough to meet this requirement. Normally, either a segment of the separation channel is used to define a fixed-volume injection plug, or auxiliary channels intersecting the separation channel can inject sample with a variable volume. The first to do this was Jeremy Glennon's group at University College Cork, in Ireland, in collaboration with the National Microelectronics Research Lab (NMRC) [47] also in Cork. Their fabricated silicon chip is shown in the inset of Fig. 9.7.5, and a schematic of the external valve connections is given in Fig. 9.7.5(a). The integrated injection loop on the chip, forming a double-tee junction with the separation channel [48–50] is seen in Fig. 9.7.5(b). This integrated injection loop gives the possibility to define sub-nanoliter injection volumes using pressure-driven flow. An injection sequence is shown in the micrographs of Fig. 9.7.5(c), where a fluorescent dye in the sample loop is shown being pushed into the separation channel to the right by a stream of mobile phase.

In deliberate considerations of process stream sampling, Robert Synovec's group at the Center for Process Analytical Chemistry (CPAC) based at the University of Washington had developed a sampling and injection strategy by routing two continuously flowing streams, one containing sample and the other mobile phase, through a double-tee junction molded on a PDMS chip [47]. Injection takes place by closing and opening valves to allow a plug of sample stream to flow down the separation channel, followed by mobile phase to perform the chromatographic analysis. Figure 9.7.6 shows the sequence, with a schematic of the channel configuration shown in the inset. This method of plug formation also allowed for more injection reproducibility in addition to moving towards a "real world" interface for process monitoring applications. As an added plus, plug size can be varied by this method. The principle drawback as reported by the authors was the difficulty in balancing pressure heads in the various streams to allow the system to perform reproducibly. This was due in part to the use of syringe pumps and split streams to deliver flows in the nanoliter per minute range. However, this was improved in a follow-up study [51]. At such low flow rates, any instability originating in the pumps can create pressure imbalances in the feed lines and chip itself, leading to unpredictable performance. Additionally, large dead volumes associated with the external tubing with respect to the microchannel volume on the chip creates excessive hold-up times for moving sample into the chip, reducing the device's ability to obtain representative samples on-line of a reaction mixture in real time.

In a further development, Schlund et al. have conceived a hydrodynamic gated injection scheme [52] similar to that developed for electrokinetic flows by Ramsey's group. In this method, an injection cross is formed by the intersection of the separation channel with a bypass channel, analogous to the analyte loading channel in 4-port CE and CEC chips. Continuous streams of sample and mobile phase are fed into the chip hydrodynamically, while at the cross the flows converge in such a way that the mobile phase stream is diverted into the separation channel, and the sample stream is forced into the bypass channel. At the confluence of the

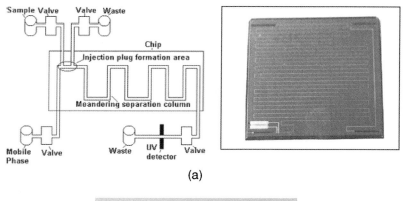

(a)

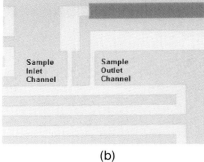

(b)

(c)

Fig. 9.7.5 Details of integrated injector and injection plug generation on LC chip from NMRC, Cork, Ireland. (a) Schematic showing external valve connections to chip for fluid control. (b) Micrograph of on-chip injection loop made by sample inlet and outlet channels forming a double tee junction with the main separation channel, with schematic of valve opening and closing to actuate injection. (c) Micrograph of a plug of florescent rhodamine dye detached from injection loop and being swept down the first segment of a separation column. Photo of glass/silicon chip shown in inset. (Reprinted from Ref. [48], ©2001, with permission from Elsevier.)

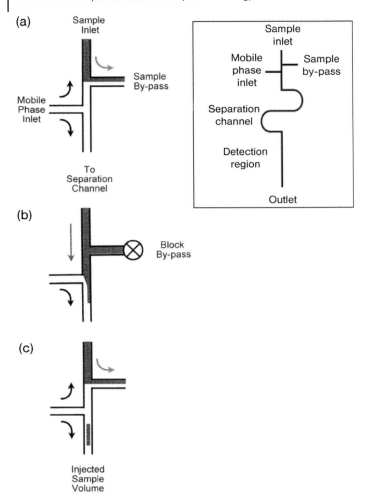

Fig. 9.7.6 Schematic of injection sequence on Synovec LC chip. (A) Chip in standby mode; sample stream entering separation channel from above, mobile phase stream flowing from left side inlet channel, both streams exiting chip in side channel leading to the right. (B) Chip in injection mode; external valve momentarily closes to stop exit stream, forcing sample to begin flowing down separation channel for short time to form a plug. (C) Chip in run mode; the external valve re-opens, re-establishing original flow pattern, this time with a plug of sample now swept down the separation channel for analysis. (Reprinted from Ref. [47], ©2000, with permission from Elsevier.)

two streams, the sample stream is "gated" from entering the separation channel inadvertently. When the flow of the mobile phase is momentarily retarded, a plug of sample stream flows into the separation channel, followed by mobile phase. The injection sequence is depicted in Fig. 9.7.7, showing a schematic of the flow of the two streams at the injection cross in the top row, a micrograph sequence of injected fluorescein dye in the middle row, and a computational fluid dynamic (FEMLAB) numerical simulation of the injection sequence in the bottom row. In Fig. 9.7.7(a), the chip is in pre-injection mode. In Fig. 9.7.7(b), a pressure perturbation momentarily interrupts the mobile phase flow, causing the analyte stream to bulge into the separation channel. Finally, in Fig. 9.7.7(c), the pressure distributions return are restored and the flows return to normal, but a plug of analyte stream is now carried into the separation channel. This injection scheme accomplishes two goals. One is that it can sample directly from a continuous flowing stream, eliminating the need for an intermediary syringe pump to sample from a process stream and subsequently inject the sample onto the chip. Secondly, the scheme is designed for on-line sampling, as the analyte stream is continuously flowing from a source onto the chip. Any change in the composition of the sample source, such as a reaction vessel, is quickly reflected in the sampling stream assuming that the flow is brought to the chip within seconds after leaving the vessel. In this way, there is no need to purge the inlet lines to the chip between analyses.

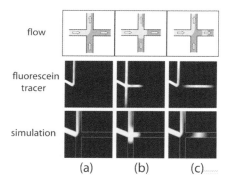

Fig. 9.7.7 Continuous sampling hydrodynamic gated injection from LC chip designed by Schlund et al. [52].
Top row: schematic of injection process shown for clarity.
Middle row: micrograph sequence of fluorescein injection.
Bottom row: results from FEMLAB computational fluid dynamic simulation of process.
(a) Standby mode: sample stream (fluorescein) enters from left, flows towards top into a bypass channel; mobile phase stream (not visible) enters from bottom, flows towards the right, entering separation channel.
(b) Injection mode: momentary reduction of mobile phase flow allows sample stream to move into separation channel, forming a plug.
(c) Run mode: re-establishment of flow pattern causes a plug of sample stream to break away, and is swept down separation channel by the mobile phase.

Column Preparation and Stationary Phases

Both open and packed columns have been demonstrated on LC and CEC chips. Strategies for making stationary phases in microchannels are few, and fall basically into three categories: (a) standard HPLC packings, (b) monolithic phases, and (c) wall coatings or bonded, each of which is now discussed in turn.

Microchannels have been filled with standard HPLC packings by flowing bead slurries into microchannels designed with microfabricated frit and weir structures, micromachined directly in the channel [53, 54], or with constrictions [55] to contain the particles. In the latter instance, 3 μm ODS-coated silica beads were packed into a channel containing constrictions (Fig. 9.7.8) under vacuum and pressure, taking advantage of the "keystone effect", whereby the initial few beads arriving at the constriction block those following to contain the packing without the use of a frit. This technique yielded a packed-column capable of generating 200 theoretical plates in 3 min for a CEC separation of two fluorescent dyes. Another approach took advantage of micromachined weirs to trap beads in chambers used for packed bed solid phase extraction or as a packed separation column [54]. In this approach, ODS-coated silica beads ranging in size from 1 to 4 μm were packed into the extraction chamber/column combination by electrokinetic transport, as seen from the top in Figs. 9.7.9(a) and (b). A side view of the chamber's weir structure is shown schematically in Fig. 9.7.9(c). The preconcentration and subsequent EOF driven elution of two fluorescent dyes was demonstrated on this device.

Bead packing has mostly been supplanted by *in situ* formation of monolithic phases, which are more successful both in terms of fabrication and column preparation. For this application, continuous polymer beds in CEC microchips

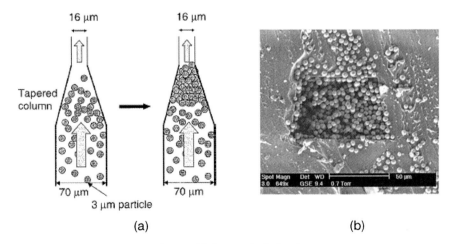

Fig. 9.7.8 (a) Diagrammatic rendering of fritless method of packing a microchannel with beads, using the "keystone effect" for self-damming by first beads to become blocked at the channel constriction.
(b) Electron micrograph of channel cross section, showing solid packing of bead bed achieved by this method.
(Reprinted with permission from Ref. [55], ©2002 American Chemical Society.)

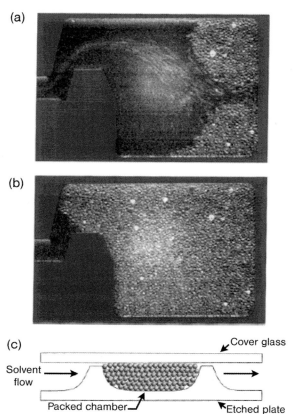

Fig. 9.7.9 Bead packing using microfabricated weirs.
(a) Micrograph view from the top of an extraction chamber on a CEC chip, showing beads flowing into and filling the chamber from a side channel by electroosmotic flow (EOF). Beads cannot escape because the weir gap is smaller than smallest bead.
(b) Micrograph of a filled chamber.
(c) Diagram of side view showing the weir structure.
(Reprinted with permission from Ref. [54], ©2000 American Chemical Society.)

have been explored by the groups of Jean Fréchet at the University of California at Berkeley [56] and Stellan Hjerten at the University of Uppsala [57, 58], both of whom worked on *in situ* polymerization of polymer monolith beds in microchannels during the same time period. The Uppsala group was the first to report on growth of continuous monolithic phases in a microchannel and use it to obtain extremely rapid yet efficient separations for both LC and CEC modes of operation [57]. These workers formed monolithic polymer phases *in situ* by a thermal polymerization process, making ionic phases for performing CEC experiments, and an anion exchange phase for pressure-driven LC separations. For the latter, the monolithic phase was designed to have a large porosity to present a low flow

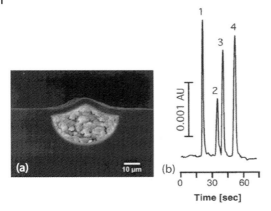

Fig. 9.7.10 First example of continuous polymer stationary phase bed (monolith) in a microchannel on LC/CEC chips. (a) Cross section of a hemi-cylindrical microchannel fabricated by isotropic dry etching on silicon substrates filled with polymer ion-exchange continuous bed stationary phase formed *in situ*. Channel is 40 μm in diameter. (b) Chromatogram of fast protein separation in an ion-exchange bed in a 4.5 cm long channel, showing high-resolution baseline separation of four proteins in less than 60 s. Gradient elution order of peaks: myoglobin (1), conalbumin (2), ovalbumin (3) and trypsin inhibitor from soybean (4). Mobile phase: phosphate buffer (20 mM, pH 7.8)/ phosphate buffer (20 mM, pH 7.8 containing 0.4 M NaCl) at 22 bar (330 psi) pressure. (Reprinted with permission from Ref. [57], ©2000 American Chemical Society.)

resistance within the channel. Fig. 9.7.10(a) is an SEM image of the monolith cross-section, showing the micropores in the polymer structure. The semi-cylindrical column diameter is 40 μm. The chip was able to achieve a remarkably rapid base-line pressure-driven chromatographic separation of proteins at relatively low pressures, using the anion exchange material. The chromatogram displayed in Fig. 9.7.10(b) demonstrates this, revealing a base-line separation of four proteins in less than 1 min in a 62 mm long column, with an effective separation length of 45 mm, using a pressure of 22 bar and a gradient elution completed in 60 s.

In a subsequent paper, diamond microchannels were fabricated by chemical vapor deposition of polycrystalline diamond on sacrificial structures in silicon [58, 59]. Touting the chemical inertness and optical and UV transparency of diamond, the authors fabricated interesting free-standing channel shapes with rectangular channel cross sections, as exemplified by the chemically vapor deposited (CVD) channel shell shown in Fig. 9.7.11(a). *In situ* monolithic phases were formed in the channels by the same process described above, and lower flow resistances were realized due to larger channel dimensions. These authors succeeded in achieving extremely rapid LC separations. The two chromatograms in Fig. 9.7.11(b) represent rapid ion exchange separations of different mixtures of proteins using gradient elution. The separation was complete in 30 s or less, the time it took for the gradient to run between its limits. The channel dimensions were 40 μm × 100 μm × 40 mm, and a pressure of 9 bar was applied. Again, the same injection technique was

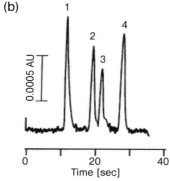
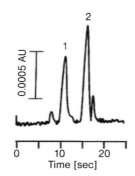

Fig. 9.7.11 (a) Electron micrograph of a chemical vapor deposited (CVD) diamond microchannel shell on the surface of a silicon substrate for microfabrication of LC chip. (b) Chromatograms of fast protein separations, showing baseline separation of the same four proteins as in the right-hand side of Fig. 9.7.10(b), this time separated in less than 30 s, and separation of larger proteins hemoglobin A_o (1) and b-lactoglobulin A (2) (right-hand side) in about 20 s. The mobile phase was the same as in Fig. 9.7.10, 9-bar operating pressure. Column dimensions: 40 μm × 100 μm × 32 mm. UV detection. (Reprinted from Ref. [58], ©2001 with permission from Elsevier.)

used, and thus reproducibility of the chromatograms was compromised. More recent publications describe the formation of a photopolymerized polymer [29] and thermally formed silica sol–gel [60] monolithic stationary phases in glass microchip. The silica skeleton with functionalized with ODS to form a reversed phase monolith. As this report was preliminary, chromatographic conditions were not optimized, but baseline separation of three catechins was shown, with efficiencies of up to 18 000 plates in a seven minute window.

The third category described above, wall coatings or bonded phases, uses simple ODS functionalization of etched channel walls in a silicon substrate [49, 50], and cationic nanoparticles for on-chip ion exchange OTLC [61]. Polyacrylate bonded coating stationary phases were developed and successfully applied for capillary OTLC in the 1990s, but as yet not have not been applied to microchips [62]. Progress in open channel bonded phases have been made in the area of CEC, both conventional and microchip-based [63].

Pumping and Fluid Handing
Fluid pumping for electro-driven devices is straightforward. A high voltage source is required with computer-controlled switches to provide orchestrated sequences of voltages to on-chip integrated electrodes, driving liquids at naturally established flow rates into the nanoliter to microliter per minute range. Reagent liquids and buffers are contained in on-chip or off-chip reservoirs. No valves, either external or internal, are needed to control the fluid handling. For pressure-driven systems this is not the case, and fluid handling is more complex. Pressure-drive systems must supply flow rates in the range of nanoliters per minute for LC microchips, and have been in the form of pressurized reservoirs, syringe pumps, or HPLC pumps using split flow. Reagents can be contained in chambers on-chip, but internal valves are needed to control the fluid handling. Air-actuated PDMS membrane internal valves (in the microchannel) have been developed in polymer chips, and solid mobile monoliths (photopolymerized *in situ*) that are liquid pressure-actuated have also been developed [26–28]. The latter can withstand much higher pressures than the former. Figure 9.7.12(a) shows a micrograph of an LC injection valve using this concept. In Fig. 9.7.12(b) depicts an injection sequence of a fluorescent dye, showing the movement of the monoliths. The fluids were mobilized at pressures between 20 and 35 bar (300–450 psi) using high pressure syringe pumps. In a very elegant and exciting development, monolithic valves were created by *in situ* photoinitiated polymerization as described for stationary phase fabrication [64]. These valves were actuated by deliberate temperature changes delivered by thermoelectric elements, enabling a phase transition at 32 °C, where desolvation and shrinkage of the polymer opens the pores in the monolith, thus allowing flow of liquid in the microchannel. The valves were capable of withstanding up to 13.8 bar (200 psi) in the closed position, with a low back pressure of only 0.3 bar (5 psi) in the open position. The response time of the monolithic valves was slow, however, taking 3.5 s to open and 5 s to close.

If external valves are used, as in hybrid systems, liquid reservoirs must be contained off-chip. The author has found that off-chip pressurized reservoirs are well suited for nanoliter flows in microchips, producing flows at constant pressure instead of constant flow rate. The former is preferred when using large volume reservoirs over constant flow rate sources, as there is a time constant in the system when the pumping system adjusts pressures to maintain constant flow rate when flow resistances change downstream. This is due to the finite compressibility of liquids, where compression of liquid in a large volume (say 10 mL) syringe pump can be on the order of several nanoliters, which is of the order of magnitude of the volume of the microchannels. The pressure change will not be felt immediately in the microchannel because the expanding liquid exiting the syringe barrel must flow for a finite time in the microchannel to allow pressure equilibration to occur,

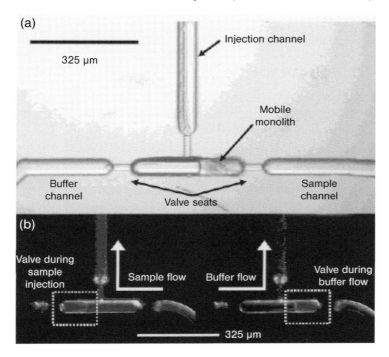

Fig. 9.7.12 Example of integrated valves produced by *in situ* polymerization of fluoropolymer mobile monoliths directly in a microchannel.
(a) Micrograph of an integrated injection valve, with central portion of horizontal channel for metering analyte aliquots into the injection channel (vertical channel). The central chamber is isolated by constrictions trapping mobile monolith, acting as a valve by blocking openings on either side to allow entrance of analyte or mobile phase (buffer) to fill the chamber. By alternating pressurization of sample and buffer sides, monolith blocks either entrance, allowing sample to fill the injection channel and buffer to push the aliquot into the separation channel.
(b) Fluorescent dye sequence showing valving action of mobile monolith. See text for details. (Reprinted with permission from Ref. [28], ©2004 American Chemical Society.)

which can take several minutes. However, constant pressure sources are passive flow delivery systems, and will not automatically compensate for any transitory or permanent blockages or changes in flow resistance downstream.

For active flow delivery, pumps must be used in pressure drive systems. Flow splitting from constant flow rate pumps, such as was done by Bjorkman et al. [58, 59], is not desirable when such small flows are demanded because of inaccuracies and fluctuations. Nowadays, mechanical pump technology has advanced to the point of providing split-less nanoliter per minute flows. Constant flow mechanical pumps, such as the Agilent 1100 series nanoflow pumps and "NanoFlow" pump from Scivex work in the nanoliter per minute range, are of course commercially available, delivering stable constant flows as low as 20 nL min^{-1}. These pumps are not amenable to miniaturization, however, and if used would need to be operated in a hybrid system. For true development of an LC sensor, highly miniaturized pumps need to be integrated within the microchip. Microfluidic MEMS versions of mechanical pumps have been developed from silicon, glass and polymer, but are not capable of generating pressures above several hundred millibars. Such low pressures are in general not useful. However, electrokinetic or electroosmotic pumps such as the EKPump™ can generate pressures up to 1000 bar, and are themselves on a microchip format, capable integration with a microchip LC unit. As such, they may present an ideal solution to the issue of miniaturization and sensorization.

Fluidic interfacing to the external world is another design issue to be considered. In earlier work, fused silica capillaries were glued to specially micromachined ports on the chip [48–50]. Kirby et al. have given an example of zero dead volume non-permanent connections to a microchip [26], gluing threaded ports machined from polyetherimide (ULTEM) onto glass LC microchips. The ports mated with HPLC fittings (nut and ferrule) to attach 360 μm OD capillary to the fluid ports of the chip. However, the most practical concept is a chip mount or chuck that makes non-permanent connections with fluid ports on the chip. Nittis et al. [65] have developed a specially designed mount for zero dead-volume high pressure non-permanent interconnects for interchanging chips, allowing for leak-proof contact between standard capillary tubing and fluid ports on a microchip. Fluids were delivered to the apparatus by using conventional HPLC fittings and tubing. In this way, chips could be easily and rapidly exchanged without the need to connect tubing directly to the chip. The author has also designed a chip mount using o-rings to seal to the ports of the chip (Fig. 9.7.13). The dead volume created by the o-rings, however, is not negligible. In Fig. 9.7.13, threaded holes for low pressure ¼″ × 28 flat-bottom fittings can be seen in the transparent (for demonstration purposes) acrylic block. These were used to connect 1/16″ or 1/8″ Teflon tubing to the block for liquid delivery.

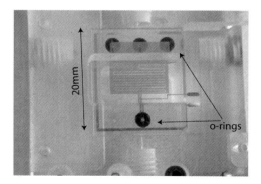

Fig. 9.7.13 Early fluidic interface to microchip designed by the author using an o-ring sealing system. Details show o-ring mating with fluid ports on chip, which is pressed to o-rings using an acrylic cover plate that is bolted to support block underneath. The chip mount is made from transparent acrylic for demonstration purposes.

Detection Methods

Detection in microseparation devices can be done by standard optical techniques, either using UV/Vis absorption or by laser-induced fluorescence for greatest sensitivity [66]. As it is a mass-sensitive technique (dependent on the absolute number of analyte molecules in the detection volume), standard UV detection becomes a problem for sensitivity due to the very short path length for light absorbance. It is possible to increase the path length by internal reflectance, but greatly complicates microfabrication. Fluorescence is more sensitive but may not be feasible for real-time applications since pre-column derivatization would be necessary. A very good alternative detection method is electrochemical detection, which is a concentration sensitive (dependent on concentration only via adsorption phenomena) detection technique, and therefore is independent of the detection cell volume, which can be made very small, the lower limit of which is determined by the working electrode size and sensitivity of the electronics. Detector electrodes in this case are microelectrodes, which with very quiet electronics can achieve a very high signal-to-noise (S/N) ratio. Electrochemical detection can be either amperometric or conductometric, and both methods have been implemented on microchips. Both methods have been used extensively in CE chips [67, 68] where the analytes are charged species, with the former method having been used in LC chips for neutral analytes [48, 52, 69, 70]. Amperometric detection involves a current measurement at constant applied potential from a sensitive potentiostat sensitive to sub-nanoampere currents. Contrary to popular belief, the polar moiety on substituted aliphatic molecules is electroactive, and the method can measure polar aliphatic as well as aromatic organic analytes. Aliphatic alcohols, acids, esters, amines, amides, aldehydes, ketones and the like are all truly electroactive in aqueous electrolytes. It can be said that they are not electroactive at dc potentials; however, the electroactivity of such intransigent molecules is exhibited when the potential is pulsed. These compounds tended to quickly foul electrodes when studied as analytes in the past, which is the origin of the misconception that they have no electroactivity. Using pulsed electrochemical detection (PED), these compounds beome responsive. This method has been successfully adapted to conventional HPLC detection using pulsed techniques, where rapid positive- and negative- pulses are applied to the working electrode to clean the surface and restore the initial state of the electrode between each measurement. LaCourse [71, 72] has shown in numerous studies that PED can detect many compounds with very good sensitivity that are difficult if not impossible by UV detection.

The CPAC group at the University of Washington has developed a refractive index optical detection method that is very sensitive and tailor-made for microchip applications. The advantage of such a technique is that it is a universal detection method, and does not depend on analyte molecules containing chromophores. In addition, it is a concentration sensitive detection method, so the detection cell volume can be made small enough to satisfy requirements for very small columns. The method, dubbed "grating light reflectance spectroscopy" (GLRS) (see Chapter 9.8), was an idea originally conceived by Anatol Brodsky and Lloyd Burgess [73, 74] at CPAC. The technique was later applied as a detector on an

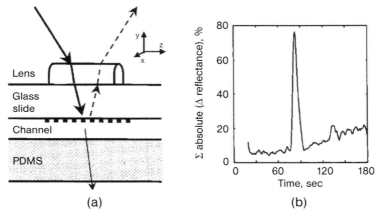

Fig. 9.7.14 Integration of GLRS (grating light reflectance spectroscopy) detection onto microchip. (a) Diagram explaining principle of GLRS technique, showing hemicylindrical lens attached to glass substrate to focus a collimated white light beam onto a 100 μm diameter spot in a microchannel (see text for details). (b) Chromatographic peak derived from the GLRS signal taken on FD&C Red food color test analyte.
(Reprinted with permission from Ref. [51], ©2002 American Chemical Society.)

LC microchip in a joint effort with the Synovec group [51]. The detection concept is schematized in Fig. 9.7.14(a). Collimated white light is focused on a metal/dielectric grating that is in contact with the fluid in a microchannel. The light is incident at a fixed angle. Light at a critical wavelength no longer is transmitted through the chip, but propagates along the interface between the grating and the fluid in the channel, satisfying the diffraction condition. About half of this incident light is reflected by the grating. The reflected light contains a critical diffraction order where light at the wavelengths near the critical wavelength forms an evanescent field at the grating interface with the sample medium (fluid), which can be observed spectroscopically. This is seen as intensity changes in spectrum of the reflected light. The critical angle is a function of the refractive index of the medium next to the grating, thus any changes in this quantity will change the wavelength and possibly the diffraction order. This technique has a demonstrated sensitivity on the order of 2×10^{-6} RIU (refractive index units), which translates to a limit of detection (LOD) on the order of 100 μM. In Fig. 9.7.14(b), a typical signal is shown for detection of FD&C Red #3 food dye injected into an LC chip similar to that described in Ref. [47]. The LOD was shown to be 170 μM for this substance, corresponding to 2.6 fmol of analyte. The light, being focused to a 100×150 μm spot, formed a detection cell of 150 pL, making this technique ideal as a detector for microseparation devices.

Future Developments for Chip-based LC
Further advances in microchip-based LC with applications on the horizon have been in development and application of revolutionary concepts in chromatography

only possible in a microchip format. An example is shear-driven chromatography, where conventional hydrodynamic pumping is replaced by literally dragging the liquid in the separation column by a moving plate [45, 75–79]. The advantage of this method of pumping is that there is no pressure drop to contend with, therefore no limitation on channel size. The performance of such devices has been shown to greatly exceed that of CEC, which is superior to pressure driven OTLC in terms of resolution and speed [78]. Thus, channels less than 0.5 µm deep have been demonstrated to yield extremely high speed and high resolution separations. One can generate very good stationary phase supports by anodizing the silicon in the channel, generating a highly porous structure approximately 500 nm deep. Combined with a sub-micron channel depth, the phase ratio in such a system can be adjusted to yield very high performance and high speed separations. The system was able to separate a coumarin C440 and C450 mixture in 50 ms, with a plate height H of 0.6 µm, corresponding to over 10^6 theoretical plates per m. Sub-second separations are routine with this technique, owing to the very large phase ratios obtainable with large capacity factor (k' of 3) stationary phases. The equivalent pressure to achieve such separation efficiencies in a conventional system would be 100 000 bar for such channel dimensions. This work demonstrates that extremely high speed separations are indeed possible while circumventing the limitations particular to pressure and electrokinetically driven systems.

9.7.3
Applications of Liquid Microseparation Devices for Process Stream Sampling and Coupling to Microreactors

9.7.3.1 Continuous-sampling CE Chips for Process Applications

As discussed above, CE and CEC techniques are highly amenable to miniaturization and adaptation as sophisticated sensors for process monitoring due to their inherent simplicity, compactness, speed and efficiency. Indeed, efforts have been made to adapt the basic technology for on-line process monitoring applications. The first CE microsystem incorporating continuous sampling dates back to the early1990s. The device was developed by the group of Andrew Ewing at Pennsylvania State University and described in a 1993 publication [80]. Although not a true microchip, the device consisted of a very wide (2.5 to 5.0 cm) rectangular separation channel composed of quartz plates separated by spacer layers ranging from 0.6 to 109 µm thick [81]. The novelty was sample introduction through a standard round capillary, where the one end of the capillary was immersed in a sample reservoir, and the other interfaced to the rectangular separation capillary by simply butting the former to the entrance slit of the latter, and translating the sample introduction capillary along the width of the separation capillary using a stepper motor, continuously injecting sample analyte into the rectangular channel electrophoretically. A diagram of the apparatus is shown in Fig. 9.7.15(a). The resulting electropherograms are three-dimensional constructs of intensity, time and lateral position within the separation channel. The data graphically show the changes in real time of the composition profile of a given mixture. In work

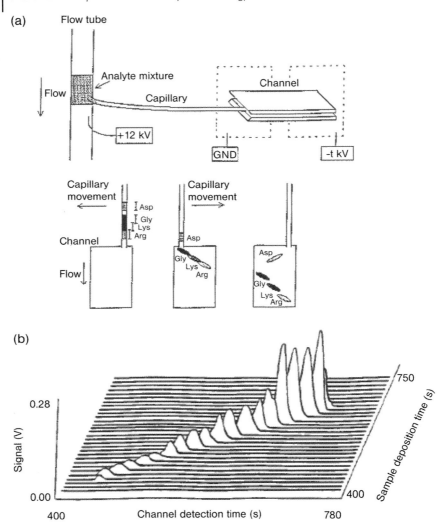

Fig. 9.7.15 First continuous sampling CE chip designed by Andrew Ewing's group at Pennsylvania State University. (a) Schematic of set-up and principle of operation for continuous flow CE chip used for on-line analysis. Capillary is rastered back and forth laterally at the gap entrance to a chip by linear translation stage actuated by stepper motors, where the analyte is pulled from a continuous flowing stream by EOF and continuously injected onto the chip. A multichannel detector is used to measure the migration of the analytes as a function of lateral position and time. The resulting electropherograms are two-dimensional, with the added dimension of lateral position (channel number), showing temporal variation of composition of the analyte stream. (b) Multichannel electropherogram of vessel contents from a recirculating sample stream, where incremental additions of arginine to the vessel are monitored in real time. (Reprinted with permission from Ref. [82], ©1996 American Chemical Society.)

demonstrating flow injection capabilities of the device for process monitoring [82], a reaction mixture was simulated by serial additions of arginine to a four-component amino acid mixture, with resulting electropherograms showing changing domains that reflected the increasing concentration of the arginine component (Fig. 9.7.15b).

The first work related to on-line sampling was published in 2001 [83]. Motivated by making a "real world" interface for a CE chip, a new form of flow-through interface was developed by the pioneering group led by Jed Harrison at the University of Alberta. The innovation was to include deep channels on the chip, through which analyte solution is pumped hydrodynamically. These deep channels, known as the sample introduction channels, or SIC, were etched in a glass substrate to a depth of 300 µm × 1 mm wide, and 20 mm long. This SIC was intersected by the electrophoresis channel network, having a cross section of 36 µm wide × 10 µm deep. Figure 9.7.16(a) shows a micrograph of the channel intersection region. With these dimensions, the electrophoretic channel network had a flow resistance over five orders of magnitude greater than the sample introduction channel. This differential allowed analyte solution to flow through the SIC without entering the electrophoretic channels. The drawback to this design was the need to externally ground the sample. Different versions of the flow-through chip were tested with batch reactors and integrated on-line precolumn microreactor for formation. Figure 9.7.16(b) shows a set-up for a chip–batch reactor interface. The reactor was a simple stirred beaker to which aliquots of fluorescein were added serially, while the contents of the beaker were pumped by a peristaltic pump through the chip via the SIC. A calibration curve derived from measurements of resulting peak areas is displayed in Fig. 9.7.16(c), showing this continuous monitoring technique to be effective in accurately quantitating reactor contents. A real batch chemical reaction was not tested. In a more complex version of the chip, a microreactor was integrated with the separation column for continuous monitoring of complexation of antigens with labeled antibodies, destined for physiological studies. Figure 9.7.17(a) shows the layout this chip, which features a junction (J2) where the sample, drawn from aliquots contained in the SIC, was mixed with reagent drawn from reservoir 6, reacted in the mixer/reaction channel, and finally injected into the separation channel from a double tee injector. In the example shown here, the device measured complexation of antibody (antiovalbumin) and hapten ovalbumin, with subsequent separation and quantitation of an antibody–hapten complex by CE. The electropherogram in Fig. 9.7.17(b) shows the separation of the ovalbumin–antiovalbumin complex from unreacted anti-ovalbumin (Ab*).

At about the same time, Lin and coworkers [84] at the National Cheng Kung University in Tainan, Taiwan published their version of a chip, which is similar to that of Harrison's group in Alberta. This group had extensive experience in flow injection analysis and on-line analysis with conventional electrophoresis. In their design, a 3 mm wide × 40 µm deep sample inlet channel (SIC) was etched for sample flow-through, which in turn was connected to the electrophoretic manifold of 80 × 40 µm cross section, in a configuration similar to the Alberta chip (see

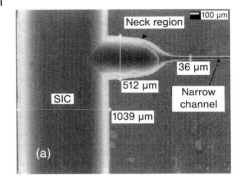
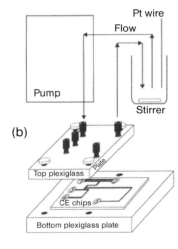
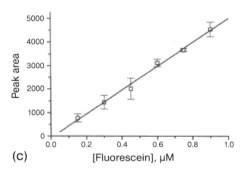

Fig. 9.7.16 Continuous sampling CE chip designed by Jed Harrison's group, University of Alberta.
(a) Micrograph of large bore sample inlet channel (SIC) coupling to loading microchannel wet etched in a glass substrate. (b) Schematic of lab set-up for semi-continuous sampling from a benchtop reaction vessel. (c) Linear calibration plot from standard additions of fluorescein to vessel, where contents are continuously recirculated through the chip by a peristaltic pump, and automatically sampled by the chip intermittently using EOF.
(Reprinted from Ref. [83], ©2001 with permission from Wiley-VCH Verlag GmbH.)

Fig. 9.7.18). The external sample source was left to float electrically (no grounding necessary), which was to make a more compatible interface with real external process sampling equipment such as a microdialysis probe head for bioreactor sampling. In addition, a gated injection method was introduced to accommodate this configuration. In this design, the buffer and sample streams flow towards each other in orthogonal directions in the microchannels, then meet at the cross in a fashion similar to Fig. 9.7.7.

Referring again to Fig. 9.7.18, the salient features of this design are that in a pre-injection state, or standby mode, a stream of analyte flows in the SIC while reservoirs D and F are grounded (Fig. 9.7.18b). Voltage is then applied to reservoir B, sample is introduced to the electrophoretic manifold and spontaneously flows to reservoir D, while buffer flows down the separation channel to reservoir F. For injection, all reservoirs are left floating momentarily, and a small plug of sample

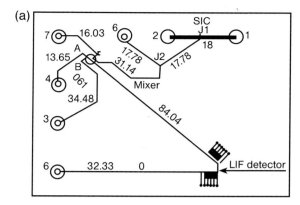

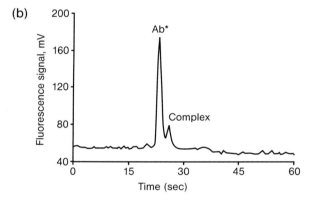

Fig. 9.7.17 Example of electrophoretic chip with integrated microreactor, also from Jed Harrison's group, University of Alberta. (a) Diagram of chip, showing complex microchannel layout for on-board reagent reservoirs, mixing (microreactor) channel and separation channel with integrated injector and detector cell. The chip is interfaced via a SIC for semi-continuous sample monitoring. (b) Example electropherogram of mixing channel (microreactor) effluent were ovalbumin and antiovalbumin were mixed. The two peaks represent the unreacted fluorescently-labeled antibody (Ab*), and the antibody–antigen complex. (Reprinted from Ref. [83], ©2001 with permission from Wiley-VCH Verlag GmbH.)

enters the separation channel (Fig. 9.7.18c). Subsequently, in run mode, the electrical connections are re-established, causing the plug to be pinched off from the main flow, and sent down the separation channel for analysis (Fig. 9.7.18d). In Fig. 9.7.19, an example electropherogram shows three series of injections of rhodamine B at a three concentration increments. The relative standard deviation (rsd) in the peaks was 7.6% by this method, with 0.64% rsd for the migration times. There was considerable influence of the hydrostatic pressure in the SIC on the injection volume. Increasing pressure from higher flow rates induced hydrodynamic flows in the sample inlet channel, perturbing the injection process.

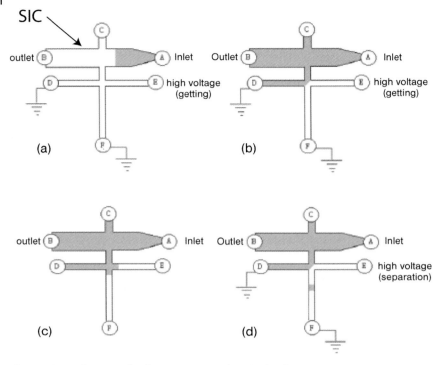

Fig. 9.7.18 Another example of continuous sampling CE chip for on-line analysis, designed by Y-H. Lin's group at the National Cheng Kung University, Tainan, Taiwan. This design allows for continuous flow through the chip via the SIC, with continuous sampling from the SIC. Gated injection is used to perform analysis at will.
See text for details on injection method.
(Reprinted from Ref. [84], ©2001 with permission from Elsevier.)

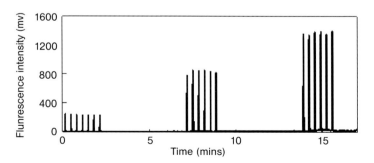

Fig. 9.7.19 Multiple-injection electropherograms showing response to three incremental additions of rhodamine dye to a reaction vessel. Contents of vessel were recirculated. See text for details.
(Reprinted from Ref. [84], ©2001 with permission from Elsevier.)

To negate this effect, the authors found it necessary to increase the gating voltage, or that imposed between inlet E and the ground points, as a function of increasing flow rate in the SIC. This problem could have been easily avoided by simply etching a deeper SIC to dramatically decrease its flow resistance with respect to that of the electrophoretic channel manifold. Attiya et al. have expounded on this point in their paper cited above [83].

Resulting from a body of work where classical flow injection coupled to CE was extensively studied for continuous monitoring [85–94], the prodigious analytical group of Zhao-Lun Fang at Zhejiang University in Hangzhou, China extended this know-how to microchip formats [90, 95–98]. Again, using a flow-through scheme, the chip design is based on a H-channel configuration for a split-flow interface with flow injection or sequential injection system, in contrast to the cross channel configuration described above. Subsequently, this group developed modified cross channel chip featuring a flow-through sample introduction reservoir on the exterior of the chip, and open to the atmosphere (Fig. 9.7.20) [98]. By use of this configuration, where sample flow was introduced from the bottom of the chip and allowed to surge into the trough glued onto the top of the chip, sample was electrokinetically transported along the sample channel and analyte injection into the separation channel was performed in the normal manner. The system was used to demonstrate its capability to measure on-line the derivatization of four amino acids, arginine, glycine, phenylanaline, and glutamic acid, with the fluorescent tag fluorescein isothiocyanate (FITC). In this experiment, reaction mixture was pumped continuously from a vessel at a rate of 0.48 mL min^{-1}, with a sample channel fill time of 30 s, and sampled intermittently using the automated method described above. The separation runs lasted 60 s each. With detection accomplished by LIF; the results are reproduced in Fig. 9.7.21, where a series of electropherograms shows the raw data for each amino acid (Fig. 9.7.21a, b), resulting in plots of the reaction kinetics (Fig. 9.7.21c).

Büttgenbach and Wilke have developed an integrated hydrodynamic injection valve for their flow-through CE microchip equipped explicitly for process analysis [99]. Realizing the short-falls of off-line batch sampling for laboratory analysis for QC and process monitoring, the authors endeavored to make a microchip with a continuous sampling flow through channel, using a hydrodynamic injection scheme akin to flow injection from a flowing stream. Building upon the designs of earlier flow-through microchips, a larger flow-through microchannel was fabricated in parallel with the smaller separation channel, both channels being connected by an orthogonally intersecting sample loading, or injection microchannel. As the device was intended for use in bioreaction monitoring, the prototype was made from biocompatible PDMS. A diagram as well as photo of the chip is shown in Fig. 9.7.22(a). During experiments where sample liquid containing analyte was continuously pumped through the flow-through channel, injection was accomplished hydrodynamically by means of an integrated passive valve placed in-line with the injection channel (Fig. 9.7.22b) When injection was desired, an external active valve placed in-line with the circulating sample flow was momentarily closed to provide a back pressure, forcing liquid to flow across

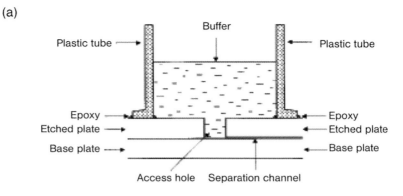
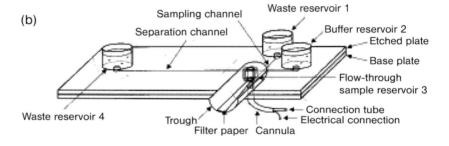
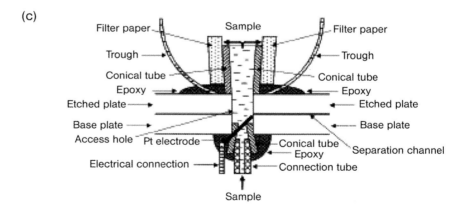

Fig. 9.7.20 Continuous sampling interface designed by Z.-L. Fang's group at Zhejiang University in Hangzhou. See text for details.
(Reprinted with permission from Ref. [98], ©2002 American Chemical Society.)

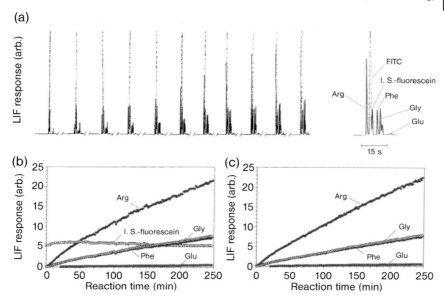

Fig. 9.7.21 Continuous sampling electropherograms from a reaction vessel wherein derivatization of four amino acids with fluorescein isothiocyanate (FITC) fluorescent tag took place, using the same chip described in Fig. 9.7.20. (a) Raw data from repeated injections over time of reaction. A zoom on a particular set of peaks from a single injection is shown on the right-hand side. (b) and (c) Plots of kinetic data on derivatization reaction for each amino acid derived from electropherogram analysis. Uncorrected data (b) and data corrected for drift (by subtracting unreacted fluorescein curve) (c). (Reprinted with permission from Ref. [98], ©2002 American Chemical Society.)

the normally closed passive valve. The external valve was closed long enough to allow the sample stream to flow to the junction of the separation channel, where a small plug of approximately 2.5 nL (average) was injected into the latter. Five consecutive electropherogram of a binary mixture of dopamine and catecholamine are shown in Fig. 9.7.22(c), showing 4–7% relative error in peak heights. No tests were made with a true reaction mixture. The separation efficiency and reproducibility degraded over time due to adsorption of impurities and the analyte itself with the PDMS capillary wall. This can be corrected by use of a glass chip.

A different on-line interfacing approach was also reported recently, where a commercially available CE chip (Micralyne) using a standard crossed microchannel configuration was reshaped by sawing off the integrated buffer reservoir and one sample reservoir to form tapered tubing fittings for attaching flexible tubing [100]. Separate lengths of tubing were used to carry both the running buffer and sample to the chip (Fig. 9.7.23a). Using a different fluid manipulation protocol, a continuous sample stream was delivered to the head of the separation column, while the buffer was delivered to a side channel which typically is used for sample loading. The sample stream contacts the entrance to the separation channel "a",

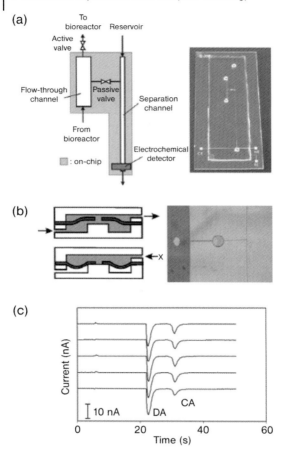

Fig. 9.7.22 Electrophoretic chip of Büttgenbach and Wilke, Technical University of Braunschweig, with a hydrodynamic injector for continuous monitoring of bioreactor stream contents.
(a) Channel layout with photo of a poly(dimethylsiloxane) (PDMS) chip.
(b) Details of PDMS membrane hydrodynamic injection system.
(c) Data from five consecutive electropherograms showing repeatability of injection method. Components are dopamine (DA) and catecholamine (CA).
(Reproduced from Figures 1, 2 and 5 from Ref. [99], ©2005 with kind permission of Springer Science and Business Media.)

but does not enter until voltage is applied to the electrode inserted in the delivery tube, and continues to flow past the channel entrance to waste. In Fig. 9.7.23(b), the injection sequence is schematized, showing (I) standby mode with buffer entering the chip from the side under EOF, and sample inlet floating, (II) analyte injected by applying 2 kV to the sample inlet while maintaining EOF of the buffer and (III) separation. The results showed no dependence of the peak heights in the electropherograms on the sample flow rate in the external line.

9.7 The Future of Microseparation Devices in Process Analytical Technology

On-line electropherograms recorded for varying concentrations in flowing streams containing the explosives trinitrobenzene (TNB) and trinitrotoluene (TNT) are shown on the left-hand side in Fig. 9.7.23(c), with 2 ppm step additions, and a mixture of phenol, 2-chlorophenol and 2,3-dichlorophenol on the right-hand side of the figure, with increases in 25 µM steps. The insets show linear calibration curves with high correlation coefficients (0.998–0.999). The authors observed that the injection volumes were insensitive to hydrodynamic flow rate in the external tubing, but, notably, due to the non-robust nature of the tubing interface, the system was limited to low sample stream flow rates and hence low hydrodynamic pressures, which apparently did not cause hydrodynamic flow on the chip.

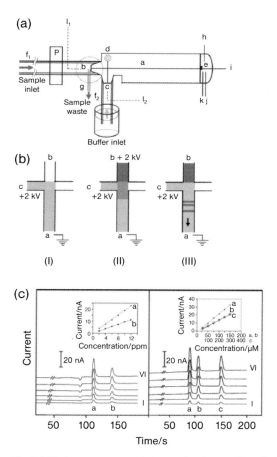

Fig. 9.7.23 A most recent variation on the theme of interfacing a CE chip to a flexible sample line for continuous monitoring. See text for details.
(From Wang et al. [100], ©2005, reproduced by permission of The Royal Society of Chemistry.)

9.7.3.2 Continuous Sampling LC Chips for Process Applications

Characteristics of Ideal LC Microchip Sensors

In consideration of LC as an on-line sensor, ideal device characteristics would be compactness, speed, very low maintenance, self-containment and low solvent consumption. Compactness is not a given here, albeit the fact that the chips themselves are very small. It is the peripheral system surrounding the chip to which attention needs to be paid in order to minimize footprint. Unlike liquid handling for CE and CEC microchips, hydrodynamic flow components involve pumps and valves, for which more effort is required to miniaturize. We have touched on current advances in these areas already, but extensive development in miniaturization and integration of hydrodynamic fluid handling components is still forthcoming. The following is envisioned.

CE and CEC are inherently faster techniques than LC. Therefore, to move towards sensor-like devices for process monitoring, columns need to be designed for high speed separations to guarantee analytical runs as short 5 min or less, with baseline or near baseline separations (resolution = 1.5) of as many as ten components. Modern high speed packed columns require very high pressures to perform, as particle sizes are 3 µm or smaller. A very viable alternative to standard packings are columns featuring monolithic stationary phases. As seen from the examples discussed above, monoliths have extremely high permeabilities and excellent mass transfer characteristics, allowing very high speed and high efficiency separations to be achieved. A very good review on modern monoliths was written recently by Svec [101]. OTLC columns would also fulfill these requirements, although this has not been demonstrated on a microchip to date. However, in standard capillary OTLC, isocratic baseline separations of multicomponent mixtures in 5 min have been demonstrated [41, 62] (i.e., seven components on a 85 cm-long, 5 µm I.D. OTLC column, with plate numbers as high as 400 000); thus the potential exists for designing microchips approaching this performance. Mass loadability is also an important consideration and, from this perspective, monoliths may be more likely preferred over OTLC columns since reaction mixtures typically contain only a few components of interest, and are fairly concentrated. Microchip formats incorporating OTLC columns have been designed with sub-nanoliter injection and detection cells, and thus would be the ideal vehicle for an OTLC approach. On the other hand, monoliths can match or outperform OTLC (in terms of separation speed) for plate numbers under 30 000 [42–44]. Since many process-related separations can be performed satisfactorily using conditions capable of generating only 1000 to 5000 plates, very short monolithic columns can also be incorporated in a microchip format for this purpose.

Sampling Interface for On-line Monitoring

Unlike the field of microchip CE, there is almost a total lack of scientific literature covering continuous sampling interfaces for chip-based LC devices. However, the author has investigated sampling methods from continuous sample streams for chip-based LC intended for on-line analysis [52]. We have independently developed a sample flow-through strategy for our LC chips, similar to those described for

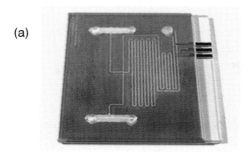

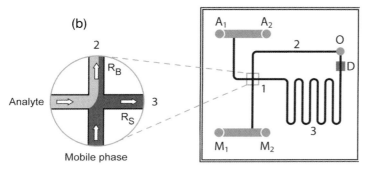

Fig. 9.7.24 All-glass LC microchip of the author's design.
(a) Chip photo, showing large bore flow-through SIC channels, microchannel system and electrochemical detector electrodes.
(b) Detail of injection cross for hydrodynamic gated injection.

the CE chips above. In this scheme, two larger-bore flow-through channels were etched into an all glass chip, and integrated with a channel system designed for a gated injection configuration, as described above in the subsection on "On-chip Hydrodynamic Sampling/Injection Strategies". Figure 9.7.24(a) shows a photo of the chip, with a description of functionalities shown in Fig. 9.7.24(b). The overall concept is to sample macroscopic flows and scale down to a microscopic flow in the chip itself. We speak of "gearing down" the flow from liters per minute to nanoliters per minute. This concept is represented by Fig. 9.7.25, showing a segment of a sample loop attached to a batch or continuous reactor containing a process stream flowing at hundreds of milliliters or even liters per minute. The LC chip is shown to sample from the main stream via a smaller tube that interfaces with the chip's SIC. In this way, the flow is split into two streams; the smaller stream passing through the chip at flow rates in the range of hundreds of microliters to milliliters per minute, driven by the pressure head in the main sample loop. The split ratio can be about 1000 : 1, where approximately 0.1% of the main flow is directed into the chip. At these flow rates, sample hold-up in the external tubing is virtually eliminated, and the system allows either for a continuous circulation directly out of a reaction vessel, through the chip, and return to the vessel, or

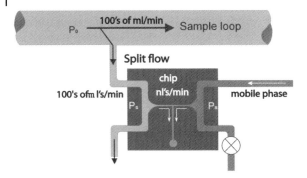

Fig. 9.7.25 The authors' principle of continuous sampling from a flowing reactor contents stream in a fast sample loop. A flow split from the main stream is caused by a smaller line connecting the chip to the sample loop, allowing the split stream, flowing on the order of hundreds of microliters per second to milliliters per second, to flow through the SIC of the chip. A pressure head in the line drives an even smaller flow into the microchannel network. The sample flows through a bypass channel (not shown), and plugs are injected at will from this stream. A $10^6 : 1$ flow reduction is realized in this manner, all the while obtaining a representative sample.

for feed from an on-line sample pretreatment stage to intermittently introduce fresh pretreated (i.e., diluted) sample into the chip. A substantial pressure head will develop in the SIC through a combination of the intermediate flow and flow resistances downstream. This intermediate pressure head forces a portion of the SIC flow into the microchannel network, which consists of a separation channel and a bypass channel. The separation channel is OTLC-like and requires a relatively low pressure of only a less than 1 bar. The column is designed to achieve efficient separations at flow rates of 10s of nanoliters per minute. Thus, a second split is made, with a split ratio again of about 1000 : 1. Overall, an eventual 1 000 000 : 1 flow split is made by this technique in a relatively facile manner for process stream sampling.

9.7.3.3 Coupling of Microseparation Devices to Microreactors

We have already shown an early example above of a simple microreactor mixing a fluorescently labeled antibody with ovalbumin (see Fig. 9.7.17), coupled to a CE column on chip [83]. Since then, virtually all work on the integration of microseparation devices with microreactors has been devoted to genomics and proteomics analyses. Owing to the complexities of the biochemistry involved, details of the microreactor components will not be given here; however, interested readers are referred to very good recent reviews [1, 102, 103]. The separation components are essentially the same as the stand-alone devices described in the above sections. In the interest of advancing this field, we show in Fig. 9.7.26 an example of a successful completion of a micro Total Analysis System (μ-TAS) using microreaction chambers for amplification (PCR) and digestion of DNA samples introduced onto the chip [104].

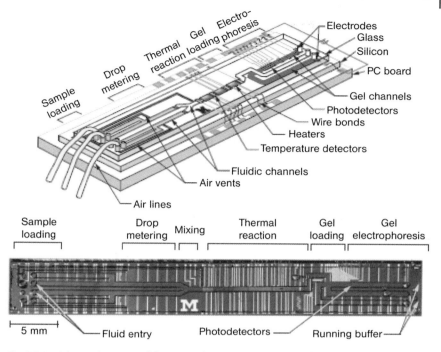

Fig. 9.7.26 Advanced DNA amplification or digestion microreactor/CE chip for DNA analysis.
(Reprinted from Ref. [104] ©1998 with kind permission from the AAAS.)

9.7.4
Potential Applications in Process Analytical Technology

9.7.4.1 Relevance of Micro-scale Device Technology

Despite the efforts cited above to develop interfacing strategies for microfluidic separation devices, there are at present no reported applications of microfluidic devices in process monitoring. The Process Analytical Technology (PAT) initiative has now been advocated by the US Food and Drug Administration (FDA) in their recent update of current Good Manufacturing Practice (cGMP) for the pharmaceutical industry [105–107]. For several economic and regulatory reasons, this initiative calls for the increasing use of multicomponent analyzers to monitor industrial scale chemical synthesis on-line, in-line or at-line, and not to rely on inference from traditional simple sensors that monitor only gases, pH, and temperature. Although PAT has been in limited practice for over 40 years by segments of the chemical industry [108], this has been done by taking conventional laboratory analytical instruments such as near-infrared and UV/visible spectrometers, process gas chromatographs and HPLC out to the plant for on-line, in-line or at-line analysis. This approach has been cumbersome, and is perhaps the main reason why PAT has not been overwhelmingly embraced by the industry. Miniaturization is now seen as the key to its future [108]. Contemporarily, microreactor technology is in its infancy, but is maturing and establishing itself in industrial settings. It may indeed become the mainstay of specialty and pharmaceutical production in the long term. As chemical processes become more scaled down to the level of microreactors, the need is clear for sensor-sized instruments that can perform multi-component analysis on-line in chemical and biochemical processes [109, 110]. However, this remains true even for conventional process chemistry. Under the umbrella of the PAT initiative, the New Sampling and Sensor Initiative (NeSSI) [25, 111, 112], advocating a modular basis for multiple micro-analyzers arrayed on a single fluidic platform, calls for the development of sensor-sized multicomponent analyzers. For techniques necessitating liquid phase separations, this would involve microfluidic lab-on-a-chip, or μ-TAS technology, to provide the necessary products. As far as other techniques are concerned, commercial successes have been met with the implementation of micromachined gas chromatographs and near-infrared spectrometry in process plants.

9.7.4.2 Potential Applications in Industrial Biotechnology and Biopharmaceutical Production

Although inroads into multicomponent analysis for on-line monitoring in industrial biotechnology and biopharma have been made with infrared spectroscopies [113], liquid chromatography in the form of HPLC remains the workhorse technique for mainly off-line, but some on-line, process monitoring [114]. This is because this technique gives a straight forward interpretation of the analytical data, whereas IR spectra require deconvolution and chemometric analysis for correct analyte determinations. Also, it is a question of tradition. However, the main impediment to widespread use for on-line HPLC is the need for sample preparation, as raw reactor contents must be at least filtered of biomass and diluted before injection. To date, there are several commercially available sampling probes that use dialysis and filtering techniques for sample clean-up but not dilution in both off-line and at-line HPLC analysis (examples of "home-built" on-line sampling probe-HPLC apparatuses are found in the literature as well [115, 116]). Groton Biosystems has marketed a fully automated sampling system for fermentation and biopharmaceutical production that performs full sample clean-up for downstream analysis, and features a built in automated CE analysis unit [117].

9.7.4.3 Sample Pretreatment

Future microfluidic LC and CE devices for on-line chemical and bioprocess monitoring will need to incorporate sample preparation unit operations such as filtering, dilution, preconcentration/solid phase extraction up-stream of the separation column, and this can done for such devices more easily than for conventional instrumentation. Using microtechnology, the sample prep unit operations can be integrated with the separation subsystem, making for a compact and even disposable unit [118, 119]. There are several examples in the literature of this level of microfluidic integration for the study of analytes of biological origin [119–126]. Purely integrated microfluidic strategies have been developed for chip-based analysis [122, 123, 126, 127], but hybrid systems can be implemented as well, incorporating off-chip sequential injection analysis techniques, offering more versatile and high throughput sample pretreatment capabilities [95, 128, 129]. Lichtenberg et al. [130] have written an extensive review on chip-based sample pretreatment systems and methods that are focused on analysis of biological targets, but some techniques are amenable to chemical process conditions.

9.7.4.4 Potential Applications in the Fields of Process Development and Microreactor Technology

The largest immediate impact of microfluidic separation devices will be in the areas of high throughput experimentation (HTE) for process development and microreactor monitoring. These areas are very amenable to the adaptation of microfluidic devices because of the very low sample consumption, compactness, speed of analysis and the ability to incorporate many chips for parallel analysis of modular batch reactor arrays and for continuous flow through monitoring of microreactors. At this time, robotic systems are available to sample the contents of individual reactors in arrays, and analyze by HPLC in a serial fashion. It has been pointed out that in general the bottleneck in high throughput experimentation is the analysis [110]. One HPLC monitoring 100 reactions at 10 min each requires at least 16 hours. Microreactors in general have similar issues [109]. For parallel liquid chromatographic analysis, Eksigent Technologies and Nanostream Inc. have products on the market that can potentially address these issues. Atofluidic Technologies is working to develop integrated sample pretreatment/LC chip cartridges for parallel analyses applied to both HTE and microreactor monitoring.

9.7.5
Future Perspectives

Hopefully, the reader has surmised from this overview of the field of microchip-based separation devices that they hold great promise for the future of the chemical process industry in general. As microreactors will likely be the key to the modernization of specialty chemical and pharmaceutical production in the short term, microtechnologies devoted to supporting microreaction chemistry will become essential as the fields develop. In fact, a co-evolution of both fields will occur as the technologies become more intertwined, providing a synergistic relationship that will benefit both by accelerating new discoveries in the respective technologies. This synergy will not only benefit process chemistry but have ramifications in the areas of biotechnology as well.

References

1 Krishnan, M., Namasivayam, V., Lin, R., Pal, R., Burns, M. A., Microfabricated reaction and separation systems. *Curr. Opin. Biotechnol.* **2001**, 12, 92–98.
2 Terry, S. C., Jerman, J. H., Angell, J. B., A Gas chromatographic air analyzer fabricated on a silicon wafer. *IEEE Trans. Electron Devices* **1979**, ED-26, 1880–1886.
3 Manz, A., Miyahara, Y., Miura, J., Watanabe, Y., Miyagi, H., Sato, K., Design of an open-tubular column liquid chromatograph using silicon chip technology. *Sens. Actuators, B* **1990**, B1(1–6), 249–255.
4 Manz, A., Graber, N., Widmer, H. M., Miniaturized total analysis systems: A novel concept for chemical sensing. *Sens. Actuators, B* **1990**, 1, 244–248.
5 Manz, A., Harrison, D. J., Verpoorte, E., Widmer, H. M., Planar chips technology for miniaturization of separation systems: A developing perspective in chemical monitoring. In *Advances in Chromatography*, ed. Brown, P. R., Grushka, E., Marcel Dekker, Inc., New York, **1993**, vol. 33, pp. 1–66.
6 Harrison, D. J., Manz, A., Fan, Z., Ludi, H., Widmers, H. M., Capillary electrophoresis and sample injection systems integrated on a planar glass chip. *Anal. Chem.* **1992**, 64, 1926–1932.
7 Manz, A., Harrison, D. J., Verpoorte, E. M. J., Fettinger, J. C., Paulus, A., Ludi, H., Widmer, H. M., Planar chips technology for miniaturization and integration of separation techniques into monitoring systems: Capillary electrophoresis on a chip. *J. Chromatogr. A* **1992**, 593(1–2), 253–258.
8 Paul, P. H., Garguilo, M. G., Rakestraw, J., Imaging of pressure- and electrokinetically driven flows through open capillaries. *Anal. Chem.* **1998**, 70, 2459–2467.
9 Auroux, P.-A., Iossifidis, D., Reyes, D. R., Manz, A., Micro total analysis systems. 2. Analytical standard operations and applications. *Anal. Chem.* **2002**, 74, 2637–2652.
10 Reyes, D. R., Iossifidis, D., Auroux, P.-A., Manz, A., Micro total analysis systems. 1. Introduction, theory and technology. *Anal. Chem.* **2002**, 74, 2623–2636.
11 Harrison, D. J., Fluri, K., Seiler, K., Fan, Z., Effenhauser, C. S., Manz, A., Micromachining a miniaturized capillary electrophoresis-based chemical analysis system on a chip. *Science* **1993**, 261, 895–897.
12 Jacobson, S. C., Hergenröder, R., Koutny, L. B., Warmack, R. J., Ramsey, J. M., Effects of injection schemes and column geometry on the performance of microchip electrophoresis devices. *Anal. Chem.* **1994**, 66, 1107–1113.
13 Culbertson, C. T., Jacobson, S. C., Ramsey, J. M., Microchip devices for high-efficiency separations. *Anal. Chem.* **2000**, 72, 5814–5819.
14 Kutter, J. P., Jacobson, S. C., Matsubara, N., Ramsey, J. M., Solvent-programmed microchip open-channel electrochromatography. *Anal. Chem.* **1998**, 70(15), 3291–3297.
15 He, B., Regnier, F., Microfabricated liquid chromatography columns based on collocated monolith support structures. *J. Pharmaceut. Biomed. Anal.* **1998**, 17, 925–932.
16 He, B., Tait, N., Regnier, F., Fabrication of nanocolumns for liquid chromatography. *Anal. Chem.* **1998**, 70(18), 3790–3797.
17 He, B., Ji, J., Regnier, F. E., Capillary electrochromatography of peptides in a microfabricated system. *J. Chromatogr. A* **1999**, 853, 257–262.
18 He, B., Tait, N., Regnier, F., Fabrication of nanocolumns for liquid chromatography. *Anal. Chem.* **1998**, 70, 3790.
19 Slentz, B. E., Penner, N. A., Regnier, F. E., Capillary electrochromatography of peptides on microfabricated poly(dimethylsiloxane) chips modified by cerium(IV)-catalyzed polymerization. *J. Chromatogr. A* **2002**, 948, 225–233.
20 Bruin, G. J., Recent developments in electrokinetically driven analysis on microfabricated devices. *Electrophoresis* **2000**, 21(18), 3931–3951.

21 Lacher, N. A., Garrison, K. E., Martin, R. S., Lunte, S. M., Microchip capillary electrophoresis/electrochemistry. *Electrophoresis* **2001**, 22, 2526–2536.
22 Verpoorte, E., Microfluidic chips for clinical and forensic analysis. *Electrophoresis* **2002**, 23(5), 677–712.
23 Khandurina, J., Guttman, A., Microscale separation and analysis. *Curr. Opin. Chem. Biol.* **2003**, 7(5), 595–602.
24 Pumera, M., Microchip-based electrochromatography: Designs and applications. *Talanta* **2005**, 66, 1045–1062.
25 http://www1.dionex.com/en-us/ic/process_analytical_DX-800/ins4551.html
26 Kirby, B. J., Reichmuth, D. S., Renzi, R. F., Shepodd, T. J., Wiedenman, B. J., Microfluidic routing of aqueous and organic flows at high pressures: Fabrication and characterization of integrated polymer microvalve elements. *Lab Chip* **2005**, 5, 184–190.
27 Kirby, B. J., Shepodd, T. J., Hasselbrink, E. F., Voltage-addressable on/off microvalves for high-pressure microchip separations. *J. Chromatogr. A* **2002**, 979, 147–154.
28 Reichmuth, D. S., Shepodd, T. J., Kirby, B. J., On-chip high-pressure picoliter injector for pressure-driven flow through porous media. *Anal. Chem.* **2004**, 76, 5063–5068.
29 Reichmuth, D. S., Sheppod, T. J., Kirby, B. J., Microchip HPLC of peptides and proteins. *Anal. Chem.* **2005**, 77, 2997–3000.
30 Ceriotti, L., Weible, K., Rooij, N. F. d., Verpoorte, E., Rectangular channels for lab-on-a-chip applications. *Microelectron. Eng.* **2003**, 67–68, 865–871.
31 Ichiki, T., Sugiyama, Y., Taura, R., Koidesawa, T., Horiike, Y., Plasma applications for biochip technology. *Thin Solid Films* **2003**, 435, 62–68.
32 Mazurczyk, R., Vieillard, J., Bouchard, A., Hannes, B., Krawczyk, S., A novel concept of the integrated fluorescence detection system and its application in a lab-on-a-chip microdevice. *Sens. Actuators, B* **2006**, 118(1–2), 11–19.
33 Park, J. H., Lee, N. E., Lee, J., Park, J. S., Park, H. D., Deep dry etching of borosilicate glass using SF6 and SF6/Ar inductively coupled plasmas. *Microelectron. Eng.* **2005**, 82(2), 119–128.
34 Lee, D.-S., Ko, J. S., Kim, Y. T., Bidirectional pumping properties of a peristaltic piezoelectric micropump with simple design and chemical resistance. *Thin Solid Films* **2004**, 468(1–2), 285–290.
35 Bu, M., Melvin, T., Ensell, G. J., Wilkinson, J. S., Evans, A. G. R., A new masking technology for deep glass etching and its microfluidic application. *Sens. Actuators A* **2004**, 115, 476–482.
36 Giddings, J. C., Chang, J. P., Myers, M. N., Davis, J. M., Caldwell, K. D., Capillary liquid chromatography in field flow fractionation-type channels. *J. Chromatogr.* **1983**, 255, 359–379.
37 Desmet, G., Baron, G. V., Chromatographic explanation for the side-wall induced band braodening in pressure-driven and shear-driven flows through cannels with high aspect-ration rectangular cross-section. *J. Chromatogr. A* **2002**, 946, 51–58.
38 Poppe, H., Mass transfer in rectangular chromatographic channels. *J. Chromatogr. A* **2002**, 948, 3–17.
39 Dutta, D., Leighton, D. T., Dispersion reduction in pressure-driven flow through microetched channels, *Anal. Chem.* **2001**, 73, 504–513.
40 Swart, R., Brouwer, S., Kraak, J. C., Poppe, H., Swelling behaviour and kinetic performance of polyacrylate stationary phases for reversed-phase and normal phase open-tubular liquid chromatography. *J. Chromatogr. A* **1996**, 732, 201–207.
41 Swart, R., Kraak, J. C., Poppe, H., Highly efficient separations on 5 mm internal diameter open tubular capillaries coated with thick polyacrylate stationary phases. *Chromatographia* **1995**, 40, 587–593.
42 Guiochon, G., Conventional packed columns vs. packed or open tubular microcolumns in liquid chromatography. *Anal. Chem.* **1981**, 53(9), 1318–1325.

43 Knox, J. H., Gilbert, M. T., Kinetic optimization of straight open-tubular liquid chromatography. *J. Chromatogr.* **1979**, 186, 405–418.

44 Poppe, H., Some reflections on speed and efficiency of modern chromatographic methods. *J. Chromatogr. A* **1997**, 778(1 + 2), 3–21.

45 Desmet, G., Baron, G. V., Simultaneous optimization of the analysis time and the concentration detectability in open-tubular liquid chromatography. *J. Chromatogr. A* **2000**, 867(1–2), 23–43.

46 Halász, I., Mass transfer in ideal and geometrically deformed open tubes II: Potential application of ideal and coiled open tubes in liquid chromatography. *J. Chromatogr.* **1979**, 173, 229–247.

47 Vahey, P. G., Park, S. H., Marquardt, B. J., Xia, Y., Burgess, L. W., Synovec, R. E., Development of a positive pressure driven microfabricated liquid chromatographic analyzer through rapid-prototyping with poly(dimethylsiloxane), optimizing chromatographic efficiency with sub-nanoliter injections. *Talanta* **2000**, 51(6), 1205–1212.

48 O'Neill, A. P., O'Brien, P., Alderman, J., Hoffman, D., McEnery, M., Murrihy, J., Glennon, J. D., On-chip definition of picolitre sample injection plugs for miniaturised liquid chromatography. *J. Chromatogr. A* **2001**, 924(1–2), 259–263.

49 McEnery, M., Tan, A., Glennon, J. D., Alderman, J., Patterson, J., O'Mathuna, S. C., Liquid chromatography on-chip: Progression towards a m-total analysis system. *Analyst (Cambridge, United Kingdom)* **2000**, 125(1), 25–27.

50 McEnery, M., Glennon, J. D., Alderman, J., O'Mathuna, S. C., Fabrication of a miniaturized liquid chromatography separation channel on silicon. *J. Capillary Electrophoresis Microchip Technol.* **1999**, 6(1 & 2), 33–36.

51 Vahey, P. G., Smith, S. A., Costin, C. D., Xia, Y., Brodsky, A., Burgess, L. W., Synovec R. E., Toward a fully integrated positive-pressure driven microfabricated liquid analyzer. *Anal. Chem.* **2002**, 74, 177–184.

52 Schlund, M., Gilbert, S. E., Schnydrig, S., Renaud, P., Continuous sampling and analysis by on-chip liquid/solid chromatography. *Sens. Actuators A* **2006**, in press.

53 Ocvirk, G., Verpoorte, E., Manz, A., Grasserbauer, M., Widmer, H. M., High performance liquid chromatography partially integrated onto a silicon chip. *Anal. Methods Instrum.* **1995**, 2(2), 74–82.

54 Oleschuk, R. D., Shultz-Lockyear, L. L., Ning, Y., Harrison, D. J., Trapping of bead-based reagents within microfluidic systems: On-chip solid-phase extraction and electrochromatography. *Anal. Chem.* **2000**, 72, 585–590.

55 Ceriotti, L., Rooij, N. F. d., Verpoorte, E., An integrated fritless column for on-chip capillary electrochromatography with conventional stationary phases. *Anal. Chem* **2002**, 74, 639–347.

56 Yu, C., Davey, M. H., Svec, F., Fréchet, J. M. J., Monolithic porous polymer for on-chip solid-phase extraction and preconcentration prepared by photoinitiated in situ polymerization within a microfluidic device. *Anal. Chem.* **2001**, 73, 5088–5096.

57 Ericson, C., Holm, J., Ericson, T., Hjerten, S., Electroosmosis- and pressure-driven chromatography in chips using continuous beds. *Anal. Chem.* **2000**, 72(1), 81–87.

58 Bjorkman, H., Ericson, C., Hjerten, S., Hjort, K., Diamond microchips for fast chromatography of proteins. *Sens. Actuators, B* **2001**, 79(1), 71–77.

59 Bjorkman, H., Ericson, C., Hjerten, S., Hjort, K., Diamond microchip capillary chromatography of proteins. Micro Total Analysis Systems **2000**, Proceedings of the mTAS Symposium, 4th, Enschede, Netherlands, May 14–18, 2000 **2000**, 187–190.

60 Ishida, A., Yoshikawa, T., Natsume, M., Kamidate, T., Reversed-phase liquid chromatography on a microchip with sample injector and monolithic silica column. *J. Chromatogr. A* **2006**, 1132, 90–98.

61 Murrihy, J. P., Breadmore, M. C., Tan, A., McEnery, M., Alderman, J., O'Mathuna, C., O'Neill, A. P., O'Brien, P., Advoldvic, N.,

Haddad, P. R., Glennon, J. D., *J. Chromatogr. A* **2001**, 924, 233.

62 Swart, R., Kraak, J. C., Poppe, H., Recent progress in open tubular liquid chromatography. *Trends Anal. Chem.* **1997**, 16, 332–342.

63 Guihen, E., Glennon, J. D., Nanoparticles in separation science – recent developments. *Anal. Lett.* **2003**, 36(15), 3309–3336.

64 Yu, C., Mutlu, S., Selvaganapathy, P., Mastrangelo, C. H., Svec, F., Fréchet, J. M. J., Flow control valves for analytical microfluidic chips without mechanical parts based on thermally responsive monolithic polymers. *Anal. Chem.* **2003**, 75, 1958–1961.

65 Nittis, V., Fortt, R., Legg, C. H., Mello, A. J. D., A high-pressure interconnect for chemical microsystem applications. *Lab Chip* **2001**, 1, 148–152.

66 Kopp, M. U., Crabtree, H. J., Manz, A., Developments in technology and applications of microsystems. *Curr. Opin. Chem. Biol.* **1997**, 1, 410–419.

67 Chen, G., Lin, Y., Wang, J., Monitoring environmental pollutants by microchip capillary electrophoresis with electrochemical detection. *Talanta* **2006**, 68, 497–503.

68 Wang, J., Electrochemical detection for microscale analytical systems: A review. *Talanta* **2002**, 56, 223–231.

69 McEnery, M. M., Aimin, T., Alderman, J., Patterson, J., O'Mathuna, C., Glennon, J. D., Liquid chromatography on-chip: Progression towards a µ-total analysis system. *The Analyst* **2000**, 125(1), 25–27.

70 Pumera, M., Merkoçi, A., Alegret, S., New materials for electrochemical sensing VII. Microfluidic chip platforms. *Trends Anal. Chem.* **2006**, 25, 219–235.

71 LaCourse, W. R., *Pulsed Electrochemical Detection in High Performance Liquid Chromatography*. Wiley Interscience, New York, **1997**.

72 LaCourse, W. R., Dasenbrock, C. O., High-performance liquid chromatography-pulsed electrochemical detection for the analysis of antibiotics. In *Advances in Chromatography*, ed. Brown, P. R., Grushka, E., Marcel Dekker, New York, **1998**, vol. 38, pp. 189–233.

73 Anderson, B. B., Brodsky, A. M., Burgess, L. W., Grating light reflection spectroscopy for determination of bulk refractive index and absorbance. *Anal. Chem.* **1996**, 68, 1081–1088.

74 Anderson, B. B., Brodsky, A. M., Burgess, L. W., Application of grating light reflection spectroscopy for analytical chemical sensing. *At-Process* **1997**, 2, 386–398.

75 Desmet, G., Vervoort, N., Clicq, D., Huau, A., Gzil, P., Baron, G. V., Shear-flow-based chromatographic separations as an alternative to pressure-driven liquid chromatography. *J. Chromatogr. A* **2002**, 948(1–2), 19–34.

76 Clicq, D., Vervoort, N., Vounckx, R., Ottevaere, H., Buijs, J., Gooijer, C., Ariese, F., Baron, G. V., Desmet, G., Sub-second liquid chromatographic separations by means of shear-driven chromatography. *J. Chromatogr. A* **2002**, 979(1–2), 33–42.

77 Desmet, G., Baron, G. V., On the possibility of shear-driven chromatography: A theoretical performance analysis. *J. Chromatogr. A* **1999**, 855(1), 57–70.

78 Clicq, D., Vankrunkelsven, S., Ranson, W., De Tandt, C., Baron, G. V., Desmet, G., High-resolution liquid chromatographic separations in 400 nm deep micro-machined silicon channels and fluorescence charge-coupled device camera detection under stopped-flow conditions. *Anal. Chim. Acta* **2004**, 507(1), 79–86.

79 Clicq, D., Tjerkstra, R. W., Gardeniers, J. G. E., van den Berg, A., Baron, G. V., Desmet, G., Porous silicon as a stationary phase for shear-driven chromatography. *J. Chromatogr. A* **2004**, 1032(1–2), 185–191.

80 Mesaros, J. M., Luo, J. G., Roeraadef, J., Ewing, A. G., Continuous electrophoretic separations in narrow channels coupled to small-bore capillaries. *Anal. Chem.* **1993**, 65, 3313–3319.

81 Bullard, K. M., Gavin, P. F., Ewing, A. G., Electrophoretic separation schemes in ultrathin channels with capillary sample introduction. *Trends Anal. Chem.* **1998**, 17, 401–410.

82 Mesaros, J. M., Gavin, P. F., Ewing, A. G., Flow injection analysis

83 Attiya, S., Jemere, A. B., Tang, T., Fitzpatrick, G., Seiler, K., Chlem, N., Harrison, D. J., Design of an interface to allow microfluidic electrophoresis chips to drink from the fire hose of the external environment. *Electrophoresis* **2001**, 22, 318–327.

84 Lin, Y.-H., Lee, G.-B., Li, C.-W., Huang, G.-R., Chen, S.-H., Flow-through sampling for electrophoresis-based microfluidic chips using hydrodynamic pumping. *J. Chromatogr. A* **2001**, 937, 115–125.

85 Chen, H.-W., Fang, Z.-L., Combination of flow injection with capillary electrophoresis. Part 4. Automated multicomponent monitoring of drug dissolution. *Anal. Chim. Acta* **1998**, 376(2), 209–220.

86 Chen, H.-W., Fang, Z.-L., Combination of flow injection with capillary electrophoresis. Part 3. Online sorption column preconcentration capillary electrophoresis system. *Anal. Chim. Acta* **1997**, 355(2–3), 135–143.

87 Chen, H.-W., Fang, Z.-L., Combination of flow injection with capillary electrophoresis. Part 5. Automated preconcentration and determination of pseudoephedrine in human plasma. *Anal. Chim. Acta* **1999**, 394(1), 13–22.

88 Liu, Z.-S., Fang, Z.-L., Combination of flow injection with capillary electrophoresis. Part 2. Chiral separation of intermediate enantiomers in chloramphenicol synthesis. *Anal. Chim. Acta* **1997**, 353(2–3), 199–205.

89 Fang, Z.-L., Liu, Z.-S., Shen, Q., Combination of flow injection with capillary electrophoresis. Part I. The basic system. *Anal. Chim. Acta* **1997**, 346(2), 135–143.

90 Huang, X. J., Pu, Q. S., Fang, Z. L., Capillary electrophoresis system with flow injection sample introduction and chemiluminescence detection on a chip platform. *The Analyst* **2001**, 126(3), 281–284.

91 Xu, Z. R., Pan, H. Y., Xu, S. K., Fang, Z. L., A sequential injection on-line column preconcentration system for determination of cadmium by electrothermal atomic absorption spectrometry. *Spectrochim. Acta, Part B* **2000**, 55(3), 213–219.

92 Fu, C. G., Fang, Z. L., Combination of flow injection with capillary electrophoresis Part 7. Microchip capillary electrophoresis system with flow injection sample introduction and amperometric detection. *Anal. Chim. Acta* **2000**, 422(1), 71–79.

93 Pu, Q. S., Fang, Z. L., Combination of flow injection with capillary electrophoresis. Part 6. A bias-free sample introduction system based on electroosmotic-flow traction. *Anal. Chim. Acta* **1999**, 398(1), 65–74.

94 Fang, Q., Shi, X. T., Sun, Y. Q., Fang, Z. L., A flow injection microdialysis sampling chemiluminescence system for in vivo on-line monitoring of glucose in intravenous and subcutaneous tissue fluid microdialysates. *Anal. Chem.* **1997**, 69(17), 3570–3577.

95 Fang, Q., Wang, F.-R., Wang, S.-L., Liu, S.-S., Xu, S.-K., Fang, Z. L., Sequential injection sample introduction microfluidic-chip based capillary electrophoresis system. *Anal. Chim. Acta* **1999**, 390(1–3), 27–37.

96 Fang, Q., Xu, G.-M., Fang, Z.-L., High throughput continuous sample introduction interfacing for microfluidic chip-based capillary electrophoresis systems. Micro Total Analysis Systems 2001, Proceedings mTAS 2001 Symposium, 5th, Monterey, CA, United States, Oct. 21–25, 2001, **2001**, 373–374.

97 Fang, Z. L., Fang, Q., Development of a low-cost microfluidic capillary-electrophoresis system coupled with flow-injection and sequential-injection sample introduction (review). *Fresenius' J. Anal. Chem.* **2001**, 370(8), 978–983.

98 Fang, Q., Xu, G.-M., Fang, Z.-L., A high-throughput continuous sample introduction interface for microfluidic chip-based capillary electrophoresis systems. *Anal. Chem.* **2002**, 74, 1223–1231.

99 Büttgenbach, S., Wilke, R., A capillary electrophoresis chip with hydrodynamic sample injection for measurements from a continuous sample flow. *Anal. Bioanal. Chem.* **2005**, 383, 733–737.

100 Wang, J., Siangproh, W., Thongngamdee, S., Chailapakul, O., Continuous monitoring with microfabricated capillary electrophoresis chip devices. *Analyst* **2005**, 130, 1390–1394.

101 Svec, F., Huber, C. G., Monolithic materials: Promises, challenges, achievements. *Anal. Chem.* **2006**, 2101–2107.

102 Vreeland, W. N., Barron, A. E., Functional materials for microscale genomic and proteomic analyses. *Curr. Opin. Biotechnol.* **2002**, 13, 87–94.

103 Weigl, B. H., Bardell, R. L., Cabrera, C. R., Lab-on-a-chip for drug development. *Adv. Drug Delivery Rev.* **2003**, 55, 349–377.

104 Burns, M. A., Johnson, B. N., Brahmasandra, S. N., Handique, K., Webster, J. R., Krishnan, M., Sammarco, T. S., Man, P. M., Jones, D., Heldsinger, D., Mastrangelo, C. H., Burke, D. T., An integrated nanoliter DNA analysis device. *Science* **1998**, 282, 484–487.

105 Marasco, C. A., Pharma's process analytical technology. *Chem. Eng. News* Feb. 21, **2005**, 201–206.

106 Pharmaceutical cGMPs for the 21st Century – A Risk-Based Approach Final Report. In Department of Health and Human Services, FDA, Ed. **2004**.

107 Guidance for Industry: PAT – A Framework for Innovative Pharmaceutical Development, Manufacturing, and Quality Assurance (Pharmaceutical cGMPs). In *Pharmaceutical cGMPs* ed., US Department of Human Services, FDA, Center for Drug Evalutation and Research (CDER), Center for Veterinary Medicine (CVM) and Office of Regulatory Affairs (ORA), Ed. **2004**.

108 Koch, M. V., Marquardt, B., Impact of micro-instrumentation on process analytical technology (PAT): A consortium approach. *J. Process Anal. Technol.* **2004**, 1(2), 12–15.

109 Thayer, A. M., Harnessing microreactions. *Chem. Eng. News* **2005**, 83(22), 43–52.

110 Thayer, A. M., Speeding up process development. *Chem. Eng. News* May 30, **2005**, 54–61.

111 Henry, C. M., Celebrating 20 years of service to industry. *Chem. Eng. News* May 17, **2004**, 40–41.

112 Workman, J., Koch, M., Veltkamp, D., Process analytical chemistry. *Anal. Chem.* **2005**, 77, 3789–3806.

113 Kornmann, H., Valentinotti, S., Duboc, P., Marison, I., Stockar, U. v., Monitoring and control of *Gluconacetobacter xylinus* fed-batch cultures using in situ mid-IR spectroscopy. *J. Biotechnol.* **2004**, 113, 231–245.

114 Schügerl, K., Progress in monitoring, modeling and control of bioprocesses during the last 20 years. *J. Biotechnol.* **2001**, 85, 149–173.

115 van de Merbel, N. C., Membrane-based sample preparation coupled on-line to chromatography or electrophoresis. *J. Chromatogr. A* **1999**, 856, 55–82.

116 van de Merbel, N. C., Lingeman, H., Brinkman, U. A. T., Sampling and analytical strategies in on-line bioprocess monitoring and control. *J. Chromatogr. A* **1996**, 725, 13–27.

117 Groton Biosystems. www.grotonbiosystems.com

118 Karwa, M., Hahn, D., Mitra, S., A sol-gel immobilization of nano an micro size sorbents in poly(dimethylsiloxane) (PDMS) microchannels for microscale solid phase extraction (SPE). *Anal. Chim. Acta* **2005**, 546, 22–29.

119 Le Gac, S., Carlier, J., Camart, J.-C., Cren-Olive, C., Rolando, C., Monoliths for microfluidic devices in proteomics. *J. Chromatogr. B* **2004**, 808(1), 3–14.

120 Steigert, J., Grumann, M., Brenner, T., Mittenbuhler, K., Nann, T., Ruhe, J., Moser, I., Haeberle, S., Riegger, L., Riegler, J., Integrated sample preparation, reaction, and detection on a high-frequency centrifugal microfluidic platform. *J. Assoc. Lab. Automation* **2005**, 10(5), 331–341.

121 Liu, W.-T., Zhu, L., Environmental microbiology-on-a-chip and its future impacts. *Trends Biotechnol.* **2005**, 23(4), 174–179.

122 Carlier, J., Arscott, S., Thomy, V., Camart, J.-C., Cren-Olive, C., Le Gac, S., Integrated microfabricated systems including a purification module and an on-chip nano electrospray ionization interface for biological analysis.

J. Chromatogr. A **2005**, 1071(1–2), 213–222.

123 Jandik, P., Weigl, B. H., Kessler, N., Cheng, J., Morris, C. J., Schulte, T., Avdalovic, N., Initial study of using a laminar fluid diffusion interface for sample preparation in high-performance liquid chromatography. *J. Chromatogr. A* **2002**, 954(1–2), 33–40.

124 Bruckner-Lea, C. J., Tsukuda, T., Dockendorff, B., Follansbee, J. C., Kingsley, M. T., Ocampo, C., Stults, J. R., Chandler, D. P., Renewable microcolumns for automated DNA purification and flow-through amplification: From sediment samples through polymerase chain reaction. *Anal. Chim. Acta* **2002**, 469(1), 129–140.

125 Laurell, T., Marko-Varga, G., Ekstrom, S., Bengtsson, M., Nilsson, J., Microfluidic components for protein characterization. *Rev. Mol. Biotechnol.* **2001**, 82(2), 161–175.

126 Schulte, T., Bardell, R., Weigl, B. H., Sample acquisition and control on-chip microfluidic sample preparation. *J. Assoc. Lab. Automation* **2000**, 5(4), 83–86.

127 Gao, J., Yin, X.-F., Fang, Z.-L., Integration of single cell injection, cell lysis, separation and detection of intracellular constituents on a microfluidic chip. *Lab Chip* **2004**, 4(1), 47–52.

128 Lee, S.-H., Sohn, O.-J., Yim, Y.-S., Han, K.-A., Hyung, G. W., Chough, S. H., Rhee, J. I., Sequential injection analysis system for on-line monitoring of L-cysteine concentration in biological processes. *Talanta* **2005**, 68, 187–192.

129 Economou, A., Sequential-injection analysis (SIA): A useful tool for on-line sample-handling and pretreatment. *Trends Anal. Chem.* **2005**, 24, 416–424.

130 Lichtenberg, J., Rooij, N. F. D., Verpoorte, E., Sample pretreatment on microfabricated devices. *Talanta* **2002**, 56, 233–266.

9.8
Grating Light Reflection Spectroscopy:
A Tool for Monitoring the Properties of Heterogeneous Matrices

Lloyd W. Burgess

9.8.1
Overview

We have developed an optical diagnostic method, Grating Light Reflection Spectroscopy (GLRS), based on measurements near critical points of intensity and phase in waves reflected from a transmission diffraction grating in contact with a diagnostic sample. The GLRS platform provides a simple, rugged, versatile method for monitoring the properties of heterogeneous matrices, and is employed as a backscattering geometry that can be applied to on-line control in the industrial processes or within a microfluidic channel for continuous micro reactor monitoring. The technique exploits anomalies in the reflection of waves from a diffraction grating at special threshold values of parameters (singular points) where one of the beams with diffraction order m_{cr} in the analyte medium transforms itself from a traveling into an evanescent wave. At these threshold parameter values, the energy of the transforming beam is redistributed among all other reflected and transmitted beams. Such redistribution has a singular character, and it can be theoretically described in a relatively simple manner, analogous to methods used in the general theory of multi-channel wave scattering. The features contained in the reflection spectrum near the thresholds allow for separation of surface and bulk effects in a sample and for simultaneous determination of the real and imaginary (absorptive) parts of the dielectric function of the sample. GLRS can provide analytically relevant information regarding mesoscopic suspensions and colloids in fluid samples, ranging from 1 nm up to 0.1 mm in diameter.

9.8.2
Introduction

Continuous process analysis demands rugged, sensitive, and selective sensing techniques that may be implemented with a minimum of sample preparation. We have developed an optical diagnostic method, Grating Light Reflection Spectroscopy (GLRS), which has shown promise as an optical sensor platform to be used in highly scattering, absorbing, and surface-active matrices. GLRS is a patented technique [1] based on measurements near critical points of intensity and phase in waves reflected from a transmission diffraction grating in contact with a diagnostic sample [2–4]. The technique exploits anomalies in the reflection of waves from a diffraction grating at special threshold values of parameters (singular points) where one of the beams with diffraction order m_{cr} in the analyte medium transforms itself from a traveling into an evanescent wave. At these threshold parameter values, the energy of the transforming beam is redistributed among all other reflected and transmitted beams. Such redistribution has a singular character, and it can be theoretically described, relatively simply, in a manner analogous to the method previously used in the general theory of multi channel wave scattering [2]. The rich information contained in the reflection spectrum near the thresholds allows for separation of surface and bulk effects in a sample and for simultaneous determination of the real and imaginary (absorptive) parts of the dielectric function of the sample.

We have demonstrated that GLRS can provide analytically relevant information regarding mesoscopic suspensions and colloids in fluid streams. Using GLRS, we can monitor changes in particle size continuously from 1 nm up to 0.1 mm in diameter. Such measurements are based on the effect of the coherence loss during wave scattering by particles and other non-uniformities in the sample matrix, where scattering events that change the properties of evanescent waves near the threshold lead to changes in the threshold characteristics. These changes manifest themselves by the appearance of a specific imaginary component in the effective dielectric constant. This result is due to coherence distortion of transmitted waves, which takes place even if the particles do not absorb waves in the exploited frequency interval. GLRS implementation involves the use of a transmission diffraction grating fabricated on an optically transparent substrate (Fig. 9.8.1). This grating consists of a micro-fabricated array of alternating strips of dielectrically contrasting materials oriented perpendicular to the incident plane of the light. The grating contacts the sample of interest, and light enters through the substrate. The incident light strikes the grating at an angle θ such

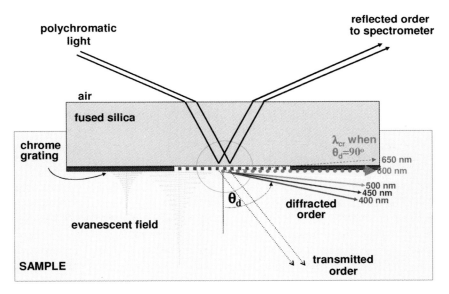

Fig. 9.8.1 The GLRS mechanism involves projecting light at the grating–sample interface and collecting the 0th order reflected light.

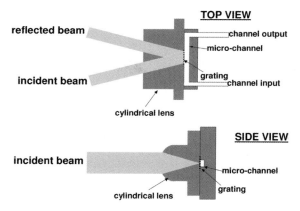

Fig. 9.8.2 Micro-channel liquid interface with GLRS detection with a sampled volume of ~180 nL.

that, depending upon the dielectric properties of the sample, a certain wavelength (the critical wavelength) within one of the diffracted orders will propagate parallel to the grating surface. Thus, the critical wavelength represents the threshold between traveling waves and evanescent waves, and exhibits a sharp discontinuity in intensity. Since all orders are coupled via the grating, this discontinuity arises in all of the other orders, including the 0th reflected order, which is the easiest to collect and analyze.

For practical applications, notably, GLRS can be used in a backscattering (180°) geometry, which can be easily fabricated into a dip-type or flow through probe for processes monitoring (Fig. 9.8.2).

9.8.3
Theory

Figure 9.8.1 also illustrates the basic GLRS sensing mechanism. Polychromatic light is projected at a transmission diffraction grating, a diffracted order produces an evanescent field, and the reflected order is analyzed with a spectrometer.

The positions of the singularities correspond to the zeros of the function of Eq. (1):

$$\delta(\omega,\theta); \delta = 0 \tag{1}$$

with δ depending upon the real part of the dielectric function of the sample, Re ε_2,

$$\delta \equiv \text{Re } \varepsilon_2 - \left(\sin\theta + m\frac{\lambda}{a}\right)^2 \tag{2}$$

where θ is the angle of incidence, m is the diffracted order, λ is the wavelength of light, and a is the grating period. The dependence of the reflected intensity,

$I(\lambda)$, on the position of the singularity, δ, and the imaginary part of the sample's dielectric function $\mathrm{Im}\,\varepsilon_2$ is given by Eq. (3):

$$I(\lambda) = c_1 \frac{2\pi}{\lambda} \sqrt{\frac{\delta + \sqrt{\delta^2 + (\mathrm{Im}\,\varepsilon_2)^2}}{2}} + c_2 \qquad (3)$$

The two constants, c_1 and c_2, in this expression represent the particular interfacial properties of a given sensor. Factors include the materials used to make the grating and its substrate as well as the incident angle, polarization state, and frequency of the incident light beam. These factors are important when considering the exact combination that will optimize the sensor performance for a given application. One can see from the terms under the square roots in Eq. (3) that the reflected intensity acquires an acute dependence on $\mathrm{Im}\,\varepsilon_2$ as δ approaches zero. Combining the features of Eqs. (1–3) shows that the spectral position of the singularity primarily depends upon the real part of the sample's dielectric function, while the magnitude/sharpness of the singularity depends on the imaginary part of the sample's dielectric function. We can use these equations to extract the real and imaginary parts of the sample's dielectric function from GLRS spectra without having to resort to indirect calibrations or multivariate statistical methods. Each of the parameters in Eq. (2) may be modulated with respect to the others to yield the vector of data containing the singularity. The most expedient experimental procedure holds both the incident angle and the grating period constant while perturbing the sample dielectric properties to generate a wavelength-dependent spectrum. The singularity then appears in this spectrum centered at a wavelength determined by Eq. (2). As with any reflection phenomenon, this technique is polarization sensitive, yielding different spectral forms for s- and p-polarized incident light.

For heterogeneous media, where we wish to relate changes in the $\mathrm{Re}\,\varepsilon_2$ and $\mathrm{Im}\,\varepsilon_2$ with the properties of particles distributed in the sample medium, we can use the coherent phase approximation [5], which states:

$$\varepsilon_2 \cong \varepsilon_m + \frac{\lambda^2}{\pi} \sum_\alpha N_\alpha A_\alpha(0) \qquad (4)$$

where ε_m is the permittivity of the suspending medium, N_α is the mean concentration of α-type particles, and $A_\alpha(0)$ is the complex forward scattering amplitude of those particles. This approximation holds in the case of not-very-high volume concentrations of particles and not-very-large dielectric contrasts between the particles and the suspending medium. For particles with dimensions less than the critical wavelength we can use the Rayleigh–Gans approximation to express the real and imaginary components of ε_2 as follows:

$$\mathrm{Re}\,\varepsilon_2 = \varepsilon_m + \rho \frac{\varepsilon_p - \varepsilon_m}{\varepsilon_p + 2\varepsilon_m} \varepsilon_m \qquad (5)$$

$$\operatorname{Im}\varepsilon_2 = \frac{3\pi}{4}\frac{\rho\overline{R}}{\lambda}(\varepsilon_m)^{\frac{3}{2}}(\varepsilon_p - \varepsilon_m)^2 \qquad (6)$$

where $\overline{R}$ is the mean particle radius, ρ is the volume fraction of the particles, and ε_p is their permittivity. Equations (5) and (6) are valid when $R/\lambda\,(\varepsilon_p - \varepsilon_m) < 1$. This value of ε_p is considered to be a particle characteristic parameter. We have assumed in Eqs. (5) and (6) that the solvent and the particles do not have appreciable absorption in the considered light frequency range. Finally, these two equations give us two relatively simple expressions that we can readily substitute into Eqs. (2) and (3) to extract the concentration and size of a scattering species suspended in solution so that, as with refractive index and absorbance, we do not have to rely on indirect calibrations. The real part of $A_\alpha(0)$ declines after reaching a maximum at $R/\lambda\,(\varepsilon_p - \varepsilon_m) \sim 1$. At the same time, the imaginary part of $A_\alpha(0)$ grows, and becomes proportional to the so-called *radar cross-section* (Fraunhofer limit) as $R/\lambda \gg 1$.

9.8.4
Results and Discussion

In the past we have presented both theoretical and experimental results demonstrating that GLRS can provide information about the concentration and average size of particles of nanometer dimensions distributed in liquid phase media [6]. To demonstrate this, we performed experiments on various concentrations of dendrimeric oligomers in water. The measurements were carried out in a continuous flow format using the micro-channel flow system illustrated above, giving the optical response shown in Fig. 9.8.3. Each size of dendrimer was sampled at ~1, 2, 3, 4, and 5% w/w; dispersion resulted in effective volume fraction dilution. The response, fitted for variable dilution and particle reference dimension (Fig. 9.8.4), demonstrates that it is clearly possible to distinguish size differences in particles distributed in water whose radii varied by less than 1 nm at concentrations of 4% w/w and above. This can be done even in the case where the characteristic length of these non-uniformities is substantially less than the scattered light wavelength, and can in fact be on the order of molecular dimensions.

Most recently we have been able to demonstrate the use of GLRS as a potential tool for on-line monitoring of milling processes over a very wide dynamic range [7]. Attrition milling is an important industrial technique used to reduce the particle size of materials, such as pharmaceutical actives, from several millimeters in size down to submicron sizes. For particles of this size regime, it is appropriate to use Van de Hulst's approximation to model the GLRS response. Figure 9.8.5 shows the first derivative modeled GLRS response, which assumes that the particles are spherical and non-absorbing, for particles with ε_p of $(1.54)^2$ in the range of ~5 μm to 100 nm in water. Samples in the small particle regime exhibit a singular response at a longer wavelength. As the particle size increases, the peak height decreases and shifts to a shorter wavelength, growing again as the particle dimension increases.

310 | *9 Selected Developments in Micro-analytical Technology*

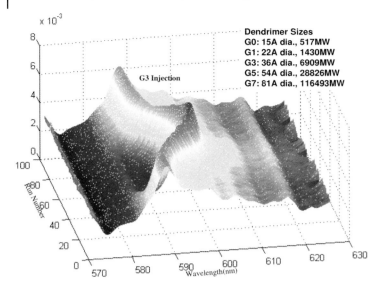

Fig. 9.8.3 Example of the spectral and temporal response of a 10-µL sample injected at a flow rate of 30 µL min^{-1}.

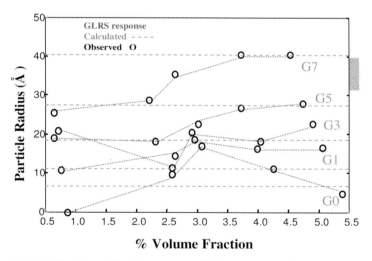

Fig. 9.8.4 Each size of dendrimer was sampled at ~1, 2, 3, 4, and 5% w/w; dispersion resulted in effective volume fraction dilution to the above concentration.

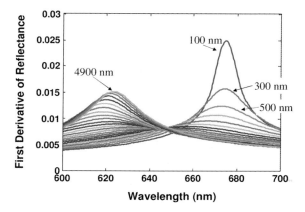

Fig. 9.8.5 Theoretical first derivative response at 30% concentration as particle size increases from 100 to 4900 nm in 200 nm increments.

Subsequently, ten randomly labeled blind samples (A–J) containing 30 (w/w)% active ingredient in water were collected at various time points during an attrition milling process and analyzed with GLRS. To analyze the samples, approximately 250 µL of each sample was injected into a GLRS flow cell and the spectra were recorded under stop flow conditions. The first derivative plots for the 30% samples are shown in Fig. 9.8.6. Since it was known that the concentration was constant, the shift in peak wavelength from the large particle scattering regime to the small particle scattering regime was observed as expected as milling progressed.

Principal components analysis (PCA), a multivariate statistical technique, was used for data reduction and pattern recognition [8]. PCA represents the variation present in many variables using a small number of principal components (PCs). PCA functions by finding a new set of axes, which more efficiently describe the variance in the data. Samples are no longer described by their intensities in

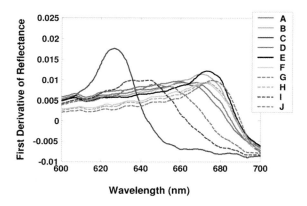

Fig. 9.8.6 GLRS first derivative plots of attrition milled samples.

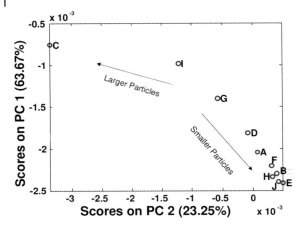

Fig. 9.8.7 Principal components analysis (PCA) scores plot of 2nd derivative data for 30% samples.

wavelength space; instead they are described by their score values on a new set of axes. PCA was performed on the second-derivative spectra obtained from the 30% samples. The PCA scores plot is shown in Fig. 9.8.7 and compared with reference results acquired using a Coulter LS 230 light scattering device at the diluted concentration of 1% (Table 9.8.1). Note that this two-component model only accounts for ~87% of the total system variance; however, the results validate that GLRS is indeed a useful technique for on-line monitoring of attrition milling processes at high concentrations.

Table 9.8.1 Reference results using a Coulter LS 230 at 1%.

Sample	Time (hours)	Size (nm)
C	0	39 270
I	1	25 500
G	2	3 263
D	4	731
A	6	438
F	8	369
H	10	232
B	12	228
J	14	199
E	16	185

9.8.5
Significance/Conclusions

The GLRS platform provides a simple, rugged, versatile method for monitoring the properties of heterogeneous matrices, and future work on improving the sensitivity, information capacity, and range of applicability of the technique will make it a powerful tool for process monitoring and high throughput experimentation. Two extensions of this effort include the investigation of an acoustic reflection embodiment of our optical work, being carried out in collaboration with Pacific Northwest National Laboratories, in Richland Washington [9], and a project to enhance the sensitivity and selectivity of the technique by applying electrochemical modulation at the metal/dielectric grating interface. This electrochemical modulation will allow the characteristics of charged colloids to be analyzed.

References

1. Anderson, B. B., Brodsky, A. M., Burgess, L. W., *U.S. Pat.* 5,502,560 (March **1996**), *U.S. Pat.* 5,610,708 (March **1997**).
2. Anderson, B. B., Brodsky, A. M., Burgess, L. W., *Phys. Rev. E* **1996**, 54, 912–921.
3. Brodsky, A. M., Burgess, L. W., Smith, S. A., *Appl. Spectrosc.* **1998**, 52, 332A–343A.
4. Anderson, B. B., Brodsky, A. M., Burgess, L. W., *Langmuir* **1997**, 13, 4273–4279.
5. Newton, R. G. *Scattering Theory of Waves and Particles*, McGraw-Hill, New York, **1966**.
6. Smith, S. A., Brodsky, A. M., Vahey, P. G., Burgess, L. W., *Anal. Chem.* **2000**, 72(18), 442.
7. Hamad, M. L., Kailasam, S., Brodsky, A. M., Han, R., Higgins, J. P., Thomas, D., Reed, R. A., Burgess, L. W., *Appl. Spectrosc.* **2005**, 59, 1.
8. Beebe, K. R., Pell, R. J., Seasholtz, M. B., *Chemometrics: A Practical Guide*, John Wiley & Sons, Inc., New York, **1998**.
9. Greenwood, M. S., Brodsky, A., Burgess, L., Bond, L. J., Hamad, M., *Ultrasonics* **2004**, 42, 531–536.

10
New Platform for Sampling and Sensor Initiative (NeSSI)

David J. Veltkamp

10.1
Introduction

This chapter describes a new type of fluidic system developed by the Chemical and Petrochemical Industries for process analyzers and their related sample handling systems. These new systems are collectively referred to, and are being developed under, the umbrella of NeSSI™ (for New Sampling/Sensor Initiative). From its beginnings, NeSSI™ has been affiliated with, and sponsored by, the Center for Process Analytical Chemistry (CPAC) at the University of Washington, Seattle, Washington, USA. Funded largely through Industrial Sponsor membership, CPAC is a multidisciplinary research center that began over twenty years ago as an NSF Industry/University Cooperative Research Center (IUCRC) directed towards fostering research in analytical sensing for the monitoring and control of industrial chemical and manufacturing processes. Now a self-sustaining research consortium still based at the University of Washington, and now involving several other universities, CPAC enjoys the support of close to 50 full and associated Sponsors from a diverse cross section of industries and National Laboratories. CPAC continues to be one of the primary focal points for pre-competitive (i.e., basic) research and open exchange on industrially relevant issues related to all aspects of process monitoring and control.

The inclusion of material on NeSSI™ is instructive in the context of this book for several reasons. First, the NeSSI™ development effort provides a good example of industry-driven efforts, and the effective use of consortia, to establish standards and hardware implementations needed to serve common industry-wide needs. Many challenges and issues related to implementation of process intensification and microreactors could benefit from a similar industry-driven consortia approach. Broad-based consortia, involving multiple industry sectors, can effectively direct research efforts on common issues and solutions. More importantly, they can serve as a unifying voice of consensus – bringing together multiple concerns and producing a single vision, or roadmap, to address those concerns. Second, the NeSSI™ efforts are resulting in commercially available

Micro Instrumentation for High Throughput Experimentation and Process Intensification – a Tool for PAT
Edited by M. V. Koch, K. M. VandenBussche and R. W. Chrisman
Copyright © 2007 WILEY-VCH Verlag GmbH & Co. KGaA, Weinheim
ISBN: 978-3-527-31425-6

hardware that could complement microreactor component developments and provide some of the fluidic interconnects and control functionality needed to help produce commercially viable microreactor systems. A common theme in this book is that microreactors (by virtue of their efficient and well-understood mass and thermal transport) can be valuable tools to understanding fundamental mechanisms in chemical processes and a means to efficiently produce materials with economies in size and waste minimization. These goals are consistent with those of process intensification. Unfortunately, commercialization of microreactors has been slow – only a few commercial systems are available, and these tend to be specialized or research systems. In contrast, many more discrete microreactor components are commercially available today. These components can be selected and combined to develop a mini chemical plant suitable for a given application. A common fluidics and sensing hardware platform, such as NeSSI™, could serve as the basis for building these types of custom microreactor systems.

Finally, the NeSSI™ efforts are facilitating the development of micro, or small-scale, chemical sensors and analyzers that could provide much of the needed sensing ability for monitoring and controlling process operations. Successful implementation of process intensification and microreactor manufacturing require the ability to measure and monitor key parameters (often requiring new analytical methods or techniques) at various stages of the process. In addition to serving as a platform for interfacing new sensing technologies to the process sample stream, NeSSI™ is also being utilized in the laboratory to support the development and testing of these new analytical tools and applications.

10.2
What is NeSSI™?

NeSSI™ (an acronym for New Sampling/Sensor Initiative) is an industry driven effort to develop and promote a new functional standard in sampling systems for industrial process analyzers. The goal of NeSSI™ is to facilitate the acceptance/ implementation of modular, miniature, and smart sample system technology to address limitations in current sampling system practices. The three keystone terms of this new technology (i.e., *modular, miniature,* and *smart*) together provide the increased functionality, ease-of-use, and flexibility needed for the sampling systems of the future and are discussed in more detail below. Notably, NeSSI™ is not just a new hardware standard (although the specification for modular surface mount hardware components based on the ANSI/ISA SP76.00.02 standard is certainly a key component of NeSSI™). Rather, NeSSI™ attempts to view all aspects of an automated sampling system (i.e., the physical sample transport, the electrical signals needed for power and communication, the processing needed for control and data handling, and its place in the plant communication and control hierarchy) and to provide specifications, or guidelines, needed for development of these systems [1].

To accomplish this, the NeSSI™ effort has focused on developing consensus among industrial end-users, technology developers, and hardware manufacturers about what is desirable and possible in sampling systems. Developing this consensus and moving the project forward is mainly carried out through regular communications and interactions of an *ad hoc* group often referred to as the NeSSI™ enthusiasts or user group. This group tries to translate the collective vision into specifications and a development roadmap. Development progress is reported, and promoted, at numerous national meetings and workshops, under the guidance of a NeSSI™ Steering Team. Wherever possible, the emphasis has been on identifying, and adapting, existing open standards that can provide the desired functionality of the NeSSI™ specifications. Where there is no suitable pre-existing technology, efforts are focused on working with standard generating committees and governing bodies to help direct future standards to include NeSSI™ specifications. Figure 10.1 shows a schematic of a complete (sometimes called a Gen II or "smart") NeSSI™ system.

The development of NeSSI™ is very much a collaborative result of end-users, manufacturers, researchers, and many others working together to modularize and miniaturize process analyzer sample systems. The technical and standardization concepts of NeSSI have been well documented through numerous presentations and workshops throughout the last four years, including yearly updates at ISA, PITTCON and FACSS conferences and updates at the biannual CPAC meetings.

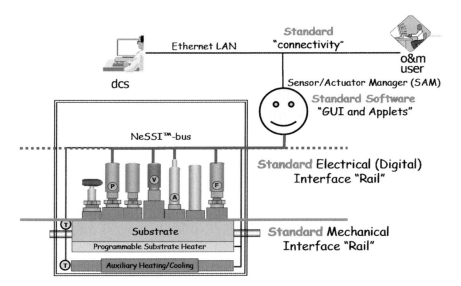

Fig. 10.1 Schematic representation of a complete NeSSI™ system, illustrating the main components and the standard interfaces outlined in the NeSSI™ specification document. Also shown are the relationship between the NeSSI™ system and the sample and data process connections.

Fig. 10.2 Example NeSSI™ systems from CIRCOR (left), Swagelok (center), and Parker-Hannifin (right).

The NeSSI™ web site (http://www.cpac.washington.edu/NeSSI/) contains a complete record of all papers, presentation, and reports related to the development of NeSSI™. These are arranged in reverse chronological order from 1999 to the present. The NeSSI™ approach has been very successful to date [2], with commercially available substrate platforms from CIRCOR Tech (Spartanburg, SC; www.circortech.com), Parker-Hannifin (Jacksonville, AL; www.parker.com), and Swagelok (Solon, OH; www.swagelok.com) and SP76 components (valves, filters, mass flow controllers, sensors, and analytical analyzers) commercially available from a growing number of vendors. Examples of NeSSI™ systems from the three major substrate vendors are shown in Fig. 10.2. Each of these NeSSI™ systems uses a different architecture for the fluidic flow between the modular SP76 surface mount component locations, but any SP76 compliant component can be interchanged between systems.

In Fig. 10.1, the box labeled "Substrate" represents the flow path of the process sample through the NeSSI™ system. Currently there are three major vendors of NeSSI™ substrates. The substrates are the MPC™ (Modular Platform Components) system from Swagelok®, the IntraFlow™ system from Parker-Hannifin, and the μMS3™ (micro Modular Substrate Sampling System) system from CIRCOR Tech. All three systems utilize the same SP76 modular surface mount standard for mounting components (shown as the parts above the substrate block in Fig. 10.1), guaranteeing interchangeability of components between systems, but use different architectures and methods of assembly for the actual flow paths between components and the process stream connections. These three NeSSI™ substrate systems, and their differentiating characteristics, are discussed in more detail below. Also shown in Fig. 10.1 are optional auxiliary heating/cooling, substrate heater, and a micro-enclosure box around the substrate and surface mount components – these may be required for different applications and are discussed in the NeSSI™ specification document. Collectively, the parts of the NeSSI™ system described in this paragraph are often referred to as Gen I, or physical, NeSSI™ and have been commercially available for several years.

Connecting the modular surface-mount components (and optional auxiliary components) in Fig. 10.1 is the NeSSI™-bus, which provides the electrical power and communication functions for controlling actuators and transmitting data to, and from, the NeSSI™ system. The NeSSI™-bus terminates at the Sensor/Actuator Manager (SAM), represented by the smiley circle in Fig. 10.1. The SAM is a combination of software and hardware responsible for handling the communication protocols for the NeSSI™-bus, processing data from sensors and analyzers, controlling actuators, and communicating status of the NeSSI™ and analyzer systems with the higher level plant distributed control system (DCS) and operations and maintenance (o&m) information systems. Both the NeSSI™-bus (a generic term coined to cover any number of open multi-drop field bus network implementations) and the SAM are important aspects of the NeSSI™ vision for adding "smarts" to the system and are described in more detail below. Together with the substrate, surface mount components, and other parts described above they represent what is often called Gen II, or "electrified", NeSSI™ system. Including standardized communication and data processing capability in the development specifications lays the foundation for validation of sample conditioning, system status reporting, data fusion, and automation of NeSSI™ systems. The goal is to achieve the same type of auto-recognition and auto-configuration of NeSSI™ devices as users are accustomed to finding with "plug-and-play" technology (such as USB devices) in modern personal computers.

There is also a huge opportunity to adapt the emerging class of "lab on a chip" sensors to a miniature/modular "smart" manifold, which could fundamentally change the way industry does process analysis. The use of NeSSI™ as a platform to stimulate development and deployment of micro instrumentation was initially described as a third generation, or Gen III, NeSSI™ system – but in reality this type of development has been going on since the first NeSSI™ hardware began to be available. Certainly, as NeSSI™ continues to evolve, its value and utility as a generic platform for innovation and implementation of new sensing technologies will also grow. NeSSI™ also lays the groundwork for the next generation of process analytical systems to be developed, featuring open connectivity architecture (Ethernet, TCP/IP, wireless, etc.), industry standard protocols (OPC, Fieldbus, Profibus, Hart, etc.), and web enabled technology (browsers, HMI, I/O servers, etc.).

NeSSI™ continues to evolve and pursue the goals of:

- Field-mounted analytical systems located in mini enclosures (boxes) near the point-of-use for the analytical result (i.e., "on-the-pipe" monitoring). This implies a move away from expensive centralized analyzer sheds with their associated long runs of sample lines.
- Easy integration of sample systems with multiple physical/chemical sensors utilizing modern multi-drop communication networks. An important aspect of the communication network (NeSSI™-bus) is "plug-n-play" (i.e., self-identification/self-configuration of sensors and actuators) interchangeability.
- Increased use of small, smart, integrated sampling, sensor, and analyzer transmitters to provide more information about the sample and the process.
- Validation of representative sample and analysis.

10.3
NeSSI™ Background

To analyze a process sample there are three choices: (a) take a sample from the process to the laboratory and run the analysis there, (b) extract a small portion of the process stream and direct that sample to flow through an analyzer located in some proximity to the process, or (c) use a probe, or other means, to insert the point of analysis into the process stream itself [3]. The first of these choices is often termed "off-line" or "grab" analysis and is still perhaps one of the most common approaches to process monitoring, especially for quality control types of analysis. There are, however, significant limitations to this type of sampling – the main one being the additional time needed to transport the sample, perform the analysis, and report the results. This tends to limit the utility of off-line analysis for providing feedback control of the process. The second choice is often termed "at-line" or "extractive" analysis and most industrial process analyzers used for controlling the process use this type of system. The advantages of at-line analysis are that the instruments used to perform the analysis can be somewhat isolated from harsh process conditions while still providing near real-time analysis results for feedback control. Of course, for at-line analysis there must be additional hardware (valves, filters, plumbing, etc.) installed to supply the extractive sample loop that carries the sample to the analyzer. This additional hardware is what is typically referred to as the "sampling system". Ideally, the sampling system will provide a representative sample of the bulk process stream and will condition the sample (e.g., filtering of particulates, adjustment of pressure and temperature, dilution, etc.) such that it is suitable for analysis by the analyzer. Finally, the third choice in process monitoring is termed "in-line" or "*in situ*" analysis. One attractive aspect of in-line analysis is that it requires no additional sampling efforts or sampling hardware – the analysis is performed inside (either through probes or windows) the process itself. That is not to say that sampling is not an issue with in-line analysis techniques – great care must be taken to insure that the environment being analyzed by the *in situ* method is representative of the bulk process. Fouling, or otherwise degrading, the *in situ* probe, or windows, by the process needs to be considered and accounted for in the design and application of in-line analyzers. Another troubling aspect of in-line analyzers is the difficulty of running calibration, or validation samples, through the analysis to assure validity of the results being reported. In extractive systems it is relatively easy to switch the sample stream with a validation stream of know properties, when the analyzer is in the process this is less easily accomplished. These three types of process analysis, broadly classified by their approach to presenting the sample for analysis, each have respective pros and cons. The emergence of Process Analytical Technology (PAT) – first in the chemical and petrochemical industries, and now the pharmaceutical and biotech industries – clearly favors either at-line or in-line analysis for their ability to produce analytical results on a time scale needed for process control and/or understanding [4]. Since NeSSI™ was primarily conceived as a replacement for traditional sampling systems employed for at-line analyzers, the remaining discussion will focus on that mode of process analysis. This is not meant to imply that extractive analysis is preferred over *in situ* analysis – in fact CPAC has been one of the leaders in promoting both types of PAT instrumentation and applications.

The idea of developing a new paradigm for how process analyzer sampling systems are designed, built, and deployed began to take shape in the late 1990s. Not surprisingly, this corresponds to a period of rapid growth in the development and application of Process Analytical Chemistry, or PAC, (an acronym that pre-dates the PAT acronym currently popularized by the FDA and pharmaceutical industries, but referring to equivalent concepts). As more attention was given to the rapid advances in analytical technology and its application to process monitoring applications, limitations in current sampling system practices began to appear more evident. Process instrumentation and analyzers were making significant improvements in stability, sensitivity, and robustness – largely due to advances in core technologies such as microcomputers, electronics, and optics. At the same time, appreciation of real-time process monitoring and advanced data handling techniques (i.e., chemometrics) were leading to more applications of on-line monitoring and the use of information from process measurements in advanced feedback control of process. Meanwhile, sampling system designs and construction have not changed significantly in the past 20+ years. There was a growing consensus among some practitioners in the field that sampling systems could benefit from a more modern approach.

A lot of the inspiration for the idea of a modular surface mount component based sampling system came from developments in the gas handling equipment used in the semiconductor industry. As you may know, semiconductor manufacturing is a process that relies on very accurate control and delivery of a large number of different high purity gases at many of the manufacturing steps. Typically, each gas line would require a pressure regulator, particulate filter, flow control and sensing, and an isolation valve – often called a "gas stick" due to the linear arrangement of the components and the associated welded tubing or fittings used to connect them together. Because space in a fabrication facility is so expensive, tool makers and suppliers have sought innovative ways to reduce the space requirements of the required gas handling equipment. By switching the gas handling components from the traditional axial connections to a down-port connection, all the components on the gas stick could be placed much closer together. Furthermore, the surface mounted down-port components could be replaced without requiring complete disassembly and reassembly of the gas stick – resulting in improvements in reliability and reduced maintenance costs. The semiconductor standards organization (Semiconductor Equipment and Materials Institute, or SEMI®, http://www.semi.org/) has incorporated these modular surface-mount gas sticks into their standards (E49) for equipment in fabrication facilities. The standard defines a standard 1½ inch footprint for surface-mount devices and, either "C" (SEMI PR 3) or "W" (SEMI PR 3.1) type metal seals for high purity applications. There is also a new de facto standard emerging for 1⅛ inch footprint gas sticks and modular surface-mount components to further shrink the size of the gas handling systems.

People looking for new ways to design sampling systems for process analyzers saw a lot to like about these new emerging standards from the semiconductor industry. The need for size reduction, although a definite benefit, was not as compelling a driver in the chemical or petrochemical industries. Rather, it was the associated reductions in sample volumes, purge times, maintenance costs, and the ability to quickly design and build a system from standardized components that made this type of architecture very appealing. Recognizing that the metal seals used in the high purity semiconductor standard were not needed, and added cost and complexity, for typical sampling system applications, these users sought to define a new standard for their industry. Roughly based on the 1½ inch footprint SEMI standard, but with elastomeric o-ring seals, work was started within the ISA-SP76 Composition Analyzers Committee to develop a modular surface-mount component specification for sampling systems.

Another main issue with traditional sampling system technology is the lack of standardization throughout the industry. In fact, the use of discrete and non-standardized designs has resulted in a "custom" design approach and a systems integration industry to specifically support custom sampling system designs. There are three problems with this traditional approach. First, the reliance on custom, or "one-off", sampling system designs directly leads to high maintenance costs – each sampling system requires unique replacement part inventories and specialized knowledge about the system operation. Secondly, the custom approach to sampling system design entails high up-front costs in the design and construction, and the resulting systems are often difficult to modify to accommodate new analyzers or conditioning requirements. Third, these traditional sample systems are characterized by multiple fittings and tubing connections, all of which are potential leak points, and a relatively large physical footprint that consumes valuable space in analyzer sheds and enclosures. As a result, sampling systems often represent a disproportionate share (relative to the analyzer itself) of the total cost of installing and operating a typical process analyzer. In particular, the cost to build for a new analyzer system is estimated at 38% for the analyzer, 30% for sampling system, and 27% for the analyzer houses where both are located. However, if one looks at the total cost of ownership of a new analyzer system over a 15 year lifetime, the actual costs to own, operate, and maintain that system is roughly twice the cost to build. Most of that cost is directly related to personnel time needed to check and repair the analyzer and sampling system. Despite the high cost associated with sampling systems, it is commonly accepted fact that roughly 80% of the problems with modern process analyzers can be attributed to the sampling system itself.

The NeSSI™ efforts at CPAC grew out of the realization that process sampling plays a critical role in the success of implementing PAT and that there were significant advantages to be realized by developing a modular, and standardized, sampling system specification. NeSSI's origins can be traced to various CPAC Focus Group sessions and a tutorial presented at the November 1999 CPAC Sponsor Meeting. From these initial meetings, a consensus was reached that most application specific sample system functions could be standardized across industry and that, by taking the appropriate steps to collectively combine best practices of the End User community, a miniature, modular and smart system can help standardize nearly 80% of the "common" designs. The remaining customization can be readily accomplished by the "plug and play" capabilities of miniature, modular components.

10.4
NeSSI™ Physical Layer (Generation I NeSSI™)

To bring a vision of a new sampling system standard to fruition, it is critically important to have the participation of manufacturers willing to supply parts and components for this initiative. In this respect the NeSSI™ effort has been very fortunate. From the beginning, some of the largest manufacturers serving the sampling system markets have provided key insights into current sampling systems, what was technically feasible in terms of new approaches, and pragmatic issues related to manufacturing, market analysis, and pricing. Without their involvement, NeSSI™ would certainly not have progressed as quickly or as smoothly as it has.

In developing new technology, there is always the risk that it will not be successful – either due to technical issues, slow progress to market, or lack of market acceptance. For NeSSI™, these risks were perhaps even greater due to the highly conservative nature of the process market. Here, the fact that NeSSI™ was coming out of the CPAC consortia provided important benefits that somewhat mitigated the risks to prospective manufacturers of the needed hardware. First, CPAC provided a common voice of end-users needs, representing a large cross section of different industries. It is much easier to justify committing resources to a single project serving 20–30 prospective customers than it is for multiple projects serving one or two companies. Secondly, CPAC provided a neutral and open forum, independent of proprietary interests, where ideas can be brought up, debated, and addressed in an academic setting. This helped to insure that the NeSSI™ development would continue to serve the needs of the process community as a whole, and not be hijacked by one or two major players in the field. Finally, CPAC provided a legal umbrella whereby different companies, who might well be competitors in the marketplace, can work together on common interests. Cooperation and collaboration to define the baseline technology, while still allowing individual companies to innovate and differentiate themselves off that baseline technology, is one of the refreshing and unique aspects offered by consortia.

Still, the enthusiasm and dedication of the hardware suppliers to the NeSSI™ vision is a little surprising – especially when one considers that NeSSI™ sales would likely be at the expense of their more traditional product offerings. Clearly, they believe there are sufficient benefits, and potential demand, to justify the investment in tooling, product development, and sales training needed to bring out new product lines for NeSSI™. One of the suppliers mentioned the reinvigoration of their company (i.e., the excitement and technical challenges of the new NeSSI™ technology in a market where the technology hasn't really changed much for many years) as a pleasant and welcome side benefit of their early adoption and support of NeSSI™.

As mentioned previously, the NeSSI™ focus has been to develop, adapt, and promote standards that will help to facilitate adaptation of the modular surface-mount sampling/sensor systems by a critical mass of manufacturers, end-users, and researchers. The physical fluidic flow layer of NeSSI™ is partially defined by the adoption of the ANSI/ISA SP76.00.02-2002 standard for modular surface-mount components. The ISA-SP76, Composition Analyzers Committee is charged with organizing, managing, and coordinating all ISA Standards and Practices Department activities related to composition analyzers (instruments or devices used for analyzing composition and characteristics of materials) for industrial monitoring. While not part of the NeSSI™ effort *per se*, there is a considerable overlap of interests and people who are members of both the ASI-SP76 Committee and the NeSSI™ user group. Work on getting the ANSI/ISA standard for modular surface mount components for sampling systems actually pre-dates the NeSSI™ formation by a few months and the progress towards establishing the standard was followed closely by the NeSSI™ group. Although it is often simply referred to as the SP76 standard, the complete title for the published standard is "ANSI/ISA-76.00.02-2002 Modular Component Interfaces for Surface-Mount Fluid Distribution Components – Part 1: Elastomeric Seals" [5]. The standard is summarized as:

> Establishes properties and physical dimensions that define the interface for surface-mount fluid distribution components with elastomeric sealing devices used within process analyzer and sample-handling systems. The interface controls the dimensions and location of the sealing surfaces to allow change of just one element of the system without modification of the entire system, making the system modular from both a design and a maintenance standpoint.

The ANSI/ISA-76.00.02-2002 document addresses only surface-mount fluid distribution components and proper sealing methods. It is also limited to sealing methods using elastomeric material for the seals. The main specification in the SP76 standard is the definition of the physical mounting footprint, mounting bolt pattern, o-ring seal dimensions, and fluidic flow channel dimensions (Fig. 10.3). Basically, it defines a modular component as being 1½ inches square with four bolts holes for attaching to the substrate and two, or more, 0.125 inch diameter fluid ports. The face of the surface-mount component is flush and is designed to seal against o-rings to provide leak proof connections to flow channels in the substrate. While relatively simple, the physical footprint standard guarantees that components from different manufacturers will be physically interchangeable.

Interestingly, "The designs of the actual system components and the flow substrate are not specified in this standard." This allows individual manufacturers to design and build NeSSI™ systems with different architectures and assembly

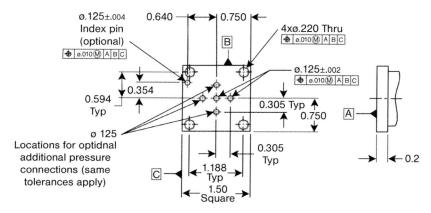

Fig. 10.3 Schematic drawing of the ANSI/ISA SP76.00.02-002 mechanical footprint and fluid ports.

methods – as long as they conform to the SP76 specification for mounting components onto their systems. This standardization effort has been embraced by many of the major fitting and sampling system providers and there are now four commercially available substrate platforms with different assembly and flow channel designs but any SP76-compliant surface-mount component will fit and function identically on any of the substrates. There are also many different SP76 modular surface-mount components (valves, filters, mass flow controllers, sensors, and analytical analyzers) commercially available from a growing number of vendors. The rest of this section will describe commercial NeSSI™ systems available from three vendors most closely associated with the NeSSI™ development effort. There are commercially available modular surface-mount systems offered by other manufacturers (and there is nothing to preclude new systems or vendors from entering the market), but our exposure and knowledge of these other systems is more limited and so we restricted ourselves to describing the systems we are most familiar with.

10.4.1
The Swagelok® MPC System

The Swagelok® Company (Solon, OH; www.swagelok.com) is a full service provider of complete NeSSI™ systems, parts including substrate flow paths and SP76 surface mount components, configuration assistance, and installation support under their MPC (Modular Platform Component) product line. The Swagelok MPC system (Fig. 10.4) was one of the first NeSSI™ systems to be commercialized. The Swagelok modular platform component system is a fluid distribution system designed for use within process analyzer and sample-handling systems. A typical MPC system consists of three layers – a substrate layer, a manifold layer, and a surface-mount layer. The substrate layer provides the main flow path between the surface-mount components. The substrate layer consists of

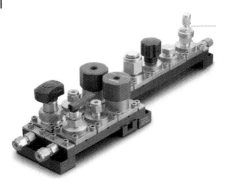

Fig. 10.4 Example of a simple Swagelok MPC NeSSI™ system.

a substrate channel and various drop-in substrate flow components. The substrate channels are available in various lengths to accommodate up to 14 surface-mount components, and adapters can extend that by mating two substrate assemblies together end to end while maintaining standard surface-mount spacing between them. The manifold layer provides the flow path between two or more parallel substrates. The manifold layer consists of a manifold channel and various drop-in manifold flow components. The manifold channels are available in various lengths to accommodate up to six parallel substrates. The substrate and manifold layers are combined to form the conduit for the system fluid, and they can be customized for any flow configuration. The surface-mount layer provides shutoff, flow control, and filtering capabilities for the system fluid. The MPC architecture is based on precision welded tube sets that carry the sample, machined substrate channels to position the tube sets and provide mounting surface for modular surface-mount components, and manifold channels and tubes to provide flow between substrate channels.

The heart of the system is the substrate flow components that carry the sample, and other fluids, through the system and provide the interconnection pathways between the surface mount components. The flow components themselves are made from one of two styles of bosses, one of which is shown in the left picture of Fig. 10.5. Each boss has a cutout for retaining the sealing o-ring, a tube end that is typically welded to the tube end of another boss, and a locating pin on the bottom that fits into a corresponding hole in the substrate channel. A completed flow component is shown in the right-hand cutaway view where two bosses are shown welded together. This is the most commonly used flow component in the MPC system. The reader will notice that the two bosses making up the flow component are slightly different – there is what is referred to as a short boss and a long boss. Also the long boss has a larger diameter locating pin and a recess for a second o-ring in the bottom of the locating pin. Both of these features are there to allow for fluidic flow to enter the manifold layer below the substrate layer by simply drilling the fluidic passage straight through in the vertical direction. This is referred to as the manifold layer of the MFC system and is mainly used to connect the flows of two, or more, substrate channels. Manifold flow components use a

Fig. 10.5 Basic flow component boss, or building block, (left) and a cutaway view of a complete flow component (right) for the Swagelok® MPC™ system. The vertical line in the cutaway view represents the weld-line joining the long and short bosses.

third type of boss similar to that shown in the left-hand view of Fig. 10.5, except there is no recess for o-ring retention – all the o-rings in the system are captured in the substrate flow components and the manifold boss simply provides a smooth sealing surface for the o-ring.

Figure 10.6 shows the cutaway view of two flow components and a surface mount component (in this case a toggle valve). There are a couple of interesting things to note about this view of the fluidic pathway. First, the surface mount component spans two flow components. The first flow component supplies fluid to the center port of the 2-port toggle valve, assuming fluidic flow from left to right. The second flow component takes the output from the side port of the valve and provides the flow path to the center port of next surface mount component. In this case the center port is also connected to the manifold layer (not shown) by virtue of the passage at the bottom of the boss). By selecting flow components with the appropriate combination of bosses (short-short, short-long, or long-long), connections to the various ports (side-side, side-center, or center-center, respectively) of the surface mount components can be controlled. There are also flow components available that skip surface mount component locations altogether (i.e., jumpers) by adding additional tube length between the two bosses. This clever arrangement accommodates all possible configurations (1-port, 2-port, and 3-port) of surface mount components and the two different possible orientations of 2-port components (side port on the left or side port on the right) with a limited number of distinct flow components. The second attractive aspect of the MPC™ design is that the alignment of the flow channel ports in the surface mount component and the substrate is accomplished using the alignment pins and corresponding holes in the substrate channel. This means that the flow components can be simply dropped into the substrate channels and the flow passages are guaranteed to line up with any components attached to the substrate. This simplifies assembly and disassembly of the MPC™ systems. Finally, all the compression force needed to deform the o-rings and provide leak tight connections at the flow passage interfaces is supplied by the four screws used to mount the surface mount component to the substrate. Since all the critical tolerances (e.g., the total height and o-ring recesses) are built into the boss machining, there is little additional care that needs to be done when assembling the system to ensure good sealing. An additional benefit of this architecture is that it requires a minimal number of sealing o-rings (typically only two per surface mount component), which also minimizes potential leak sources due to o-ring failure.

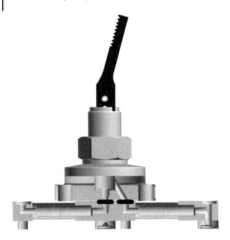

Fig. 10.6 Cutaway view of the flow path for a single surface mount component (in this case a 2-port toggle valve) in the Swagelok® MPC™ system.

Finally, the cutaway view in Fig. 10.7 shows the flow paths of all three layers (manifold, substrate, and surface mount) assembled together into a MPC system. A simple flow diagram for the system is also shown at the bottom of Fig. 10.7. The flow generally runs in the substrate between surface mount components and goes up into one surface mount component and then back done into the substrate – this is typical for all NeSSI™ systems. Notably, various substrate flow components are used to accommodate the different flow connection requirements of the various surface mount components. This is also typical in a NeSSI™ system – there is no specification that dictates which ports of the SP76 configuration are to be designated as inputs or outputs, or which ports will be present in a SP76 component. This allows component designers the flexibility to pick the most advantageous flow path for a given surface mount component (some will have a preferred, or required, flow direction) and the substrate designers need to accommodate this diversity. Another typical feature of NeSSI™ systems is the use of standard fittings to connect the NeSSI™ system to the process flow – in this case a compression fitting on the flow component on the left of Fig. 10.7. The availability of standard fitting connectors and adapters allows hybrid NeSSI™ systems to be constructed where modular surface mount and traditional sampling system components can be combined. NeSSI™ systems are typically panel mounted in industrial settings and the MPC system provides mounting blocks (feet and supports) for this purpose. The Swagelok® MPC™ system is assembled with a single size 10-32 hex socket cap screw, a single size standard o-ring, and requires only a single size hex wrench. Surface-mount components can easily be serviced from the top of the assembly without disturbing any other components, removal of the system from its enclosure, or requiring disassembly of the system. All surface-mount components, adapters, and caps are easily interchangeable on any surface-mount position because of the modularity of components and the

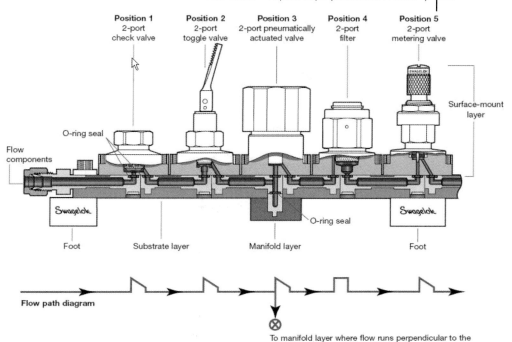

Fig. 10.7 Cutaway view of a five-position Swagelok® MPC™ assembly with a manifold layer assembled below Position 3. At this position, flow is diverted in two directions – up to the surface-mount component and down into the manifold layer.

use of the standard ANSI/ISA 76.00.02 interface. Again, these are features that are shared by all the commercially available NeSSI™ systems.

The photograph in Fig. 10.8 shows a partially assembled Swagelok® MPC™ NeSSI™ system. In the photo is the long anodized aluminum substrate channel with some surface mount components mounted and some flow components visible in the un-mounted positions (one is lying on top of the substrate waiting to be dropped into the channel and one is lying along side just below the substrate). Near the top of the photo is a two-position manifold channel and flow component that will be attached under the substrate channel (using hex bolts inserted into the countersunk holes visible in the substrate) to allow flow between this substrate stream and another (not shown). Also notice the grooves machined into the substrate top on either side of the hole used for the drop-down connection to the manifold layer (about the center of the surface mount component, where the flow components butt up against each other) – this allows for leak detection (using a "leak sniffer" for gases or visually for liquids). Finally, notice the process connection flow component at the end of the substrate (lower left corner of Fig. 10.8), which are available in various standard fitting sizes.

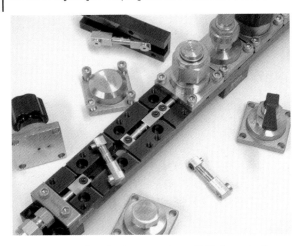

Fig. 10.8 Examples of substrate and manifold channels, flow components, and surface mount components for the Swagelok® MPC™ NeSSI™ system.

Swagelok also provides a software tool to aid the user in MPC system component selection and the assembly process. This tool, called the MPC System Configurator, is a CD-ROM software program and is freely available from the Swagelok web site or through local vendors. The Configurator allows the user to create a customized system by defining, placing, and connecting surface-mount components on a layout grid using a simple drag-and-drop user interface. Once the layout is complete, the Configurator identifies the MPC series flow connectors (including substrates, manifolds, seals, and assembly hardware) that are necessary to build the complete system. A bill of material is generated for ease of ordering components, and an assembly diagram is produced to facilitate assembly. Figure 10.9 shows a screen shot of the MPC Configurator layout screen (left) and the generated Bill of Materials (right). This tool makes it very easy to design, configure, and order MPC systems – even for users with a limited knowledge of NeSSI™ or modular surface mount sampling systems. Of course, technical assistance is also available from Swagelok to help users design, assemble, and install their MPC systems.

10.4.2
The CIRCOR Tech μMS3™ (Micro Modular Substrate Sampling System) System

The CIRCOR Tech (Spartanburg, SC; www.circortech.com) is also a full service provider of NeSSI™ systems and components. The company (through its associated subsidiaries – Circle Seal, GO Regulators, Hoke, and others) has a long history of serving the traditional sampling system markets in the chemical and petrochemical industries. Their NeSSI™ offering is called the micro Modular Substrate Sampling System (μMS^3™) (Fig. 10.10) and is described on their web site as:

10.4 NeSSI™ Physical Layer (Generation I NeSSI™)

[it] was developed specifically for the Process Analytical Instrumentation market, and incorporates a building block and flow tubeset architecture for simplicity. The μMS³™ is the latest generation sampling system, an evolution of our original popular MS³ robust design. The μMS³™ System is a unique design that focuses on maximizing efficiency and minimizing costs. The flexibility of the block-and-flow-tubeset architecture makes it the perfect solution for any gas or liquid sample conditioning system, single or multistream.

CIRCOR Tech provides the parts needed to construct a μMS3™ system, "build to order" complete systems, and a wide range of SP76 surface mount components for NeSSI™ systems. They also provide services for designing and installing their μMS3™ systems according to customer specifications.

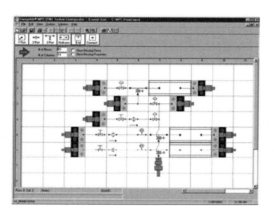 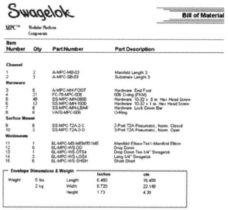

Fig. 10.9 Swagelok MPC Configurator layout screen (left), where the user defines the system with drag-and-drop blocks, and a generated bill of materials (right) used to order the system.

Fig. 10.10 Example of a μMS³™ (micro Modular Substrate Sampling System) NeSSI™ System from CIRCOR Tech.

Fig. 10.11 Basic building block (left) and a selection of flow tubesets (right) for the CIRCOR Tech µMS³ system.

The µMS³™ system is built around a single machined building block (shown on the left-hand side of Fig. 10.11), which provides the mounting for surface mount components on top and attachment of flow tubes beneath the block. Each building block accommodates a single SP76 surface mount component using the screw holes at the extreme corners of the block top. The flanges on either end of the block allow the blocks to be screwed together to form substrate "sticks". Each block has passages drilled for all three of the flow ports specified in the SP76 standard and machined recesses for capturing the sealing o-rings (visible in the center of the building block). On either side of the flow passages are counterbored holes for screws used to attach the fluid flow tubesets (examples of which are shown on the right of Fig. 10.11). The tubesets are available in various pre-welded lengths and consist of machined mounting flanges (with the threaded holes for mounting screws and grooves for capturing the sealing o-rings) electron-beam welded to tubing. Note that the tubesets have the flanges welded either perpendicular to the tube (for connecting passages between adjacent blocks in a stream) or in-line with the tubes (for connections between adjacent streams). Interestingly, CIRCOR provides its tubesets with a passivating coating (Restek Silcosteel®) as a standard treatment to help insure low maintenance and long life in various applications (Restek Sulfinert® coatings are also available as an option).

Using this arrangement, a µMS3™ system is assembled in the following way. First the basic building blocks are screwed together (using the flanges and screw holes at either end of the block, as shown in Fig. 10.12A) to provide the basic substrate for mounting the surface mount components. Next, the tubesets are screwed to the underside of the blocks (Fig. 10.12B) to achieve the desired fluidic interconnects between the blocks (unused passages in the blocks are simply filled using supplied plugs) and between substrates. The next step is to attach the feet and tubesets that provide panel mounting and connections to the external process streams (Fig. 10.12C). Finally, the surface mount components are attached (Fig. 10.12D) to the top of the blocks using the four mounting screws per component (as per the SP76 standard).

The perceived advantages of the µMS3™ system are its inherent simplicity (only one type of substrate, the basic building block, used throughout and standard collection of tubesets), ease of assembly (only two hex-head screw sizes and same size o-ring for all seals), easy detection of leaks (due to visible tubeset connections under the system), and easy adaptation to any flow configuration required by customer's application.

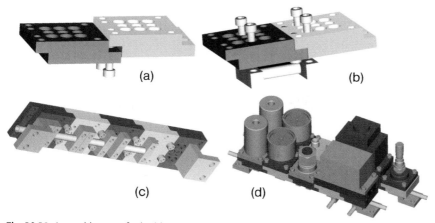

Fig. 10.12 Assembly steps for building a μMS3™ system:
(A) joining of the basic building blocks, (B) attachment of a tubeset,
(C) the underside view showing the tubesets and mounting feet attached,
and (D) with the surface mount components mounted.

Fig. 10.13 The μMS3™ NeSSI™ system used as the basis for the International Student Games at ISA Expo2005. (Courtesy of Robert Sherman, CIRCOR Tech.)

To demonstrate the ease of putting together a NeSSI™ system, Robert Sherman, CIRCOR Tech recently sponsored a practical exercise as the International Student Games at the ISA EXPO2005 held in Chicago, IL. In this exercise, teams of three or four students (Juniors or Seniors in either Engineering or Chemistry) were given the task of assembling a working μMS3™ system from supplied schematic flow diagram, some drawings of a system, and a box of the necessary hardware parts. Even though most of the student participants had little, or no, experience with sampling systems in general (and certainly none with NeSSI™ systems), most of the teams were able to successfully assemble the parts and produce a working NeSSI™ system within the 45 minute time limit! Figure 10.13 shows a photo of the completed NeSSI™ system used for this student competition.

10.4.3
The Parker-Hannifin Intraflow™ System

The Intraflow™ System by Parker-Hannifin (Jacksonville, AL; www.parker.com) is actually the second generation of NeSSI™ systems offered by the company. Their first system, called the Parker Integrated Conditioning System (PICS), was the only commercial offering to support the 2.25 inch, five-stream surface mount specification originally included in the SP76 draft standard. This system, as shown in Fig. 10.14, was designed with five flow channels specifically for compatibility with most existing sample conditioning systems. It supported both the draft 1½″ and a 2¼″ SP76 elastomeric interface specifications. The larger footprint was designed to utilize more existing industrial devices and integrate additional auxiliary flow streams (waste, actuator supply air, vent, steam, etc.) into the sampling system. As shown in the inserts of Fig. 10.14, the additional size of the PICS blocks allows the 1.5 inch port cross arrangement to be extended to include additional ports for the five possible streams flowing through the block. In this system all the fluidic flows are contained within passages drilled into the blocks. The blocks are bolted together using the shown special bolts and o-ring seals are used to seal the passage connections between the blocks. The reader will notice that the PICS system is quite a bit larger and more massive than other systems described so far. It is also an inherently linear system – any connections between

Fig. 10.14 Initial NeSSI™ offering from Parker-Hannifin; the Integrated Conditioning System (PICS) could accommodate both 1½″ and 2¼″ surface mount components based on the original draft ANSI/ASI SP76 specification.

PICS rails are accomplished using traditional tube and fittings (although with five possible fluid streams, the need for spanning sticks is somewhat reduced). In the end, the additional complexity in manufacturing and assembly needed to support the additional streams, and an end-user desire to standardize on the 1½ inch SP76 standard, led to discontinuing this product line and dropping the 2¼ inch surface mount specification from the final SP76 standard.

The Parker-Hannifin PICS system was replaced by the Intraflow™ System as their main NeSSI™ commercial product line. Of course, their previous experience with the PICS system helped them to produce a second-generation NeSSI™ system with unique architecture and some interesting abilities. Also, the large number of 1½ inch SP76 surface mount components they have developed work with the new system as well. The Parker-Hannifin web site describes the Intraflow™ System as:

> Parker IntraFlow™ substrate fittings have been developed specifically for analytical, lab and other complex general purpose instrumentation flow control systems. ISA/ANSI SP76.00.02 compliant, Parker IntraFlow™ fittings provide maximum flexibility with minimal space requirements. All flowpaths, regardless of direction, are maintained on a single plane within the system as there are no lower level manifold blocks required. Contiguous fitting flow paths are intra-connected with slip fit pressure connectors, while a threaded pegboard provides connection force and rigidity when fittings are mounted with cap screws. System assembly couldn't be simpler: a 5/32″ hex head wrench is all you need to build with Parker IntraFlow™.

Figure 10.15 shows an example of the Parker-Hannifin Intraflow™ system. In this case it is a basic fast loop with sample coming in the left top connection, going through a shutoff valve, a check valve, a pressure gauge, and a filter. The filtered sample exits to the analyzer at the bottom fitting and excess unfiltered sample returns through the rotometer and the top right connection. At first glance it may appear that the surface mount components are simply bolted to the pegboard without any visible flow tubes or channels. Looking closer at the left-hand image, one will notice that the surface mount components appear to have thicker bases than examples shown earlier. In the right-hand image, most of the surface mount components have been removed – revealing little mounting blocks, or Intraflow™ fittings, which are mounted to the pegboard. Still, it is not obvious how fluids would flow from the obvious external process connections through the system and through the blocks to the surface mount components. Parker has come up with a clever way to tie the flows together; allowing the Intraflow™ fittings to handle even complex flow patterns in a single plane of blocks.

The Intraflow™ fitting (Parker calls it a fitting rather than a block or substrate) is the basic building block of the new Parker NeSSI™ system. Figure 10.16 shows an example of an Intraflow™ fitting, and its associated slip-fit pressure connection. The fittings themselves are monolithic stainless steel blocks with several features machined into them. At the outer corners of the fitting are threaded holes for

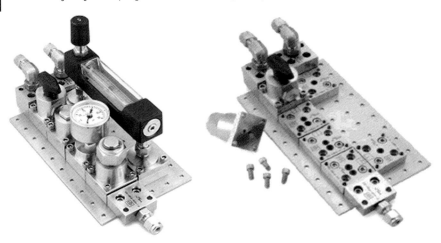

Fig. 10.15 Example of a basic fast loop Parker-Hannifin Intraflow™ system.

attaching SP76 surface mount components. Just inside of that are counterbored holes where the screws used to secure the fitting to the pegboard are inserted. Near the center of the fitting are the passages and o-ring retaining recesses for the fluidic interconnect to the surface mount component. Depending on the desired flow pattern, one or more sides of the fitting will have passages for connecting flows to other Intraflow™ fittings. These passages are slip-fit counterbores into which slide Pressure Connectors (the other key component in building Intraflow™ systems, shown just to the right of the fitting in Fig. 10.16). The fitting's internal flow paths are ⅛-inch diameter and are precision machined at the intersections to minimize dead volume, as shown in the schematic on the right-hand side of Fig. 10.16. The slip-fit Pressure Connectors provide the flowpath intra-connectivity between Parker IntraFlow™ fittings in the system. Pressure Connectors are supplied as fully assembled parts with standard size elastomer o-rings in the outer groves and an internal orifice of ⅛ inch. They are available in both standard and jumper lengths (for systems with flow requirements between non-contiguous fittings).

Fig. 10.16 The Parker Intraflow™ fitting and slip-fit Pressure Connector. On the right-hand side is a schematic of the flows within the Intraflow™ fitting.

10.4 NeSSI™ Physical Layer (Generation I NeSSI™)

All Parker IntraFlow™ fittings are permanently marked with the part number, flow schematic and material heat traceability code for easy identification during assembly and when servicing in the field. So, looking at the Intraflow™ fitting in Fig. 10.16, we can see that it is a T-fitting with the center port connected to passages on the right and bottom sides of the fitting. Notice that there is nothing in the symmetric design of the Intraflow™ fitting (other than the orientation and number of passages to the SP76 surface mount component) that restricts its orientation relative to other fittings in the system – all orientations are possible. This feature leads to a very high degree of flexibility to accommodate complex flow requirements using a small inventory of only 16 different Intraflow™ fitting configurations. This also means that all flow paths are on the same plane – there is no need for lower level manifolds to provide bridges between streams. This is touted as an advantage of the Intraflow™ design to improve recognition of flows in the system, minimize system weight, and eliminate additional leak points. More importantly, the Intraflow™ architecture makes it possible to implement low dead volume stream switching (using a 3-port fitting and a 3-way surface mount valve) in a single position – something that is difficult to accomplish with other NeSSI™ systems.

Assembling an Intraflow™ system is very simple (Fig. 10.17). The first step is to select and layout the Intraflow™ fittings according to the desired flow paths. Next, the Pressure Connectors are inserted into the fittings to complete the flow paths. Then the Intraflow™ fittings are secured to the threaded pegboard using the four hex head screws. This fixes the relative position of the fittings and insures the Pressure Connections are captured and cannot leak. Parker offers a wide range of heated and non-heated pegboard sizes to build the IntraFlow™ system. For the heated pegboards, heat is applied through conduction directly to the base of the fitting (within 5/32" of the actual fluid flowpath), increasing heat transfer efficiency. The heating can be provided either by commercially available standard general purpose heater cartridges inserted directly into the pegboard or by steam heat channeled pegboards. Finally, the SP76 surface mount components are attached to the fittings using another four hex head screws. Note, it is also possible to build the system up sequentially (as shown), completing each component location before moving on to the next component. However, for complex flow patterns, it may be easier to complete the flow interconnects before the Intraflow™ fittings are bolted down (plus it takes advantage of the flow markings printed on the fittings to visualize the layout). One inherent limitation of the Intraflow™ system is that it can be difficult to modify a system, or change out a fitting block, without disassembling a large part of the layout from the pegboard.

Parker-Hannifin also offers a web-based software tool, called Design-Pro™, for designing and specifying your Intraflow™ NeSSI™ system. Design-Pro™ is a drag-and-drop type tool that utilizes the ISA SP76 symbol library for components and produces Bill-of-Materials, fitting layout, system layout, and system definition drawing output files. Of course, technical assistance from Parker-Hannifin is also available to help the user in all aspects of design, build, and installation of Intraflow™ NeSSI™ systems.

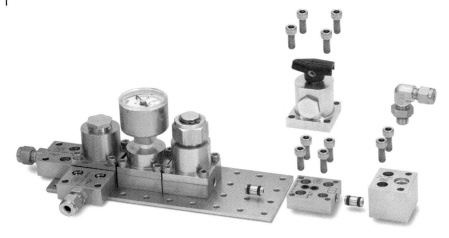

Fig. 10.17 Picture showing how SP76 modular surface-mount components are attached to one of the commercially available substrates.

10.4.4
General Characteristics

For all three NeSSI™ systems described above, the general performance specifications are quite similar. The high pressure ratings of these systems are 500 psig (34 bar) for the Parker system and 1000 psi (69.4 bar) for the CIRCOR Tech and Swagelok systems. Temperature ratings for the systems mostly reflect the elastomeric material selected for the o-rings in the systems. The standard o-ring material is fluorocarbon rubber for the Parker and Swagelok systems and Viton® for the CIRCOR system – although almost any o-ring material can be specified by the user. Since standard o-ring sizes are used by all manufacturers, compatible o-ring selection should not be an issue. For most NeSSI™ systems the default material of construction for the wetted parts is 316L stainless steel, although most vendors also offer more exotic materials for demanding, or corrosive, applications.

The biggest question about NeSSI™ system performance is related to flow rates and pressure drops in the systems, especially in liquid applications. People look at the SP76 standard port dimensions and wonder what that translates to in terms of viscosity and throughput limits for liquids. Here, the hard numbers have been a little slow becoming publicized. Swagelok commissioned a study (unpublished) and found that the pressure drops of the individual surface mount components were much more significant in determining the overall C_v of the system than the substrate flow components. The port orifice dimensions are 0.090-inch ID for Swagelok's MPC components (0.100″ ID for the substrate), 0.110-inch ID for CIRCOR's μMS3™ components (0.120″ for their transfer tubes), and 0.115-inch ID for Parker's Intraflow™ substrate and components. To put these numbers into context, a ⅛″ OD × 0.028″ wall tubing has an ID of 0.069″ and thick walled (0.065″) ¼″ OD tubing has 0.120″ ID. So, the NeSSI™ system port dimensions are really not that different than the tubing they may be replacing in traditional sampling systems. Experience in installed applications indicates that diesel fuels (and other liquids with similar viscosities) can be routinely processed in NeSSI™ systems without difficulty.

10.5
NeSSI™ Electrical Layer (Generation II NeSSI™)

The other main specifications promoted by NeSSI™ are directed at providing automation and "smarts" to NeSSI™ systems. There are guidelines for a standardized electrical and communication "plug & play" field bus (also called distributed I/O or multi-drop serial communication) to be adopted. This NeSSI-bus would provide power and bi-directional electrical signal lines for devices (SP76 compliant sensors and actuators) on the NeSSI™ substrate, as well as auxiliary equipment such as heaters and environmental sensors that may be present in the sampling system enclosures. It was deemed important that the electrical signals be carried over a modern field bus network protocol (as opposed to traditional 4-20 ma wiring) to take advantage of self-defining and self-diagnosing capabilities of the network protocols and to reduce installation and maintenance costs. Regardless of the actual field bus network and protocol selected, it was important that this be a commercially available, open, and supported protocol to reduce costs and facilitate implementation.

The NeSSI™ specification also describes a Sensor Actuator Manager (called SAM) that would identify installed components, configure them for the application, verify their operation, and provide integration of their data signals (in engineering units) for transmission to higher level process monitoring and control networks. The SAM is a combination of software and hardware layers that function as the host to the sensor/actuator communication bus, the gateway to the plant communication and control networks (Ethernet LAN, DCS, OPC, etc.), an environment for running software applets for repetitive tasks and diagnostics, and an open development space for custom software provided by the system manufacturers or end-users. The SAM represents a standardized way for the NeSSI™ system to interact with the plant information and maintenance infrastructure to increase confidence in results coming from process analyzers.

With these capabilities, the NeSSI™ system will provide a common physical connection to the sample (the SP76 mechanical rail) and a common electrical and communication connection to the user (the NeSSI-bus electrical rail). Between these two rails reside the transducers that interact with the sample and return either data or status information. While not as far in their development or widespread adoption by industry, these standardization efforts to provide automation and communication are an important aspect in the eventual utility of NeSSI™. As the automation and communication aspects of NeSSI™ increases, it will be possible to not only analyze a sample stream and provide an *in situ* measurement of a validation standard, but each analysis will also have associated sensor readings to demonstrate that the sample was conditioned correctly and the system was working as designed – this is the holy grail of a "smart" sampling/sensor system.

The interesting aspect of the NeSSI™ rail topology is that designing new sensors or actuators for the NeSSI™ system should be much simpler – the developer has only to concentrate on the transducer functionality since the physical and electrical interfaces are provided by the NeSSI™ specifications. This provides the springboard for what has been termed the Generation III NeSSI™ – incorporation of micro-analytical devices (i.e., analyzers, spectrometers, and chromatograms) directly onto NeSSI™ systems as another surface mount component. Interestingly, the move to incorporate analytical, and even micro-analytical, devices on the NeSSI™ system is already taking place.

10.5.1
The NeSSI™ Bus

When the NeSSI™ group began writing the specification for the automated NeSSI™ system, they wanted to provide a forward-looking vision that would set a target for increased reliability, ease-of-use, and functionality beyond what current sampling systems could provide. Clearly, an important aspect of that vision was the ability of the NeSSI™ system to monitor its status and provide automated, or remote, control of its operation. To accomplish this goal, the systems needed to have sensors and actuators and a means of communicating the monitoring and control signals. This means wiring. Certainly, it would be possible to achieve this monitoring and control functionality using conventional wiring practices such as 4–20 milliamp signals individually wired to discrete terminals of a distribution panel or DCS. However, this approach carries many of the same disadvantages in the wiring that the NeSSI™ physical layer sought to eliminate in the plumbing. For example, traditional wiring is labor intensive to design and implement, it is labor intensive to troubleshoot and maintain, it can be difficult to adapt the wiring to accommodate changes such as adding new sensors or actuators, and it generally moves the generation of the control signals and the processing of the sensor data upstream in the plant infrastructure and away from the point-of-use at the sampling system and analyzer.

The NeSSI™ team wanted a better solution that was more consistent with the modularity and standardization they had defined in the physical layer. It obviously lessens the value of having a system where components can be easily installed, replaced, or reconfigured if it just moves the productivity bottleneck to the wiring and communication part of the system. Their vision for the electrical layer was a single multi-drop cable and a compact single plug-in connector for each sensor or actuator. This multi-strand cable would provide power to components that needed it and the control and data signals for operating the components. Furthermore, the system would be able to recognize the individual components connected to the cable and know how to communicate with them. For actuators, it would know what signals are needed to control them and be able to verify the completion of the control commands. For sensors, it would know the format of the returned data, what is required to translate the signals into engineering units, and any diagnostic or reliability metrics associated with the sensor readings. If something needed to be replaced or added to the system, it should not require any extensive modifications to the existing wiring or communication logic. In short, the electrical power and communication should be as transparent and easy to use as possible, requiring minimal intervention on the part of the installation and maintenance personnel.

One only needs to look at the examples of "plug & play" simplicity provided by the universal serial bus (USB) and modern Ethernet networks in personal computers to realize that technology exists that could make the NeSSI™ vision a reality. However, that is not to say that an off-the-shelf solution exists. The NeSSI™ group has been working on defining and evaluating solutions that would fit into the somewhat unique constraints imposed by the chemical and petrochemical processing industries that are the main consumers for NeSSI™ systems.

Within the chemical industry process analyzers, and therefore sampling systems, are often viewed as a necessary evil – they are installed only if the data they provide can be shown to lead to better products, increased process safety, or increased process efficiency. For new technology, such as NeSSI™, to be adapted it must demonstrate that it provides better information or a lower cost way of doing things. The cost of ownership advantages demonstrated by the NeSSI™ physical layer, relative to traditional sampling system plumbing, is just now being accepted as a better, and justifiable, technology. It has taken roughly five years to turn that corner. For the electrical layer, it is a more radical departure from traditional practices and, despite good arguments for the advantages it will provide, it remains to be seen how quickly it will be adopted by the industry. Therefore, the NeSSI™ effort has been very careful to pay attention to the pragmatic issues of cost and industry acceptance. That is why the focus has been to look most closely at existing communication protocols and open, or non-proprietary, networking solutions that already have some use and acceptance by industry.

An equally important constraint, and one that is more specific to the industries targeted by NeSSI™, has to do with the issue of safety related to electrical signals in the processing environment. Chemical and petrochemical processing plants are often inherently dangerous places with the risk of fires and explosions from the materials being processed. Strict safety guidelines and regulations are in place for equipment placed in these plants with different classifications, or zones, established depending on the inherent risk posed by the materials being processed. According to the NeSSI™ specification:

> The majority of analytical systems required for the petrochemical and refining industries are located in Hazardous Electrical Locations. Consequently all NeSSI components – including transducers, wiring systems, optical systems, enclosures – must be designed to meet the various global Hazardous Location standards. The NeSSI specification favors the use of Division 1 (Zone 1) rated equipment inside equipment enclosures which handle hazardous (flammable) fluids. Designing to this capability, although more stringent than

Division 2/Zone 2, will provide increased design flexibility, lower cost to build, simpler maintenance, optimum safety and suitability for the widest range of applications and requirements. Division 1 (Zone 1) will allow the use of analyzer optical windows, as well as dynamic and static elastomer seals, without the need for additional electrical protection methods such as purging, continuous dilution or pressurization.

In this context a hazardous area is designated as any location in which a combustible material is or may be present in the atmosphere in sufficient concentration to produce an ignitable mixture.

So the specification recognizes that operation of NeSSI™ systems are likely to occur in hazardous locations and should be designed to meet the tough requirements of Division 1 (Zone 1) classification. [IEC uses the Zone system. North America has used the Division system; however, current and emerging North American regulations will now be built on the Zone system. (Canada accepts the Zone regulation for new installations.)] Notice that this applies to all aspects of the NeSSI™ system, not just the electrical connections. The specification document goes into more detail about the various classifications and recommendations for meeting the requirements. Table 10.1 shows how the IEC symbol relates to the zone classification and various approved methods for satisfying the requirements for different zones. Notably, the decision to specify such a tough hazardous area qualification for NeSSI™ was not without dissention among the NeSSI™ user group. There was a minority opinion that calling out such a specification would have a detrimental effect on manufacturers having to design to that standard and would unnecessarily increase the cost to those with applications in less hazardous areas. In the end, it was difficult to prove that the spec would inherently lead to higher prices and that the flexibility afforded by having only one NeSSI™ compliant standard was worth the additional design effort. After all, once a component was certified for a hazardous area classification, it could certainly be

Table 10.1 IEC recognized methods of protection (for reference).

Technique	IEC symbol	Permitted zone
Oil immersion	Ex o	1 & 2
Pressurization	Ex p	1 & 2
Powder filling	Ex q	1 & 2
Flameproofing (explosion proof)	Ex d	1 & 2
Increased safety	Ex e	1 & 2
Intrinsic safety (2 fault tolerant)	Ex ia	0, 1 & 2
Intrinsic safety (1 fault tolerant)	Ex ib	1 & 2
Non-incendive	Ex n	2
Encapsulation	Ex m	1 & 2

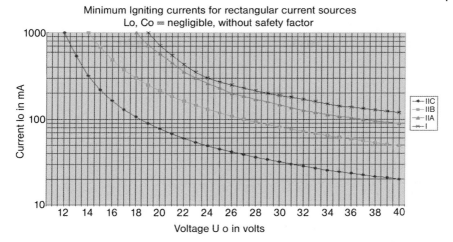

Fig. 10.18 Intrinsically safe operating region.
(Courtesy of Dr. U. Johannsmeyer, PTB, from IEC Subcommittee SC31G.)

used in non-hazardous areas. In hindsight, deciding to stick to the more rigorous specification certainly delayed development of the NeSSI-bus (as discussed below) but will probably result in a safer and more universally accepted product.

For the electrical layer itself, the NeSSI™ specification calls for using intrinsically safe wiring practices as the preferred method of meeting the Division 1 (Zone 1) requirements. Intrinsically safe is a term used to describe the situation where, in the simplest terms, if there is a grounding or break in the wire or connection, there is not enough electrical power to generate a spark that could cause ignition. It is an operational definition that is related to the voltage and amperage carried in the wiring. Figure 10.18 illustrates the safe operating region for intrinsic safety. The NeSSI™ specification for the intrinsically safe NeSSI-bus says:

> As an intrinsically safe operations environment basically limits the available power, careful consideration shall be given to the power requirements for the sensors/actuators (transducers) and the associated NeSSI-bus communication components. "As a minimum, the intrinsically safe NeSSI-bus design shall meet *Ex ib* (optional *Ex ia*) requirements according to EN50020 or IEC 79-11, the alternative FISCO (Fieldbus Intrinsically Safe Concept) Model shall be evaluated as the basis for designing and connecting the NeSSI IS bus and components. In addition, an intrinsically safe sensor bus is preferred to avoid any wiring connections or disconnection during "plug and play" swapping of sensors and actuators on any substrate position. Using intrinsically safe cables/connectors allows for live connections and disconnection in hazardous areas without danger of explosion or the need for a gas test and a safe work permit. Additional savings will be realized by eliminating the need for conduit, condulets and conduit/cable seals and glands.

Most modern field bus communication networks (i.e., Foundation Fieldbus, CANbus, PROFIBUS, SDS, ControlNet, LONworks, etc), inherently provide much of the "plug & play" functionality described in the NeSSI™ specification. Most of the "plug & play" magic is accomplished by requiring each device that will be connected to the network to have an associated electronic description of its capabilities, identity, and requirements. This description might be embedded in the device itself (like USB devices) or in a separate file available to the software running the network protocol. When connected to the field bus network there is an exchange of recognition packets that identify the device and calls up its electronic description. The network protocol software then configures its communication to the device depending on its description. This allows components to be installed or switched without having to manually reconfigure the system wiring or software.

Initially, it was anticipated that the NeSSI™ community would simply adopt one of these field bus networks as the basis for the NeSSI-bus. The DeviceNet™ protocol, running in a controller area network (CAN), was initially selected as the communication protocol since it met most of the requirements for the NeSSI-bus. It has a large installed base (although not so much in the process analyzer market), is fairly inexpensive, and the chipset easily fit the physical size constraints for mounting on top of SP76 components. A working NeSSI™ system was actually built and demonstrated in 2003 by Swagelok using DeviceNet™ as the network and Labview as the software. Unfortunately, DeviceNet™ was not certified as intrinsically safe and therefore was not suitable. In July of 2003, members of the NeSSI™ group met with OVDA (the governing body for DeviceNet™) to explore the possibility of extending it to an intrinsically safe physical specification. The meeting results were that while it was certainly possible to develop and certify an intrinsically safe specification, the funding to do so would be prohibitive for the NeSSI™ effort.

In fact the only commercially available field bus network certified as being intrinsically safe was Foundation™ Fieldbus (and by extension Profibus, since it runs on the same physical layer and chipset). Promoted internally at CPAC by Honeywell, an announcement was made in February 2005 that Foundation™ Fieldbus had been adopted as the communication standard for NeSSI™. But Foundation™ Fieldbus was not without problems as the right choice for the NeSSI-bus. The main limitation was that the chipset and circuit board needed to implement the protocol was too large for devices that must fit within the 1½-inch footprint for SP76 devices. The chipset and circuit board could be redesigned and shrunk to fit the size constraints of NeSSI™ devices, but again the costs of leading that effort were beyond CPAC and NeSSI™. Also there were concerns about the added cost associated with manufacturing Foundation™ Fieldbus devices.

Recently, several things have pointed to a resolution of the NeSSI-bus issues. First, for the past few years there has been an effort under way at NIST called IEEE 1451 (http://www.motion.aptd.nist.gov/). The IEEE 1451, a family of Smart Transducer Interface Standards, describes a set of open, common, network-independent communication interfaces for connecting transducers (sensors or actuators) to microprocessors, instrumentation systems, and control/field networks. The key feature of these standards is the definition of Transducer Electronic Data Sheets (TEDS). The TEDS is a memory device attached to the transducer, which stores transducer identification, calibration, correction data, measurement range, and manufacture-related information, etc. The goal of 1451 is to allow the access of transducer data through a common set of interfaces whether the transducers are connected to systems or networks via a wired or wireless means. The family of IEEE 1451 standards are sponsored by the IEEE Instrumentation and Measurement Society's Sensor Technology Technical Committee. The first standard to be published was IEEE 1451.2, which was approved in September of 1997, which is a specification of an electronic data sheet and a digital interface to access that data sheet, read sensors, and set actuators. The interesting thing about the IEEE 1451 effort is that is not another field network; it is a collection of open standards that may be used with multiple networks. For the NeSSI™ effort the most important effort of the IEEE 1451 family of standards is the IEEE 1451.6. The standard defines the mapping of IEEE 1451 TEDS to the CANopen dictionary entries as well as communication messages, process data, configuration parameter, and diagnosis information. It adopts the CANopen device profile for measuring devices and closed-loop controllers. This project defines an intrinsically safe (IS) CAN physical layer. There are already many implementations of CANopen network interfaces based on the CAN in Automation (CiA) DS 404 device profile, which has been available for over 10 years. CiA has assigned the rights to DS 104 over to IEEE 1451 and added definitions needed for intrinsically safe operation. Finally, it looks like a CAN based network that is intrinsically safe will be available within a relatively short time.

On other fronts, Emerson has offered to make its Foundation™ Fieldbus chipset and circuit board available to NeSSI™ device manufacturers wishing the utilize that protocol and IS network for their NeSSI-bus. This is significant since the Emerson hardware is very close to the dimensions needed to fit NeSSI™ devices. Also, it appears that the high cost associated with Foundation™ Fieldbus may be something of an urban legend – latest indications are that the cost adder over other field bus networks is minimal. There are also efforts underway by ABB to adapt the I^2C network protocol used in its process analyzers to an IS NeSSI-bus. So, the situation has changed quite dramatically. Five years ago (when NeSSI™ first started looking for a communication network) there were no viable choices that supported intrinsic safety and a small enough physical footprint to be selected as the NeSSI-bus. Now it looks like that within the next two years there may be three viable networks to choose from.

10.5.2
The NeSSI™ Sensor Actuator Manager (SAM)

The draft NeSSI™ Gen II specification describes many required and desirable aspects of the SAM but will leave the development and implementation to individual manufacturers. To date, this important aspect of the Gen II draft specification has not been demonstrated and development of a functional SAM remains an opportunity for research and commercialization. But there are also indications that progress is being made. In the NeSSI™ specification there are provisions for two types of SAM implementation. In one embodiment, the SAM is a stand-alone device, such as a PC computer, a programmed logic controller (PCL), or a compact "hockey-puck" industrial microprocessor controller acting as a gateway between the NeSSI-bus and the plant Ethernet networks. In the other implementation, the SAM is actually embedded in the controller of a process analyzer. Recently, three companies, ABB, Emerson, and Siemens, have agreed to develop NeSSI™ SAM functionality embedded in their process gas chromatography analyzers. This is a interesting and welcome development as it will help demonstrate the perceived advantages of a "smart" sampling system in conjunction with the most commonly used process analyzer. To help define the important software functionality for the SAM, a recent NeSSI™ workshop focused on defining the highest priority monitoring and alerting tasks among prospective end-users of NeSSI™ systems.

10.6
Advantages of NeSSI™ in Laboratory Applications

While the NeSSI™ concepts were originally conceived and developed for industrial process stream sampling applications, this same technology could be used as a general platform architecture supporting analytical measurements and other research activities in the laboratory. This is an area that CPAC is actively pursuing in its research program and promoting within academia. Notably, despite its growing acceptance by industry, the modular surface-mount hardware, related NeSSI™ specifications, and the industrial development efforts are still fairly young and continuing to evolve. The adoption and development of this technology for laboratory applications will complement, and benefit from, the continuing development efforts of industrial end-users and vendors. Successful adaptation of NeSSI™ systems for laboratory use will help demonstrate and evaluate the technology in more diverse applications, and with less associated risk, than would be possible in industrial process settings. Expanding the demonstrated application base of NeSSI™ to include laboratory uses will also broaden the potential markets, help reduce the costs, and stimulate additional development within both industry and academia.

Several advantages can be envisioned from adaptations of the NeSSI™ technology to laboratory applications. While some of these advantages could be achieved with currently available laboratory flow systems, the ease of assembly, built in communication, rugged construction, self-diagnostic abilities, and automated control functionality of the NeSSI™ approach represents entirely new capabilities not available in current commercial products. Interestingly, the emergence of the modular surface-mount sampling/sensor systems may be signaling a reversal of recent trends in the analytical community toward analytical probe, or *in situ* measurement, development. While originally favored as an answer to problems with traditional extractive sampling technology, *in situ* probes have their own problems with calibration and validation that have led many to abandon that approach in critical monitoring situations.

10.6.1
Provide a Consistent Physical Environment for making Analytical Measurements

One of the most compelling advantages of NeSSI™ is that it provides a consistent, reproducible, and verifiable fluidics environment for making analytical measurements. The importance of stability in the sampling system is well understood in industry (where it is estimated that 80% of the problems with on-line analyzers are due to problems with the sampling system) but is less appreciated in academic and laboratory settings. As regulatory requirements and an increased emphasis on hazardous agent detection and pharmaceutical monitoring become more predominant in academic research, the ability to produce verified analytical measurements is becoming more important. By providing integrated stream switching on the modular sampling/sensor system manifold, calibration and/or validation samples can be introduced and analyzed along with the sample stream. This will improve analytical confidence and traceability and reduce, or eliminate, analytical errors due to sensing element placement and/or sample conditioning variability between calibration and analysis steps.

10.6.2
Easily Constructed and Configured Interfaces for Analytical Instrumentation

The modular nature of the surface-mount components makes it very easy to assemble or reconfigure a fluidic system to meet the requirements of a particular analyzer or sensor. Figure 10.19 shows three simple examples of surface-mount component systems being used to provide all, or part, of an analytical interface.

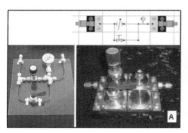

Fig. 10.19 Examples of surface-mount instrument interfaces. (A) A simple gas chromatograph helium gas metering system, showing the traditional assembly made from tubing and fittings, a four-component surface-mount replacement, and the surface-mount layout and flow diagram at the top. (B) A simple two-component surface-mount interface for an oxygen sensor (Teledyne Analytical) with a standard rotameter on the inlet, the sensor in the first component positions, and a metering valve in the second component position before the outlet. (C) An at-line system for introducing a process gas sample into the inlet stream of a process GC; a diagram of the surface-mount layout and flow is shown above.

While these examples could have been assembled using traditional tubing and fittings, the surface-mount systems are more compact, easier to maintain (e.g., to remove a surface mount component from the substrate involves removing only four screws), and provide a path forward where adding automated components requires only swapping out the SP76 mechanical component for an SP76 actuator. This flexibility and ease of assembling prototype interfaces is something that conventional laboratory systems cannot match and becomes even more important in the face of the dwindling skilled support resources (machinists, glassblowers, and electronics technicians) facing most academic (and industrial) institutions.

10.6.3
Facilitate Reduced Sensor or Analyzer Development Costs

The systems being promoted by NeSSI™ will also make it easier and less costly to develop new analytical technology through the use of "plug and play" components with well-defined sample interfaces and standardized connectivity protocols – freeing the researcher to concentrate on analytical development. By incorporating appropriate sample conditioning in the surface-mount system, the range of potential sample conditions that a sensor or analyzer must accommodate is reduced. This can lead to simpler transduction mechanisms and/or analysis protocols. For example, many gas phase sensors are sensitive to humidity, particulates, and temperature – by controlling these variables with surface-mount components (dryers, filters, and substrate heaters) upstream of the sensor, the researcher can focus on the analytical technology rather than "protecting" his sensor from detrimental environmental or sample conditions. The modular nature of these systems makes it possible to specify certain combinations of surface-mount conditioning components for different application environments.

Furthermore, the reduced size and internal volumes of the modular surface-mount sampling/sensor systems is a good scale match with many of the MEMS and microfabricated miniature analytical systems being developed in research laboratories around the world. Despite the immense efforts to develop the sensing capability of the micro-analytical devices, little attention has been focused on how the sample is going to presented to the very small devices in a reproducible way without clogging the very small flow channels of these devices.

Similarly, the NeSSI™ efforts to adopt a standardized communication bus, or network, and the hardware and software needed to poll and collect data from sensing elements will help free the researcher from developing their own custom electronic data I/O and user interface software. The standardization on a commercially available field bus network protocol (a largely transparent software layer) and hardware (chip-set level electronic interface for I/O) means that researchers can easily (using a common skill-set and widely available tools) integrate their analytical devices into an existing logging and reporting network and be insured that their device will be "plug & play" compatible with other sensors and actuators on that network. Figure 10.20 shows some of the analytical technologies that have been developed or are under development for NeSSI™

NeSSI™ Sensing Technologies

- Gas Chromatography
  - Thermal Desorption (?)
  - Trap/Purge Injection (+)
- Dielectric (√)
- Spectroscopies
  - IR (?), NIR (√)
  - UV- Vis (√)
  - Raman (√)
  - Fluorescence (+)
- Impedance (+)
- Conductivity (√)
- Vapochromic Sensors (+)
- GLRS (+)
- Particle Sizing
  - Light scattering (?)
- Turbidity (+)
- pH (√)
- RGA (+)
- Mass Spectrometry (√)
- LC, SEC, IC (+)
- Terahertz (?)
- Refractive Index (√)

Fig. 10.20 Survey of analytical technologies available (√), demonstrated (+), or mentioned (?) for NeSSI™.

systems. Items with a check mark (✓) are commercially available as SP76 components or interfaces, those with a plus (+) sign are known to be demonstrated, and those with a question (?) mark have been mentioned as planned or underway developments.

10.6.4
Increased Ability to Integrate Multiple Sensing Elements

The small footprint of the NeSSI™ systems, relative to conventional systems, means one is not likely to be constrained to limit the number of sensing elements one could incorporate in a single application stream. This means that multiple sensors, or even redundant sensors for critical monitoring situations, can be integrated into a single system. One area expected to provide large benefits and new analytical capability is the formation of sensor arrays or "smart" systems to more fully characterize complex samples and verify consistent sample conditioning. The ability to tie different analytical techniques together on the same modular sampling/sensor system substrate could lead to the creation of new hyphenated instrumentation. Similarly, the development of analytical "clusters" incorporating the multiple analysis and sample conditioning functionality required for complete assays, or routine analysis, of common samples (e.g., water quality monitoring) in a single unified structure is already being considered.

10.6.5
Ease Adoption of Developed Analytical Solutions to Industrial Applications

Since the NeSSI™ components used in laboratory systems would be available as components of industrial systems, sensing technology developed and tested using these systems could be easily transferred. Adaptations needed to accommodate differences in stream characteristics (flow, pressure, etc.) between laboratory and process applications can be easily implemented by switching modular components. This is distinctly different than with other fluidic systems commonly used in laboratory research – often these laboratory systems are not suitable, or require substantial modification, for process monitoring applications, which leads to slower adoption of the developed sensing technology. Conversely, there are many innovative technologies and monitoring solutions developed by industrial researchers that never get exposed to academia. Having a common measurement platform whereby these practices could be exchanged and implemented would be mutually beneficial and lead to increased cross-fertilization of ideas. Furthermore, the adaptation of these surface-mount sampling/sensor systems would help train analytical chemists in the more industrially relevant techniques of stream sampling and analysis rather than the traditional "beaker" or "test tube" analysis commonly taught in academia.

10.6.6
Applications to Other Laboratory Tasks

While most of the discussion above was related to the advantages of NeSSI™ in the area of supporting analytical measurements, there are also many other potential applications of this technology as general purpose laboratory fluid delivery systems. In general, fluidic systems already exist that are probably better suited for these tasks than systems built from currently available modular surface-mount components, but two brief examples will be given to show what might be possible. Increased development and adaptation of computer controlled surface-mount fluidic components and automated manipulations could be used to replace traditional volumetric sample preparation. Recent developments in high precision flow controllers for gas and liquid applications are rapidly being adapted to the modular surface-mount format, leading to increased precision and reproducibility in liquid dispensing, mixing, and formulation applications. The ability to incorporate surface-mount sensing components can help to insure you are dispensing what you want. There is also considerable interest, and a CPAC funded project, related to combining NeSSI™ with modular micromixing, microreactors, and micro-unit-ops components being developed in Europe to produce micro-scale chemical plants. Currently, the NeSSI™ systems are being investigated as general fluidic delivery system for solvents, adducts, and reagents going into microreactor and for monitoring product formation. However, eventually the microreactor could become just another surface-mount component and one could perform reactions or synthesis within the modular sampling/sensor system itself.

10.6.7
Cost, Convenience, and Safety

Besides the technical advantages listed above, there are also some less direct benefits to adopting and promoting NeSSI™ in the laboratory. Unfortunately, much of the time spent conducting research involves assembling systems and apparatus needed to support the planned experimentation. While some of this setup effort is unavoidable due to the nature of research, having a standardized toolkit of fluidic and sensing components that could easily be assembled for specific tasks or rapid prototyping would certainly reduce the burden and increase researcher productivity. The "plug and play" nature of NeSSI™ implies that a relatively small toolkit of such components could be reassembled and reconfigured by students to perform various analytical tasks typically taught in the laboratory. Also, the commercial-grade construction of modular sampling/sensor system components would make them more durable than glassware commonly used in teaching laboratories, thereby reducing long-term costs associated with laboratory curriculum. Finally, the reduced volumes and closed, leak-free, nature of a NeSSI™ system may reduce exposure risks and waste generation, thereby increasing overall laboratory safety.

One potential barrier to the widespread adoption of NeSSI™ hardware by the laboratory chemist is the generally higher cost typically associated with process instrumentation and fittings, relative to conventional laboratory equipment. When the industrial NeSSI™ systems were first introduced, the cost multiplier was estimated at 1.5 times the cost of a conventional process sampling system. Currently, estimates from manufacturers who deal in both types of sampling systems place the multiplier at about 1.1 times the conventional system costs and estimates predict the multiplier will be 0.7 within three years – so the cost is approaching parity with conventional laboratory costs. Given that NeSSI™ systems are inherently more durable and robust than most laboratory fluidic systems, and the increased value derived from standardization and communication features mentioned above, hardware cost are not likely to be a serious impediment to the adoption and use of modular sampling/sensor systems in the laboratory.

10.7
Conclusions

This chapter has sought to introduce the philosophy and commercial implementations for new sampling systems being developed under the NeSSI™ program. Other chapters make references to NeSSI™ in several different contexts. It is important to remember that NeSSI™, and efforts by others that pre-dated NeSSI™, was mainly trying to improve the practice of designing, installing, and maintaining sampling systems. As NeSSI™ has evolved and gained in popularity there seems to be many other applications of this technology beyond just supporting process monitoring analyzers. Certainly, much of the success of the NeSSI™ program can be directly attributed to the vision and tireless efforts of the many people who helped define, nurture, and promote the program. There was never any direct reward for all their efforts, just the satisfaction of knowing that they got to play a part in defining a new direction for an entire industry. Also keep in mind that the participation would neither have been as broad or as effective across the industry without the neutral meeting ground provided by the academic consortia of CPAC – it should serve as a reminder to support, and be involved in as much as possible, consortia-based opportunities that you may come across in your companies interactions with universities and competitors.

Finally, the reader may lament a certain lack of recent results and examples of NeSSI™ applications in process settings. This is partly because things are happening at a time scale too quick to not be outdated in a book. For recent NeSSI™ news, please visit the NeSSI™ web site at http://www.cpac.washington.edu/NeSSI/, which maintains a complete history of papers, presentations, and workshop results related to the topic.

References

1 R. Dubois, P. van Vuuen, *NeSSI™ (New Sampling/Sensor Initative) Generation II Specification*, Center for Process Analytical Chemistry (CPAC), University of Washington, Seattle WA, **2003**, www.cpac.washington.edu/NeSSI.

2 D. M. Simko, "NeSSI – concept to reality", *Technical Papers of ISA*, 428 (Proc. – Anal. Division Symp., 2002), **2002**, 122–130.

3 J. B. Callis, D. L. Illman, B. R. Kowalski, Process analytical chemistry. *Anal. Chem.* **1987**, 59(9), 624A–626A, 628A, 630A, 632A, 635A, 637A.

4 F. McLennan, B. Kowalski (eds.), *Process Analytical Chemistry*, Blackie, London, **1995**.

5 *ANSI/ISA SP76.00.02-2002 Modular Component Interfaces for Surface-Mount Fluid Distribution Components – Part1: Elastomeric Seals*, Instrumentation, Systems, and Automation Society (ISA), Compositional Analyzers Committee, **2002**, www.isa.org.

11
Catalyst Characterization for Gas Phase Processes

Michelle J. Cohn and Douglas B. Galloway

11.1
Introduction

Characterization tools are critical to technology development and to the advancement of basic scientific knowledge for materials, catalysts, adsorbents and process technologies. An increase in fundamental knowledge and better predictive tools will reduce overall technology delivery time by identifying and measuring key properties and correlating critical properties with performance. These tools need to be deployed in all phases of product/process development, from idea validation to commercialization, and, if necessary, troubleshooting.

Tools for examining molecular and long-range structures, chemistry and bonding are widely applied to both new and existing molecular sieves, catalysts and adsorbents. Properties of materials, including reactivity, thermochemistry, acidity and diffusion, can also be readily measured. As much work as possible work is done using realistic *in situ* environments to closely mimic actual process conditions and chemistries.

A suite of tools is necessary to accelerate catalyst and process development. These tools will be used determine key catalyst functions such as:

1. acidity
2. metallic sites
3. acid–metal balance
4. acid–metal proximity
5. pore geometry of microporous and mesoporous materials
6. effect of contaminants on catalyst acid and metal functions
7. determination of reaction mechanisms
8. measurement of reaction kinetics
9. determine catalyst coking and deactivation
10. metallurgical failure analysis.

Table 11.1 gives examples of specific useful tools.

Micro Instrumentation for High Throughput Experimentation and Process Intensification – a Tool for PAT
Edited by M. V. Koch, K. M. VandenBussche and R. W. Chrisman
Copyright © 2007 WILEY-VCH Verlag GmbH & Co. KGaA, Weinheim
ISBN: 978-3-527-31425-6

Table 11.1 Some key tools for characterization.

Characterization	Tool
Micro and chemical structural analysis, including porosity, bonding, coking	X-ray diffraction (XRD) Nuclear magnetic resonance (NMR) Fourier-transform infrared (FTIR) Scanning electron microscopy (SEM) Transmission electron microscopy (TEM) Atomic force microscopy (AFM) Raman spectroscopy Small probe molecule volumetric and gravimetric adsorption
Reactivity and acidity	Ammonia temperature-programmed desorption (NH_3TPD) Calorimetry: liquid–liquid, liquid–solid and gas–solid Probe molecule microreactivity testing
Kinetics and diffusion	Steady-state isotopic transient kinetic analysis (SSITKA) Temporal analysis of products (TAP) Tapered element oscillating microbalance (TEOM) Temperature scanning reactor (TSR) Zero length chromatography (ZLC) Pulsed field gradient NMR
Metallic sites	Temperature-programmed reduction (TPR) X-ray adsorption near-edge spectroscopy (XANES) Extended X-ray adsorption fine structure (EXAFS) Mössbauer spectroscopy X-ray photoelectron spectroscopy (XPS) Electron paramagnetic spectroscopy (EPR)

Many of these tools can give information at the nano or atomic scale. In addition, some can be performed at realistic operating conditions (temperature, pressure, flow, chemical composition), which result in the most immediately transferable understanding.

The next section describes in detail many of the less well-known tools for kinetics, reactivity and diffusion.

11.2
Characterizing Reactivity, Reaction Mechanisms, and Diffusion

As customer demands for the development and optimization of chemical processes continue to increase, the need to improve the fundamental understanding of a given process also grows. By better understanding and quantifying the mechanisms for diffusion, adsorption, and reaction of molecules in catalytic materials, the rate of developing a novel chemical process is greatly enhanced. To gain the needed insight, techniques need to be developed and implemented within a company that allow scientists and engineers to obtain information about reaction mechanisms and pathways. Likewise, characterization methods to measure fundamental kinetic constants such as adsorption constants, activation energies, heats of adsorption, and diffusion constants are also needed for process development. This information not only impacts the selection and development of new catalyst systems but also plays a key role in providing some of the fundamental constants required to model and scale-up new processes.

11.2.1
Reactivity Testing

Historically, reactivity testing has formed the basis for most catalyst screening and process development work. Typically, this work has involved screening, testing, and optimizing catalytic materials in pilot plant scale reactor systems. Although these systems can utilize commercial feeds, and provide valuable information at commercially relevant process conditions, they require several days to test a single sample and normally involve a large commitment in resources. More recently, developments in informatics, automation, and instrumentation have lead to the integration of combinatorial techniques into material research and catalytic testing [1]. By utilizing a combinatorial format, hundreds of catalyst can be tested and evaluated in one-tenth the amount of time required by conventional methods. These advancements have already made a large impact in reducing the cycle time for testing and screening ideas and catalytic materials.

The concept of combinatorial screening can be further divided into two subsystems, focus level screening and discovery screening [2]. While the focus system provides more detailed information such as kinetics and selectivity on many catalyst samples per test, a discovery system can screen thousands of catalytic materials with sufficient resolution to identify promising hits. An example of a discovery assay for reactivity testing that has been successfully implemented at UOP LLC is the laser activated membrane introduction mass spectrometry (LAMIMS) system [2, 3].

The LAMIMS system was developed as a modification of membrane introduction mass spectrometry, and uses a silicone membrane to separate the process stream from the vacuum of a quadrupole mass spectrometer. Figure 11.1 shows the details of the reactor system used in a LAMIMS analysis system. Once a catalyst array is loaded in the reactor, the system uses a CO_2 laser to activate individual catalyst samples while a reactant stream passes through the sample holder. By monitoring the proper mass, the integrated signal from the mass spectrometer is used to evaluate the reactivity of a given catalyst sample. The use of large arrays, combined with the requirement of less the 90 s to test a single catalyst sample, enables this technique to screen catalyst samples at a rate of more than 1000 samples per day.

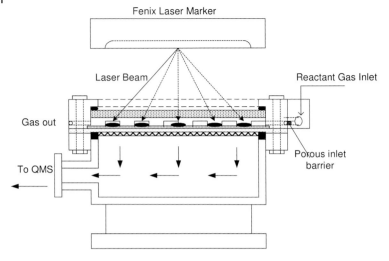

Fig. 11.1 Schematic of the LAMIMS reactor system.

11.2.2
Kinetic Tools

While combinatorial reactivity testing provides the ability to quickly test and screen catalyst systems and catalyst samples, other kinetic tools are needed to provide deeper understanding. One such tool is the temperature scanning reactor (TSR) system developed by Wojciechowski [4, 5, 6]. The TSR concept allows for the rapid collection of kinetic data. By ramping the reactor temperature at several space velocities, the TSR can cover a wide range of process conditions in a single experiment that would take many runs in a conventional reactor system. Figure 11.2 illustrates a conversion surface consisting of conversion, space-time, temperature (χ, τ, T) triplets generated from a TSR run used to study transalkylation reactions. While the ability to quickly generate reactivity data over a wide range of conditions helps to make this technique valuable for process development, even further value can be added using TSR methodology for data processing [7]. Taking the numerical derivate of the (χ, τ, T) surface generates a three-dimensional kinetic surface of rate as a function of space-time and temperature.

In the past, process development and kinetic modeling involved collecting pilot plant data from several runs. Owing to the resources involved with pilot plant testing, kinetic data is typically collected on only commercially viable prototypes. Since TSR equipment and methodology allow for the collection and generation of kinetic data quickly and efficiently, meaningful kinetic surfaces can be generated for prototype samples. This data allows for process development teams to evaluate process concepts and novel catalyst formulations in much greater detail. Along with evaluating samples quickly, the deeper understanding provided by techniques such as the TSR allows for better decision making, leads to a reduction in the

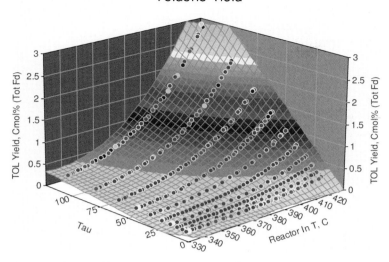

Fig. 11.2 Toluene yield versus reactor temperature for various space-times.

time required for process development, and provides kinetic constants that can be used in process models.

Another useful tool for characterizing catalyst systems is steady-state isotopic transient kinetic analysis (SSITKA) [7] Originally developed by Happel [8], Bennett [9], and Bileon [10], SSITKA is based on the fact that the isotope effect is minimal for heavier atoms such as carbon. By measuring the isotopic composition of a product stream, the technique uses a step change between isotopes for two reactant streams to determine the concentration of surface intermediates and their corresponding average residence time. Since the isotopic step change has no impact on the energetics of a catalyst system, SSITKA allows for kinetic measurements to be made under steady-state conditions.

Most SSITKA measurements have been made using a mass spectrometer to monitor the isotopic composition of the product stream. Although this limits the utility of the technique, it has been used successfully to characterize many reactions, including methanation, methanol synthesis, ammonia synthesis, Fischer–Tropsch synthesis, and ethane hydrogenolysis [11]. More recently, GC separation and hydrogenolysis of the product stream have been incorporated into a SSITKA system, and applied to the characterization of n-butane isomerization over sulfated zirconia [12]. By analyzing the product stream during a step change from unlabeled to labeled n-butane, the number and average lifetime of surface intermediates leading to isobutene were studied as a function of H_2 addition and Pt promotion. As an example of this work, Fig. 11.3 shows the large increase in surface intermediates as a result of the addition of a Pt promoter to sulfated zirconia in the presence of H_2. During the development of new processes and testing of novel catalyst formation, the information given by techniques such as

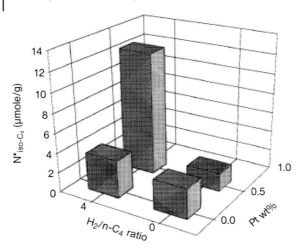

Fig. 11.3 Effects of Pt and H_2 on the initial concentration of surface intermediates. (From Ref. [12], reprinted with permission.)

SSITKA is extremely valuable in gaining insight into novel reaction pathways, and quantifying the specific impact of different types of active sites on catalytic reaction mechanisms.

For many catalytic processes involving hydrocarbons the major deactivation pathway comes from the formation and deposition of carbonaceous species. Often referred to as coke, on many catalytic materials its deposition leads to transport resistances for reactants and products and blocks access to active sites. Understanding its rate of formation and impact on catalyst deactivation is of great importance in developing and modeling new catalytic processes. Developing a fundamental understanding of coke deposition enables the proper selection of catalytic materials and process conditions to minimize the impact of deactivation on a process.

Although coke deposition can be modeled in many ways, a strong case can be made that it is best to relate catalyst deactivation to its coke content, and use the gas composition, temperature and catalyst activity as parameters to describe coke formation [13]. Although gravimetric measurements have been useful for the study of coke deposition, gravimetric balances have limited utility for kinetic studies due to the large amount of reactant bypass of the catalyst bed [14]. Lacking the ability to characterize catalyst samples under plug flow conditions, precise kinetic measurements are very difficult to make using gravimetric balances. However, the development of the tapered element oscillating microbalance (TEOM) has proven to be a major improvement in the ability to characterize coke deposition [15]. Using a hollow oscillating element with the catalyst bed located in one end, the catalyst mass is determined by monitoring the natural frequency of the element. Since the reactant stream passes through the hollow element and then the catalyst bed, the catalyst bed mass can be monitored under plug flow conditions. Simultaneously

Effect of Coke on Catalyst Activity and Selectivity

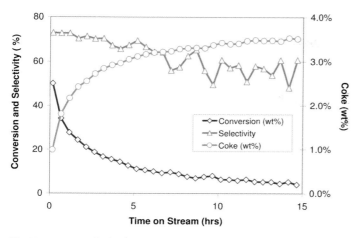

Fig. 11.4 Impact of coke deposition on conversion and selectivity.

measuring the changes in the catalyst mass along with catalyst activity (Fig. 11.4) gives direct insight into the impact of coke levels on catalyst activity. This crucial information, gained with minimal assumptions, is needed to properly model catalyst deactivation for many processes.

11.2.3
Diffusion Tools

Owing to several unique properties, molecular sieves are utilized in many commercial catalyst and adsorption processes. While providing the possibility of many types of active catalytic, or adsorption, sites due to their typically high capacity for ion exchange, some of the most unique properties of molecular sieves are a result of their microporous structure. Typically exhibiting structures with pore diameters in the same range as the cross-sectional size of many hydrocarbons of industrial interest, molecular sieves show such useful properties as shape-selectivity and molecular sieving capabilities. However, the close relationship between the size of a reactant or product molecule in a catalyzed reaction or the size of an adsorbate molecule in a separation process often leads to diffusion having a large or even dominant impact on the rate of a process. Likewise, secondary processes such as coking or hydrothermal treatments can easily modify a material to the point that a given process becomes diffusion controlled.

Clearly, techniques to characterize diffusion can provide important and useful insight into many catalytic and separation processes that employ molecular sieves. Fortunately, many techniques have been developed to determine diffusion coefficients on numerous molecular sieves. Several of the more successful methods that have been developed over the past two decades include pulsed field gradient NMR, tracer desorption NMR, single crystal membrane, frequency response, and zero length chromatography (ZLC). While all of these techniques offer many advantages and disadvantages, zero bed chromatography, more commonly known as ZLC, offers a relatively simple yet effective method for characterizing diffusion. First demonstrated by Eic and Ruthven [16], the technique has been used extensively for characterizing many combinations of microporous materials and adsorbates.

The basic features of the ZLC system consist of a gas chromatograph with a flame ionization detector (FID), a ZLC column housed within the GC, a switching valve, two mass flow controllers, and a data acquisition PC. After loading a small quantity of sample into a modified zero-volume union, the union is attached directly to the FID of the GC system. Following pretreatment of the sample, the sample is equilibrated with the adsorbate stream at a low partial pressure. After the equilibration period, the switching valve is activated to allow the argon purge stream to flow across the sample. By evaluating the normalized desorption curve, C/C_o, as a function of time in the long-time region, the intracrystalline diffusion constant, D, can be determined from Eq. (1):

$$\ln \frac{C}{C_o} = \ln\left(\frac{2}{L}\right) - \pi^2 \frac{D}{R^2} t \tag{1}$$

where R is defined as the effective crystal radius and L is defined by Eq. (2):

$$L = \frac{F R^2}{3 V_{cat} K D} \tag{2}$$

with the purge flow rate given by F, the sample volume given by V_{cat}, and the equilibrium ratio given by K.

Greater insight into the impact of mass transport limitations on activity and selectivity is gained by combining information from techniques such as ZLC with results from activity tests. As an example, Fig. 11.5 shows ZLC results for the measured diffusion constants of propane and butane on four separate catalysts. As expected, for a given sample the diffusion constant for propane is higher than that for n-butane. More importantly, however, a clear positive correlation between catalyst selectivity and the measured diffusion constant is observed. By developing such correlations between reactivity measurements and diffusion, potential catalytic materials can be screened quickly and efficiently. A more effective use of techniques for measuring diffusion, though, occurs when they are used to better understand fundamental differences in performance for a set of catalyst samples. Differences in hydrothermal treatments and other factors can have a dramatic effect on the accessibility of an active site on a given catalyst. Without the availability of techniques such as ZLC, it becomes very hard to gain insight into such factors.

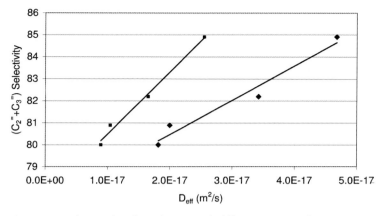

Fig. 11.5 Correlation of catalyst selectivity with diffusion constants for propane (■) and n-butane (♦) measured on four catalyst samples.

11.3 Summary

Although activity and selectivity measurements remain a major resource for generating information for process development, a wide array of tools are now available for characterizing reactivity, reaction mechanisms, and diffusion. While combinatorial techniques such as LAMIMS can screen up to 1000 catalyst samples per day in a single apparatus, the use of other kinetic tools during catalyst development is crucial. While techniques such as LAMIMS allow the screening of milligram quantities of catalysts during the initial phase of process development, many other tools are needed to provide information as the process is scaled up to a commercial unit that could easily utilize up to 100 000 kg of catalyst. To successfully scale a catalytic process that will involve an increase in scale of nearly ten orders of magnitude, as much fundamental information as possible needs to be measured. Techniques such as TSR and SSITKA provide kinetic constants that give researchers critical insight into reaction mechanisms, and provide information necessary to model a catalytic process. Other techniques such as the TEOM and ZLC measurements also provide important information regarding the impact of carbon deposition, hydrothermal treatments, and other factors on catalyst deactivation and active site accessibility. All of these techniques form an important set of tools that help to provide a better understanding of catalytic processes, and improve the speed of process development.

References

1. M. L. Bricker, J. W. A. Sachtler, R. D. Gillespie, C. P. McGonegal, H. V. D. S. Bem, J. S. Holmgren, *Appl. Surf. Sci.* **2004**, 223, 109.
2. A. Nayar, R. Liu, R. J. Allen, M. J. McCall, R. R. Willis, E. S. Smotkin, *Anal. Chem.* **2002**, 74, 1933.
3. A. Nayar, Y. T. Kim, J. Rodriguez, R. R. Willis, D. B. Galloway, F. Falih, E. S. Smotkin, *Appl. Surf. Sci.* **2004**, 223, 118.
4. B. W. Wojciechowski, *Catal. Today* **1997**, 36, 167.
5. N. M. Rice, B. W. Wojciechowski, *Catal. Today* **1997**, 36, 191.
6. S. P. Asprey, N. M. Rice, B. W. Wojciechowski, *Catal. Today* **1997**, 36, 209.
7. S. L. Shannon, J. G. Goodwin, Jr., *Chem. Rev.* **1995**, 95, 677.
8. J. Happel, *Chem. Eng. Sci.* **1978**, 33, 1567.
9. C. O. Bennett, in *Catalysis Under Transient Conditions*, ed. A. T. Bell, L. L. Heeedus, ACS Svmposium Series, American Chemical Society, Washington, DC, **1982**, vol. 178, p. 1.
10. P. Biloen, *J. Mol. Catal.* **1983**, 21, 17.
11. S. L. Shannon, J. G. Goodwin, Jr., *Chem. Rev.* **1995**, 95, 677.
12. S. Y. Kim, J. G. Goodwin, Jr., D. Galloway, *Catal. Today* **2000**, 63, 21.
13. G. B. Froment, *Proceedings of the Sixth International Congress of Catalysis*, **1976**, vol. 1, 10.
14. G. F. Froment, J. D. Meyer, E. G. Derouane, *J. Catal.* **1990**, 124, 391.
15. F. Hershkowitz, P. D. Madiara, *Ind. Eng. Chem. Res.* **1993**, 32, 2969.
16. M. Eic, D. M. Ruthven, *Zeolites* **1988**, 8, 40.

12
Integrated Microreactor System for Gas Phase Reactions

David J. Quiram, Klavs F. Jensen, Martin A. Schmidt, Patrick L. Mills,
James F. Ryley, Mark D. Wetzel, and Daniel J. Kraus

12.1
Overview

Microfabrication technology provides a platform for invention of highly instrumented, microscale reactor systems that incorporate new innovative analytical capabilities, allow point-of-use synthesis and small-scale manufacturing, or are amendable for implementation of novel high-throughput screening methods. However, the use of integrated circuit-like reaction devices, such as the MIT thin-film microreactor, also gives rise to a spectrum of new engineering challenges, such as reactor integration, process robustness, operation at realistic process conditions, and ease of use.

A prototype design for parallel gas-phase microreactors is described that combines electrical, mechanical, and chemical engineering principles to create an integrated, packaged system. It includes specialized components for microfluidics, gas-phase catalytic reaction, temperature and pressure control, thermal energy management, on-line process analysis, process automation, and process safety management in a single unit. The system platform has operational features that are equivalent to a conventional laboratory-scale reactor system, but the footprint requires less than 10% of the space.

The heart of the system is a microreactor packaging scheme that is based upon a commercially available microchip socket. This approach allows the silicon-based reactor die, which contains dual parallel reaction channels with more than 100 electrical contacts, to be installed and removed in a straightforward fashion without removing any fluidic and electronic connections. Various supporting microreactor functions, such as gas feed flow control, gas feed mixing, and various temperature control systems, are mounted on standard CompactPCI® electronic boards. The boards are subsequently installed in a commercially available computer chassis. Electrical connections between the boards are achieved through a standard backplane and custom-built input-output PC boards. A National Instruments® embedded real-time processor is used to provide closed-loop process control and to monitor system alarms. A host PC is used as the human–machine interface for operator interaction and for historical data logging.

To demonstrate feasibility and to evaluate system performance, the gas-phase oxidation of methane to synthesis gas and ammonia oxidation to nitrogen oxides over heterogeneous catalysts were utilized as test reactions. Both reactions were successfully operated under steady-state conditions for a limited period of on-stream time. Methane oxidation was more challenging to study than ammonia oxidation since it required temperatures on the order of 600 °C for appreciable reactant conversion. At this temperature, the platinum heater lines in the silicon reactor chip experienced a loss in electrical continuity. Ammonia oxidation was operative at lower temperatures since NO production varied from less than 5% selectivity at 250 °C to about 30% selectivity at 500 °C. Similarly, N_2O production was detected at about 400 °C and approached a selectivity of 20% at 500 °C. These results were in good agreement with previous experiments using the T-microreactor at MIT as well as conventional ammonia oxidation reactors reported in the literature. Besides demonstrating parallel gas-phase reactions on a chip, another key result was the ability of the process control scheme to perform closed-loop PID control on 24 microreactor heaters at a maximum data rate of 500 Hz. Other details of the system design and challenges associated with this MEMS-like reactor system are also described.

12.2
Introduction

Microreactor technology (MRT) has evolved over the past decade into an established discipline that has applications ranging from high-throughput catalyst screening for chemicals (Younes-Metzler et al., 2005), petroleum intermediates (Huybrechts et al., 2003), and fine chemicals (Chen et al., 2005) to small-scale manufacturing of agricultural chemicals and other classes of organic chemicals (Hessel and Löwe, 2003a, 2003b, 2003c, 2005; Watts, 2004). Microscale reactor technologies are also being used to accelerate the synthesis of new liquid crystals (Styring, 2004), to study biological substrate oxidation kinetics (Doig et al., 2002; Puskeiler et al., 2005), and to generate nanoparticles (Shchukin and Sukhorukov, 2004). MRT is also driving new developments in synergistic technologies, such as microscale analytical separations (Palmer and Remcho, 2002), combinatorial microfluidics (Wiese et al., 2005), and multiphase-flow micro-hydrodynamics (Tovbin, 2002). To construct MRT components, new materials of construction (Dietrich et al., 2005), fabrication tools (Bong-Bang et al., 2005), and fabrication techniques (Truckenmüller et al., 2006) are undergoing development and implementation. Miniaturization of process components has also led to the invention of various special-purpose process components, such as microstructured mixers (Pfeifer et al., 2005), heat exchangers (Alm et al., 2005), miniature gas–liquid reactors (Chambers et al., 2005; Commenge et al., 2005), and fluid–fluid separators (e.g., Benz et al., 2001; Majors, 2006; Sato and Goto, 2004). The evolution in both the hardware and applications of MRT over the past decade has been largely captured through the annual International Conference on Microreaction Technology (IMRET) that was initiated with IMRET 1 (Ehrfeld et al., 1997). Since this first conference, eight IMRET conferences have been organized on an alternating basis by DECHEMA in Europe and by the AIChE in North America, with the latest gathering being IMRET 9 (Schütte et al., 2006). It is also generally agreed that microscale systems represent a definitive methodology for process intensification (Charpentier, 2005). The latter is also consistent with the often-cited advantages of microreaction systems when compared with conventional reactor systems, which include (a) safer operation; (b) simplified scale-up; (c) improved heat and mass transfer characteristics; and (d) application flexibility (Jensen, 1999; Ehrfeld et al., 2000).

Microreaction technology within DuPont evolved according to the event versus time schedule described in Table 12.1. Research activities have spanned nearly two decades since they were initiated in the latter part of the 1980s. Early projects were driven by the desire to produce hazardous chemicals upon demand, thereby eliminating the need for transportation and storage (Anonymous, 2003). Target molecules were inorganic or organic compounds that contained functional groups needed for synthesis of high-valued added molecules, such as agricultural chemicals. The reactions of interest generally involved the use of heterogeneous catalysts or fast acid–base chemistries, were highly energetic (e.g., $\Delta H_r > -50$ kcal mol^{-1}), generated corrosive reaction media (such as aqueous HCl), and could potentially result in explosive mixtures due to the presence of a hydrocarbon and O_2. For these reasons, the microreactor design effort considered various methods for on-board reactant mixing, approaches for thermal energy management and heat integration, techniques for micro-machining methods using various materials of construction, concepts for controlling reaction light-off, and methodologies for incorporation of solid catalytic materials (Ryley, 1999). To meet these requirements in a compact package, various components were integrated by creating multi-layered laminate structures (Ashmead et al., 1996, 1997). However, these efforts did not address the problem of connecting arrays of reactors to increase overall throughput, or integration of microreactors into other system components with automation to create a single platform system.

Referring to Table 12.1, DuPont and MIT initiated a collaboration in the late 1990s that was part of the Microfluidic Molecular Systems (Microflumes) Program at the Defense Advanced Program Agency (Lee, 1999). The goal was to transform fluid-based protocols and operations, such as those used for chemical or biochemical reaction, fluid transport, synthesis, and engineering, from the laboratory to the field. This was to be accomplished by first developing chip-scale fluid-handling capability to replace macroscale fluidic components, such as pumps, valves, and reactors, and then integrating these new microscale components with automated sample capture and on-line analytical. These capabilities would provide the soldier with programmable micromachines that could perform hundreds of fluid-based process sequences to meet dynamic requirements in chemical and biological analysis as well as chemical synthesis.

Table 12.1 Evolution of microreactor technology in DuPont.

Year	Research and development activity
1987–1994	DuPont develops microreactor concepts and prototype microreactors so that selected target chemistries can be demonstrated in an automated fashion. • New hazardous materials laws limit transport of certain chemicals • Quantities needed for testing are typically small • Disposal of unwanted chemicals can be costly • Certain large scale chemical reactions are dangerous • Some chemicals have limited shelf-life
1994–1997	DuPont funds collaborative effort with MIT to develop accurate transport-kinetic models for microscale chemical reactors. • Microreactors consist of well-defined geometries, but multi-layered structures when compared to larger-scale conventional reactors • Existing reactor modeling packages were not predictive of microreactors
1994–1999	DuPont proves that different reactions can be conducted and accurately measured on a single chip • DuPont develops multi-layered structure and microarray format • DuPont develops coupled microreactor–analytical system for direct monitoring of reaction performance
1997–2000	DARPA (Defense Advanced Research Program Agency) provides additional funding for DuPont/MIT collaborative research effort to prove that a microreactor system can be designed for safe, parallel operation • Microreactor could have application for different types of chemistries and different types of users (e.g., DuPont Pharmaceuticals) • Interest in using microreactor to synthesize different types of reactions in parallel exists
2001	DuPont divests DuPont Pharmaceuticals to Bristol-Myers-Squibb
2002	DuPont identifies microreactor technology for potential venture and develops MIT alliance
2002–2006	Novel parallel microreactor systems for fermentation are demonstrated in a new program initiative in the DuPont/MIT alliance

The primary objective of this work was to develop a stand-alone gas-phase microreactor system whose functionality satisfied the objectives of the DARPA Microflumes program. Emphasis was placed on designing and constructing a MEMS-based system from first principles that had the same functionalities that are present in a more conventional laboratory-scale system, such as the DuPont MARS system (Mills and Nicole, 2005). The key challenges in microreactor integration as well as demonstrating prototype packaging schemes are addressed. No previous effort has demonstrated multiple MEMS-like reactors running in parallel with an integrated fluid handling and control system for each reactor. A unique feature is that key process functions consist of modular components that fit inside a commercially-available computer chassis. The components are designed and fabricated from first principles as prototype electro-mechanical boards that are configured to perform one or more process-specific functions. To demonstrate feasibility, the oxidation of CH_4 and NH_3 are used as test reactions. Operation of these reactions is also illustrated under commercially relevant conditions using integrated MEMS components.

12.3
Microreactor Packaging

The microreactor design was constrained by the requirement that all electrical and fluidic connections between the microreactor and other system components be implemented using so-called *packaging* methods that were easy-to-use while maintaining process robustness. The microreactor and any supporting components had to be heated so they could safely operate above the dew points of the feed and product stream compositions, thereby preventing any product condensation from the gas-phase. For these reasons, a simplified methodology was required that facilitated microreactor removal and replacement.

12.3.1
Microreactor Packaging Hardware

Previous microreactor packing schemes developed at both MIT and DuPont could not be used since the electrical and fluidic interfaces required significant effort to implement. The microreactor packaging problem was eventually solved by using the DieMate™ Known Good Die (KGD) socket manufactured by Texas Instruments (Texas Instruments Incorporated, 34 Forest Street, P.O. Box 2964, Attleboro, MA).

Figure 12.1 shows photographs of the Texas Instruments DieMate™ Known Good Die socket used for the final microreactor packaging scheme. The particular DieMate™ socket chosen for the microreactor scale-up system was the 110-pin version (Texas Instruments, Dallas, TX, P/N CBC110 004 K) since this could accommodate the required sensors.

12.3.2
DieMate™ Electrical Interconnect Details

As shown in Fig. 12.1(a), the DieMate™ socket holds the reactor die in place and makes electrical contact through a row of pins on each side of the socket. The pins are forced into contact with the die by four springs that are placed on each corner of the socket. These springs are visible in the side view shown in Fig. 12.1(b). The springs force the upper piece of the socket away from the lower socket body. This force is redirected by the pins to press on the die in the socket. The latching force

12 Integrated Microreactor System for Gas Phase Reactions

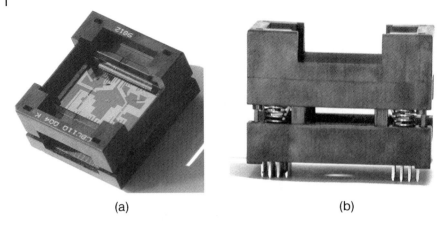

(a) (b)

Fig. 12.1 Texas Instruments DieMate™ Known Good Die socket used for microreactor packaging in the scale-up system.
(a) Perspective view of the TI socket with the reactor chip in place.
(b) Side view of the TI socket, showing the springs used for creating pin force.

of the pins is more than sufficient to form a gas-tight seal for atmospheric pressure reactions using O-rings. By pressing on the top piece of the socket, the springs are compressed downward and the pins of the socket are lifted off and away from the microreactor die. This action allows the microreactor die to be easily replaced. The bond pads of the microreactor die and the pins of the DieMate™ socket are self-aligning in this system, which simplifies installation.

Figure 12.2 is a rendered version of the scale-up microreactor die that shows the metallization pattern. Note that a large amount of the silicon area is covered by the lead lines. However, some of the leads on the die are still narrow. This was performed to make the resistances in all the leads to be approximately the same. This also simplified the design of the electrical circuits used for controlling the microheaters and micro-temperature sensors.

12.3.3
DieMate™ Fluidic Interconnect Details

Modifications were required to accommodate the fluidic interconnections since the DieMate™ socket was originally designed for use with integrated circuits. The solid bottom of the socket was removed so that a gas manifold could access the microreactor die from underneath. The final microreactor packaging scheme is shown in Fig. 12.3. The DieMate™ manifold extends up into the socket where it forms a seal with the Pyrex® bottom of the microreactor die using Kalrez® O-rings. The sealing force is created by the pins pressing on the top of the microreactor, and also by the socket head cap screws pushing the DieMate™ manifold upwards into the socket. Feed and product gas transfer lines constructed from 1/16-inch-o.d. type 316L stainless steel are connected to the DieMate™ manifold.

12.3 Microreactor Packaging

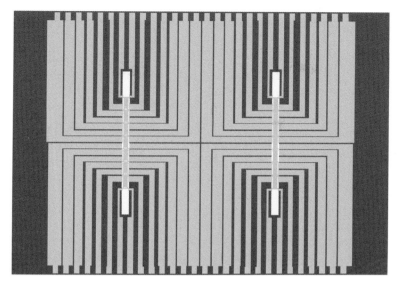

Fig. 12.2 Rendered top view of the scale-up microreactor die. Metallization is shown in gray, silicon substrate is blue, and exposed membrane is in yellow.

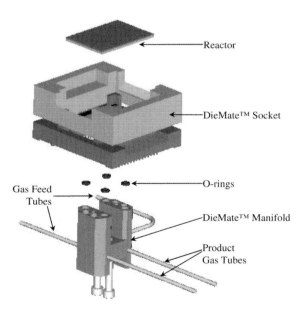

Fig. 12.3 Exploded view of the DieMate™ packaging assembly.

12.3.4
DieMate™ Manifold Heating

The DieMate™ manifold also functioned as the microreactor die heater since it had four drilled-through holes to accommodate ⅛-inch-o.d. cartridge heaters where each provided 15 watt of power (60 W total). The holes are visible adjacent to the O-rings in Fig. 12.3. The temperature of the DieMate™ manifold was monitored using two 1/16-inch-o.d. × 1-inch 100 Ω platinum Resistive Temperature Devices (RTDs).

12.3.5
DieMate™ Manifold Fabrication

The DieMate™ manifold (Fig. 12.3) was fabricated from type 316L stainless steel using conventional machining techniques. This material of construction was inert to the anticipated components present in the reaction mixture and was also suitable for operation at the design temperature of 200 °C. The DieMate™ manifold nut plate was also machined from type 316L stainless steel.

12.3.6
DieMate™ Manifold Insulation

The DieMate™ manifold is insulated to improve thermal uniformity and to prevent exposure of the DieMate socket to excess temperature. (The DieMate™ socket maximum temperature specification is 150 °C.) Figure 12.4 shows the insulated DieMate™ manifold before attachment of the 1/16-inch transfer lines. The insulation consists of a ceramic fiber strip that has a maximum operating temperature of 2300 °F (1260° C).

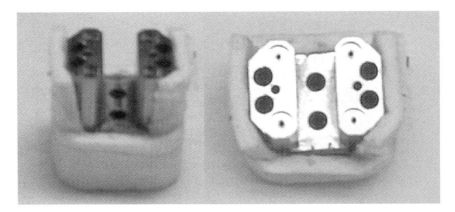

Fig. 12.4 Insulated DieMate™ manifold: top isometric view (left) and top view (right).

12.4
Microreactor System Design

The Automated Integrated Microreactor System (AIMS) was designed to be functionally equivalent to a conventional heterogeneous gas-phase catalyst testing system with parallel reactors, such as the Multiple Automated Reactor System or MARS (Mills and Nicole, 2005). One key difference between the AIMS and MARS-like reactor systems is the use of MicroElectricalMechanical System (MEMS) components, such as microvalves and micro-MFCs. Another key difference is the use of a computer chassis as the platform for mounting the various sub-components that perform various key process functions, such as flow control, temperature control, and chemical reaction. These sub-components consist of various custom-designed boards that resemble those used in various common types of electronic equipment. Conventional on-line gas sampling valves and GC analytical equipment were utilized since a micro-scale analytical system would have required detailed development due to the need for multiple-columns with a column switching scheme. This aspect is a possible future topic.

The various gas flow manifolds and microreactors are packed by mounting the devices on standard 6U CompactPCI boards (6U designates the height of the board since the CompactPCI standard supports many different board heights). The boards are then inserted into the computer chassis and the electrical connections between the boards are then made through the backplane, ribbon cables, and rear I/O boards. Key design aspects are highlighted in the following sections.

12.4.1
System Chassis

The AIMS is housed in a standard CompactPCI chassis manufactured by Kaparel (Kaparel Corporation, 97 Randall Dr., Waterloo, Ontario, N2V 1C5, Canada). Figure 12.5 shows front and rear views of the final AIMS chassis with all of the boards installed. This particular chassis is an integrated sub-rack that measures 19-inch wide × 15¾-inch high × 11¾-inch deep. It has a maximum capacity of fourteen 6U, 64 or 32-bit CompactPCI cards and up to two double-wide (8HP) 6U power supplies. The chassis includes two SK 3344012 RiCool 12VDC blowers, which are located in the top compartment of the chassis. They are used to cool the system components and to minimize the accumulation of process gases inside the chassis. These particular fans are rated for 440 CFM of air and are manufactured by Rittal Corporation (Rittal Corporation, 1 Rittal Place, Springfield, OH 45504).

Electrical interconnections between the boards are made using a backplane and ribbon cables between rear I/O cards. Ribbon cables on the front side of the chassis connect the reactor boards to the heater driver circuit boards. The rear I/O cards are custom-built PC boards that are used to transfer signals from cables connected to the control computer to the system boards on the front. Ribbon cables are used to transfer connections between the rear I/O cards because there were not enough lines available in the backplane to make all of the required connections.

372 | 12 Integrated Microreactor System for Gas Phase Reactions

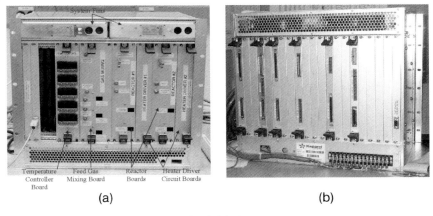

Fig. 12.5 Views of the computer chassis that holds the various AIMS process card. Front (a) and rear (b) views of the AIMS without any external fluidic connections.

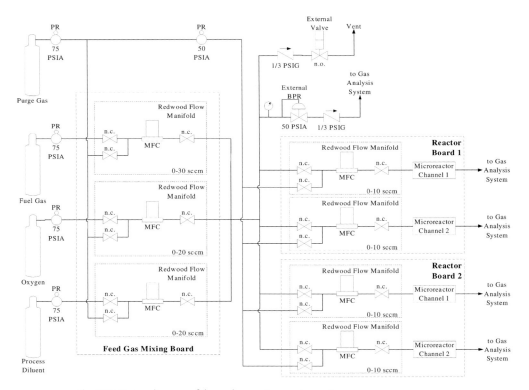

Fig. 12.6 Piping diagram of the scale-up microreactor system.

The backplane is a H110 CompactPCI telephony backplane also obtained from Kaparel. Two PS4400 backplane boards were used along with one PS1150 board. The backplane PCBs were slightly modified by removing selected chip capacitors that were not compatible with the amount of current being carried on the backplane. Only analog signals were transmitted by the backplane, so the digital communications capability of the backplane bus was not utilized.

Details of all of the electrical connections are beyond the scope of this chapter. It suffices to say that detailed I/O net lists were made of the connections that were needed between the boards and the location of each end of the connection. In addition, these lists describe all of the connections to the system control computer and any external electrical signals.

Figure 12.6 shows a process flow diagram of the AIMS (some details concerning gas delivery to the system has been excluded for simplification). The key components include (a) a feed gas manifold, which consists of a pressure regulators and a bank of mass flow controllers (MFCs) for on-line generation of the reaction feed gas mixture; (b) a reactor manifold, which consists of dedicated MFCs that meter the feed gas mixture to individual parallel-operating microreactors; and (c) an on-line reactor feed and product gas analysis system. A brief description of the various components that perform these various functions is given below.

12.4.2
Primary Feed Gas Manifold

The primary feed gas manifold generates a steady-state composition containing a fuel, an oxidizer, and a process diluent to the downstream reactor boards. The fuel gas is typically a light alkane or olefin in the C_1 to C_4 range or ammonia (NH_3). However, any other reactant could be used provided it can be supplied in a gas or vapor state. If the vapor is generated from a non-volatile liquid, a special-purpose vaporizer board would be required. The reactants used here were supplied from a commercial cylinder as a binary mixture diluted with either inert helium (He) or nitrogen (N_2). When NH_3 is used as one of the reactants, krypton (Kr) is used as the cylinder gas diluent since N_2 is one of the oxidation reaction products. The process diluent is either He or N_2 when a light hydrocarbon is used as the reactant, or He when NH_3 is used as the reactant. These diluents allow the resulting fuel/oxidizer/inert composition to exist below the lower flammability limit for safe operation. The oxidant is pure oxygen (O_2). All of these gases enter the system at a secondary regulator supply pressure of 75 psia.

Metering of the individual reactor feed gas constituents to generate the desired feed gas composition actually occurs on dedicated feed gas mixing boards located within the AIMS chassis. The process elements contained on this board are shown within the dashed sub-region in Fig. 12.6.

The feed gas mixing board indicated in Fig. 12.6 actually contains three parallel manifolds that control the process gas flow rates using a Redwood Microsystems (Redwood MicroSystems, Inc., 959 Hamilton Ave., Menlo Park, CA, 94025) MEMS-Flow® MFC and three Redwood Microsystems normally closed Shut-Off

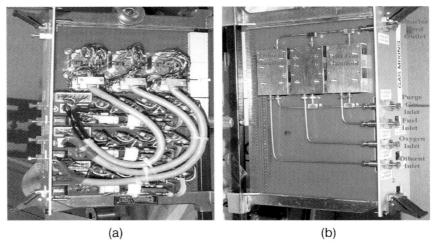

Fig. 12.7 (a) Right-hand side of the feed gas mixing board, showing the cables for the feed gas controls. (b) Left-hand side of the feed gas mixing board, which shows the various feed gas manifolds.

Valves (SOVs) (this assembly is referred to as the Redwood flow manifold). The composition of the reactor feed gas mixture is defined by setting the flow rates for each of the process gases using these dedicated mass flow controllers (MFCs).

A right-side view of a feed gas mixing board that shows the various electronic assemblies and wiring is shown in Fig. 12.7(a). The system feed gas inlet connections for fuel, oxygen, diluent, and purge gas are visible in the left side view shown in Fig. 12.7(b). The tubing assemblies were constructed from 1/16-inch type 316L stainless steel tubing. Serial communication to the Redwood flow manifolds is available for diagnostic purposes through three, four-conductor RJ-11 jacks on the front of the card.

12.4.3
Reactor Feed Gas Manifold

Once the fuel, oxidant, and process diluent gases exit their respective legs of the various feed gas mixing boards, they converge to a single line so the resulting feed gas mixture can be directed to a common manifold. The manifold serves as a feed gas source for the downstream reactor boards, the feed gas analysis system, and for venting excess feed gas during regular operation or during an emergency situation.

12.4.4
Reactor Board

The reactor board is a key part of the feed gas manifold. Two reactor boards are used, but this number is not a limitation since additional boards can be added as long as the AIMS chassis contains enough slots to accommodate them. Each reactor board has two separate connections where reactor feed gas can be introduced from the feed manifold. These connections serve as independent feed inlet points for two parallel Redwood Flow Manifolds (RFMs) that are used to control the gas flow rate to each channel of the microreactor die. A total of four microreactors can be operated since each die has two independently controlled reaction channels. The RFMs are identical to those used for the feed gas mixing board except for their inlet and exit operating pressures.

An isometric view of a reactor board is shown in Fig. 12.8. It contains the electronics for the Redwood flow manifolds as daughter and granddaughter cards. The electronics needed to operate six temperature sensors on each microreactor channel are located underneath these cards. The ribbon connector, which is also visible in Fig. 12.8, is used to transfer electrical signals directly from the reactor board to the heater circuit board (described below). Serial communications for the Redwood flow manifolds is provided by two, four-conductor RJ-11 jacks on the front. The front of the board has gas inlets for the reactor feed and the purge gas. In addition, there are two gas outlets, one for each reaction channel. The inlet and outlet fittings are 1/16-inch type 316L stainless steel. The tubing assembly component having the greatest pressure sensitivity is the microreactor membrane. This will rupture at ca. 7 psig (Firebaugh, 1998).

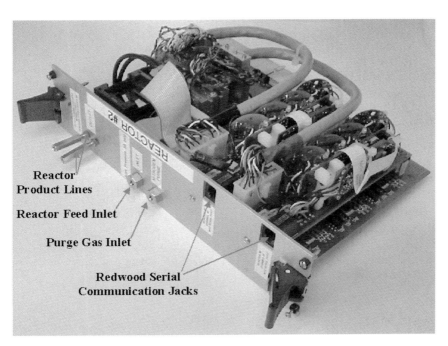

Fig. 12.8 Isometric view of a reactor board.

Fig. 12.9 Clam shell heater used for the product transfer lines from the DieMate™ manifold to the front of the board.

12.4.5
Transport Tube Heating

Tubing that is used for transfer of product gases between the DieMate™ manifold to the on-line GC analysis system must be heated to at least 200 °C to prevent product condensation and to minimize any potential for line blockage. The product transfer lines from the microreactor are heated by conduction through the heating cartridges in the DieMate™ manifold using a clam shell heater that encases the tubing to the front of the card. Fig. 12.9 shows a picture of the final clamshell heater design.

A flexible Kapton®-coated heater is wrapped around the clamshell and is powered using a 24 VDC input. The temperature of the clamshell is monitored using two 100 Ω platinum Resistive Temperature Devices (RTD's). Both the DieMate™ manifold and the clamshell heater are insulated using ceramic fiber strip obtained from a commercial source with a maximum temperature rating of 2300 °F.

12.4.6
Temperature Controller Board

The temperature controller board (Fig. 12.10) consists of five CAL (CAL Control Inc., 1580 S.Milwaukee Ave., Libertyville, IL 60048) 3300 temperature controllers mounted on a standard 6U CompactPCI boards. These devices are used to control the temperature of the DieMate™ manifolds and the product transfer lines by manipulating the power input to their respective heaters. Two controllers are

Fig. 12.10 Temperature controller board.

used for the DieMate™ manifolds themselves and another two are used for the transfer lines from the DieMate™ manifold to the front panel. Two controllers per DieMate™ manifold are required since each reactor board requires a dedicated controller.

12.4.7
Heater Driver Circuit Boards

There are two heater driver circuit boards in the AIMS – one for each reactor board. The left-most board is shown in Fig. 12.11 and is referred to as Heater Driver Circuit Board 1. It contains the circuitry necessary to power twelve microheaters on a microreactor die. Both of the cards used in the system are identical and interchangeable.

The heater circuits are voltage-controlled so the power input to the microreactor heaters can be reduced as the local reaction temperature increases. However, increasing the temperature causes the resistance of the microheaters to also increase, so the current supplied to the microheaters must be reduced to maintain

Fig. 12.11 Heater driver circuit board.

a constant voltage. This is an important control feature since local reaction ignition in a reaction channel can generate temperature increases in excess of 100 °C. Another complication is that this increase can occur over a period of several milliseconds so the sensor and controller response times must be capable of detecting and responding to it.

12.4.8
Testing Procedures

The various AIMS components were subjected to a series of detailed tests as they were assembled to ensure that they functioned properly. The individual system boards were also closely examined to ensure that the board manufacturer had followed the specified board net lists. Fortunately, no major problems were discovered during the initial electrical testing. Instead, only some minor setbacks were encountered, such as a lead needing to be tied to ground and a capacitor that was inserted incorrectly. Most of the effort was spent on the Redwood mass flow controllers owing to a factory design problem that was uncovered during the troubleshooting, which was later corrected. Overall, the results were very encouraging since no problems were discovered that required a redesign of one of the printed circuit boards or any other major changes. Additional details are provided in Chapter 6 of Quiram's thesis (2002).

12.5
System Control and Process Monitoring

The control hardware and software presented a challenge that went beyond the limits of traditional laboratory systems and involved the use of components in β-form, or components that were just commercially released. The primary challenges included managing the rather large number of input/output signals and meeting the high control loop cycle rate without degradation of performance. In addition, the control computer was responsible for safety monitoring and to engage interlocks as needed. As in most industrial reactor systems, the control computer only executed the process control loop, and another system provided the operator interface.

This section provides an overview of the system hardware, the design principles associated with the process control software, and a summary of the various process control human–machine interfaces.

12.5.1
System Hardware

The AIMS control and monitoring system includes two major components: (1) the Programmable Logic Controller (PLC) and (2) the Human-Machine Interface (HMI). A PLC was used since the operating system was especially robust and not prone to failure as often seen in personal computers running in the Microsoft Windows™ environment. Because the PLC is responsible for process safety aspects, reliability is the most important factor. This eliminates any machine operating under a Windows™ OS.

The PLC used for the AIMS is a National Instruments (NI) PXI-1010 chassis with an embedded controller. PXI is the NI version of the Compact PCI standard. This computer can be booted under Windows™ or a proprietary NI real-time operating system. The real-time operating system is used for process control of the AIMS. This computer is responsible for monitoring all of the discrete and analog inputs from the AIMS as well as providing the discrete and analog outputs for process control.

The AIMS PLC is actually composed of two NI PXI chassis that are referred to here as a PXI-1010 and a PXI-1000B. Two chassis were required since the total number of boards exceeded the number of slots available in a single chassis. Figure 12.12 shows the two chassis with the boards used for controlling the AIMS. A MXI-3 bus extender cable connects these chassis by extending the bus of the system beyond the backplane of the PXI-1010 chassis. This is performed using two NI PXI-8330 boards where one is mounted in each chassis.

The PXI-1010 chassis contains the embedded PXI controller board, a NI PXI-8170, which uses a 700 MHz Pentium™ III processor as the CPU. This processor runs independently of the computer running the HMI. This particular CPU was not available with an Ethernet connection port so a MAXI 100L Ethernet card made by Nexcom was installed (NEX Computers, Inc., 46560 Fremont Blvd. #401, Fremont, CA 94538). In addition, the 1010 chassis has one PXI-6527 digital I/O board that provides the required 16 digital output signals and the 20 digital input channels. This board has 24 optically isolated digital input channels and 24 solid-state relay output channels. All of the digital outputs were 0 VDC in the low state and 24 VDC in the high state. The Acopian® power supplies provided the external power needed for these channels. The digital input channels were 0 VDC in the low state and either 5 or 24 VDC in the high state. A 100 pin NI SH100-100-F cable connects this board to the AIMS chassis.

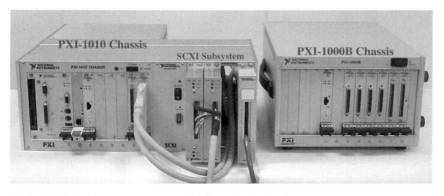

Fig. 12.12 National Instruments control hardware for the AIMS.

Table 12.2 Boards in the primary PXI-1010 chassis.

Slot no.[a]	Module[b]	Description	No. of channels used	Cable[c]
1	8170	Pentium™ III 700 MHz embedded computer	N/A	N/A
2	MAXI 100L[d]	LAN Ethernet connectivity module	N/A	Ethernet
3	8330	MXI-3 bus extender module	N/A	MXI-3
7	6527	Digital I/O module	20 I/16 O	SH100-100-F
8	6025E	Multifunction I/O module for reading the system RTDs	1	SH100-100
SCXI-1	1121/1321	RTD measurement	4	Custom
SCXI-2	1121/1321	RTD measurement	1	Custom
SCXI-4	1160/1324	General purpose relay module	6	Custom

a) Skipped slot numbers indicate empty slots in the chassis.
b) Unless otherwise noted, the boards in the PXI-1010 chassis are manufactured by National Instruments.
c) All cables are manufactured by National Instruments except the Ethernet and custom cables.
d) Manufactured by Nexcom.

Table 12.2 summarizes the boards used in the primary PXI-1010 chassis that control the AIMS.

Referring to Table 12.2, the PXI-1010 chassis also contains a four-slot SCXI subsystem for three SCXI boards that were not available in the PXI form. The SCXI subsystem contains two SCXI-1121 boards that are used to measure the temperatures of five RTDs in the AIMS. These RTD signals originate from the DieMate™ manifold and the product heater line. Each board can measure temperatures from four RTDs with an accuracy of ±1 °C. The RTDs are wired in a four-wire configuration. A NI SCXI-1321 high voltage terminal block is used to connect the RTD signals to the SCXI-1121 module. A NI PXI-6025E multifunction I/O board is used to read the voltage signals from the SCXI-1121 boards with the multiplexing done on channel 0 of the 6025E. A 100-pin NI SH100-100 cable connects the 6025E to the SCXI subsystem. (Normally, the signals from these boards would be read directly through the bus link from the SCXI subsystem to the PXI chassis; however, a bug in the software prevented this while the system was operating in real-time mode.) A NI SCXI-1160 general-purpose relay module is used to provide the six needed power relays for the AIMS. This board consists of 16 isolated Single-Pole Double-Throw (SPDT) latching relays. A NI SCXI-1324 high voltage terminal block is used to connect the power signals to the SCXI-1160 module.

Table 12.3 Boards in the PXI-1000B chassis.

Slot no.[a]	Module[b]	Description	No. of channels used	Cable[c]
1	8330	MXI-3 bus extender module	N/A	MXI-3
3	6713	Analog output module	8	SH68-68-EP
4	6713	Analog output module	8	SH68-68-EP
5	6713	Analog output module	7	SH68-68-EP
6	6713	Analog output module	8	SH68-68-EP
7	6071E	Analog input module	29	SH100-100
8	6071E	Analog input module	26	SH100-100

a) Skipped slot numbers indicate empty slots in the chassis.
b) All of the boards in the PXI-1000B chassis are manufactured by National Instruments.
c) All of the cables are manufactured by National Instruments.

Table 12.3 summarizes the boards located in the PXI-1000B chassis. This chassis contains four PXI-6713 analog output boards that provide the thirty-one needed analog output signals from the PLC to the AIMS. Each PXI-6713 board has eight analog output channels with 12-bit resolution. A 68-pin shielded NI SH68-68-EP cable connects each of these boards to the AIMS chassis. The PXI-1000B chassis also contains two PXI-6071E multifunction I/O boards that provide the 55 needed analog input channels. Each board can read 32 differential analog inputs with 12-bit resolution in various signal gain configurations. Each 6071E module is connected to the AIMS with a 100 pin SH100-100 cable.

12.5.2
Process Control Software

The control program for the PLC was written in G using LabVIEW version 6.03. G is the proprietary graphical programming language of National Instruments. Programs written in G can be run in the Windows™ operating environment or in the real-time operating environment of the embedded controller. Development work was performed on a separate computer that downloads the control VI's to the real-time embedded controller. The control loop can be monitored through the development computer, but it does not have to be operating for the controller to function.

The control program executing on the PLC is responsible for five tasks: (1) monitoring process variables and transmitting them to the HMI; (2) examining the process variables for abnormal conditions; (3) executing interlocks based on the status of the process variables; (4) performing closed-loop control of the microreactor heaters (24 heaters in total); and (5) receiving and implementing operator input from the HMI. All of these actions are considered part of the control loop and must be executed within a specified time period for the controller to be operating in real time. Thus, the speed of the backplane, processor, and communications are all factors in determining the maximum loop execution rate.

The HMI connects the AIMS hardware to the user. The HMI is actually an interface program that runs on a computer connected to the PLC. For the AIMS, the HMI was coded in G using LabVIEW 6.02. During testing of the AIMS, the HMI was run on a older Dell model OptiPlex GX110 with a Pentium™ III processor and 256 Mbytes of RAM operating under Windows™ NT 4.0 Service Pack 5. LabVIEW is available for several operating environments, including LINUX, so these machine specifications are not restrictive. (Windows NT, 2000, or XP is recommended for the operating system due to its improved stability. A failure in the HMI computer will not prevent the controller from operating, but will result in the loss of data.) A 10 Mb s^{-1} Ethernet connection was established between the HMI computer and the NI embedded controller. Both machines were capable of 100 Mb s^{-1} transmission speeds, but the routers in the subnet limited operation to only 10 Mb s^{-1}. In either case, the data transmission rate was still much faster than the HMI computer could either display or store information, so much of the data was not seen by the user. (The data was not lost. Instead, the display update rate on the HMI was set to around 10 Hz, so all of the data points were not displayed. The data logging software recorded data at a maximum rate of 4 Hz.) However, this did not affect the performance of the AIMS since the real-time controller executes all the necessary control functions. Instead, only the display of data and its logging is affected.

The HMI is responsible for displaying process sensor and controller data to the user, and for historical logging of data. These actions allow the user to make decisions so new process set points can be entered. The data displays are achieved through numerical displays and other visual indicators, e.g., a valve colored green would be open. The HMI design must carefully consider the appropriate display methods and the organization of the data. With over 200 process variables in the AIMS, it would have been easy to overwhelm the operator with data. Thus, most control actions are done pictorially by clicking on a valve or another object. In this way, the data is displayed in close proximity to where the user can enter new process set points.

The other task of the HMI is to log all of the process variables at a designated time interval. For this task, the LabVIEW Data Logging and Supervisory Control Module was installed on the HMI computer. This add-on to LabVIEW gives the developer the ability to establish automatic data logging, features for working with applications having a large I/O channel count, and alarm management and event logging capabilities. In addition, this module provides the developer with OPC connectivity tools, which were used in communicating with the CAL 3300 temperature controllers. The built-in data logging features of this package were used by establishing tags for the process variables to be recorded. The properties of the tags could then be adjusted to the users needs. For example, the dead band for data logging can be changed to help minimize database size. Data logging was

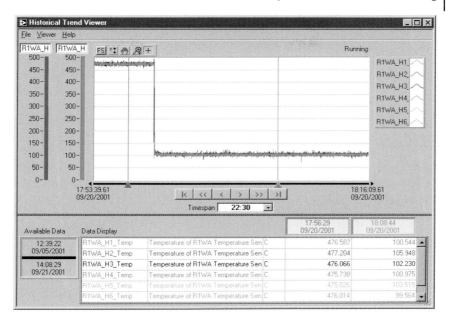

Fig. 12.13 LabVIEW historical trend viewer GUI.

initiated as soon as the HMI program was started, and storage was performed using a National Instruments Citadel database. Data in this format can be accessed through the LabVIEW historical trend viewer or through third-party software, such as Microsoft's Excel using the Citadel Open Database Connectivity (ODBC) driver. This driver allows other applications to access the data directly using Structured Query Language (SQL) queries. The LabVIEW historical trend viewer allows the user to access process variables by specifying a time window to display and the desired process variables. Data retrieval from the database is transparent in the sense that the appropriate database file is determined by LabVIEW, and the process variable values are then retrieved from the database for the given time window.

Figure 12.13 shows a screen shot of the LabVIEW HMI with some sample data from the AIMS. The data displayed by the historical trend viewer can be exported to standard text file formats that can be read by spreadsheet applications.

12.5.3
Process Control Interface

The AIMS HMI is organized around a central control panel that allows the operator to quickly view many of the process variables and to easily access more detailed panels. This is performed by selecting various tabs that are associated with the operator interface. A detailed description of these various tabs along with actual screens is outside the scope of this chapter. A few brief illustrations are given below to convey the design of a typical tab.

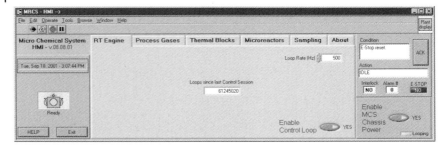

Fig. 12.14 AIMS HMI main control panel with the RT Engine tab selected.

12.5.3.1 Run Time (RT) Engine tab

The real-time control program on the AIMS PLC is started on the *Run Time (RT) Engine* tab of the central panel. Figure 12.14 shows a screen shot of this panel. This panel allows the operator to select the desired loop rate of the PLC and to initiate operation of the PLC control loop. It also displays the number of loops that have been completed since the last enable command was given. The other tabs on this central panel allow the operator to quickly navigate to other sections of the system and to examine the process conditions. These tabs include process gases, thermal blocks, microreactors, and sampling. Further sub-panels can be opened from these primary tabs on this central panel. The tabs are ordered from left to right according to the flow of gas in the system, except for the first tab. The particular one shown below is used to enable the PLC control loop.

12.5.3.2 Microreactors Tab

The operation of the microreactors is monitored through the *Microreactors* tab on the main control panel. This panel displays the position of the SOVs, the feed gas flow rate to each microreactor channel, and whether or not the microreactor heaters are enabled. Additional information on the operation of each of the microreactor channels can be obtained by clicking on the *Open Panel* button next to the reactor name. In addition, pressing the *Configure Reactor Control* button opens a sub-panel where the operator can configure the temperature control mode for each of the microreactor channels to be either manual or automatic PID control. Some salient aspects of the reactor control panel are given below since this is the key system component.

The various *Reactor Control* panels allow the operator to review much more detailed information on the operation of an individual microreactor channel. It also allows the operator to make changes to the process set points. Figure 12.15 shows the *Reactor Control* panel for channel A of the microreactor on Reactor Board 1. Here, the operator can choose the gas stream flowing to the microreactor (either feed gas or purge gas), specify the gas flow rate, and define the set points for the microreactor heaters. The *Temperature Reading* indicators at the top of the channel block in the middle of the panel show the measured temperatures of the first six microreactor heaters in the channel using the microreactor RTDs (as

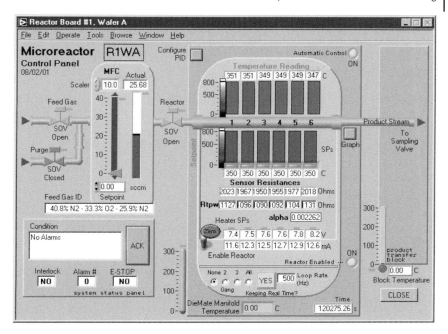

Fig. 12.15 Microreactor Control sub-panel for the AIMS HMI.

stated previously, only the first six of the seven microreactor heaters are utilized in the scale-up system). These temperatures are also indicated graphically with the color bars below each of the displayed values. The particular panel shown in Fig. 12.15 is in the automatic control mode, so the controls for the numerical and graphical temperature set points are also displayed below the temperature indicators. Here, the operator can enter new values for the temperature set points and observe the current set points. Located below these controls are the temperature sensor resistance indicators, which display the raw resistance data before it is converted into a temperature. The measured temperatures are calculated using the resistance at the triple point of water, R_{tpw}, for each of the microreactor RTDs and the platinum layer temperature coefficient of resistivity, α (alpha in the display). Both of these controls are located below the displayed sensor resistances. The operator must enter values into these fields for the temperature reading to be calculated for the microreactor RTDs. Information on the heater performance is below these controls with indicators that display the voltage and current being applied to each of the microheaters in the reactor channel. In manual control mode, the operator enters the voltage to be applied across the microreactor heaters directly. In automatic control mode, this value is calculated by the PID control loop to maintain the set point temperature. The toggle switch located at the left of the microheater indicators allows the user to enable or disable the microreactor heaters. When the heaters are disabled, a relay on the Heater Circuit Board directs the heater power to a fixed resistor on the circuit board (designed to simulate the load of a microreactor heater). This relay was added as a safety measure to enable the system to quickly remove power from the microreactor heaters and to aid in testing by preventing unwanted power surges during startup from destroying the microreactor membranes. The lighted *Reactor Enabled* indicator on the right of this block simply displays whether the microreactor heater is actually enabled (the position of the relay on the Heater Circuit Board is sensed by the control electronics). In the section at the bottom of this block, the operator has the ability to gang the microreactor heaters together. This feature allows the user to change the set points of multiple heaters by entering a single value. Indicators also tell the operator whether or not the control loop is maintaining real-time and the current loop rate. This information is used for analyzing performance of the control loop.

12.5.4
Automation Testing Procedures

The hardware of the control system of the AIMS was tested simultaneously with the AIMS hardware described in Section 12.3. The AIMS PLC control program and HMI were both developed during this same period and were tested when the assembly of the AIMS and its control hardware was completed. Debugging and testing was an ongoing process that was performed simultaneously with the creation of the PLC control program and the HMI. The final software validation was performed when the AIMS was operated with inert gases flowing through the system and then with reacting gases. Owing to the complexity of these two computer codes, debugging actually occurred throughout the entire testing phase of the scale-up system. Modifications to the software were made to add functionality and improve ease-of-use for the control program and HMI.

12.6
Microreactor Process Safety

12.6.1
Overview

The construction and testing of the AIMS microreactor system followed the Process Safety Management (PSM) guidelines of DuPont for their Wilmington, Delaware facilities. These guidelines are specifically addressed in DuPont Corporate SHE Standard S21A. In general, the AIMS was designed and tested to minimize the possibility of any safety incidents. This included the release of unacceptable levels of toxic gases, explosions, and deflagrations. Electrical hazards were also considered in system design due to the extensive power network in the AIMS.

The DuPont hazards analysis roadmap was consulted to determine the level of safety audit needed for the AIMS. A Research Safety Review (RSR) was determined to be sufficient because the system was not designed for unattended operation. The RSR required documentation for all process safety hazards using the RSR checklist as a guideline. A separate Process Hazards Analysis (PHA) was prepared to address in detail any items requiring documentation in the various RSR checklists. In addition, an Equipment Safety Audit (ESA) was also completed as required by DuPont PSM guidelines.

12.6.2
Process Hazards

The major hazards associated with the AIMS were feed and product flammability, chemical exposure, and elevated process temperatures. The risk of fire in the system was reduced by the use of diluted feed streams that did not allow the formation of flammable mixtures in the system. In addition, system interlocks prevented the uncontrolled release of process gases in the event of an incident. These interlocks also minimized the risk of exposure of the system operators to process gases. In addition, the process was entirely contained in a fume hood, so following the system Standard Operating Procedures (SOPs) helped prevent such exposure. The risk of injury due to high temperatures in and around the system was also be minimized by closely following the system SOPs during operation and system maintenance. A detailed description of all process hazards is omitted here for brevity.

In the event that an explosive composition was reached during operation, the small volumes associated with the reactors, valves, and transfer lines minimized the potential for injury to personnel and damage to equipment. The entire system gas volume inside the chassis was 3.4 cm^3. As a worst case scenario during operation, a stoichiometric composition of CH_4 and O_2 was assumed to be present in all of the gas lines. Further, it was assumed that the system was pressurized to 150 psig, which was well above the point at which the pressure relief devices would activate. If this entire volume of gas underwent complete combustion, it would release the equivalent energy of 0.00023 lbs of TNT (assuming the gas is at 20 C and employing a value of 490 kcal lb^{-1} for TNT). Thus, the only explosion hazard presented by this system was the accumulation of flammable mixtures outside the system tubing.

Any notable accumulation of gas was unlikely since the two fans inside the chassis create a flow rate of approximately 180 CFM of air through the system. This corresponded to more than 95 complete air changes or turnovers every minute (Heck and Manning, 2000). The most likely zone of gas escape would be above the microreactor due to a membrane failure. If this occurs, the control system should have interlocked and shutoff the flow of combustible gas to that reaction channel. The flammable gas that does escape would have been immediately diluted by air flowing over the microreactor at an estimated rate of 120 ft min^{-1} (Heck and Manning, 2000). To provide a more detailed analysis of gas mixing in the immediate vicinity of a microreactor die, a computational fluid dynamics (CFD) model was constructed to simulate the gas flow hydrodynamics. This simulation quantifies that there is a recirculation zone above the reactor with an airflow rate as low as 20 ft min^{-1} (Heck and Manning, 2000). Nonetheless, this was deemed to be sufficient to dilute any escaping gas.

12.6.3
AIMS Safety Features

The AIMS was designed to fail to a safe state. System faults were normally handled by the National Instruments controller. The LabVIEW® code for this controller was running under Pharlap, which is a deterministic operating system. Thus, the reliability problems of LabVIEW® running under a Windows environment were not present in this system. In addition, a Brentek model P8-WDT24/PLC watchdog timer (Brentek International, Inc., Ridgewood Rd., York, PA 17402) was in place to detect any failure in the NI control system. The watchdog timer requires a signal from the controller once every second, or when a relay interrupts power to the system chassis.

Under normal operating conditions, the National Instruments controller monitors for process faults and takes corrective action upon identification. The system will override operator commands issued at the HMI during a safety interlock. The unsafe conditions that result in a process interlock are:

- The feed gas composition falls within the lower and upper flammability regime.
- Mechanical failure of a microreactor.
- An over-temperature of a heated part or the sampling valve oven.
- An under-temperature of a heated part or the sampling valve oven.
- Fume hood exhaust failure.
- Failure of one of the two CompactPCI chassis exhaust fans.
- Power failure.
- Controller timeout detected by the watchdog timer.

Most of the system interlocks are handled by the embedded NI controller. This is similar to that used in other laboratory-scale reactor systems (Mills and Nicole, 2005) except the Siemens 545 PLCs have been replaced with the National Instruments system. The National Instruments controller monitors the status of the AIMS process variables and takes appropriate action if any interlock conditions are found. In addition to these interlocks, the external heater controllers that are used for the system heating tapes and sampling valve box contain hardwired over-temperature interlocks built-in to these controllers. However, these interlocks only interrupt power to the affected heating device, so the NI controller must still take action to shutdown the rest of the process.

Power failure is considered a system interlock, but the NI controller cannot take any active measures to return to a safe state. Instead, the system was designed so that in case of a power failure the system shuts down by relieving any pressure in the process lines. However, if this occurs, the process lines are not purged with an inert gas. Thus, when any system maintenance was performed, such as replacing a microreactor, it was necessary to purge all the process lines with an inert gas. Ideally, a combination of normally open and normally closed valves would have allowed the system to be automatically purged with an inert gas during power failure. Unfortunately, the Redwood MicroSystems SOVs were not available in a normally open configuration, so this was not possible. However, an external normally open valve was placed on the reactor feed line to vent excess pressure in the system during a power failure.

Similarly, the hood status was monitored by a discrete contact closure that was connected to the power circuits of the hood. Thus, a failure of the hood exhaust effectively caused a system interlock by removing power to the AIMS. The NI controller was not connected to one of the affected circuits, so it remained operational. Nonetheless, without power in the AIMS, it could not take further action with the system.

12.7
Microreactor Experimental Methods

12.7.1
Overview

The AIMS performance was evaluated using methane oxidation and ammonia oxidation as the test reactions in separate experiments. The catalysis and reaction engineering of these reactions have been the subject of extensive research on length scales ranging from single crystals to commercial-scale reactor systems. Consequently, the purpose of the AIMS testing was not to identify some new scientific behavior or engineering technology. Instead, the goal was to assess the functionality of the AIMS in comparison with a more conventional laboratory reactor system currently being used at DuPont, such as the MARS system (Mills and Nicole, 2005). By showing that the AIMS produced data that compared favorably with the MARS, a first step in the miniaturization of complex reactor systems would be achieved. To accomplish this objective, it was first necessary to demonstrate that the MEMS components could effectively control gas flow and other reaction variables in the system while providing the same level of safety management designed into conventional equipment. The ability of the microreactor to reproduce the behavior of conventional laboratory-scale tubular reactors was not examined since the membrane microreactor design did not duplicate the same type or magnitude of transport-kinetic interactions that occurred in the conventional system. However, the membrane microreactor did have a few unique operational features that were not available in traditional tube reactors, such as the ability to control the catalyst temperature in the reaction zone over a length scale on the order of 1 mm or less.

Comparison of the AIMS operation with a conventional laboratory-scale reactor system operation, such as the MARS reactor (Mills and Nicole, 2005), is briefly discussed here by describing the procedure followed during a typical experimental run. This discussion is less detailed than the system Standard Operating Procedures (SOPs), but focuses more on the experimental protocol. Following this, an overview of the GC analysis methods used is given. The methods used to extract reactor performance information from the raw GC data are omitted since the details are provided elsewhere (Quiram, 2002).

12.7.2
Experimental Protocol

The experimental procedures for conducting a reaction experiment with the AIMS were generally adopted directly from those previously developed for the MARS units at DuPont (Mills and Nicole, 2005). The basic GC analysis techniques were also borrowed from the MARS so that the protocols for determining the feed and product gas compositions were essentially identical. However, an important difference between the AIMS and the MARS was that this latter reactor system was designed for unattended automated catalyst performance evaluations. One of these protocols involved long-term catalyst life studies, sometimes on the order of several thousand hours. The required automation and safety features for unattended operation of the AIMS had been incorporated into the system design, but overall system robustness for this particular operating mode had not been demonstrated. Future work should demonstrate that the AIMS is capable of long-term unattended operation, which is defined here as an uninterrupted on-stream time of at least 500 hours.

Reaction testing was initiated by enabling the microreactor heaters and setting the heaters either in manual or automatic control mode. In the manual control mode, the operator set the microreactor heater voltages. While in the automatic control mode, the operator could set the desired microreactor heater temperature. The product gas compositions were determined from each of the four reactor channels.

The GC analysis time was the limiting step in the data collection process. The microreactor heaters reached equilibrium extremely quickly (on the order of 10 ms), so transient product data was not be observed by the GC. Another limiting factor on the maximum testing speed of the system was the tubing volume between the microreactor exit and the multi-position valve. The gas residence time that was required to thoroughly displace the tubing volume over the typical range of feed flow rates was less than 1 min. In most experiments, the flow rates at the microreactor inlet were unchanged, but the temperature of the microreactor heaters was varied. Since the GC analysis required about 15 min cycle^{-1}, only five to ten different temperature settings were explored in a single run. During longer runs, additional reactor feed gas injections were taken to determine if variation in the feed gas composition had occurred.

At the end of an experiment, the microreactor channels were purged with N_2 or He (depending on the purge gas connected to the system at the time) and additional GC injections of feed gas were taken. During a few runs, a microreactor failed through membrane rupture or the heaters becoming inactive. At this point, the microreactor channel was purged for 1 min and that line was isolated from the rest of the system by closing its SOVs. After the additional feed gas injections were completed, the system was shut-down by purging all the lines with inert gas and removing power from the chassis. The heated components outside of the AIMS chassis were also turned off, except for the sampling valve box to avoid thermal cycling of these valves, which could create leaks between the valve rotor and seal.

12.7.3
GC Method

The GC techniques used for analyzing the AIMS product and feed gas were adopted from the multidimensional methods developed at DuPont for the MARS (Mills and Guise, 1994, 1996; Delaney and Mills, 1999; Nicole et al., 2000; Mills and Nicole, 2005). Although the GC columns and temperature programs for the CH_4 and NH_3 oxidations were specifically selected to both resolve and quantify the products from these particular reactions, the physical setup of the sampling system was the same. Details are omitted since they are mainly provided in the above-cited references. The gas chromatograph used was an Agilent 5890 Series II Plus (Agilent Technologies, 395 Page Mill Rd., Palo Alto, CA 94303). Data collection was performed with Agilents' GC ChemStation software revision A.06.03. Although ChemStation software is in-use at DuPont, the MARS units typically used more traditional Laboratory Information Management Systems (LIMS) for GC data management.

The raw GC data was converted into compositional data using the internal standard method with either N_2 or Kr as the reference species. Once the species compositions were determined, values for the reactant conversion, product selectivities, and product yields were evaluated using the standard expressions. Errors in these parameters were calculated by closing the mass balance on the reaction. This was done by comparing the feed gas composition with the measured product gas compositions. Large errors in the mass balance (> 5%) were indicative of physical problems with the system, which were typically attributed to leaks. Smaller errors (< 5%) were caused by uncertainties in the GC molar response factors as well as fluctuations in flow rates of the MFCs on the Feed Gas Mixing Board.

12.8
AIMS Testing Results and Discussion

12.8.1
Overview

Operating the AIMS under the reaction conditions was similar to other catalyst test systems at DuPont that utilize reactors run in parallel, such as the MARS system (Mills and Nicole, 2005). Considerable care was needed to setup the system so as to obtain quality, reproducible results. However, after this step, operation was straightforward. This was mainly due to the advanced automation used to operate both the AIMS and the MARS. Specifically, the operator could quickly scan all of the relevant process variables using the HMI, and automatic system interlocks ensured that they were maintained within safe bounds during a run. Automated GC sampling protocols, taken directly from experience with the MARS, required minimal operator intervention to acquire reaction data. The most involved part was analyzing the GC results. However, this aspect could also be automated, as had been done to a certain extent with the various MARS units.

The following sections summarize the testing results of the AIMS. Section 12.8.2 gives an overview of the operation of the AIMS and provides qualitative comparisons between the AIMS and the MARS units. Additional information is given on the operation of the AIMS and related to issues. Subsections 12.8.2.6 and 12.8.2.7 discuss in detail the operation of the Redwood Microsystems components used in the AIMS. Because of the number of problems associated with these devices, they were generally considered the major obstacle in operation of the AIMS. Section 12.8.3 describes the performance of the process automation system and selected process sensors. Reaction performance results are described in Section 12.8.4. Some reaction results obtained from operating the AIMS are then given for the CH_4 (Section 12.8.4.1) and NH_3 (Section 12.8.4.2) oxidation reactions. Finally, the key findings from the system testing are summarized.

12.8.2
AIMS vs. MARS Setups

This section summarizes the key qualitative comparisons between the AIMS and MARS systems from several practical perspectives.

12.8.2.1 System Size
The main difference between the AIMS and the MARS units was the relative size of the various process components. Although miniaturization of reaction systems was an objective of this project, this also led to some unexpected problems. Specifically, debugging and repairing hardware problems in the AIMS was usually more difficult than with the MARS. In particular, the closed-system chassis prevented direct monitoring of the individual components during operation, e.g., probing an on-board component with a gas leak detector or a multimeter. Since debugging is easiest to perform under energized conditions, this particular aspect presented a challenge. However, the difficulty of maintenance should not be viewed as an overriding disadvantage. Future versions are expected to contain more robust components, and could incorporate features allowing easier diagnosis, such as placing electrical test points near the front of the system boards to monitor important voltages.

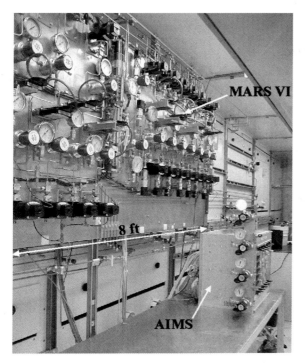

Fig. 12.16 Size comparison between AIMS and MARS Version VI.

12.8.2.2 Gas Flow Manifold

One advantage of miniaturization is that the entire system could be easily moved to another location for maintenance. Figure 12.16 compares the AIMS to the flow manifold for a MARS Version VI. An oven that houses the MARS parallel reactor bank was removed from the hood to provide a clearer perspective view. The front area of the AIMS occupied about one-tenth of the area of the MARS flow manifold. Obviously, the AIMS offers a significant advantage by reducing the space requirements in a fume hood. In addition, an operator can easily access all of the parts of the AIMS without needing a footstool or stepping inside the hood.

12.8.2.3 Process Automation

Although the size of the flow manifold and reactor components have been substantially reduced in the AIMS, no such reduction has occurred in the control system since both the MARS and AIMS rely on components that have already been miniaturized. Figure 12.17 shows both the Siemens S7-300 controller for the MARS Version VI and the National Instruments controller for the AIMS. Their footprints are essentially the same, with both having a similar number of inputs and outputs. The Siemens controller used for the MARS Version VI could have been used for operating the AIMS, but it was not capable of PID control loop rates faster than 20 Hz.

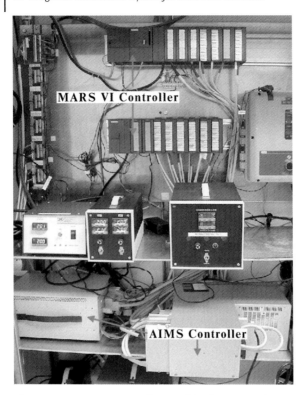

Fig. 12.17 Automation controllers used for the AIMS and MARS Version VI.

12.8.2.4 Electrical Wiring

One notable advantage of the AIMS was the simplicity of setting up the wiring. Because all of the components in the AIMS chassis were wired internally to connectors on the rear of the chassis, there was no need to connect individual wires from the components to the controllers. This is necessary for the MARS units, which adds a significant cost to their construction. For example, wiring the MARS Version VI took an experienced instrumentation technician about one week to assemble the field wires and then physically connect them between the PLC boards and the field instruments. The wiring for the AIMS was performed through cables connecting the AIMS chassis to the NI PXI chassis. This took only about ½ hour to complete when the AIMS was moved into the fume hood for testing. Future versions of miniaturized reactor systems could improve on this even further by converting the analog signals into a digital format so communications could be done through a standard digital communications protocol, such as TCP/IP. This would require customized electronics in the reactor system chassis, but it would reduce the complexity of the external controller substantially by eliminating many of the boards. In addition, the proximity of the controller to the reactor system would not be as important since analog signals would no longer need routing.

12.8.2.5 Reactor Failure and Replacement

The routine operation of the AIMS offered no particular issues except for operation of the Redwood components. The microreactors almost always functioned properly in the experiments. During one experiment, a substantial leak was found at the O-ring seal between the DieMate™ manifold and one of the reaction channels. This line was manually isolated from the system before the feed gas was introduced to the reactors, so no danger was posed. This was caused by an improper adjustment of the socket head cap screws on the DieMate™ manifold. This was avoided in subsequent runs by setting the torque on these screws to a consistent value. In addition, the microreactor die failed when the heaters produced an excessive temperature. In many cases, the heaters simply stopped functioning, but the membrane remained intact. The failure was usually due to morphological changes in the platinum layer that interrupted the continuity of electricity to the heater lines. However, the membrane did rupture on many of the reactors, but this had been previously observed for this type of reactor design (Srinivasan et al., 1997; Srinivasan, 1998). When this occurred, the reactor line was automatically purged with inert gas and then it was isolated from the remaining active reactor lines.

12.8.2.6 Redwood Microvalve Operation

The Achilles heel of the AIMS system was the Redwood microvalves that were used for positive flow shut-off. Failures generally occurred when the system was first powered up, which suggested that power transients were the cause of many of the problems. Specifically, these valves were thermally actuated, so opening one of the SOVs produced a large transient current draw of about one amp, but for less than ½ second. Adjustments were made in the LabVIEW control program to prevent the simultaneous actuation of multiple valves. In particular, the valves were not allowed to actuate when power was first applied to the chassis.

Another difficulty was that the normally-closed valves SOVs would sometimes fail to open when actuated. This was traced to an inherent design issue, so it was necessary to simply replace the Redwood flow manifold with another manifold. Fortunately, some spare manifolds had been ordered that could be used while the components were repaired by Redwood.

12.8.2.7 Redwood Mass Flow Controllers

The Redwood Mass Flow Controllers (MFCs) were able to control mass flow, but with some limitations. In particular, great care had to be exercised when calibrating the MFCs since the calibration curve sometimes exhibited an inflection point in the flow rate versus set point response. A few MFCs did have linear calibration curves, but many exhibited this nonlinearity to varying extents. This was probably due to the flow measurement and control techniques of these MFCs, which operated by controlling the pressure drop across an orifice. Discussions with Redwood staff indicated that adjusting the internal flow control and measurement parameters for these controllers required a considerable amount of work.

12.8.2.8 Other Process Components

The only item that needed replacing was a DieMate™ socket whose pins had been damaged by improper handling. All other components inside the AIMS chassis functioned without any failure during the tests.

12.8.3
Automation and Hardware Evaluations

This section summarizes key results and findings on the prototype temperature measurement elements and process control systems. Generally, most HMI and PLC control program components operated according to the design specifications. However, a few items remained problematic throughout the system evaluation. In some cases, the problem was probably due to electronics and not the software. Difficulties with operating the Redwood SOVs and MFCs are discussed above.

12.8.3.1 Temperature Sensors

The temperature measurements generated by the various system RTDs were almost always suspect from the onset of startup. Specifically, the temperatures measured by the RTDs using the SCXI-1121 boards, which were displayed in the *Thermal Blocks* tab of the main control panel, were nearly always inaccurate. This was attributed to poor electrical connections inside the AIMS chassis. In particular, the connectors on the front system boards to the backplane were determined to be the source of difficulty. It was noted during testing that removing a board and simply reinserting it could dramatically improve or degrade one of the RTD values. The backplane connectors themselves are not completely at fault since the frequency of board removal and replacement in the AIMS was much higher than anticipated by the project team. In fact, most of the system boards were probably removed and replaced many hundreds of times throughout the project lifetime. This operation was always a somewhat complicated procedure and a good deal of care was needed to simply avoid damaging the pins on the backplane connecters.

Notably, the RTD temperatures measured by the CAL controllers in the AIMS chassis were deemed to be quite reliable. Occasionally, a temperature measurement would be clearly inaccurate. However, during most of the testing, the RTDs appeared accurate, as verified by independent temperature probe readings. (Although difficult to verify by probing directly, other indicators such as the power output by the controllers to maintain a set point temperature could be monitored to infer actual temperature.) These controllers were more reliable since fewer connections were required to connect the temperature controller boards to the reactor boards. Thermocouples were not used for temperature measurements since the various metals used in the backplane and PCB would not have been compatible with their use. In this case, it would have been necessary to place a voltage amplifier on the thermal couple signals so that their signals could have been measured as an analog input.

12.8.3.2 Temperature Controllers

The CAL controllers exhibited some other operational problems. In particular, the OPC connection to the HMI was not very reliable. It was often difficult to initiate communication with the CAL controllers and incorrect controller parameters would sometimes be transmitted, e.g., set points would be changed to incorrect values or other controller parameters would be unintentionally altered. Much time was spent trying to debug these communication problems by adjusting parameters on the controllers and the RS232/485 converter. Unfortunately, it was not possible to resolve all the problems completely. The difficulty probably was due to the particular converter design that was used, but poor connections in the serial links could also have been a source of difficulty. Electrical noise should not have been a problem since RS-485 communications typically work well in these types of environments. Because of these problems, it was preferred to input operating setpoints and other controller parameters directly to the CAL controllers without using the HMI.

12.8.3.3 Process Control Loops

Various values for the control loop rate were used to evaluate the ability of the system to maintain real-time PID temperature control of the microreactor heaters. It was found that the control loop could successfully operate at a rate of 500 Hz with all 24 microreactor heaters being under closed-loop PID control. At faster rates, the processor had insufficient time to complete the TCP/IP communications with the HMI, which resulted in data loss. Somewhat conservative loop parameters were used because of the noise in the temperature measurement signal, but temperature control was quite stable under a wide variety of both reactive and non-reactive conditions. The dynamics of the closed loop control algorithm were not explored since the data logging rate (4 Hz) was not fast enough to perform these types of evaluations. This would be possible with program modifications, but this issue was not a project objective. It could be considered as a topic for future investigation.

The few problems that were not solved really had minimal effect on the ability to use the system and did not detract at all from its safety. The most impressive finding was the ability of the PLC to perform closed-loop PID control on 24 microreactor heaters at a maximum rate of 500 Hz. Although this high loop rate is achievable in specialized control systems, this rate of control for hardware whose cost was less than $20k is a notable accomplishment.

12.8.4
Reaction Performance Results

This section compares typical AIMS microreactor performance results for both CH_4 oxidation and NH_3 oxidation in separate sequences of experiments. The oxidation of NH_3 reaction placed lower demands on the microreactor design and control system since nearly complete conversion could be obtained at less than 300 °C. However, the GC analysis of the NH_3 oxidation products was considerably more difficult. Open-loop temperature control was initially used for the reactor system testing, but closed-loop temperature control was later investigated. Typical GC product analysis data, microreactor performance data, and PID controller performance results for the microreactor heaters are summarized below.

12.8.4.1 Methane Oxidation

Methane oxidation was the first reaction studied using the AIMS system. The goal was to operate the microreactor under fuel-lean conditions with millisecond contact times (Williams et al., 2005) while achieving high degrees of conversion for O_2, which was the limiting reactant.

Figure 12.18 shows a typical chromatogram obtained from an on-line product gas analysis for the oxidation of CH_4. The chromatograms obtained from the TCD were typically characterized by steady baselines with well-resolved peaks. The product spectrum indicates that both partial and total combustion reactions of CH_4 are present. Both CO_2 and H_2O are baseline-resolved using a Hayesep® R column while the O_2, N_2, CH_4, and CO are baseline-resolved on a molecular sieve column. The overall analysis time requires about 9 min. Some gaps exist in the retention times for the various components, but no attempt was made to reduce these by additional optimization of the temperature program. The overall analysis time could have been reduced if the dual-oven method had been used as described elsewhere (Delaney and Mills, 1999; Nicole et al., 2000).

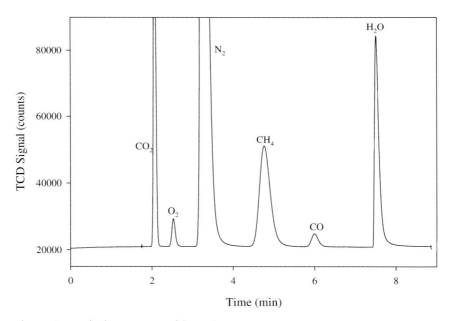

Fig. 12.18 Sample chromatogram of the methane oxidation reaction products.

Figure 12.19 shows typical CH_4 and O_2 conversion versus CO selectivity data at various reactor temperatures obtained from the AIMS. In this case, the microreactor was successfully operated up to an upper temperature limit of about 575 °C. (Temperatures measured above 500 °C by the microreactor RTDs are questionable since the platinum layer morphology starts changing rapidly at high temperatures. This subsequently changes the behavior of the platinum temperatures sensors.) The large error bars for the CO selectivity values are due to the low conversion of CH_4. Reduction of the error bars would require more precise measurements of the small (< 0.5 vol%) CO compositions observed. The accuracy could be increased by using a larger sample loop volume, which would increase the signal-to-noise ratio for CO. However, the resulting TCD output signal for N_2, which is the component present at the highest concentration, would then saturate the A to D converter. As a result, the N_2 peak could not be used as the internal standard so quantification of the CO peak could not be performed using this approach. Similarly, the error bars on the O_2 conversion are large because of the differential levels of conversion. Consequently, small errors in quantification of O_2 in the feed and product gas are magnified when the conversion, which is based upon a difference in the composition, is evaluated. Another problem is that the relative molar response factor for H_2O was not precisely measured, so the relative molar response factor value of NH_3 was used as an estimate. The error introduced by this estimate is reflected in the errors on the calculated O_2 conversion.

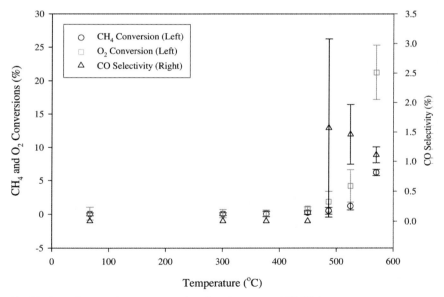

Fig. 12.19 Methane oxidation reaction data for microreactor ACT-G2-4 channel B with a feed gas flow rate of 10.1 mL min^{-1}. The feed composition was 14.9% methane, 10.3% oxygen, 39.8% nitrogen, and balance helium. This microreactor was placed in Reactor Board 2. Error bars represent the 95% confidence interval of each value.

Microreactors operating under the same reaction conditions were compared, not only to examine possible problems with the AIMS, but also to quantify any differences in performance between the microreactor channels. Figure 12.20 compares the O_2 conversion versus temperature data for three different reaction channels operating with the same feed gas composition and feed gas flow rate. The CH_4 conversion data are similar and omitted for brevity. The conversions below 550 °C are less than 1% and lie well within the calculated error bounds of the values. However, the error bars increase as the temperature increase, which becomes particularly apparent as the temperature approaches 600 °C, as the O_2 conversions approach 20%. The error bars increase with increasing temperature because of the changing structure of the platinum used for the catalyst and the temperature sensors. The change of the platinum catalyst structure was indicated by large fluctuations in the product gas composition during the analysis (the microreactors usually did not last long enough under these conditions to find a steady-state condition.). In addition, further annealing of the platinum lines of the temperature sensors reduced the accuracy of the temperature measurements.

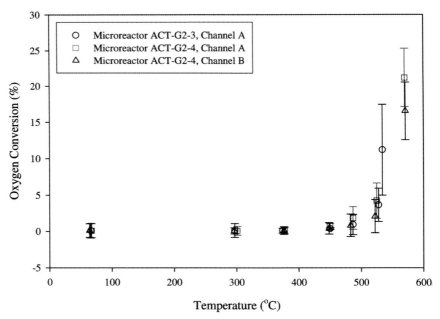

Fig. 12.20 O_2 conversion data versus temperature for the oxidation of CH_4 for three microreactor channels. Open-loop temperature control was used in this experiment with each heater set to the same heater voltage. The temperature used for the plot is the average of the measured RTD temperatures. These results come from one AIMS experiment with three microreactors running simultaneously. (There was a gas leak in Reactor Board 1 for microreactor channel B, so this channel was not utilized.) Feed gas composition: 14.9% CH_4, 10.3% O_2, 39.8% N_2, and balance He. Feed gas flow rate to each microreactor channel was 10 mL min^{-1}. Error bars represent the 95% confidence interval of each value.

High temperatures caused the resistance of the temperature sensor to irreversibly decrease to a lower value. Hence, temperatures much higher than 600 °C produce a false resistance reading close to 600 °C because of the change in platinum structure.

In summary, the oxidation of CH_4 was more difficult to study in the AIMS microreactor design due to the high reaction temperatures required for integral levels of conversion. Generally, differential level conversions (< 10%) of CH_4 were observed. The main challenge was the degradation of the platinum microreactor heaters. The heater lines usually lost continuity when operated above 600 °C. To a lesser extent, membrane failure also prevented the microreactors from reaching these high temperatures. Because high temperatures could not be sustained for a long period (> 10 min), very limited reaction data were available at these temperature points since a GC analysis took about 10 min to complete. Very low selectivities, less than 10%, to CO were observed, and none or extremely small amounts of H_2 (< 0.5%) were observed because of this temperature limit. Unfortunately, the high temperature required (> 700 °C) for the partial oxidation reaction was barely achievable, and the reactors generally failed when the temperature became high enough to produce significant amounts of carbon monoxide. Only on one occasion was any H_2 detected in the product stream. Despite the above-mentioned difficulties, CH_4 oxidation was the first reaction that demonstrated the AIMS could be successfully operated with multiple microreactors at unique operating conditions for each one.

12.8.4.2 Ammonia Oxidation

Because of the challenges in obtaining high conversion for the methane oxidation reaction, the oxidation of NH_3 was subsequently studied. For safety reasons, NH_3 was maintained as the limiting reagent.

Analysis of the NH_3 oxidation reaction products was complicated by the need to identify a GC column packing that produced both symmetrical peak shapes for both NH_3 and H_2O while maintaining a reasonable overall analysis time. The best GC column packing identified was a ⅛-inch OD × 6 ft type 316 SS column packed with 100/120 Hayesep® T porous polymer. The packing was treated with a 5% KOH solution to deactivate some of the NH_3 adsorption sites. Figure 12.21 shows a typical chromatogram generated from the AIMS for the oxidation of NH_3. Complete analysis required 15 min, which limits the sample throughput to four per hour. Nitric oxide, NH_3, and H_2O were resolved on the Hayesep® T column while Ar, O_2, N_2, and Kr were resolved on the molecular sieve column. Ar and O_2 co-elute as a composite peak. Baseline resolution of these two species in the molecular sieve column would require a cryogenically-cooled column. The water peak is unsymmetrical and elutes about 8 min after the peak for NH_3. Despite these apparent deficiencies, the water peak could be reliably quantified.

Figure 12.22 shows an example of the reaction results for testing the AIMS with ammonia oxidation. Conversion below 200 °C was not measurable, whereas conversion above 300 °C was complete. In between these two values, the product gas composition exiting the microreactor was unstable. This resulted in the large

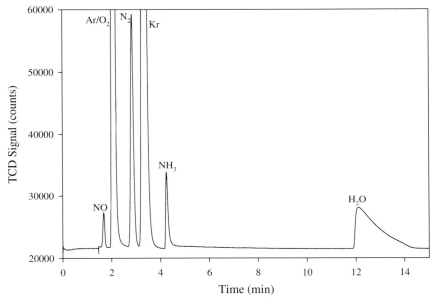

Fig. 12.21 Sample chromatogram of the products of ammonia oxidation.

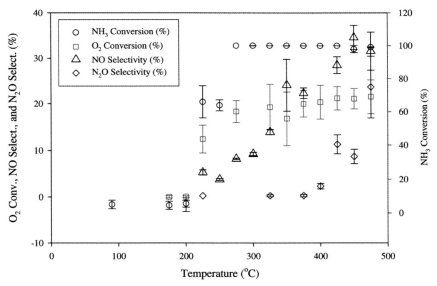

Fig. 12.22 Ammonia oxidation reaction data versus temperature for microreactor ACT-G3-7 channel A with a feed gas flow rate of 4.0 mL min^{-1}. (Closed-loop temperature control was used for this experiment, and each heater was given the same set point temperature.)
The feed composition was 8.8% NH_3, 25.6% O_2, 6.9% Ar, 14.3% Kr, and the balance He. This microreactor was placed in Reactor Board 1. Error bars represent the 95% confidence interval of each value.

error bounds for the NH_3 and O_2 conversion values in this region. Unfortunately, the cycle time for the GC method was not fast enough to investigate this phenomenon, but it is believed to be related to the large amount of heat released by the reaction.

The selectivity data shown in Fig. 12.22 indicate that significant amounts of NO are produced by the microreactor during the oxidation of NH_3. The data indicate that selectivity improves dramatically, from almost zero at 250 °C to approximately 30% at 500 °C.

Similarly, small amounts of N_2O are detected at approximately 400 °C, and this product was observed during most of the ammonia oxidation trials. Srinivasan et al. (1997) and Srinivasan (1998) did not report the presence of N_2O during studies on the oxidation of NH_3 in a prototype MIT microreactor design. This may have been due to an overlap with CO_2 since these species have the same molecular weights.

Figure 12.23 shows the heater power applied to the microreactor heaters for 200, 250, and 300 °C. For 200 °C, there is very little heat contribution from the reaction so the power profile is almost flat. However, there is significant heat generated by the reaction for 300 °C, so each heater steadily needs more power as there is less reaction occurring to provide the heat as the gas travels down the length of the heated segment. The 250 °C set point is almost in between these two examples, with the profile rising linearly at the beginning and then steadying out for the later heaters.

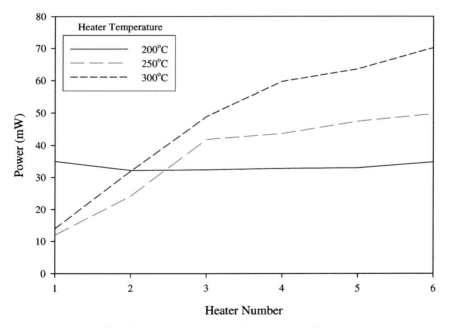

Fig. 12.23 Power profile of the microreactor heaters for 200, 250, and 300 °C.

12.9
Summary and Conclusions

The testing of the AIMS revealed some minor problems in the mechanical layout of the components. This made diagnosing and repairing problems difficult since direct access to most components was limited. However, the most significant challenge was operation of the Redwood Microsystems MEMS valves. These valves were not very robust and required component replacement on many occasions. In addition, the MFCs, when they were functioning, did not behave similarly to conventional MFCs. This increased the difficulty of setting up the system for testing. Despite these obstacles, the system automation made the AIMS both safe and easy to use. It was successfully tested using the methane oxidation and ammonia oxidation reactions. Because of temperature limitations of the microreactors used in this scale-up system, methane conversion was generally less than 10%. Subsequently, ammonia oxidation was studied because of prior success in working with this reaction. This allowed exploration of the AIMS over a wide variety of conversions of ammonia. Closed-loop PID temperature control was successfully tested on this reaction system.

Overall, the AIMS functioned comparably to the conventional catalyst test stations used at DuPont. Individual reactors could be started or shut-down independently of the other reactors in the system. In addition, the safety systems responded appropriately during testing and prevented the occurrence of any safety incidents during this project.

References

Alm, B., Knitter, R., Haußelt, J., Development of a ceramic micro heat exchanger – Design, construction, and testing, *Chem. Eng. Technol.*, 28(12), **2005**, 1554–1560.

Anonymous, DuPont researchers receive award for improving chemical safety, http://www.scienceblog.com/community/older/2003/C/2003411.html

Ashmead, J. W., Blaisdell, C. T., Johnson, M. H., Nyquist, J. K., Perrotto, J. A., Ryley, Jr., J. F. (inventors), Integrated chemical processing apparatus and processes for the preparation thereof, *US Pat.* 5 534 328 assigned to E. I. du Pont de Nemours, July 9, **1996**.

Ashmead, J. W., Blaisdell, C. T., Johnson, M. H., Nyquist, J. K., Perrotto, J. A., Ryley, Jr., J. F. (inventors), Integrated chemical processing apparatus and processes for the preparation thereof, *US Pat.* 5 690 763 assigned to E. I. du Pont de Nemours, November 25, **1997**.

Benz, K., Jäckel, K.-P,. Regenauer, K.-J., Schiewe, J., Drese, K., Ehrfeld, W., Hessel, V., Löwe, H., Utilization of micromixers for extraction processes, *Chem. Eng. Technol.*, 24(1), **2001**, 11–17.

Bong-Bang, Y., Lee, K-M., Oh, S., 5-axis micro milling machine for machining micro parts, *Int. J. Adv. Manuf. Technol.* 25, **2005**, 888–894.

Chambers, R. D., Fox, M. A., Holling, D., Nakano, T. T., Okazoe, G., Versatile gas/liquid microreactors for industry, *Chem. Eng. Technol.*, 28(3), **2005**, 344–352.

Charpentier, J.-C., Process intensification by miniaturization, *Chem. Eng. Technol.*, 28(3), **2005**, 255–258.

Chen, B., Dingerdissen, U., Krauter, J. G. E., Lansink Rotgerink, H. G. J,. Möbus, K., Ostgard, D., Panster, J. Riermeier, T. H., Seebald, S., Tacke, T., Trauthwein, H.,

New developments in hydrogenation catalysis particularly in synthesis of fine and intermediate chemicals, *Appl. Catal. A: General*, 280(1), **2005**, 17–46.

Commenge, J.-M., Falk, L., Corriou, J.-P., Matlosz, M., Analysis of microstructured reactor characteristics for process miniaturization and intensification, *Chem. Eng. Technol.*, 28(4), **2005**, 446–458.

Delaney, T. M., Mills, P. L. "Multidimensional process gas chromatography based on a single GC with dual ovens", Presented at 38th Annual Eastern Analytical Symposium, Somerset, NJ, 15, **1999**, Gas Chromatography Poster Session, Poster 71.

Dietrich, T. R., Freitag, A., Scholz, R., Production and characteristics of microreactors made from glass, *Chem. Eng. Technol.*, 28(4), **2005**, 477–483.

Doig, S. D., Pickering, S. C. R., Lye, G. J., Woodley, J. W., The use of microscale processing technologies for quantification of biocatalytic Baeyer-Villiger oxidation kinetics, *Biotechnol. Bioeng.*, 80(1), **2002**, 42–49.

Ehrfeld, W., Hessel, V., Lehr, H., Microreactors for chemical synthesis and biotechnology – Current developments and future applications in Microsystem Technology in Chemistry and Life Sciences, A. Manz and A. Becker (Eds.), 194, **1997**, 233–252.

Ehrfeld, W., Hessel, V., Löwe, H., *Microreactors: New Technology for Modern Chemistry*, Wiley-VCH, Weinheim, 1st ed., **2000**.

Firebaugh, S. "Improving reactor yield," Internal memo, Department of Chemical Engineering, Massachusetts Institute of Technology, Feb. 9, **1998**.

Heck, M., Manning, A. *Flotherm Analysis and Modeling of Airflow Patterns and Temperature Distribution in the Kaperel Card Cage*, Flomerics, Inc., Marlborough, MA, Report No. USE0400-11, **2000**.

Hessel, V., Löwe, H., Microchemical engineering: Components, plant concepts, user acceptance – Part I, *Chem. Eng. Technol.*, 26(1), **2003a**, 13–24.

Hessel, V., Löwe, H., Microchemical engineering: Components, plant concepts, user acceptance – Part II, *Chem. Eng. Technol.*, 26(4), **2003b**, 391–408.

Hessel, V., Löwe, H., Microchemical engineering: Components, plant concepts, user acceptance – Part III, *Chem. Eng. Technol.*, 26(5), **2003c**, 531–544.

Hessel, V., Löwe, H., Organic synthesis with microstructured reactors, *Chem. Eng. Technol.*, 28(3), **2005**, 267–284.

Huybrechts, W., Mijoin, J., Jacobs, P. A., Martens, J. A., Development of a fixed-bed continuous-flow high-throughput reactor for long-chain n-alkane hydroconversion, *Appl. Catal. A: General*, 243(1)1, **2003**, 1–13.

Jensen, K. J., Microchemical systems: Status, challenges, and opportunities, *AIChE J.*, 45(10), **1999**, 2051–2054.

Lee, A., "Microfluidic molecular systems program at DARPA", presentation given at the NIST Labs, Microanalytical Lab/MEMS Seminar Series, Gaithersburg, MD, October 28, **1999**.

Majors, R. E., Minaturized approaches to conventional liquid-liquid extraction, *LC/GC North Am.*, 24(2), **2006**, 118–130.

Mills, P. L., Guise, W. E. "A multidimensional gas chromatographic method for on-line process analysis of isobutane or isobutylene partial oxidation reaction products", Presented at 33rd Eastern Analytical Symposium and Exposition, Somerset, NJ, November 15, **1994**, Session on Gas Chromatography.

Mills, P. L., Guise, W. E., A multidimensional gas chromatography method for analysis of n-butane oxidation reaction products, *J. Chromatogr. Sci.*, 34(10), **1996**, 431–459.

Mills, P. L., Nicole, J. F., Multiple automated reactor system (MARS). 1. A novel reactor system for detailed testing of gas-phase heterogeneous oxidation catalysts, *Ind. Eng. Chem. Res.*, 44, **2005**, 6435–6452.

Nicole, J. F., Mills, P. L., Delaney, T. M., "Multidimensional on-line process gas chromatography based on a single GC with two independently controlled ovens", Presented at AIChE 2000 Annual Meeting, Los Angeles, November 14, **2000**, Session on Lab and Pilot Scale Instrumentation, Paper 276d.

Palmer, C., Remcho, V. T., Microscale liquid-phase separations, *Anal. Bioanal. Chem.*, 372(1), **2002**, 35–36.

Pfeifer, P., Bohn, L., Görke, O., Haas-Santo, K., Schubert, K., Microstructured components for hydrogen production from various hydrocarbons, *Chem. Eng. Technol.*, 28(4), **2005**, 474–476.

Puskeiler, R., Kaufmann, K., Weuster-Botz, D., Development, parallelization, and automation of a gas-inducing milliliter-scale bioreactor for high-throughput bioprocess design (HTBD), *Biotechnol. Bioeng.*, 89(5), **2005**, 512–523.

Quiram, D. J., *Characterization and Systems Integration of Microreactors*, Sc.D. Thesis, Department of Chemical Engineering, Massachusetts Institute of Technology, February **2002**.

Ryley, J. F. "Microchemical system applications – DuPont experience", Presented at Microchemical Systems and Their Applications, June 16–18, **1999**, Reston, VA.

Sato, M., Goto, M., Note: Gas absorption in water with microchannel devices, *Sep. Sci. Technol.*, 39, **2004**, 3163–3167.

Shchukin, D. G., Sukhorukov, G. B., Nanoparticle synthesis in engineered organic nanoscale reactors, *Adv. Mater.*, 16(8), **2004**, 671–682.

Schütte, R., Matlosz, M., Renken, A., Liauw, M., Langer, O.-U. (organizers), The Ninth International Conference on Microreaction Technology (IMRET 9), Potsdam/Berlin, Germany, September 6–9, **2006**.

Srinivasan, R., Hsing, I.-M., Berger, P. E., Jensen, K. F., Firebaugh, S. L., Schmidt, M. A., Harold, M. P., Lerou, J. J., Ryley, J. F., Micromachined reactors for catalytic partial oxidation reactions, *AIChE J.*, 43(11), **1997**, 3059–3069.

Srinivasan, R., *Microfabricated Reactors for Partial Oxidation Reactions*, Sc.D. Dissertation, Massachusetts Institute of Technology, Cambridge, MA, **1998**.

Styring, P., Towards liquid crystal synthesis using high throughput and micro reactor technologies, *Mol. Cryst. Liq. Cryst.*, 411, **2004**, 1059–1070.

Tovbin, Y. K. Gas and liquid transfer in narrow pores, *Theor. Foundations Chem. Eng.*, 36(3), **2002**, 214–221.

Truckenmüller, R., Cheng, Y., Ahrens, R., Bahrs, H., Fischer, G., Lehrmann, J., Micro ultrasonic welding: Joining of chemically inert polymer microparts for single material fluidic components and systems, *Microsyst. Technol.*, 12(10–11), **2006**, 1027–1029.

Watts, P., Chemical synthesis in micro reactors, *Chem. Ing. Technik*, 76(5), **2004**, 555–559.

Younes-Metzler, O., Svagin, J., Jensen, S., Christensen, C. H., Hansen, O., Quaade, U., Microfabricated high-temperature reactor for catalytic partial oxidation of methane, *Appl. Catal. A: General*, 284(1–2), **2005**, 5–10.

Wiese, M., Schneider, G., Kappe C. O. (Eds.), *QSAR & Combinatorial Sci.*, Special issue on Combinatorial Fluidics, 24(6), **2005**, 695–803.

Williams, K., LeClerc, C. A., Schmidt, L. D., Rapid lightoff of syngas from methane: A transient product analysis, *AIChE J.*, 51(1), **2005**, 247–260.

13
Liquid Phase Process Characterization

Daniel A. Hickman and Daniel D. Sobeck

13.1
Overview

This chapter describes the development of a system to characterize homogeneous liquid phase reactions using serial screening in a tubular reactor. The system used a single microchannel reactor with reactants injected as finite pulses. Fundamental reaction engineering principles (e.g., equations and constraints for heat transfer, mixing, axial dispersion, pressure drop) were used to design or select the system components to enable analysis of the reactor effluent at undiluted, steady-state concentrations while injecting only microliters of the reactant solutions per experiment. These reaction engineering principles and their application to the design of this system are summarized below. This chapter also shows experimental data that validate the expected low axial dispersion of injected reactants. An error analysis demonstrates the impact of uncertainties in system variables on the accuracy and precision of reaction kinetic parameters derived from experiments in this system.

13.2
Background

The concept of a highly automated scale-up process is enticing. One vision of such a process for homogeneous (liquid phase, non-catalytic) reactions starts with little more than a list of reactants, solvents, and desired products. Given this information as well as constraints imposed by economic, environmental, safety, and practical factors, a highly automated system could include both software and hardware components to generate an optimal reactor design. The necessary software components would:

1. generate potential reaction mechanisms;
2. convert proposed mechanisms into reaction rate expressions;

3. insert these rate expressions into a model of the experimental reactor system, including at least material balances and perhaps energy and momentum balances;
4. perform parameter estimation to determine the kinetic parameters for the rate expressions in each of several candidate models;
5. perform model discrimination to rank the candidate models;
6. construct and execute model-based experimental design, enabling efficient convergence to an optimized model and parameters, with the experimental design, parameter estimation, and model discrimination steps completed in an iterative fashion;
7. define the optimal reactor configuration and operating conditions to maximize the chosen objective function (e.g., economic profit).

The above list essentially summarizes the tasks performed by a reaction engineer in the development of a new reaction process. The quality of the final result of this reaction engineering methodology, whether automated or not, is intrinsically linked to a necessary hardware component of this process: the experimental reactor system. Any shortcomings of this physical system would be manifest as errors or uncertainties in the final reactor design.

High throughput systems for the development of chemical reaction processes are becoming increasingly common. For homogeneous liquid phase reaction systems, the commercially available systems are almost all parallel, multireactor systems [1, 2, 3, 4]. In these systems, high throughput experimental testing is achieved by operating multiple batch reactors simultaneously. An alternative approach that has not been given as much attention is high throughput operation in a serial fashion. A few examples of serial high throughput screening in the literature include:

1. Sequential screening of heterogeneous catalysts [5].
2. A simple microfluidic system that transports solutions with rapid mixing and no axial dispersion by using networks of microchannels with rectangular cross sections, hydrophobic surfaces fabricated using rapid prototyping in polydimethylsiloxane, and elimination of axial dispersion by localizing the reactants within aqueous plugs separated by a water-immiscible oil [6].
3. A micromixer-tube system for serial transient screening of organometallic catalysts by sequential injection of pulses of catalysts and reactants [7].
4. Temperature scanning reactors (TSRs) for high throughput serial screening of reaction conditions [8, 9, 10].

In recent years, commercially available microreactor systems have also been designed for process development and small scale production. These systems are offered with varying levels of automation, integrated sensors, and on-line analysis. For example, systems offered by Siemens [11], CPC-Cellular Process Chemistry Systems GmbH [12], and mikroglas chemtech GmbH enable systematic screening of process parameters for the design and optimization of liquid phase chemical processes as well as production of kilogram quantities of materials. These examples all utilize a continuous flow of the reactant mixtures through a microchannel microreactor.

This chapter explores one extension of these continuous automated reactor system concepts. This extension combines the concept of highly automated chemical process development in microscale laboratory equipment with the concept of flow injection analysis [13] to achieve high throughput serial screening in a microchannel device. Using this approach, small volumes of reactants are injected into a continuously flowing stream, with multiple experiments performed as sequential injections. The frequency of such experiments is typically limited by the requirements of the product analysis methodology. In this particular application, the system design is limited to homogeneous liquid phase reactions that occur between reactants introduced in two separate miscible streams. The next section provides the design basis for this system and also defines the system capabilities.

13.3
System Design Basis

Figure 13.1 gives a conceptual schematic of this system. In this process, a solvent stream is fed continually by two pumps through two sample injection valves. The two sample loops in these valves are filled with the two desired reactant mixtures and simultaneously injected into the flowing carrier streams. The two finite pulses of reactant mixtures reach a mixing tee at the inlet of the reactor, and the combined reactant mixture is transported through the reactor to the analytical sampling valve.

A key aspect of the design strategy for this reactor system is to enable quantification of reaction kinetics with a steady-state mass balance model of the reactor. This is enabled only if the injected finite pulses are adequately large in volume to provide a volume at the point of analytical sampling that contains the reactants and products at the steady-state concentrations (i.e., not diluted by axial dispersion into the carrier stream before and after the finite pulse). Acceptable and unacceptable levels of axial dispersion are illustrated in Fig. 13.2, a plot of concentration (y-axis) as a function of distance traveled by the pulse (x-axis). Given an adequately large

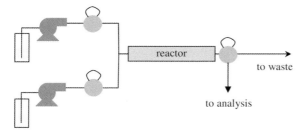

Fig. 13.1 Conceptual schematic of high throughput serial screening reactor system.

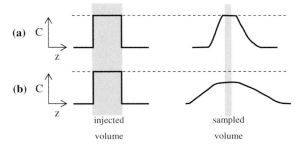

Fig. 13.2 Inlet and outlet concentration profiles:
(a) acceptably low axial dispersion, (b) unacceptably high axial dispersion.

injection volume, or a flow path with minimized axial dispersion, one can achieve a sampled volume that is not diluted as a result of axial dispersion into the carrier stream (Fig. 13.2a). However, higher levels of axial dispersion, or smaller injected volumes, will result in dilution of the entire dispersed pulse (Fig. 13.2b). In this case, none of the reactor effluent will contain reactants at their equivalent steady-state concentrations, and a more complex dynamic mass balance is required to extract reaction rate parameters, giving larger uncertainties in the final rate parameters.

To design a system that achieves acceptably low axial dispersion (Fig. 13.2a), a few fundamental principles must be applied. These principles of reaction engineering are described in the following paragraphs. They lead naturally to an optimal system design when combined with the following objectives: (a) maximize the experimental throughput, (b) minimize the amount of materials to be consumed, (c) minimize the uncertainty in the data, and (d) maximize the available ranges of residence times, temperatures, pressures, and concentrations.

To eliminate uncertainties associated with distribution of the reaction mixture to multiple parallel microchannels, and to eliminate associated axial dispersion in the necessary distribution and collection headers, a single channel microreactor was constructed of commercially available tubing. The criteria that governed the selection of the reactor and other tubing dimensions included:

1. minimum achievable residence time of less than 10 s;
2. maximum achievable residence time on the order of several hours;
3. reactor volume at least 20 times that in the mixing tee;
4. reaction fluid essentially isothermal throughout the reactor;
5. characteristic mixing time less than one-fourth of the minimum residence time (based on two colaminar streams at the reactor inlet);
6. pressure drop at the minimum residence time less than 100 psi;
7. contacting pattern accurately represented using a one-dimensional (1D) axial dispersion model;
8. minimum time of 60 s for at least 99% of the steady-state concentration available for analytical sampling.

The first criterion was defined to enable quantitative studies of relatively fast liquid phase reactions. The second was defined to enable quantitative studies of relatively slow reactions by striking a balance between reactor volume and the lower end of the feed pump flow rate ranges. The third criterion was chosen to ensure that the mathematical modeling of the reactor could neglect reactions occurring in the inlet mixing tee. This constraint can be immediately dismissed by using a low dead volume connector (e.g., Valco part number MY1XCS6 with 0.15-inch diameter holes in a Y-configuration, having a total estimated internal volume of 0.1 µL). The remaining five criteria are further developed below.

13.3.1
Heat Transfer in Laminar Pipe Flow

The fourth reactor design criterion (the reaction fluid must be essentially isothermal throughout the reactor) enables the complexity of the reactor model and the associated potential for errors to be minimized. Following the example of Mears for a fixed bed reactor [14], "isothermal" was defined as giving a reaction rate at the maximum radial mean temperature that is no more than 5% higher than the rate achieved at the wall temperature. At steady-state, neglecting axial heat conduction, the heat generated in a differential reactor volume must equal the heat removed through the walls:

$$\Re(-\Delta H)\, dV = U\, (T_b - T_w)\, dA \tag{1}$$

For a cylindrical channel with all of the heat transfer resistance assumed to be on the reaction fluid side ($U \sim 3.5\, k/d_t$ at low Reynolds numbers [15]), this equation becomes:

$$\frac{\Re(-\Delta H)\, d_t}{4} = \frac{3.5\, k\, (T_b - T_w)}{d_t} \tag{2}$$

Assuming the temperature dependence of the reaction rate follows the Arrhenius form, the ratio of the maximum rates at the maximum radial mean (bulk) temperature and the wall temperature must satisfy the following constraint defined earlier:

$$\ln\left(\frac{\Re_{max}(T_b)}{\Re_{max}(T_w)}\right) = -\frac{E_a}{R_g}\left(\frac{1}{T_b} - \frac{1}{T_w}\right) \le \ln(1.05) \cong 0.05 \tag{3}$$

Combining the last two expressions and eliminating T_b, the criterion for isothermal performance gives the following constraint on the channel diameter:

$$d_t^2 \le \frac{14\, k\, R_g\, T_w^2}{\Re_{max}(-\Delta H)\, (20\, E_a - R_g\, T_w)} \tag{4}$$

Table 13.1 Parameter values used for evaluating design constraints.

Parameter	Value	Units
k	0.6038	W/m-K
T_w	293	K
$\Re_{max}$	75.3	mol/m³-s
$(-\Delta H)$	126	kJ/mol
E_a	126	kJ/mol
D_m	1.16×10^{-9}	m²/s

Conservatively (for a positive order reaction), the maximum rate corresponds to the rate at the channel inlet, where the concentration driving force is the highest. Using the parameter values listed in Table 13.1, which are based on dilute acetone in water at 20 °C, the maximum channel diameter is 0.5 mm (0.02 inch). For this calculation, the value of $\Re_{max}$ was based on the following assumptions, which give the most conservative (highest) value for the maximum rate: first order reaction, initial concentration of 5000 mol m^{-3} of the rate controlling reactant, and 10% conversion of that reactant at 7 s residence time.

13.3.2
Mixing at the Reactor Inlet Tee

The fifth reactor design criterion requires that the measured reaction rates not be significantly affected by the rate of mixing at the reactor inlet. This constraint can be defined either by ensuring that the reactor residence time exceeds the characteristic mixing time by an acceptable margin, or by using two-dimensional (2D) simulations to calculate the error introduced by assuming instantaneous mixing at the channel inlet. Both of these approaches will be illustrated.

13.3.2.1 Characteristic Mixing Time Analysis

The mixing mechanism in this laminar flow system must be diffusion because the mixing must occur normal to the laminar flow streamlines. Laminar flow prevails in systems with these small dimensions. Thus, the characteristic mixing time is the characteristic diffusion time, defined for two colaminar streams mixing in a cylindrical channel as:

$$t_D = \frac{R^2}{4 D_m} \qquad (5)$$

The design constraint listed earlier requires that $\tau_{min} \geq 4\, t_D$, or $\tau_{min} \geq R^2/D_m$. (The basis for the choice of the factor of 4 is arbitrary but is supported by simulations described later.) Thus, to achieve the desired minimum residence time, the channel diameter must be less than 0.21 mm (0.0085 inch). This value is more limiting than the value of 0.5 mm imposed by the heat transfer analysis.

13.3.2.2 Two-dimensional Simulation

A second method used to determine the effect of inlet mixing on the reactor performance is to simulate a 2D cylindrical channel with reaction. This approach compares the fractional conversion predicted for the case with perfect initial mixing to that predicted for the case where two reactant streams are initially fed as two distinct colaminar streams. In the latter case, the inlet boundary condition is defined in order to give a symmetric boundary condition as defined below, where the two reactants are labeled "1" and "2":

$$c_1(z=0) = \begin{cases} c_{1i} \text{ for } 0 \leq r < R_1 \\ 0 \text{ for } R_1 \leq r < R \end{cases} \tag{6}$$

$$c_2(z=0) = \begin{cases} 0 \text{ for } 0 \leq r < R_1 \\ c_{2i} \text{ for } R_1 \leq r < R \end{cases} \tag{7}$$

The component mass balance for each of the two reactants is given below:

$$2\bar{v}_z\left[1-\left(\frac{r}{R}\right)^2\right]\frac{dc_i}{dz} = D_{m,i}\left(\frac{d^2c_i}{dz^2} + \frac{d^2c_i}{dr^2} + \frac{1}{r}\frac{dc_i}{dr}\right) - k_r\, c_1\, c_2 \tag{8}$$

For these simulations, the discretization of the mass balance equations used first order backward finite differences over a uniform grid of 100 intervals in the axial direction and a third order orthogonal collocation over 50 finite elements in the radial direction [16].

Figure 13.3 shows the concentration profile along the axis (centerline), and along the wall, of a cylindrical channel 0.18 mm (0.007 inch) in diameter for the case where $c_{1i} = 100$ mol m^{-3} and $c_{2i} = 0$ mol m^{-3} at a total flow rate of 3.55 μL s^{-1}. This corresponds to a residence time of 7 s in a 1-m long channel that is 0.007" in diameter. The molecular diffusivity was assumed to be 1.16×10^{-9} m^2 s^{-1}, the value for acetone in water at infinite dilution and 25 °C [17]. The characteristic mixing time calculated earlier for this scenario is 1.7 s, corresponding to an axial position of 0.24 m. Thus, the characteristic mixing time calculation represents a residence time at which the mixing is approaching complete, with the ratio of the centerline (maximum) to wall (minimum) concentration at 1.15.

To quantify the effect of the incomplete mixing on reaction rates in the front of the reactor channel, this same simulation was repeated assuming second order kinetics (first order in each of the two components) and $c_{1i} = c_{2i} = 100$ mol m^{-3}. A rate constant k_r of 1.0×10^{-3} m^3 mol^{-1} s^{-1} was used to give an intermediate level of conversion (near 25%). This case can be compared with a simulation in which the inlet boundary conditions were changed to assume complete mixing (50 mol m^{-3} of each component across the entire inlet cross section). The axial fractional conversion profiles for these two cases (unmixed and premixed feeds) are shown in Fig. 13.4, where the unmixed feed curve is the average of the calculated values for the two components. The computed conversions for the two components were

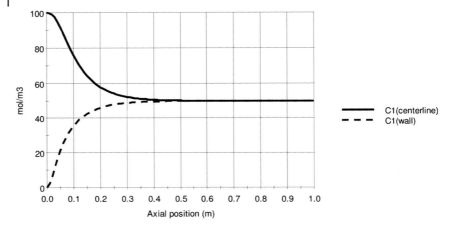

Fig. 13.3 Steady-state concentration profile of a tracer of component 1 injected in a stream on the centerline at 50% of the total flow.

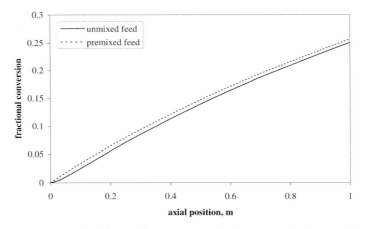

Fig. 13.4 Simulated fractional conversion profiles for premixed and unmixed feeds.

slightly different because of numerical errors resulting from the very steep radial gradients at the interface between the two feeds at the reactor inlet.

As shown in Fig. 13.4, the difference in the overall conversion is the consequence of a lower conversion rate at the reactor inlet, as expected. Although the characteristic mixing time corresponds to a distance of 0.24 m, the reaction rate (slope of the conversion profile) of the unmixed feed case approaches that of the premixed feed case by 0.1 m, or 10% of the total reactor volume. Furthermore, the predicted conversion for the unmixed and premixed feeds is 25.1% and 25.7%, respectively. This corresponds to a relative difference of 2%, an acceptable level of error that will only decrease as the reactor residence time increases. This analysis validates the reactor design criterion that the characteristic mixing time based on two colaminar streams at the reactor inlet must be less than one-fourth of the minimum residence time.

13.3.3
Pressure Drop in Laminar Pipe Flow

The sixth reactor design criterion requires that the pressure drop at the minimum residence time be less than 100 psi. For a small diameter channel, the flow through that channel will be laminar for all flow rates of interest for this particular application. Neglecting end effects, the solutions to the equations of continuity and of motion for steady-state laminar flow of an incompressible Newtonian fluid are well-known, yielding a parabolic velocity distribution and the Hagen–Poiseuille equation for pressure drop, as given in Eqs. (9) and (10):

$$v_z(r) = \frac{8F}{\pi d_t^2}\left[1-\left(\frac{r}{R}\right)^2\right] = 2\bar{v}_z\left[1-\left(\frac{r}{R}\right)^2\right] \quad (9)$$

$$\frac{\Delta P}{L} = \frac{128 F \mu}{\pi d_t^4} \quad (10)$$

For the case considered earlier (7 s residence time in a 0.18 mm, or 0.007-inch, diameter by 1-m long tube), using the viscosity of water at ambient temperature, the predicted pressure drop across the reactor channel is 23 psi.

13.3.4
Axial Dispersion in Laminar Pipe Flow

The seventh reactor design criterion requires that the contacting pattern be adequately represented using a 1D axial dispersion model. Neglecting the irrelevant case of extremely low flow rates, the single component mass balance in laminar pipe flow can be accurately described by one of three mathematical models, depending on the residence time in the tube [18]. If the residence time is sufficiently short, then molecular diffusion in the radial direction has a negligible impact on the residence time distribution of the fluid. This is called the pure convection regime. If the residence time is sufficiently long, a 1D axial dispersion model provides an accurate representation of the residence time distribution of the fluid. Between these two regimes, the mass balance must be represented using a 2D model, or an approximation of the complete 2D representation.

To simplify the extraction of reaction rate parameters from experimental data, the tube was designed so that the 1D axial dispersion model would apply over the entire range of expected residence times. This model is given in Eq. (11):

$$\bar{v}_z \frac{dc_i}{dz} = D_z \frac{d^2 c_i}{dz^2} + \sum_j v_{ij} \Re_j \quad (11)$$

where the axial dispersion coefficient (D_z) is a function of the molecular diffusivity, flow rate, and tube diameter.

$$D_z = D_m + \frac{\bar{v}_z^2 d_t^2}{192 D_m} \tag{12}$$

For reactor channel aspect ratios (L/D) that exceed 10, the boundary defining the limit of applicability of the axial dispersion model is approximated by Eq. (13) [18]:

$$\log_{10} Bo = \log_{10}\left(\frac{\bar{v}_z d_t}{D_m}\right) < 1.025 \log_{10}\left(\frac{L}{d_t}\right) + 0.65 \tag{13}$$

Equation (13) defines the maximum flow rate (minimum residence time) in a given tube that can be modeled using the 1D axial dispersion model.

13.3.4.1 Plug Flow Criterion

Using the methodology of Mears [19], the effect of axial dispersion on the performance of a tubular reactor can be ignored if the constraint given in Eq. (14) is satisfied:

$$0.05 > \frac{\bar{v}_z}{k_r L} \ln\left(1 + \frac{D_z k_r^2 L}{\bar{v}_z^3}\right) \tag{14}$$

This constraint was derived for a first order reaction by choosing to define "plug flow" as the case where the additional length needed to match the performance of a truly plug flow reactor is less than 5%. If this inequality is satisfied, the axial dispersion in the reactor can be ignored, enabling the experimental reaction rate data to be extracted using the plug flow reactor mass balance equation:

$$\bar{v}_z \frac{dc_i}{dz} = \sum_j v_{ij} \Re_j \tag{15}$$

For example, consider the short residence time case presented earlier (7 s residence time, 1-m long channel with 0.007-inch diameter). For a first order reaction with $k_r = 0.05$ and an inlet concentration of 100 mol m^{-3}, about 30% conversion would be achieved. The right-hand side of the dimensionless inequality above, assuming a molecular diffusivity of 1.16×10^{-9} m^2 s^{-1}, is 0.007. Thus, the plug flow criterion is easily satisfied. For rate constants lower than the value in this example, the above criterion is even more readily satisfied. Thus, the channel dimensions used in this example will enable a plug flow mass balance to be the basis for all reaction rate parameter calculations, simplifying the data analysis.

13.3.5
Injector Loop Sizing

The eighth reactor design criterion requires that 99% of the steady-state concentration be achieved in the analytical sample loop for at least 60 s. To achieve a sampled volume that is not diluted as a result of axial dispersion into the carrier stream, the injector loop volumes combined must be enough larger than the analytical sample loop to (a) compensate for the axial dispersion that occurs and (b) provide an adequate time window for the analytical sample to be removed from the process stream and injected into the analytical system. Furthermore, the two injector loops should be chosen of an adequately small diameter to minimize the axial dispersion at the tail of the injected finite pulse. By experimental trial and error, the smallest feasible injector loop inner diameter was 0.5 mm (0.02 inch). While smaller diameter tubing is available, the excessive pressure drop encountered during filling of such loops prevented robot-automated purging and filling of these loops.

13.4
System Capabilities

Our ultimate vision was to automate the reactor scale-up process, from experimental design through the development of mathematical models of reaction kinetics. To achieve this vision, we chose to first demonstrate that a custom, single channel microreactor could be used to generate quantitative scale-up data.

13.4.1
System Overview

The mechanical system used for this evaluation consisted of several automated components: feed control, reactants preparation, synchronized injected reactant pulses, reactor temperature and pressure and residence time control, on-line analytical, fraction collector for product sample collection, as well as programming for multi-system coordination.

Figure 13.5 gives a screen capture of the programming schematic of this system.

13.4.1.1 Feed System
The feed system contained an integrated balance to quantify mass flow rate, multiple pressurized feed containers to supply material to the process, a stream select valve to choose which feed container supplied material, and a feed pump to transfer the material from the feed container through the reactor. Syringe pumps provided accurate and stable flow rates, ranging from a few $\mu L\ min^{-1}$ to over 1 $mL\ min^{-1}$, at pressures as high as 1000 psi. These flow rate and pressure capabilities exceed limitations imposed by other system components.

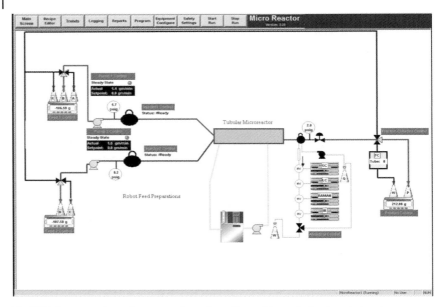

Fig. 13.5 Screen capture of programming schematic used to control automated system.

Material was injected into the process streams as they flowed from the feed containers to the reactor. Synchronizing the injecting of these two valves allowed for simultaneous injections of reactant A into feed stream A and reactant B into feed stream B.

13.4.1.2 Robotics Feed Preparation

The injected pulses could be prepared and loaded into the loop manually or automatically. A robotics system enabled automated sample preparation and loop loading. The feed preparation robot accurately metered reactants and solvents into feed bottles over a wide range of target masses. Four individual containers could be used to feed solvent to the robotics system. A 5-mL syringe was used to deliver quantities greater than 1-mL, while a 1-mL syringe was used for enhanced accuracy when delivering quantities below 1-mL. The robotics 1-mL syringe accurately delivered materials down to 0.06 grams. Concentrated reactants were stored in specified vials of the vial rack. The Teflon™ vial rack held a total of 24 vials at 25-mL volume each. The concentrated reactants were typically stored in vials 1–X, with the remaining vials used for the diluted samples that were being injected into the process streams as finite pulses. The robot activities were coordinated via software programming. This scheduler programming instructed the robot which sample to prepare or load next.

13.4.1.3 Reactor

We designed this system to allow various ~1 mL scale continuous process equipment types (e.g., reactors, separators) to be modularly inserted. The microreactor jacket was connected to a heat transfer system that consisted of a heating and cooling bath connected to an external recirculating pump. The downstream pressure was controlled using a pressure control valve coupled with a pressure transducer.

13.4.1.4 On-line GC Analysis

The product stream exiting the reactor flowed through an analytical sampling valve. Between analyses, the analytical system circulated the transfer solvent. As the apex of the sample plug passed through the sample injection valve, the valve was switched to the inject position for separation and analysis by GC or LC. Evaluation of the overall processing cycle time showed that decreasing the volume of the analytical sample loop from 100 to 10 µL could reduce the overall processing cycle time by up to 5 hours per sample. The product disposition valve could be used to capture a product sample in a vial by directing flow to the fraction collector, to recycle flow back to feed container R, or to direct flow to a weighed product or waste container.

13.4.1.5 Process Control

A combination of computer hardware, software, and programming were used for process control, data acquisition, and operator inputs. The primary process control resided on a Control PC with set-points being delivered to the local equipment controllers. Connections between the Control PC and servers on the network enabled downloading of files from the Internet, backing up data and files to the Intranet, and remote process monitoring. Process monitoring occurred by monitoring trend screens locally or remotely. Trend screens were preprogrammed graphs of key process variables as a function of time (e.g., temperature, pressure, flow rate, and residence time). The entire process could be controlled by entering a single command at a time or could be automatically controlled via a recipe. Recipe control consisted of executing preprogrammed commands used to control the automated equipment (e.g., pumps, baths, robotics, injectors, analytical equipment, and feed/product container selections).

13.4.2
Experimental Evaluation

Two alcohols (hereafter referred to as Tracer 1 and Tracer 2) were chosen to be the tracers for a first round of tests, using water as the solvent. Subsequent experiments varied the lag time between the injection of the tracers (Tracer 1 from one injector and Tracer 2 from the other injector) and the sampling of the reactor effluent. The objective was to verify that the chosen reactor volume (31.4 µL using 0.007" ID tubing), injector loop volumes (30.4 µL using 0.02" ID tubing), feed tubing volumes (2.5 µL each using 0.007" ID tubing), analytical sample loop volume (10 µL using 0.010" ID tubing), and target flow rate (20 µL min^{-1}) would provide steady-state concentrations in the reactor effluent. The delay time between feed injection and effluent analysis was varied by systematically changing the reactor

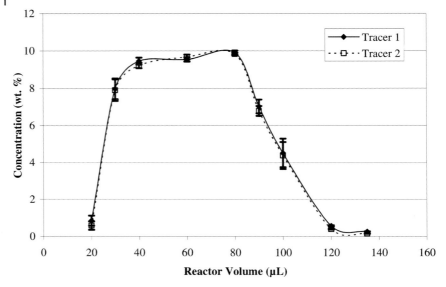

Fig. 13.6 Flat top experiment, stainless steel reactor.

volume entered into the Control PC. The Control PC then calculated the desired injection time based on the time for the center of the injected pulse to reach the center of the analytical loop. Thus, the lag time was the sum of the reactor volume, half of the combined injector loop volumes, the feed tubing volumes, and half of the analytical sample loop volume divided by the total flow rate, and entering a smaller reactor volume resulted in a shorter lag time. Figure 13.6 displays the results from this experiment.

As shown in Fig. 13.6, the mean lag time corresponds to a ~60 µL reactor volume, which is significantly higher than the 31 µL reactor volume calculated using the nominal tubing diameter. The inside diameter tubing tolerance for 0.007″ tubing is ±0.001″. Thus, the maximum expected reactor volume is 41 µL, which is still smaller than the ~60 µL observed in the lag time experiments.

The results of Fig. 13.6 were compared with a 1D axial dispersion model of the same process configuration (i.e., two injector loops, two lines from the loops to the mixing tee, the reactor tubing, and the sample loop tubing with the dimensions cited earlier). The output for the case of a 10 µL min^{-1} flow rate from each pump, with 10 wt.% Tracer 1 injected simultaneously from each injection loop, is shown in Fig. 13.7. The model predicted analytical loop peak concentration (i.e., exceeding 99% of the inlet injector concentration) arrived earlier than the peak concentration resulting from the actual process data.

Figure 13.8 displays the same data compared with the 1D axial dispersion model with the reactor tube diameter increased to match the mean residence time achieved in the experiments. Increasing the theoretical tubing diameter to 0.0096″ resulted in a reactor volume increase from 31.4 to 58.5 µL. This diameter is outside the ±0.001″ range of uncertainty provided by the manufacturer. While

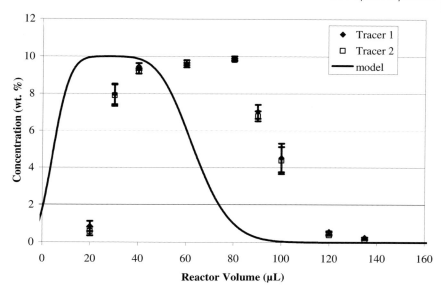

Fig. 13.7 One-dimensional axial dispersion model prediction compared to figure 6 data.

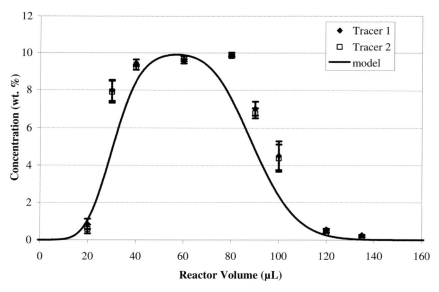

Fig. 13.8 One-dimensional axial dispersion model prediction with reactor volume fit to figure 6 data.

it is possible that the tubing is larger than the nominal diameter provided by the manufacturer (published at 0.007″), the following section identifies an alternative explanation.

Figure 13.8 also shows that the area under the curve is greater for the experimental data than that predicted by the model. This indicates a mass balance problem with the experimental procedure. This mass balance problem is also addressed in the next section.

13.4.2.1 Materials of Construction Effects

Glass-lined Stainless Steel Tubing

For the tests described in this section, an acetonitrile solvent and two new tracers (Tracers 3 and 4) were used. The stainless steel reactor from the previous experiments was replaced with 51.25″ of glass-lined steel tubing having an inside diameter between 0.008″ and 0.010″. This corresponded to a new reactor volume ranging from 42.2 to 65.7 µL. As before, we designed experiments to verify that this new combination of tubing and conditions (i.e., new glass-lined reactor volume, injector volume, analytical sample volume, and flow rate) would produce finite pulses of sufficient duration to allow the reactant and product concentrations in the effluent stream being sampled to attain non-dispersed concentrations (i.e., steady-state "flat top"). Our experiments varied the delay time between the reactant pulse injections and the analytical sample gathering by systematically varying the reactor volume entered into the Control PC. Figure 13.9 shows the results from this experiment.

For the Tracer 4 curve, the flat top occurred from approximately 122 to 162 µL. Surprisingly, this volume was not near the geometrically calculated 42.2–65.7 µL volume. The 40 µL width of the flat top was in the range of the sum of the two injector volumes, approximately 60 µL. However, the "full width at half maximum,"

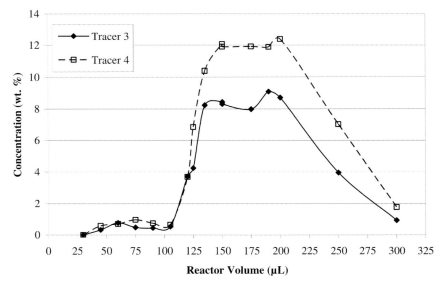

Fig. 13.9 Flat top experiment, glass lined reactor, teflon™ injectors and transfer line.

which should coincide with the sum of the injector volumes, corresponded to 100 μL. Integration of the two curves in Fig. 13.9 gives 218% and 150%, respectively, of the mass of Tracers 4 and 3 originally loaded in the injectors. These high values corroborate the higher than expected peak widths but are not easily explained. One hypothesis is that the polymeric injector loops have a significant capacity for these components within the tubing walls. Thus, during the loading of these loops, extra material can be absorbed into the polymer phase. This extra material then elutes when the injector loops are purged by the flowing solvent into the downstream reactor tubing.

Stainless Steel Tubing

If an absorption phenomenon is occurring, changing the materials of construction of the wetted surfaces could impact the results. The best reactor results (in which the reactor volume based on the tracer analysis most closely matches the geometrically calculated reactor volume) were attained with stainless steel materials of construction. The injector loops, tubing from the injectors to the mixing tee, and the reactor were all changed to stainless steel materials of construction of equivalent volumes. The flat top experiments were repeated for this configuration; Fig. 13.10 shows the results.

As shown, the stainless steel system gave the maximum concentration from approximately 50 to 94 μL. This was much closer to the geometrically calculated 42.2 μL to 65.7 μL, although it was still slightly larger than expected. This experiment conclusively demonstrated that materials of construction impact the lag time, supporting the absorption hypothesis. Integration of the two curves in Fig. 13.10 gives 159% and 114%, respectively, of the mass of Tracers 4 and

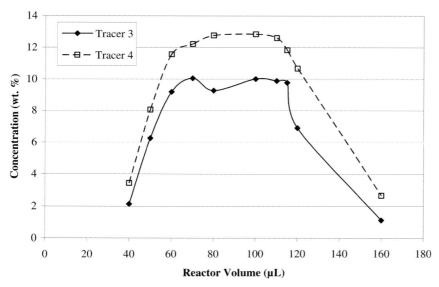

Fig. 13.10 Flat top experiment, all stainless steel wetted materials of construction.

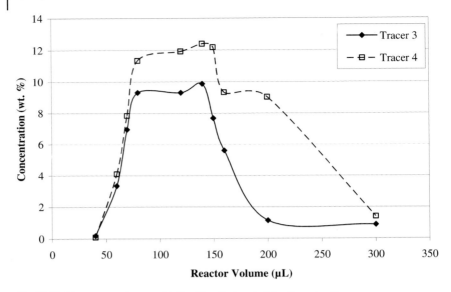

Fig. 13.11 Flat top experiment, all teflon™ wetted materials of construction.

3 originally loaded in the injectors. The Tracer 3 mass balance is within the uncertainties of the various measurements. The Tracer 4 mass balance may be primarily the result of a shift in the LC response factor.

Teflon™ Tubing
Next, we replaced the injector loops, tubing from the injectors to the mixing tee, and reactor with Teflon™ tubing of equivalent volumes. The flat top experiment was then repeated, with the results from this experiment plotted in Fig. 13.11.

The system with all Teflon™ tubing gave the maximum concentration from approximately 66 to 122 μL. Unexpectedly, this is a significantly longer lag time than obtained with the stainless steel system. Integration of the two curves in Fig. 13.11 gives 266% and 148%, respectively, of the mass of Tracers 4 and 3 originally loaded in the injectors. These high values provide further support for the hypothesis that these materials are being absorbed by the Teflon™ tubing.

13.4.2.2 Summary
The preceding discussion shows that the custom single channel reactor gave the expected low axial dispersion of injected reactants, which is consistent with the analysis of axial dispersion of Newtonian fluids in circular tubes as documented by Aris [18, 20]. However, experiments with different materials of construction demonstrated a variable lag time between the injection of a reactant pulse and its arrival at the analytical sample point as a function of the material of construction. The impact of this observation is that only certain materials of construction may be used for high throughput serial screening by injection of reactant pulses, dependent upon the chemistry employed.

13.4.3
Quantitative Measurements of Reaction Kinetics

To validate the utility of this reactor system for quantitative measurements of reaction kinetics, we evaluated the cumulative impact of errors introduced by the various system components. For this illustration, the variabilities in the pump flow rates, the initial reactant sample concentrations, and the reactor effluent analysis are included. Uncertainties in other process parameters (such as the reaction temperature) that may affect the measured rate are assumed to be negligible.

According to the literature for the syringe pumps, the flow reproducibility is ±5%. Assuming that this corresponds to the 95% confidence interval based on a set of five measurements with a normal distribution, the standard deviation for this distribution corresponds to 4.0% of the mean flow rate. For this analysis, the mean measured in those reproducibility tests is assumed to coincide with the set point.

We quantified the accuracy and precision of the automated sample preparation system by preparing samples of various quantities of water and weighing those samples after each addition of water. The data for the amount metered using the 1-mL syringe (typically used to deliver reactant during feed mixture preparation) are summarized in Table 13.2, where the number of samples taken for each target mass was four or five. Table 13.3 provides the results of similar experiments with the larger 5-mL syringe (used to deliver solvent during feed mixture preparation). As indicated by these two tables, the greatest sample concentration variability will result during preparation of dilute mixtures of reactant with the 1-mL syringe. For this error propagation analysis, this worst case variability will be used. The inaccuracy of the weight delivered by the robot syringe will be ignored because this is a systematic error that can be eliminated by calibration of the sample preparation system.

Finally, five replicate LC analyses with the two LC test components gave sample standard deviations of 5.6% and 5.4% for the two components. Assuming this

Table 13.2 Weight variability of replicate samples delivered through the 1 mL syringe.

Target mass (g)	0.06	0.08	0.1	1	4
Error of sample mean relative to target	−1.5%	−7.6%	−2.4%	−2.3%	−0.7%
Sample standard deviation relative to sample mean	3.5%	3.3%	0.4%	0.2%	0.2%

Table 13.3 Weight variability of replicate samples delivered through the 5 mL syringe.

Target mass (g)	0.4	0.7	1	10
Error of sample mean relative to target	−3.3%	−1.9%	−1.4%	−0.4%
Sample standard deviation relative to sample mean	0.3%	0.1%	0.1%	0.02%

variability is solely the result of error in the analysis and not in the sample preparation, the error propagation analysis uses a relative standard deviation of 5.5% for the measured concentrations.

For this illustration, a first order reaction of a single component will be chosen, where only the feed component concentration is measured. For a first order reaction in a plug flow reactor, the rate constant can be calculated from a single measurement as follows:

$$k_r = \frac{F \ln\left(\frac{c_{1f}}{c_{1i}}\right)}{V} \tag{16}$$

Since the initial concentration is actually the result of the combination of two streams from two different pumps, the form of this equation that reflects all of the system variables is as follows:

$$k_r = \frac{(F_1 + F_2) \ln\left[\frac{c_{1f}(F_1 + F_2)}{c_{1u} F_1}\right]}{V} \tag{17}$$

where c_{1u} is the concentration of component 1 in its feed injector loop. Thus, the expected standard deviation associated with the rate constant k_r can be calculated by performing an error propagation analysis [21]. Equation (18) is then the resulting equation representing the variance ($s_{k_r}^2$, the square of the standard deviation) of the computed result.

$$\frac{s_{k_r}^2}{k_r^2} = \frac{s_{F_1}^2 + s_{F_2}^2}{(F_1 + F_2)^2} + \frac{\left(\frac{s_{c_{1f}}}{c_{1f}}\right)^2 + \left(\frac{s_{c_{1u}}}{c_{1u}}\right)^2 + \left(\frac{s_{F_1}}{F_1}\right)^2 + \frac{s_{F_1}^2 + s_{F_2}^2}{(F_1 + F_2)^2}}{\left[\ln\left(\frac{c_{1f}(F_1 + F_2)}{c_{1u} F_1}\right)\right]^2} \tag{18}$$

Using the relative standard deviations listed earlier ($s_{F_1}/F_1 = s_{F_2}/F_2 = 0.04$, $s_{c_{1u}}/c_{1u} = 0.035$, and $s_{c_{1f}}/c_{1f} = 0.055$), the error in the computed value of k_r is $s_{k_r}/k_r = 0.12$. This error is significant if this value is to be applied to the design of a commercial scale reactor. If a rate constant with this error were used to design a full-scale reactor, the reactor would have to be over-designed relative to the volume predicted based on the computed value of k_r to compensate for this uncertainty. In this single reaction example, the full scale reactor volume is inversely related to the rate constant as indicated by Eq. (18). Thus, to achieve 95% confidence that the reactor volume will be large enough to achieve the desired conversion, the design volume must be set 31% larger than the volume calculated based on the mean computed value of k_r.

This analysis highlights the importance of quantifying and minimizing the variability in each system component. If those variabilities were vanishingly small, the true value of k_r could be computed based on a single experiment. However, if those variabilities are significant, such as in this example, multiple experiments will be required to determine the value of k_r with acceptable uncertainty. If experiments in this system were used to discriminate between competing mechanistic models, minimizing the system variability would increase the efficiency and reliability of the model discrimination process.

Although not illustrated here, systematic errors in the system components must be eliminated or known because those systematic errors translate directly to an error in the final computed result, such as the reaction rate constant. For example, if the pumps actually deliver only 90% of their set point value, the computed value of k_r will be $1/0.9 = 111\%$ of its actual value.

13.5
Conclusions

The custom single channel reactor was shown to give the expected low axial dispersion of injected reactants, which is consistent with the analysis of axial dispersion of Newtonian fluids in circular tubes as documented by Aris [18, 20]. However, experiments with different materials of construction demonstrated a variable lag time between the injection of a reactant pulse and its arrival at the analytical sample point as a function of the material of construction. We hypothesize that this effect is the result of adsorption or absorption of the reactants on or in the tubing walls. The impact of this observation is that only some materials of construction may be used for high throughput serial screening by injection of reactant pulses, dependent upon the chemistry employed.

For this high throughput serial screening approach to succeed, the observed discrepancy between geometrically calculated and experimentally measured reactor volumes must be resolved. If this issue cannot be resolved, quantitative reaction rate data cannot be generated using the pulse injection mode of operation.

Nomenclature

A	heat transfer area (m^2)
Bo	Bodenstein number
c_{1f}	effluent concentration of component 1 (mol m^{-3})
c_{1i}	inlet concentration of component 1 (mol m^{-3})
c_{1u}	undiluted concentration component 1 in its injector loop (mol m^{-3})
c_i	concentration of species i (mol m^{-3})
$D_{m,i}$	molecular diffusivity (of component i) (m^2 s^{-1})
d_t	reactor channel diameter (m)
D_z	axial dispersion coefficient (m^2 s^{-1})

E_a activation energy (J mol^{-1})
F_i volumetric flow rate (of stream i) (m^3 s^{-1})
k thermal conductivity of the reaction fluid (W m^{-1} K^{-1})
k_r reaction rate constant; m^3 mol^{-1} s^{-1} if second order, s^{-1} if first order
L reactor channel length (m)
$\mathfrak{R}_{max}$ reaction rate maximum (mol m^{-3} s^{-1})
$\mathfrak{R}$ reaction rate (mol m^{-3} s^{-1})
r radial coordinate (m)
R reactor channel radius (m)
R_1 radial coordinate of annular interface between colaminar feed streams (m)
R_g universal gas constant, 8.314 J mol^{-1} K^{-1}
s_j standard deviation of variable j
T_b bulk temperature in the reaction fluid (K)
t_D characteristic diffusion time (s)
T_w wall temperature (K)
U overall heat transfer coefficient (W m^{-2} K^{-1})
V reactor volume (m^3)
$v_z(r)$ axial component of the velocity at radial coordinate r (m s^{-1})
$\bar{v}_z$ mean axial velocity (m s^{-1})
z axial coordinate (m)
ΔH heat of reaction (J mol^{-1})
ΔP pressure drop (Pa)
μ viscosity (kg m^{-1} s^{-1})
ν_{ij} stoichiometric coefficient of component i in reaction j
τ_{min} minimum residence time (s)

References

1 Symyx Parallel Polymerization Reactor, http://www.symyx.com/images/pdfs/Polyolefin.pdf.
2 Argonaut Endeavor® Catalyst Screening System, http://www.argotech.com/products/chemical/endeavor.html.
3 Chemspeed AutoPlant A100 (http://www.chemspeed.com/content/products/products_autoplant_a100.shtml) and Accelerator SLT100 Synthesizer (http://www.chemspeed.com/content/products/products_slt100.shtml).
4 Mettler-Toledo MiniBlock XT Solution Phase Synthesizer, http://www.bohdan.com/miniblockxt.pdf.
5 P. L. Mills, J. F. Nicole, A novel reactor for high-throughput screening of gas–solid catalyzed reactions, *Chem. Eng. Sci.* **2004**, 59, 5345–5354.
6 H. Song, J. D. Tice, R. F. Ismagilov, A microfluidic system for controlling reaction networks in time, *Angew. Chem. Int. Ed.* **2003**, 42, 767.
7 H. Pennemann et al., Investigations on pulse broadening for catalyst screening in gas/liquid systems, *Chem. Eng. Sci.* **2004**, 50, 1814–1823.
8 B. W. Wojciechowski, The temperature scanning reactor I: Reactor types and modes of operation, *Catal. Today* **1997**, 36, 167–190.
9 N. M. Rice, B. W. Wojciechowski, The temperature scanning reactor II: Theory of operation, *Catal. Today* **1997**, 36, 191–207.

10 S. P. Asprey, N. M. Rice, B. W. Wojciechowski, The temperature scanning reactor III: Experimental procedures and data processing, *Catal. Today* **1997**, 36, 209–226.

11 W. Ferstl et al., Development of an automated microreaction system with integrated sensorics for process screening and production, *Chem. Eng. J.* **2004**, 101, 431–438.

12 S. Taghavi-Moghadam, A. Kleemann, K. G. Golbig, Microreaction technology as a novel approach to drug design, process development and reliability, *Org. Proc. Res. Dev.* **2001**, 5, 652–658.

13 J. Ruzicka, Discovering flow injection. from sample to live cell and from solution to suspension, *The Analyst* **1994**, 119, 1925.

14 D. E. Mears, Diagnostic criteria for heat transport limitations in fixed bed reactors, *J. Catal.* **1971**, 20, 127–131.

15 See Figure 22-3, p. 348 in C. O. Bennett, J. E. Myers, *Momentum, Heat, and Mass Transfer*, 3rd ed., McGraw-Hill, New York, **1982**.

16 B. A. Finlayson et al., Mathematics, in *Perry's Chemical Engineers' Handbook*, 7th ed., Eds. R. H. Perry, D. W. Green, McGraw-Hill, New York, **1997**.

17 E. L. Cussler, *Diffusion: Mass Transfer in Fluid Systems*, Cambridge University Press, Cambridge, **1984**, p. 116.

18 O. Levenspiel, *The Chemical Reactor Omnibook*, OSU Book Stores, Inc., Corvallis, Oregon, **1993**.

19 D. E. Mears, The role of axial dispersion in trickle-flow laboratory reactors, *Chem. Eng. Sci.* **1971**, 26, 1361–1366.

20 R. Aris, *Proc. R. Soc.* **1956**, 235A, 67.

21 D. A. Skoog, D. M. West, F. J. Holler, *Fundamentals of Analytical Chemistry*, Saunders College Publishing, Philadelphia, **1996**, p. A-34.

14
Novel Systems for New Chemistry Exploration

Paul Watts

14.1
Introduction

The success of the pharmaceutical industry depends largely on the synthesis and screening of novel chemical entities and their ability to optimize selected leads into marketable drugs. In a market where development costs are exceptionally high, the ability to shorten optimization times would be desirable to reduce this cost. Furthermore, new technology that would enable a cost-effective upward step-change in the number of lead candidates that could be prepared would provide an obvious competitive advantage, as this would potentially enable the discovery of a better lead with enhanced therapeutic properties, with reduced side effects.

So how can a chemist make more compounds for evaluation? Clearly the pharmaceutical industry has invested heavily in automation; however, one of the slowest steps is still the synthesis, purification and isolation of potential drug candidates. Several pharmaceutical companies have acknowledged that new technology, such as microreactors, offer many fundamental and practical advantages of relevance to automated high throughput synthesis and product purification.

Microreactors consist of a network of micron-sized channels etched into a solid substrate [1, 2]. They may be fabricated from a range of materials, including glass, silicon, quartz, metals and polymers, using various fabrication techniques, including photolithography, hot embossing, powder blasting, injection molding and laser micro forming [3]. For glass microreactors, which are the most common for synthetic applications involving the use of organic solvents, photolithographic fabrication of channel networks is generally performed [4, 5].

For chemical synthesis the channel networks are connected to a series of reservoirs containing chemical reagents to form the complete device with overall dimensions of a few centimeters. Reagents can be brought together in a specific sequence, mixed and allowed to react for a specified time in a controlled region of the channel network using various methods, including electrokinetic or hydrodynamic pumping. For a newcomer to the field of microreactor technology,

Fig. 14.1 Glass microreactor suitable for hydrodynamic pumping.

the easiest way to operate a device is by hydrodynamic control, using syringe-type pumps to maneuver solutions around the channel network. If hydrodynamic pumping is required it is possible to thermally bond ceramic HPLC-type adaptors to the glass device (Fig. 14.1). Connecting syringe pumps to such devices is relatively easy and enables the reactor to be integrated with a HPLC system, for example. This type of microreactor is ideal for reaction optimization.

Nevertheless, this approach has the disadvantage of requiring either large external pumps or the complex fabrication of small moving parts within the reactor itself. Notably, although this approach is relatively easy if reacting just two solutions, it becomes far more complex to accurately control the fluidics when introducing more than three reagents into the device, which is required when conducting multistep reactions; in these situations much more care is required when designing the exact dimensions of the reactor channels.

A more elegant way of pumping solutions around a channel network is by electroosmotic flow (EOF) [6–9], using voltage sequences applied via electrodes placed within the reagent reservoirs (Fig. 14.2). This method has several significant advantages over hydrodynamic based pumping methods, as it can be easily miniaturized as no moving parts are involved and the required voltage sequences can be readily applied under automated computer control.

For a glass microreactor, the channel wall–solution interface normally has a negative charge, arising from ionization of surface silanol (SiOH) groups. This immobile surface charge (SiO–) attracts a diffuse layer (on the order of nm thick) of mobile, oppositely charged counter-ions in the solution adjacent to the channel wall (cations for a negatively-charged glass channel wall). As shown schematically in Fig. 14.3, application of an electric field along the channel length causes the nm thick layer of mobile cations to move towards the more negative electrode, which drags all of the intervening solution in the bulk of the channel with it; notably, however, this only works effectively in channels less than 300 μm in diameter. An important feature of EOF is that the liquid velocity is constant across the channel, except in the nm thick regions of the diffuse layer of counter-ions very close to

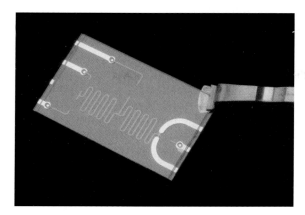

Fig. 14.2 Glass microreactor with integrated electrodes for EOF (LioniX BV).

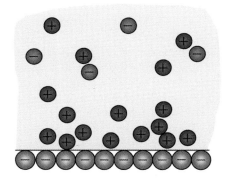 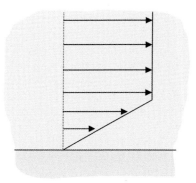

Fig. 14.3 Voltage-driven movement of the cations at the channel wall (left-hand side) produces a flat EOF velocity profile across the channel except within the nm thick diffuse counter-ion layer (right-hand side).

the wall. Unlike EOF, pressure-driven flow produces a parabolic velocity profile with high velocities in the channel centre and slow velocities near to the wall, giving rise to increased blurring of reagent zones along a channel length. The flat velocity profile associated with EOF is advantageous when *in situ* separation is required, as discussed later.

For EOF to be achieved, polar solvent types must be used. The EOF fluid velocity v_{eof} is given by Eq. (1):

$$v_{eof} = -\frac{E \varepsilon \varepsilon_0 \zeta}{\eta} \quad (1)$$

where E is the electric field (voltage divided by electrode separation), ε is the relative dielectric constant of the liquid, ε_0 is the permittivity of free space, ζ is the zeta potential of the channel wall–solution interface and η is the liquid viscosity.

Table 14.1 Relationship between magnitude of EOF and solvent properties.

Solvent	Dielectric constant	Viscosity (cP)	Polarity index (P)	Flow rate (μL min^{-1})
MeCN	37.5 (20 °C)	0.38	5.8	5.30
DMF	36.7 (25 °C)	0.92	6.4	1.67
EtOH	24.6 (25 °C)	1.10	5.2	0.90
THF	7.58 (25 °C)	0.55	4.0	1.00

Consequently, solvents that possess a high dielectric constant (i.e., polar solvents) and low viscosity (η) will have a higher flow rate (Table 14.1 and Fig. 14.4).

Figure 14.4 shows that the solvent flow rate is directly proportional to the field strength applied; as a result the flow rates within the channels can be easily controlled. Clearly, this limitation prevents nonpolar solvents such as hexane and dichloromethane from being used in EOF controlled microreactors.

Notably, however, under EOF control, charged solutes move with an electrophoretic velocity in addition to the electroosmotic velocity of the solvent. An elegant example of this was demonstrated by Fletcher et al. who reacted Ni^{2+} ions [from Ni(NO$_3$)$_2$] with pyridine-2-azo-p-dimethylaniline (PADA) within an EOF based microreactor to produce a [NiPADA]$^{2+}$ complex [10]. The authors reacted 2 mM Ni^{2+} with 2 mM PADA, which in a batch reaction would produce product (assuming a 100% conversion) of 1 mM concentration; however, within the microreactor they report that the product was produced in 12 mM concentration. This is because the positively charged Ni^{2+} ions move with a higher electrophoretic velocity than the neutral PADA molecules, hence the Ni^{2+} ions move through the PADA solution, leading to preconcentration of the product within the channel. The importance of this is that, within EOF-based reactors, concentration gradients are non-uniform, which may be exploited in chemical processes, as discussed later.

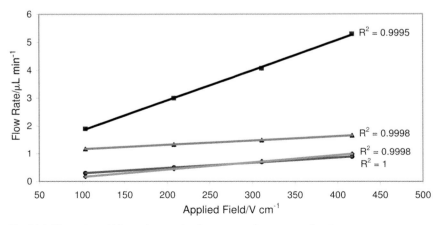

Fig. 14.4 Flow rates within a microreactor for a range of common solvents.

Table 14.2 Comparison of reaction rates and yields for batch and microreactions.

Reaction	In flask		In microreactor channel	
	Time	Conversion (%)	Time	Conversion (%)
Suzuki	8 h	70	6 s	78
Wittig	3 h	48	3 s	60

A concerted effort is now underway to establish the benefits that microreactors can bring to the field of reaction chemistry. Many reactions (reviewed in detail in Refs. [1, 2, 11–13]) have been demonstrated to show enhanced reactivity, product yield and selectivity when performed in microreactors as compared with conventional bench-top glassware. But in the pharmaceutical industry the speed at which candidates can be prepared (and screened) is clearly most critical and this is where microreactors offer a real advantage. For instance, in the Wittig reaction the yield was of the order of 70% for both batch and micro-reactions; however, in the microreactor the product was generated in approximately 6 s compared with several hours for the batch reaction [14]. Comparable results were observed in the Suzuki reaction (Table 14.2) [15]; clearly this would enable more compounds to be prepared in a given period of time, which is one of the aims if the pharmaceutical industry wishes to screen more compounds.

In addition to the enhanced speed of synthesis, a microchannel system also provides a potential separation column and a non-turbulent environment for partition between solvents. Integration of a microreactor device, via purification, to one of the many highly sensitive microchannel-based biological assay systems would enable the compounds to be screened. Apart from the greatly reduced reaction times demonstrated for the microreactors, handling times to assay and chemical reagent costs would be virtually eliminated; as shown diagrammatically in Fig. 14.5.

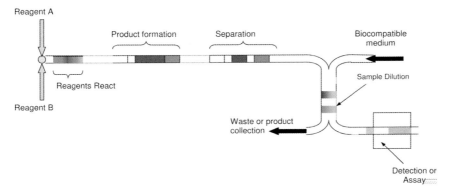

Fig. 14.5 Integration of a microreactor to an assay system.

In lead optimization using conventional batch technology, irrespective of whether parallel or iterative mode is chosen, validation and optimization of reactions tends to be one of the major rate-limiting steps. Based on the model described in Fig. 14.5, however, it can be seen that if the biological assay was replaced by a chemical assay and the conditions not the reagents are varied, then an auto-optimization could be carried out. Replication of the optimized channel as a parallel bundle could then provide a means of amplifying the amounts of material generated.

14.2
Chemical Synthesis in Microreactors

14.2.1
Synthesis of Pyrazoles

To develop the above methodology, Garcia-Egido et al. have recently reported the synthesis of a combinatorial library of pyrazoles within a glass microreactor system operated using hydrodynamic control [16]. A T-shaped microreactor was used to react seven 1,3-dicarbonyl compounds **1** sequentially with three hydrazine derivatives **2** to produce a 21-member library of pyrazoles **3** (Scheme 14.1). The automated system consisted of an autosampler to introduce small volumes of reagents into the chip, a pump to move the reagents through the microreactor, and a dilution system to enable a small sample to be diverted to an LC-MS instrument for analysis. In most cases the pyrazoles **3** were obtained in 99% conversion, but clearly an additional chromatography step would allow products to be further purified. However, to obtain these high yields the authors found it necessary to use 80 equivalents of the hydrazine reagent. To achieve their final aim, the purified products would need to be analyzed using an integrated flow-through bioassay system to enable *in situ* screening to be performed.

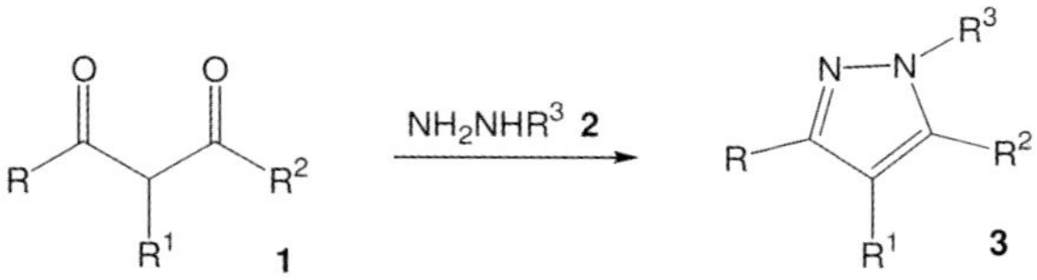

Scheme 14.1 Synthesis of pyrazoles.

For completeness, readers may be interested to compare the synthesis of pyrazoles in an EOF based system where Wiles et al. have reacted a series of 1,3-diketones with hydrazines, to prepare pyrazoles in > 98% conversion using only stoichiometric equivalents of the reagents [17], although a detailed study to compare the difference in reaction efficiency has not been conducted.

Clearly, this advanced type of system is suitable for drug discovery applications and could be applied to a range of other reactions performed within microreactors

14.2 Chemical Synthesis in Microreactors

under hydrodynamic control [18–24]. Although the above system is by far the most sophisticated development to date, cynics argue that the overall system is hardly miniaturized; the microreactor itself is tiny but the overall system is still composed of several pieces of large bench-top instrumentation. This is where EOF-based systems are potentially advantageous as external pumps are not necessary and purification could be achieved using on-chip electrophoretic separation, rather than using large external instrumentation, such as the HPLC/MS described above.

14.2.2
Peptide Synthesis

To develop an EOF based system, Watts et al. have conducted an extensive study on peptide synthesis, where they prepared a library of peptide derivatives within a computer-controlled microreactor system operating under EOF [25–28]. The authors demonstrated that dipeptides could be prepared from pre-activated carboxylic acids. They optimized the reaction using the pentafluorophenyl (PFP) ester of Fmoc-β-alanine **4** with amine **5** to give dipeptide **6** quantitatively in 20 min (Scheme 14.2). This represented a significant increase in yield compared with the traditional batch synthesis, where only a 50% yield was obtained in 24 h. The authors then used the methodology developed to consecutively react other pentafluorophenyl esters with amine **5** to produce a library of dipeptides. Other amines could also be used in a similar way, demonstrating the versatility of the technique to very rapidly prepare numerous related compounds [26].

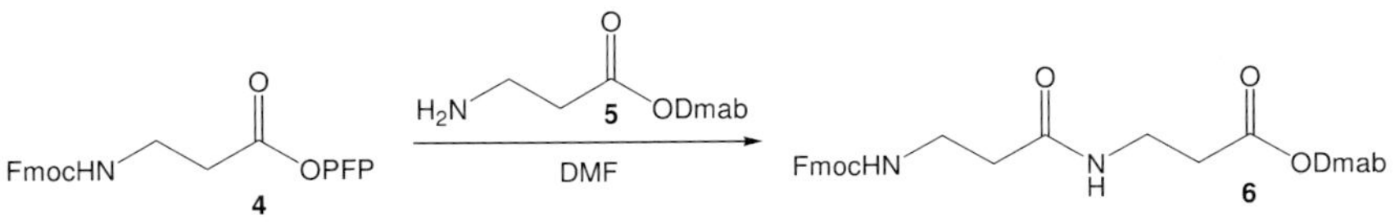

Scheme 14.2 Peptide synthesis.

More significantly, the first example of a multistep synthesis in an EOF-based microreactor was demonstrated, where the authors extended the approach to the synthesis of tripeptide **8** [26]. Pentafluorophenyl ester **4** reacted with amine **5** to form dipeptide **6**, which was reacted *in situ* with DBU to effect Fmoc deprotection. The amine **7** was then reacted *in situ* with another equivalent of **4** to prepare tripeptide **8** in 30% overall conversion (Scheme 14.3). Clearly this approach could be applied to a wide range of amino acids to rapidly produce much larger libraries of peptides. More generally, this approach could be used to generate highly unstable or highly toxic reagents *in situ* within a reactor, enabling reactions to be conducted more safely.

Although all dipeptide bond-forming reactions produced the appropriate dipeptide in 100% conversion, based on consumption of the pentafluorophenyl ester **10**, the product was still contaminated with residual amine **4** and pentafluorophenol (the by-product of the reaction). George et al. have reported that the dipeptide **9** may be separated from the reaction mixture using the reactor manifold

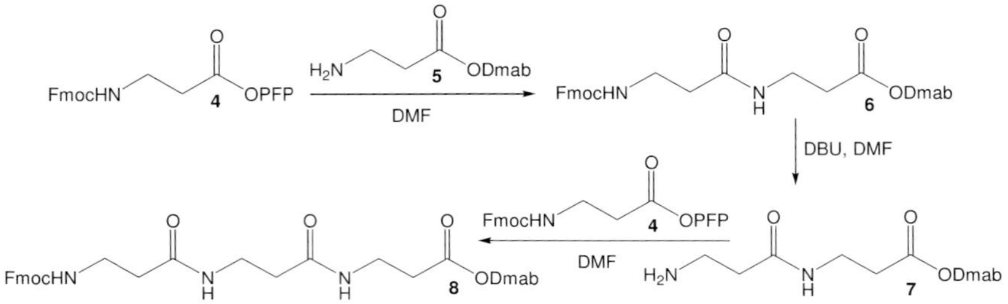

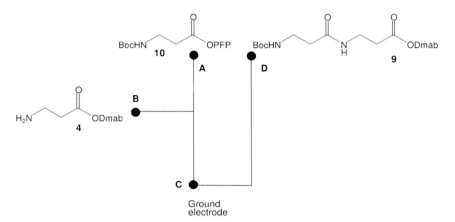

Scheme 14.3 Multistep peptide synthesis.

Fig. 14.6 Microreactor manifold for synthesis and electrophoretic purification of peptides.

shown schematically in Fig. 14.6, where the reaction mixture was collected in the ground reservoir during the synthesis and then the peptide is purified by electrophoresis and collected in reservoir D [28]. Hence this methodology enabled the synthesis and separation to be efficiently conducted within an integrated microreactor without the need for large peripheral equipment attached. However, further research is still needed to investigate integration of bioassay devices into this type of system. Environmentally, such methodology is advantageous as no additional solvent is used in the purification procedure; hence such technology is of interest in the area of "green" chemistry.

Having demonstrated that amide and peptide bonds could be successfully formed using microreactor technology, the authors then investigated racemization in reactions involving α-amino acids [27]. Reaction of the pentafluorophenyl ester of (R)-2-phenylbutyric acid **11** with α-methylbenzylamine **12**, gave the product **13** in quantitative conversion with 4.2% racemization (Scheme 14.4). Importantly this represented less racemization than observed in the batch reaction at the same concentration and temperature. The reduced level of racemization was attributed to the reduced reaction times observed within the microreactors; this clearly demonstrates that there are other advantages associated with the technology.

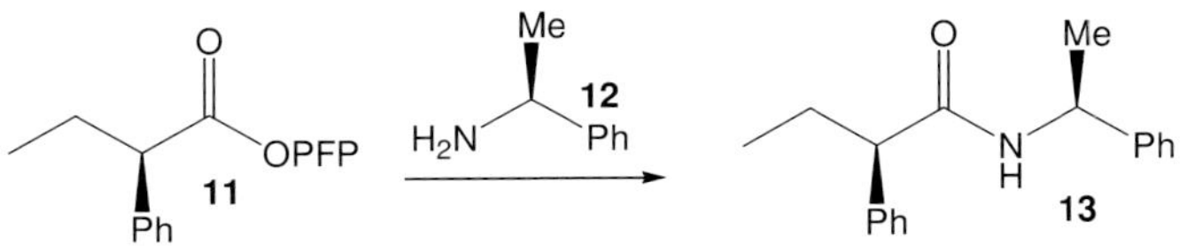

Scheme 14.4 Racemization in the synthesis of amides.

Naturally, the above method of integrating the chemical reaction and purification, through electrophoretic separation, could be applied to a range of other reactions that have been performed within EOF pumped devices [29–33]. However, when conducting multistep reactions it needs to be determined if it is necessary to sequentially purify intermediates as they are formed or whether it is possible to simply purify the product at the end of the reaction sequence.

14.2.3
Reaction Optimization

Another key area that microreactor technology is ideal for is reaction optimization. Wiles et al. [34] have demonstrated this with regard to carbanion chemistry, which is one of the most common methods of C–C bond formation used in the chemical and pharmaceutical industry. In such reactions, large volumes of highly pyrophoric bases are frequently employed, with the consequence that large quantities of heat are generated, meaning that careful temperature control is required to prevent by-product formation. Hence, microreactors have a considerable attraction for these reactions because the reactor enables excellent temperature control of the reaction. An elegant example of reaction optimization is given by the preparation of enolates from a series of 1,3-diketones using an organic base and their subsequent reaction with various Michael acceptors such as **14** to afford 1,4-addition products within a microreactor (Scheme 14.5) [34].

When using a continuous flow of the reagents **14** and **15**, only 15% conversion into the adduct **17** was observed, compared with 56% when the diketone **16** was reacted with **14** to form the Michael adduct **18**. The authors demonstrated enhancements in conversions through the application of a stopped-flow technique.

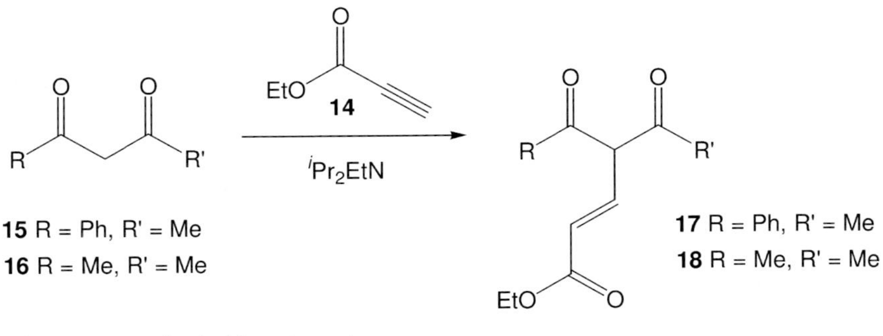

15 R = Ph, R' = Me
16 R = Me, R' = Me

17 R = Ph, R' = Me
18 R = Me, R' = Me

Scheme 14.5 Michael addition in a microreactor.

This procedure involved the mobilization of reagents through the device for a designated period, using an applied field, and the flow was subsequently paused by the removal of the applied field, prior to re-applying the field. Using the regime of 2.5 s on and 5 s off, the conversion to the product **17** was improved to 34%, while lengthening the stopped-flow period to 10 s, resulted in a further increase to 100%. This optimization process would have taken typically 30 min, far less than if batch reactions were being evaluated. This was compared with the preparation of **18**, in which the regime of 2.5 s on and 5 s off resulted in an increase in conversion to 95%. This demonstrated that the enolate of 2,4-pentanedione **16** was more reactive than the corresponding enolate of benzoyl acetone **15**. The authors propose that the observed increase in conversion, when using the technique of stopped flow, was due to an effective increase in residence time within the device, corresponding to the different kinetics associated with these reactions. This approach is clearly relevant to those wishing to study reaction kinetics of such reactions. Furthermore, when the reactions were conducted in batch a side product was formed (in up to 20% yield) from the reaction of the base (^{i}Pr$_2$EtN) with the Michael acceptor **14**; because the reagents are added spatially within the microreactor no by-products were detected and as a result the desired products were obtained in higher yield.

14.2.4
Stereochemistry

Microreactor technology has also been applied in some reactions to be able to more accurately control the stereoselectivity of reactions. Skelton and coworkers have reported the application of microreactors for the Wittig reaction [14, 35]. The authors used the microreactor to prepare the *cis*- and *trans*-nitrostilbene esters **19** and **20** using the Wittig reaction (Scheme 14.6). Several features such as stoichiometry and stereochemistry were investigated. When two equivalents of the aldehyde **22** to the phosphonium salt **21** were used in the reaction, a conversion of 70% was achieved. The microreactor demonstrated an increase in reaction efficiency of 10% over the traditional batch synthesis. The reaction stoichiometry

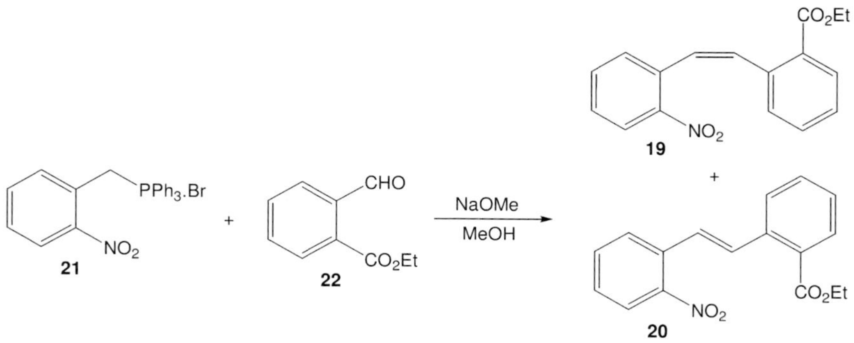

Scheme 14.6 Wittig reaction.

was subsequently reduced to 1 : 1, but using a continuous flow of reagents, as above, the conversion was poor (39%). The conversion was increased to 59% using an injection technique, where "slugs" of the phosphonium salt **21** were injected into a continuous flow of the aldehyde **22**.

The research was further extended to investigate the stereochemistry of the reaction. The ratio of isomers **19** and **20** was controlled by altering the voltages applied to the reagent reservoirs, which in turn affected the EOF and electrophoretic mobility of the individual reagents. The variation in the external voltage subsequently altered the relative reagent concentrations within the device, producing cis/trans ratios in the region 0.57 to 5.21. In comparison, the authors report that a traditional batch synthesis, based on the same reaction time, concentration, solvent and stoichiometry, afforded a cis/trans ratio of approximately 3 : 1 in all cases. This demonstrated that significant control is possible in a microreactor operated by EOF compared with batch reactions.

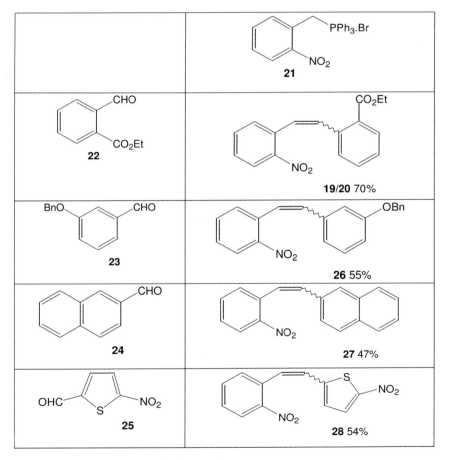

Fig. 14.7 Library of stilbenes produced using the Wittig reaction.

The authors then prepared a small library of compounds by sequentially substituting different aldehydes in the reaction. Reaction of 2-nitrobenzyl(triphenylphosphonium) bromide **21** with 3-benzyloxybenzaldehyde **23**, 2-naphthaldehyde **24** and 5-nitrothiophene-2-carboxaldehyde **25** produced stilbenes **26–28** in 55%, 47% and 54% overall conversion, respectively (Fig. 14.7). These results show that for a series of aromatic aldehydes of comparable reactivity, the products are obtained in similar conversions, which demonstrates the versatility of the methodology for the preparation of larger libraries of compounds.

14.3
Chemical Synthesis in Flow Reactors

Although different approaches to compound purification have been evaluated within microreactors, such as the HPLC purification and electrophoretic separation as discussed above, the question remains as to if it necessary to formerly "purify" all reactions? Wiles et al. have reported the solution phase Knoevenagel reaction within an EOF based microreactor, using the reaction manifold illustrated in Fig. 14.8. Reaction of diketone **29** with dimethylamine **30** formed the enolate, which was reacted *in situ* with one equivalent of aldehyde **31** to give the Knoevenagel adduct **32** in quantitative conversion. However, the product was still contaminated with dimethylamine. The consequence of this, and indeed the vast majority of reactions conducted in microreactors, is that a batch-type purification is then required; this clearly detracts from the many advantages that can be achieved through the technology.

This problem could, evidently, be circumvented by incorporating a supported catalyst within the microreactor. To create a catalyst bed within a microreactor, the device was fabricated from a top block and two etched plates (placed back to back), which enabled a deeper catalyst bed to be achieved (Fig. 14.9). The main channels were 130 µm wide and 50 µm deep and the catalyst bed was 800 µm wide, 100 µm deep and 10 mm long.

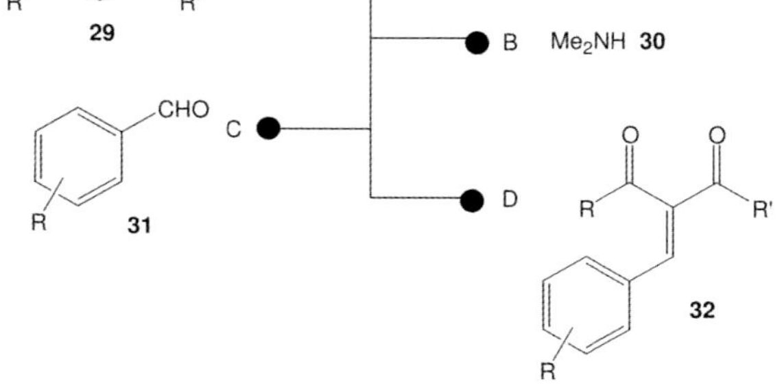

Fig. 14.8 Microreactor manifold for solution-phase Knoevenagel reactions.

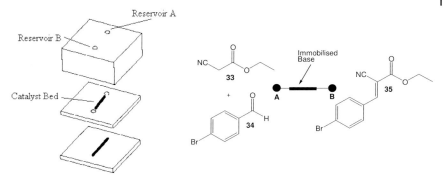

Fig. 14.9 Schematic of reaction manifold for Knoevenagel reactions.

The catalyst bed was subsequently packed with piperazine functionalized silica, and ethyl cyanoacetate **33** and 4-bromobenzaldehyde **34** were reacted in acetonitrile to produce ethyl 3-(4-bromophenyl)-2-cyanoacrylate **35** [36]. The reaction was investigated at a range of field strengths (i.e., flow rates) to optimize the conditions. As expected, as the field strength is reduced the conversion into product increases. After optimizing the reaction, a series of other Knoevenagel reactions were conducted using other aldehydes [36]. Reaction of ethyl propiolate with benzaldehyde, 3,5-dimethoxybenzaldehyde and 4-benzyloxybenzaldehyde generated products in 82 ± 2, 83 ± 4 and 51 ± 7% conversion, respectively.

In an extension of the study the group conducted an investigation using solid-supported bases in capillaries of various sizes to see if greater quantities of compound could be produced. They found that capillaries in the range of 500 μm to 5 mm in diameter (3 cm long) could be packed with various solid support bases and used in the Knoevenagel reaction (Fig. 14.10). Using EOF to move regents through the catalyst bed it was possible to prepare the product **32** in greater than 99% conversion and in 99% purity [37]. Potentially this approach to synthesis would mean that the product could be screened for biological activity without having to formally purify the reaction mixture, which would make the overall process even more efficient.

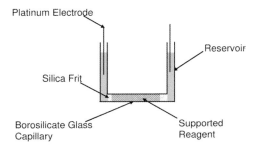

Fig. 14.10 Schematic of apparatus used in solid-supported Knoevenagel reactions.

Importantly, the authors demonstrated that the immobilized reagents suffered less physical damage within the flow-through system compared with batch reactions. They conducted fourteen sequential reactions, obtaining an average conversion of 99.1% with an RSD of 0.65. Furthermore, significantly, EOF pumping means that no backpressures are generated, making it is far easier to use this approach than a hydrodynamically pumped system. Another advantage of this type of "microreactor" is that larger quantities of compound can be readily prepared. In the 500 μm capillary 28 mg was prepared in 4 h, while in the 5 mm capillary 500 mg was synthesized in just 15 min. Notably, in open (unpacked) capillaries EOF is restricted to channels of less than 300 μm; however, these results demonstrate that EOF can be used in much larger systems when the capillary is packed with a reagent. Clearly, this approach to synthesis allows more moderate quantities of compounds to be prepared if required, and full instrumental characterization may be easily performed using traditional techniques such as NMR, IR etc. The ability to prepare modest quantities of analytically pure compounds without laborious purification procedures provides a cost-efficient tool for the synthesis of intermediates.

Having elucidated the optimum conditions for the Knoevenagel reaction in a flow reactor, a range of other reactions using different activated methylene derivatives and aldehydes (Table 14.3) was conducted. In all cases excellent product purities and yields were obtained. The reaction of benzaldehyde and ethyl cyanoacetate was also performed using 3-(dimethylamino)propyl-functionalized silica gel, 3-aminopropyl-functionalised silica gel, 3-(1,3,4,6,7,8-hexahydro-2H-pyrimido[1,2-1]pyrimidino)propyl-functionalized silica gel and polymer-supported diazabicyclo[2.2.2]octane, whereby excellent conversions were obtained (> 99.0%) in all cases [37].

Table 14.3 Synthesis of α,β-unsaturated compounds in a continuous flow reactor.

Aldehyde	Activated methylene	Conversion (%)	Yield (%)
Benzaldehyde	Ethyl cyanoacetate	99.98	99.70
4-Bromobenzaldehyde	Ethyl cyanoacetate	99.96	99.35
Methyl-4-formyl benzoate	Ethyl cyanoacetate	100.0	99.80
3,5-Dimethoxybenzaldehyde	Ethyl cyanoacetate	99.86	99.89
4-Benzyloxybenzaldehyde	Ethyl cyanoacetate	99.99	99.67
Benzaldehyde	Malononitrile	99.98	99.40
4-Bromobenzaldehyde	Malononitrile	99.89	99.79
Methyl-4-formyl benzoate	Malononitrile	100.0	98.84
3,5-Dimethoxybenzaldehyde	Malononitrile	99.87	99.17
4-Benzyloxybenzaldehyde	Malononitrile	100.0	99.73

Table 14.4 Synthesis of dimethyl acetals in a continuous flow reactor.

Starting Material	Conversion (%)	Yield (%)
4-Bromobenzaldehyde	99.60	99.73
Benzaldehyde	100.0	99.21
4-Cyanobenzaldehyde	99.84	99.24
4-Chlorobenzaldehyde	99.96	99.45
4-Benzyloxybenzaldehyde	99.93	99.65
3,5-Dimethoxybenzaldehyde	99.93	99.44
2-Naphthaldehyde	99.94	99.84
Methyl-4-formyl benzoate	99.92	99.42
5-Nitro-2-thiophenecarboxaldehyde	99.95	99.80
trans-Cinnamaldehyde	99.93	99.44
Propiophenone	99.93	99.88

The methodology was then applied to acid-catalyzed reactions, namely the synthesis of acetals [38] from a premixed solution of aldehyde or ketone and trimethylorthoformate. The reaction products were subsequently collected and concentrated *in vacuo* before analysis of the crude product by NMR spectroscopy. As Table 14.4 illustrates, using this approach, excellent product purities and yields were again obtained.

Having demonstrated the preparation of an array of small organic compounds in high purity and yield, multistep reactions were then conducted by spatially incorporating two supported reagents into a continuous flow reactor. As Fig. 14.11 illustrates, the first step of the reaction consists of an acid-catalyzed acetal deprotection to afford the respective aldehyde and the second step involves the base-catalyzed condensation of the aldehyde with an activated methylene. Using this approach, 100% deprotection of the acetal to aldehyde was observed and 99% conversion of the aldehyde to the desired unsaturated product was observed to an give analytically pure product in 99.6% yield.

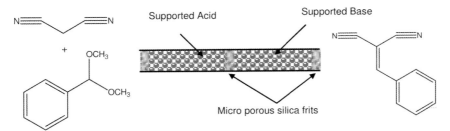

Fig. 14.11 Schematic of apparatus used for multistep reactions.

Compared with the use of pressure-driven flow, electroosmotic flow (EOF) is advantageous as it is simple to use, employs no mechanical parts and generates minimal back-pressure; the latter is particularly important with respect to the use of supported reagents as it prevents reagent degradation. Consequently, as no mechanical pumps are required the size of the operating system is greatly reduced to a power supply and the reactor. In addition, the reactor(s) can be computer controlled and operated remotely from a safe working distance, reducing the amount of valuable fume cupboard or bench space required.

14.3.1
Large-scale Manufacture

Current production technology is based on the scaling-up of successful laboratory scale reactions by firstly constructing a pilot plant, followed by a final increase in scale to enable production. This approach is, however, fundamentally flawed, as at each stage changes are made to the overall surface-to-volume ratio of the reactors, which in turn affects mass and heat transfer processes. These variations in reactor conditions, therefore, result in changes to the process, meaning that it is necessary to evaluate the process and reoptimize it at each stage of scale up process. Consequently, the route from bench to large scale production is both costly and time consuming. It is therefore postulated that through the application of microreactor technology, the transfer from laboratory to production would be both rapid and cost-effective as processes would initially be optimized on a single device; to increase the production capacity more devices would be employed – a technique referred to as "numbering up" or "scale out". With the number of techniques amenable to mass production increasing, the commercial availability of parallel reactors is starting to be realized. Along with the ability to reduce the transfer time between initial discovery and production, the scale out approach is also advantageous as it enables access to an array of features not commonly used in traditional scale out approaches, such as reduced reaction times and the ability to work in the explosive limit.

From a production perspective, the scale out approach is advantageous as it enables changes in production volume by simply increasing or decreasing the number of devices employed, therefore meeting the customer demand. Also, through the use of generic reactor designs, custom syntheses could be performed with relative ease. Compared with a production plant whereby reactors are configured for a single function, this flexibility is both advantageous and cost-effective.

Very few papers exist in the open literature where pharmaceutical companies describe examples of specific chemical reactions that have been performed within microreactors. A recent paper by Zhang et al. from Johnson and Johnson Pharmaceuticals [39] describes a series of reactions performed using the CPC microreactor system [13]. The paper details a wide variety of reactions that have been used successfully to prepare kilogram quantities of products. The system has been used to conduct exothermic reactions, high temperature reactions, reactions involving unstable intermediates and reactions involving unstable reagents.

14.4
Conclusions

Microreactor chemistry is currently showing great promise as a novel method on which to build new chemical technology and processes. Reactions performed in a microreactor invariably generate relatively pure products in high yield, in comparison with equivalent bulk reactions, in much shorter times. One immediate and obvious application is, therefore, in combinatorial chemistry and drug discovery, where the generation of compounds either with different reagents or under variable conditions is an essential factor. An interesting twist to the chemistry carried out to date is the opportunity to separate reactants and products in real time, which would enable rapid screening to be facilitated. As an extension of the technology it has been demonstrated that kilogram quantities of products can be produced without having to "scale up" the synthesis.

References

1. W. Ehrfeld, V. Hessel, H. Löwe, *Microreactors: New Technology for Modern Chemistry*, Wiley-VCH, Weinheim, **2000**.
2. R. E. Oosterbroek, A. van den Berg, *Lab-on-a-Chip: Miniaturised Systems for (Bio)Chemical Analysis and Synthesis*, Elsevier, Amsterdam, **2003**.
3. M. Madou, *Fundamentals of Microfabrication*, CRC Press, Boca Raton, FL, **1997**.
4. T. McCreedy, *TrAC* **2000**, 19, 396–401.
5. T. McCreedy, *Anal. Chim. Acta* **2001**, 427, 39–43.
6. J. Th.G. Overbeek, *Electrokinetic Phenomena In Colloid Science*, Vol. 1. ed. H. R. Kruyt, Elsevier, Amsterdam, **1952**, pp. 195–244.
7. C. L. Rice, R. Whitehead, *J. Phys. Chem.* **1965**, 69, 4017–4024.
8. R. J. Hunter, *Zeta Potential in Colloid Science*, Academic Press, London, **1981**.
9. J. Jednacak, V. Pravdic, W. Haller, *J. Colloid Interface Sci.* **1974**, 49, 16–23.
10. P. D. I. Fletcher, S. J. Haswell, X. Zhang, *Lab Chip* **2002**, 2, 102–112.
11. P. D. I. Fletcher, S. J. Haswell, E. Pombo-Villar, B. H. Warrington, P. Watts, S. Y. F. Wong, X. Zhang, *Tetrahedron*, **2002**, 58, 4735–4757.
12. K. Jahnisch, V. Hessel, H. Lowe, M. Baerns, *Angew. Chem. Int. Ed.* **2004**, 43, 406–446.
13. T. Schwalbe, V. Autze, G. Wille, *Chimia* **2002**, 56, 636–646.
14. V. Skelton, G. M. Greenway, S. J. Haswell, P. Styring, D. O. Morgan, B. Warrington, S. Y. F. Wong, *Analyst* **2001**, 126, 7–10.
15. G. M. Greenway, S. J. Haswell, D. O. Morgan, V. Skelton, P. Styring, *Sens. Actuators B* **2000**, 63, 153–158.
16. E. Garcia-Egido, V. Spikmans, S. Y. F. Wong, B. H. Warrington, *Lab Chip* **2003**, 3, 73–76.
17. C. Wiles, P. Watts, S. J. Haswell, *Org. Proc. Res. Develop.* **2004**, 8, 28–32.
18. E. Garcia-Egido, S. Y. F. Wong, B. H. Warrington, *Lab Chip* **2002**, 2, 170.
19. H. Lu, M. A. Schmidt, K. F. Jenson, *Lab Chip* **2001**, 1, 22.
20. R. C. R. Wootton, R. Fortt, A. J. de Mello, *Org. Proc. Res. Develop.* **2002**, 6, 187.
21. H. Hisamoto, T. Saito, M. Tokeshi, A. Hibara, T. Kitamori, *Chem. Commun.* **2001**, 2662.
22. S. Suga, M. Okajima, K. Fujiwara, J. Yoshida, *J. Am. Chem. Soc.* **2001**, 123, 7941.
23. J. Yoshida, S. Suga, *Chem. Eur. J.* **2002**, 8, 2651.
24. K. Kanno, H. Maeda, S. Izumo, M. Ikuno, K. Takeshita, A. Tashiro, M. Fujii, *Lab Chip* **2002**, 2, 15.

25 P. Watts, C. Wiles, S. J. Haswell, E. Pombo-Villar, P. Styring, *Chem. Commun.* **2001**, 990–991.
26 P. Watts, C. Wiles, S. J. Haswell, E. Pombo-Villar, *Tetrahedron* **2002**, 58, 5427–5439.
27 P. Watts, C. Wiles, S. J. Haswell, E. Pombo-Villar, *Lab Chip* **2002**, 2, 141–144.
28 V. George, P. Watts, S. J. Haswell, E. Pombo-Villar, *Chem. Commun.* **2003**, 2886–2887.
29 G. N. Doku, S. J. Haswell, T. McCreedy, G. M. Greenway, *Analyst* **2001**, 126, 14–20.
30 M. Sands, S. J. Haswell, S. M. Kelly, V. Skelton, D. O. Morgan, P. Styring, B. H. Warrington, *Lab Chip* **2001**, 1, 64–65.
31 N. G. Wilson, T. McCreedy, *Chem. Commun.* **2000**, 733–734.
32 C. Wiles, P. Watts, S. J. Haswell, E. Pombo-Villar, *Lab Chip* **2001**, 1, 100–101.
33 C. Wiles, P. Watts, S. J. Haswell, *Chem. Commun.* **2002**, 1034–1035.
34 C. Wiles, P. Watts, S. J. Haswell, E. Pombo-Villar, *Lab Chip* **2002**, 2, 62–64.
35 V. Skelton, G. M. Greenway, S. J. Haswell, P. Styring, D. O. Morgan, B. Warrington, S. Y. F. Wong, *Analyst* **2001**, 126, 11–13.
36 N. Nikbin, P. Watts, *Org. Proc. Res. Develop.* **2004**, 8, 942–944.
37 C. Wiles, P. Watts, S. J. Haswell, *Tetrahedron* **2004**, 60, 8421–8427.
38 C. Wiles, P. Watts, S. J. Haswell, *Tetrahedron* **2005**, 61, 5209–5217.
39 X. Zhang, S. Stefanick, F. J. Villani, *Org. Proc. Res. Develop.* **2004**, 8, 455–460.

15
Going from Laboratory to Pilot Plant to Production using Microreactors

Michael Grund, Michael Häberl, Dirk Schmalz, and Hanns Wurziger

15.1
Introduction

For some years now microreactors [1, 2] have been attracting the interest of the scientific community. Since microfabrication technology made microstructured fluidic devices available they have been studied for their scope and limitations. There have also been attempts to introduce such systems into the industrial setting [3].

Many aspects of microreaction technology have been reviewed recently [4]. The unique properties of miniaturized reaction systems have been discussed in the context of production in the multi-kilogram up to ton scale. We at Merck KGaA have tried to apply this technology to the production of fine chemicals [5], triggered by our very early access to a proprietary silicon micromixer [6] with unique mixing properties.

15.2
Nitration

15.2.1
General Remarks

The nitration of aromatic compounds is a fundamental reaction [7] of utmost importance to the chemical industry. Many different regimens for this unit-process are known [8]. Nitrations have been described in microreactors [9–11] and during our own work with microreactors we have also gained experience with nitrations [12]. We have shown that it is possible to generate, in the laboratory, smaller amounts of chemicals using microreactors, exemplified by the continuous nitration of 8.6 g of N-methoxycarbonyl-1,2,3,4-tetrahydro-isoquinoline over 6 full days. In an unlimited period of time one could produce unlimited amounts of chemicals with a single microsystem. Since this is unrealistic we are not convinced that microreactors are the means for the production of larger quantities of chemicals. Rather, we think the small dimensions and hold-up volumes of microreaction systems are of advantage in the optimization [13] of reactions at an early stage, when only small amounts of materials are available. In addition, we were intrigued by the idea of shortening the upscaling process.

We wanted to apply this idea to EMD 503982, a factor Xa-inhibitor [14] under development at Merck KGaA as an antithrombotic agent. Phase I has been finished recently and phase II is scheduled. A key component of this compound is 4-(4-aminophenyl)morpholin-3-one (3). This aniline can be generated by hydrogenation from the corresponding 4-(4-nitrophenyl)morpholin-3-one (2) [15], which in turn is accessible by nitration of the precursor 4-phenylmorpholin-3-one (1) [16] (Scheme 15.1).

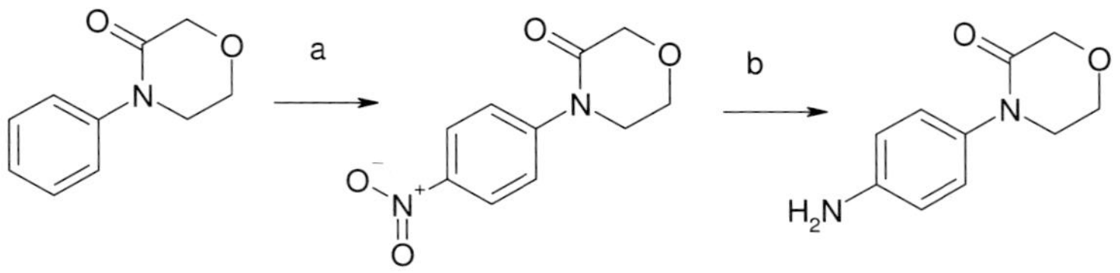

a: nitration
b: hydrogenation

Scheme 15.1 (a) Nitration; (b) hydrogenation.

For the phase 2 development work, larger kilogram amounts of the key component are needed, and we thought a continuous nitration process could be very advantageous for several reasons.

We wanted to find the optimal conditions for the nitration in the laboratory on a small scale and subsequently transfer these conditions directly into the much larger scale of a pilot plant.

Classical nitrations are performed in a batch reactor, where the nitrating species is added carefully to the bulk of substrate in a suitable solvent. The larger the batches, the longer the addition of the nitrating agent lasts. In the large-scale batch-nitration of 4-(phenyl)morpholin-3-one, increasing amounts of ring-opened by-products were observed. To minimize the formation of such difficult to remove byproducts the batches had to be kept below 500 L.

Another aspect is the safety of nitrations. Batches of several hundred liters bear a substantial oxidation hazard. Very often the nitration products are explosives. By using microreactors the holdup volume is reduced to the micro- or milliliter scale and even in the production scale this volume ranges in the lower liter scale.

For nitrations of liquids these are brought into contact with the nitrating agent directly. When solids are nitrated these normally have to be dissolved in a suitable solvent, which itself is not nitrated. Concentrated sulfuric acid is often used as a solvent for nitrations. It is a mild solvent (as long as it does not get in touch with water) with an excellent dissolving power and often enhances the speed of nitration reactions. However, it has some disadvantages in the workup phase. Sulfuric acid has to be diluted heavily with water and neutralized prior to disposal. During this process large amounts of heat have to be removed.

Acetic acid is also known as a solvent for nitrations. Unlike sulfuric acid this can advantageously be removed by distillation after the reaction is finished.

15.2.2
Orienting Laboratory Nitrations

15.2.2.1 Acetic Acid

Because of the expected better workup and disposal procedures, nitration in acetic acid was tried first. Stock solutions of 4-(phenyl)morpholin-3-one in glacial acetic acid (2 M) and nitric acid (65%) in concentrated sulfuric acid (4.8 M) were used. For the reaction optimization, two disposable polypropylene syringes (3 mL capacity) were filled with the different stock solutions and mounted on a syringe pump. The syringes were connected with the silicon micromixer that was connected with a residence time unit, a Teflon tube of known inner diameter and length. The flow of the syringe pump was adjusted according to the residence time required, the only variable. For simplicity, nitrations were carried out at room temperature. The whole system was allowed to reach equilibrium before 1 drop of efflux was collected in a HPLC autosampler-vial. This vial was charged with 2 mL of a 1 : 1 water–acetonitrile mixture to make sure the nitration is quenched immediately. The effectiveness of the quench was easily proven by recording a new HPLC chromatogram from the same vial 24 h after the first analysis: the different chromatograms were superimposable. In acetic acid the nitration seems to be rather sluggish (Table 15.1). The starting material is consumed fairly slowly with a low yield of the desired meta/para mixture. From the increase of the unwanted ortho isomer over time one can state that the selectivity is rather low. To find out whether nitration at a higher temperature influenced the results, the experiment was repeated at 52 °C (Table 15.2). Now the reaction rate was increased, but the ratio of the ortho isomer versus the meta/para isomers was even worse.

For the next experiment a 1 M solution of 4-(phenyl)morpholin-3-one in acetic acid was treated with neat fuming nitric acid. Then the nitrating species was used in roughly 24-fold molar excess. Now (Table 15.3) the reaction proceeds much faster.

Table 15.1 Nitration with 65% nitric acid in acetic acid at room temperature.

Time (min)	Educt (%)	Ortho (%)	Meta/para (%)
4.0	79.6	4.5	14.9
8.0	49.4	11.7	36.9
16.0	46.1	13.4	38.4
32.0	37.7	19.9	43.4

Table 15.2 Nitration with 65% nitric acid in acetic acid at 52 °C.

Time (min)	Educt (%)	Ortho (%)	Meta/para (%)
8.0	25.6	29.6	42.3
16.0	8.8	34.6	53.1

Table 15.3 Nitration with fuming nitric acid in acetic acid.

Time (min)	Educt (%)	Ortho (%)	Meta/para (%)
8.0	9.4	48.4	40.5
10.6	5.7	48.0	44.8
15.0	0.6	50.6	47.4

After 15 min reaction time the starting material is almost completely consumed but, disappointingly, the ratio of the positional isomers has changed again: the ortho isomer now is formed in excess over the meta/para isomers.

15.2.2.2 Acetic Anhydride
When the acetic acid was replaced by acetic acid anhydride under otherwise unchanged conditions, the reaction became extremely fast. Within a one-minute reaction time the starting material was consumed completely and only a small amount of the wanted para-mononitro compound was found besides a higher nitrated compound as the major product. Even if a more favorable selectivity would have been observed, and although a nitration with nitric acid in acetic anhydride can be carried out safely [17] with the micromixer, these conditions have to be avoided because the entire reaction heat is evolved immediately and very probably cannot be controlled on a larger scale.

15.2.2.3 Laboratory Optimization
For economic reasons it would be desirable to use reagents in equimolar amount to the substrate. In the laboratory setting, however, only one equivalent of nitrating species was not sufficient. The reaction was rather slow and starting material always remained. A greater than twofold excess of nitrating species accelerated the nitration reaction, but caused more side products.

From all the different test-reactions we learnt that the nitration should be sufficiently fast and para-selective. From all the preliminary experiments we decided to use a double molar excess of nitrating species. In the laboratory setup we used 1 M 4-(phenyl)morpholin-3-one in concentrated sulfuric acid and 2 M 65% nitric acid in concentrated sulfuric acid. It turned out that this solution was stable at least 3 months at room temperature, as evidenced by HPLC.

The optimal conditions (Table 15.4) for nitration were found in a residence time of 6 min and a double molar excess of nitrating species.

Table 15.4 Nitration with 65% nitric acid in concentrated sulfuric acid.

Time (min)	Educt (%)	Ortho (%)	Meta/para (%)
3.0	4.2	9.7	83.8
4.0	1.4	8.8	87.7
5.3	0.0	9.1	89.8
6.4	0.0	8.5	89.8

The raw HPLC data are not corrected, but it can be easily derived from the figures that the nitration in concentrated sulfuric acid is fairly fast. The starting material is consumed quickly with an almost constant ortho/meta-para ratio. Under the HPLC conditions used the meta and para isomer peaks were unresolved. The meta isomer appears with a retention time of 2.65 min. Figure 15.1 shows the chromatogram for the effluent of the nitration. There is some remaining starting material eluting at 2.5 min, followed by some ortho isomer at 2.57 min. At higher retention times higher nitrated products appear. Figure 15.2 shows the chromatogram of recrystallized material from isopropanol. Here the hump for the meta isomer is gone.

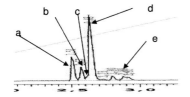

a: starting material
b: ortho-isomer
c: meta-isomer
d: para-isomer
e: higher nitrated products

Fig. 15.1 HPLC chromatogram of the crude reaction mixture.

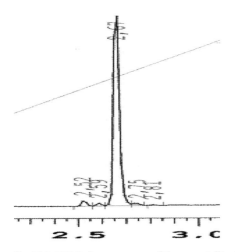

Fig. 15.2 HPLC chromatogram of the recrystallized product.

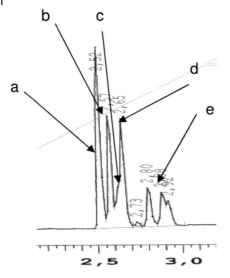

Fig. 15.3 HPLC chromatogram of the mother liquor.
(a) Starting-material enriched;
(b) ortho-isomer enriched;
(c) meta-isomer enriched;
(d) para-isomer depleted;
(e) higher nitrated products enriched.

Figure 15.3 shows the chromatogram of the mother liquor. Here the unwanted isomers are enriched and the peak of the desired para-product is depleted. These results show that the combination of nitrating conditions and workup-recrystallization can deliver the desired material with high purity (99.2%). In carrying out the experiments with the silicon micromixer the materials to be used could be reduced tenfold, and a time saving of 75% was achieved. In the larger laboratory-scale, where for safety reasons the reaction has to be optimized and larger amounts of product are desired, this advantage is lost [18].

15.3
Microreaction System "MICROTAUROS"

15.3.1
First Prototype Silicon Micromixer

The first prototype micromixer was a roughly 1 mm thick silicon plate with the dimensions of ca. 17 × 30 mm (Fig. 15.4). The first trivial task to be solved was how to introduce reagents and organic solvents into this micromixer. The first approach was to glue a hypodermic needle onto the silicon chip. The resulting construction was fairly fragile and had to be stabilized by a polymer cast.

Fig. 15.4 Silicon micromixer with glued-in hypodermic needles and cast stabilization.

This setup worked for aqueous solutions but not for organic solvents or aggressive reagents, which dissolved the glue or attacked the cast and the system became leaky. The cast was only applied in the inlet and outlet regions. The rest of the silicon body was not stabilized with the danger of braking.

15.3.2
Second Prototype Silicon Micromixer with Connector

The second prototype silicon mixer (Fig. 15.5) was slightly bigger and more robust. It had two rectangular slits next to the inlet and outlet holes. Through these slits the legs of an U-shaped plastic part were mounted to hold an adapter. This adapter was fitted with an olive as a connector for soft tubing. The adapter was forced against the silicon body of the micromixer using an excenter (Fig. 15.6). A leak-proof liquid connection was realized with an elastic O-ring. However, this connecting device seemed too fragile to withstand the harsh conditions we wanted to apply.

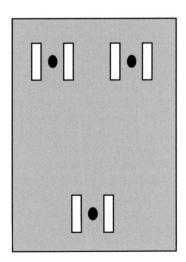

Fig. 15.5 Second mixer prototype with etched-through slits next to the inlet and outlet openings.

Fig. 15.6 Mounted connection system with U-shaped legs.

15.3.3
Optimized Connection System

From our experience with medium-pressure chromatography we were familiar with the Omnifit [19] tube fittings for universal Teflon® tubing. Therefore, we constructed a plastic cube with a threaded hole and small legs, which were fitted through the rectangular apertures in the silicon micromixer. A stabilizing counterplate was applied and the whole setup fixed by an attachment device (Figs. 15.7 and 15.8). By carefully hand screwing the Omnifit fitting into the threaded connecting cube we were able to establish a stable and tight connection up to 20 bar.

This connection system [20] ensures that chemicals come in contact with only silicon and Teflon®. Chemicals and reagents were applied with disposable syringes

Fig. 15.7 New connection system for etched-through silicon mixers, front view.

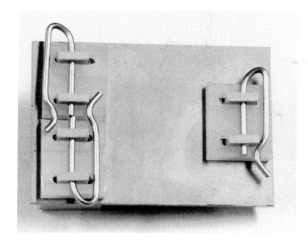

Fig. 15.8 New connection system for etched-through silicon mixers, back view.

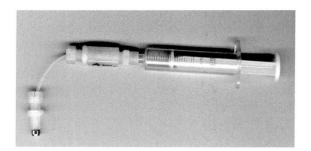

Fig. 15.9 Complete Omnifit connection system with disposable syringe.

that were connected by commercial Luer fittings (Fig. 15.9). For controlled flow rates stepper-motor driven syringe pumps [21] were used. The syringes were operated manually only for cleaning purposes.

With the prerequisite of safe handling of organic compounds we carried out several chemical reactions in the flow-through mode.

15.3.4
First Summary

At this point we summarize some experiences with the micromixing system: the reagents, starting materials and products must be homogeneously dissolved during passage through the system. Very often the products are less soluble than the starting materials in a given solvent. This may be overcome by the use of different reagents, solvents or dilution conditions. Bases such as triethylamine form hydrochlorides that are insoluble in many solvents, e.g., dichloromethane.

Pyridine as a base can often do the same job, but its hydrochloride is soluble. Polymerizations must be avoided. The reactions performed are often much faster than anticipated from the literature. Owing to the precise control of the reaction conditions, the yield and purity of products are generally better than in corresponding batch reactions.

15.3.5
Optimized Micromixer with an Advanced Connection System

At higher flow rates we observed a drastic rise in the backpressure of the micromixer, which eventually destroyed the silicon body of the mixer. This problem could be overcome by a new mixer-design with parallel channels (Fig. 15.10). The individual mixing structures (Fig. 15.11) were arranged in rows with six "split-

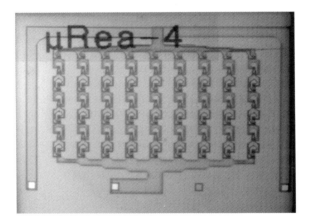

Fig. 15.10 New mixer design with parallel mixing cascades.

Fig. 15.11 Enlarged single mixing element, etched in silicon.

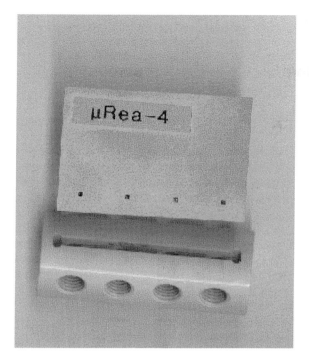

Fig. 15.12 New, single part connector.

and-recombine" mixing structures. Manifolds connected the parallel rows. The inlet manifold was fed by two channels that functioned as heat exchangers. Using this design all inlet and outlet openings could be positioned at the same side. This made it possible to construct a new simplified single-piece connector (Fig. 15.12) [22] for the Omnifit fittings. A precise trench with the length and width of the silicon mixer body was cut in a polymer piece and threaded holes were drilled in to address the openings.

15.4
Automated Reaction Optimization

15.4.1
The Principle

The optimization process as described above can be further improved. One has to bring the system into equilibrium, collect a sample and analyze it by HPLC. After a set of different experiments the individual chromatograms have to be analyzed to generate the yield–residence time curve. If several parameters have to be varied (residence time, temperature and stoichiometry) an automatic system in

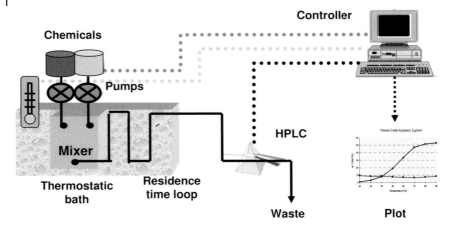

Fig. 15.13 Setup of the automatic reaction optimizer "MICROTAUROS".

combination with a statistic-planning tool would be helpful. Figure 15.13 shows the computer-operated setup, consisting of two syringe pumps, a cryostat/thermostat, fraction collector, HPLC-system with autosampler and the program "Cornerstone" for statistical planning of the individual experiments. Since each individual experiment is allocated to a unique position in the fraction collector/autosampler rack this rack can be used to track the individual experiments by the autosampler of the HPLC instrument. The fraction collector rack is charged with HPLC vials prefilled with 1 mL of water–acetonitrile (1 : 1) quench solution.

After all experiments are finished the whole fraction collector/auto sampler rack is transferred manually to the HPLC autosampler. When the HPLC runs are finished the chromatograms are sent back to the controlling computer for analysis and generation of the required reports.

15.4.2
First Prototype Reaction Optimizer

For the laboratory prototype a Gilson fraction collector was used. The whole setup as depicted was named "MICROTAUROS" (Fig. 15.14) (microreactor for automated reaction optimisation) and worked fairly well with only a few drawbacks. The length of the tube necessary for reaching every position on the fraction collector rack precluded very short reaction times. Higher pump rates would compromise the advantage of a laboratory system with the rather small amounts of materials and small syringes. A much more severe drawback is the fact that a three-way solenoid valve had to be used. In the equilibrium phase the material stream is switched to waste and only diverted for collection of analytical samples.

A given undispensed volume remains in the dispensing needle and is unavoidably dispensed later together with fresh material of the following experiment.

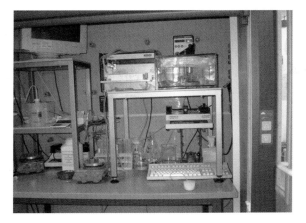

Fig. 15.14 First prototype of MICROTAUROS.

15.4.3
Sampling System without Cross-contamination

This cross-contamination issue was finally solved by simple means [23]. The end of the residence time unit was placed directly over the receiving vial. Instead of moving the dispensing needle over the stationary rack, the rack was moved below the stationary dispenser. This was possible by mounting the receiving HPLC autosampler rack on an X-Z stage (Fig. 15.15). To avoid cross-contamination the steady pump flow was not diverted by a valve during equilibration. A vacuum source, which sucked away the complete stream, was placed next to the dispenser mouth. Only for collection of an analytical sample was the vacuum interrupted to allow a droplet to fall undisturbed down into the HPLC vial. Figure 15.16 shows the operating principle. To guarantee the same temperatures in the silicon

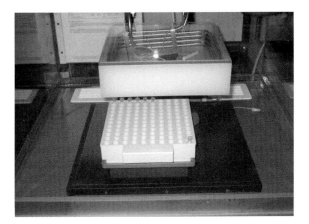

Fig. 15.15 Thermostated bath mounted above a moving autosampler tray.

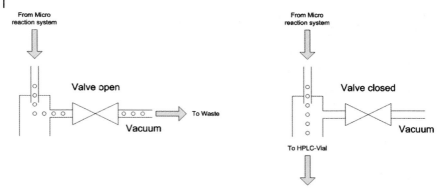

Fig. 15.16 Principle of the new contamination-free sampling device.

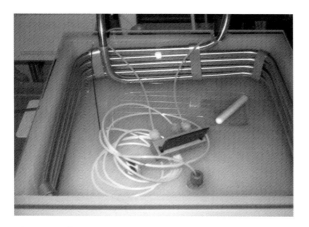

Fig. 15.17 Thermostated bath of MICROTAUROS with a micromixer and residence-time loop.

micromixer and the subsequent residence time unit they were both immersed into a thermostatic bath (Fig. 15.17).

15.5
Upscale in Larger Laboratory Scale

15.5.1
Laboratory Modules for Process Development

The modular design of the reaction system allows multistep syntheses in a larger scale. Figure 15.18 shows two dosing modules with pumps, a reaction module and a control unit. The modular reaction system allows the rapid assembly of dedicated reaction equipment. It can easily be modified and failing units can be replaced without dismantling the whole setup.

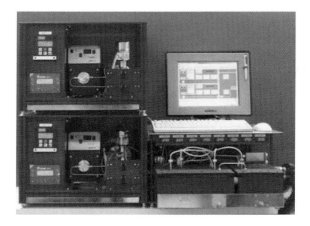

Fig. 15.18 Modules for process development.

15.5.2
Nitration with 65% Nitric Acid in Concentrated Sulfuric Acid

The homemade continuous reaction system in the large-scale laboratory consisted of a static mixer with a dead volume of 1.67 mL (Kenics) [24] and a 3-m residence time loop of 2 mm inner diameter with a total volume of 9.42 mL. Geometries of this scale have proven to be of value. They allow the removal of the large amount of heat and reduce clogging, both of which can be problematic in very small systems. Clogging can occur during the reaction by precipitation or because of particles in starting materials of technical grade.

A statistical experimental plan was generated with variation of reaction temperature, residence time and stoichiometry. The optimal parameters from the laboratory experiments were one factor within the experimental plan. The result of this cycle was quite surprising. As can be seen from the Table 15.5, experiment 9 gave the best result. This is very close to the best conditions found in the laboratory optimization. As previously seen, there was no great dependence of the isomer ratio on the residence time.

The initial mixing of the components seems to be the commanding step in the reaction. The different results of this larger scale laboratory experiment compared with the laboratory experiments with the silicon micromixer may be due to the different masses of substrate and nitrating species. Possibly, the differing stoichiometries may be of influence. In the laboratory setup, on average 50 µL of a 1 M solution of 4-(phenyl)morpholin-3-one was pumped through the system. This is roughly 0.5 g h^{-1}, whereas in the pilot plant under comparable conditions 130 g h^{-1} are processed. The different masses led to quite different heats evolved. In the miniature scale precise temperature control at room temperature was guaranteed. In this larger scale temperatures of more than 50 °C were observed, which were possibly responsible for the improved results.

Table 15.5 Nitration on a larger laboratory scale.

Exp.	T (°C)	Time (min)	Stoichio- metric ratio	Educt (%)	Meta (%)	Ortho (%)	Para (%)	By-product (%)	Selectivity (%)
1	10	8	1	0.6	10.5	15.8	70.5	1.8	71.5
2	10	2	1	0.3	10.3	15.5	70.0	2.5	71.2
3	10	2	3	0.3	11.1	15.1	70.4	2.0	71.5
4	20	5	2	0.2	11.2	15.2	61.2	9.1	63.3
5	30	2	3	0.3	12.0	15.4	62.3	7.4	64.2
6	30	8	3	0.2	10.6	15.6	64.9	6.6	66.4
7	30	2	1	0.2	12.3	17.7	68.4	0.7	69.0
8	20	5	2	0.2	11.0	16.1	70.9	1.2	71.5
9	20	5	2	0.2	10.6	16.5	71.3	0.7	72.0
10	30	3	1	0.2	11.5	16.6	69.6	1.7	70.0
11	10	8	3	0.2	10.1	14.3	70.2	4.0	71.2

15.5.3
Nitration with Neat 65% Nitric Acid

So far, in all experiments the nitrating agent was prepared by dissolving the appropriate amounts of 65% nitric acid in concentrated sulfuric acid. This is an additional process in the whole procedure. Therefore, it was decided to add the 65% nitric acid directly to the stream of the 1 M solution of 4-(phenyl)morpholin-3-one in sulfuric acid. In this case the molar ratio of the nitrating species versus 4-(phenyl)morpholin-3-one is around 10 : 1 and residence times longer than 2.5 min would require rather long tubing. However, quantitative conversion could be brought about with low molarities and short residence times. Therefore, the residence time of the experimental plan was kept unchanged at 2.5 min (Table 15.6).

Using this experimental plan only four experiments were necessary to prove a complete conversion.

Table 15.6 Experimental plan for the nitration of 4-(phenyl)morpholin-3-one with neat 65% nitric acid.

T (°C)	Residence time (min)	Concentration of 4-(phenyl)-morpholin-3-one in conc. sulfuric acid (mol kg^{-1})	Excess nitric acid vs. 4-(phenyl)-morpholin-3-one (mol mol^{-1})	Starting material (area-%)	Ortho product (%)	Meta product (%)	Para product (%)	Open chain by-product (%)	Conversion (%)	Selectivity (%)	Yield (%)
10	2.5	0.53	1.0	0.25	12.55	14.42	69.80	0.90	99.74	71.28	71.10
10	2.5	0.53	2.0	0.00	13.10	14.25	69.50	1.10	100.00	70.95	70.95
10	2.5	0.83	1.0	2.10	14.00	12.80	68.60	0.70	97.86	69.86	68.36
10	2.5	0.83	2.0	0.50	15.30	12.10	69.50	0.80	99.49	70.77	70.41

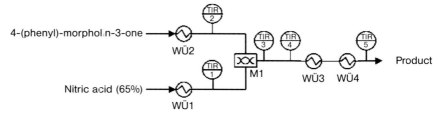

Fig. 15.19 Single-step arrangement for the nitration of 4-(phenyl)morpholin-3-one.

15.6
Upscale in a Pilot Plant

The high viscosity of concentrated sulfuric acid causes a fairly high backpressure, which was mandatory for residence times as short as possible. The homemade pilot plant continuous reaction system consisted of a static mixer and a residence time loop with a total volume of 565 mL (Fig. 15.19). The maximum temperature of the whole system was limited to 60 °C, because it was found that the complete heat of reaction was liberated within one second during the flow through the static mixer. To control the reaction temperature in this scale correctly dimensioned heat sinks are mandatory. With these precautions the same results were obtained as before. Using this equipment over 200 kg of 4-(phenyl)morpholin-3-one were nitrated within 50 h.

15.7
A Concept for the Future

Presently, pharmaceuticals are produced batchwise in multipurpose plants. For different reasons increasingly dedicated mono-production plants [25] are recommended. The integration of microreaction concepts into existing production plants is still ongoing. Continuous flow reactions are less flexible than batch reactions, particularly during workup steps such as phase separations, crystallizations and other operations.

It is suggested that batch production plants be combined with modules for continuous-flow reaction to optimize existing processes most economically.

15.8
Conclusion

The inherent danger of a batch nitration in reactors with more than 500-L capacity was the rationale to develop a continuous process. Larger amounts of unwanted by-products due to long dosing times can be reduced. In this continuous nitration the reaction volume is diminished by about two orders of magnitude, which markedly increased the inherent safety.

In our systems the nitration gave good conversions and by avoiding the ring-opened by-products the selectivity was superior to the batch process.

Minimal effort for upscaling was necessary to perform quality nitrations in a larger laboratory scale or even semi-technical scale. In addition, the limitations of the batch-scale can be avoided. The subsequent hydrolysis is performed in a continuous process as well, thus reducing the oxidative potential and eliminating the necessity for extended stirring times in an acidic environment.

As well as the nitration-specific advantages, the time, the amount of materials necessary for the process validation, and the risk of scaling-up are all reduced considerably.

The procedures described and the limited amounts of materials to be produced apply to multi-task reactors only. Mono-task units lack the necessary economic basis [26] for amounts in the 100-kg range.

References

1 P. D. Fletcher, S. J. Haswell, E. Pombo-Villar, B. H. Warrington, P. Watts, S. Y. F. Wong, X. Zhang, *Tetrahedron* **2002**, 58, 4735–4757.

2 K. Jähnisch, V. Hessel H. Löwe, M. Baerns, *Angew. Chem.* **2004**, 116, 410–451.

3 T. Bayer, D. Pysall, O. Wachsen in *Microreaction Technology: Industrial Aspects; Proceedings of the Third International Conference on Microreaction Technology* (ed. W. Ehrfeldt), Springer Verlag, Heidelberg, **2001**, pp. 165–170; *Chem. Abstr.* 133: 282144.

4 *Chem.-Ing.-Tech.* **2004**, 76 Special Vol. 5.

5 H. Krummradt, U. Koop, J. Stoldt in *Microreaction Technology: Industrial Aspects; Proceedings of the Third International Conference on Microreaction Technology* (ed. W. Ehrfeldt), Springer Verlag, Heidelberg, **2001**, pp. 181–186; *Chem. Abstr.* 133: 176817.

6 N. Schwesinger, T. Frank, *Ger. Offen.* DE 19511603 A1 19961002.

7 G. A. Olah, R. Malhotra, C. Narang, *Nitration: Methods and Mechanisms*, Wiley Publishers, New York, **1989**.

8 W. Seidenfaden, D. Pawellek in *Houben-Weyl, Methoden der Organischen Chemie*, Vol.10/1, Georg Thieme Verlag, Stuttgart, **1971**.

9 G. Panke, T. Schwalbe, W. Stirner, S. Taghavi-Moghadam, G. Wille, *Synthesis* **2003**, 2827–2830.

10 H. Pennemann, P. Watts, S. J. Haswell, V. Hessel, H. Löwe, *Org. Proc. Res. Dev.* **2004**, 8, 422–439 and literature cited therein.

11 J. R. Burns, C. Ramshaw in *Proceedings Fourth Int. Conf. on Process Intensification for the Chemical Industry: Better Processes for Better Products*, Brugge, Belgium, **2001**, BRH Group Ltd., Cranfield, UK, *Chem. Abstr.* 136: 371400.

12 H. Wurziger, F. Stoldt, K. Fabian, N. Schwesinger, *Ger. Offen.* DE 19935692 A1 20010102; *Chem. Abstr.* 134: 149276.

13 S. Löbbecke, W. Ferstl, A. Lohf, A. Steckenborn, J. Hassel, M. Häberl, D. Schmalz, H. Muntermann, T. Bayer, M. Kinzl, I. Leipprand, *Chem.-Ing.-Tech.* **2004**, 76, 637–640.

14 C. Tsaklakidis, D. Dorsch, W. Mederski, B. Cezanne, J. Gleitz, WO 2004087646 A2 20041014; *Chem. Abstr.* 141: 350179.

15 W. Mederski et al. *Bioorg. Med. Chem. Lett.* **2004**, 14, 5817–5822.

16 C. Thomas, M. Berwe, A. Straub, WO 2005026135 A1 20050324; *Chem. Abstr.* 142: 316448.

17 W. König, *Angew. Chem.* **1955**, 67, 157–158.

18 D. Schmalz, M. Häberl, N. Oldenburg, M. Grund, H. Muntermann, U. Kunz, Chem.-Ing.-Tech., **2005**, 77, 859–868.
19 Omnifit Limited, 2 College Park, Coldhams Lane, Cambridge CB1 3HD, England.
20 G. Brenner, M. Schmelz, H. Wurziger, N. Schwesinger, Ger. Offen. DE 19746585 A1 19990429.
21 Standard Pump 22, Harvard Apparatus, Inc., 84 October Hill Rd., Halliston, MA 01746.
22 M. Hohmann, M. Schmelz, H. Wurziger, N. Schwesinger, Ger. Offen. DE 19854096 A1 20000525.
23 M. Schmelz, G. Brenner, T. Greve, H. Wurziger, M. Arndt, WO 2005003733 A1 20050113; Chem. Abstr. 142: 106179.
24 Kenics, Chemineer, Inc., P. O. Box 1123, Dayton, OH 45401.
25 C. Ewers, Pharma Supply Chain, ECV-Editio-Cantor-Verlag, Aulendorf, **2002**.
26 D. Schmalz, Dissertation, TU Clausthal, Clausthal-Zellerfeld, Germany, **2005**.

Part III
A Summary and Path Forward

16
Concluding Remarks

Melvin V. Koch, Ray W. Chrisman, and Kurt M. VandenBussche

16.1
Summary

The preceding chapters in Parts I and II show that the value of micro-instrumentation is being demonstrated broadly and that recent research results continue to improve on the ability of these devices to play an increasingly important role in both research and manufacturing environments. As the value of micro-unit operations is realized, there will be an increased evaluation of these devices to replace macro-units or to be incorporated within macro-unit operations. The superior mass and heat transfer capability frequently leads to significant quality, energy and environmental benefits.

In addition, we are learning that the rapid heat and mass transfer characteristics of micro-instrumentation can provide significant advantages in the early R&D stages of material development. Historically, reactions run in the laboratory have been carried out in reaction equipment that is poorly characterized. The result has been that information developed in the early R&D stages is a convolution of the intrinsic chemistry and the characteristics of the reaction equipment. Thus as reaction equipment is changed during scale up the measured reaction characteristics can change in unexpected ways such as in the formation of unwanted impurities or dramatically longer reaction times with lower yields. The use of microscale instrumentation enables the relatively easy modeling of the reaction equipment to extract its impact on the measured reaction parameters for the determination of intrinsic reaction rates.

The impact of this new ability to more precisely measure reaction chemistry is still in its infancy. Since intrinsic chemistry is often very fast and microreaction equipment is very compatible with real-time analysis it is often possible to extract very large and precise quantities of reaction information very rapidly. Researchers are only just beginning to understand how to treat this wealth of information. As reported in the preceding chapters, it is possible to gather high quality rate expressions from laboratory scale experimentation, which enables much more rapid, precise and dramatically lower cost scale up of chemical reactions. Given the high automation potential of this equipment it is reasonable to expect that in the not too distant future it will be possible to generate chemical rate expressions in the time it use to take to develop material for larger scale testing. In fact, chemical kinetics might also be developed rapidly at an early stage, which should allow synthesis chemists to develop much better reaction chemistry to insure the most efficient reaction routes reach the commercialization stage.

However, as noted by Hickman and Sobeck in Part II, Chapter 13, at this very early stage, care must be exercised to fully understand the performance of the reaction equipment. As reactions are performed in microscale equipment the researcher must be careful not to assume equipment performance characteristics. While high quality data can be generated, researchers will need to be aware that the high surface-to-volume ratio micro-equipment can interact with materials in ways that can surprise researchers new to the field. Thus, standard equipment evaluation tests are and need to continue to be developed to insure that the correct equipment and materials of construction are used to insure that precise reaction parameters are developed.

Another concept that has evolved is that within a corporate function like discovery, process development, or process optimization, it is important to again refer to the Development Cycle (Fig. 16.1) to emphasize the value of gathering data and interpreting it (converting it into process knowledge) to move ahead to the next function's cycle. Eventually, the Development Cycle of the production function is responsible for setting the control of the process via this approach. Effective data gathering in the discovery and early process development studies will minimize the need to run extensive and expensive studies in the later process development and process optimization stages. It is a risk assessment calculation as to whether to improve the process of a potential new product before the product has been chosen for production. The potential cost savings in terms of time and capital spending are significant if the potential new product becomes a product and is moved into production.

Once it is recognized that each of the above corporate functions has the opportunity to operate with a Development Cycle, the next step is to minimize the

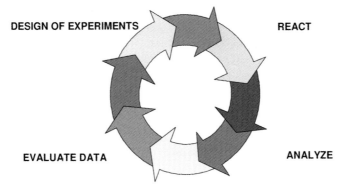

Fig. 16.1 The development cycle.

barriers in moving between cycles. Historically, there are significant barriers to moving projects between functions. Often there are too many duplicative efforts or alternate approaches to gathering data. Thus, another concept that has arisen is that cross functional cooperation within an organization is a worthwhile exercise. Accomplishing this will increase productivity on the target project and probably result in subsequent positive interactions between the functions on future projects. It is beneficial for the early phase process development personnel to learn what type of information is needed by the later development phase engineers. There are additional benefits when the engineers become familiar with the capabilities of the discovery research and the early process development functions to gather data.

16.2
The Path Forward

- Success is now being demonstrated in various areas of the pharmaceutical industry via the use of advances in micro-instrumentation. Combinatorial chemistry in discovery and reaction optimization, coupled with micro-analytical tools, is showing value to the industry. This effective demonstration of how PAT (process analytical technology) can be useful in pharmaceutical processes has resulted in an increasing interest by the biotechnology industry to evaluate measurement tools for monitoring the bio-reactors. This approach will help to follow the formation of desired product and undesirable by-products. The ability to learn more about the affect of processing conditions on composition and concentration will improve productivity and quality. Regulatory agencies like the FDA will encourage the use of PAT tools in the biological fields to define the operating space and to obtain the desired process understanding on which to base process control – for improved products. There is also continued progress in the use of micro-instrumentation in the discovery and process development phases of biotechnology. There are examples now where microreactors can be used to study various process and nutrient related process variables and assist in process optimization studies.
- It is anticipated that micro-instrumentation will allow for effective evaluation and eventual use of continuous processes in areas where batch approaches have been traditional. The expected impact of this will be significant on energy savings and in environmental and productivity aspects of commercial processing. The areas where this will show early benefits are in pharmaceuticals, but it is envisioned that biotechnology will incorporate the advantages of continuous processing. This will have a broad impact on medicinal bio-products as well as biochemical and biofuel production.
- The value of standardization of the components of sampling systems is being shown with the success of the NeSSI platforms. As the cost related to design, installation, and maintenance of individual one-of-a-kind sampling systems increases, it made sense to consider the use of a standard platform that could be flexible in mounting components unique to the needs to the process stream. It

was an added benefit when it was demonstrated that the NeSSI platform could also function as a base for analytical instrumentation. NeSSI with analytical tools can conduct an array of analytical techniques on a sample stream that is uniformly conditioned. As a result the data from the various techniques can be studied with chemometric approaches for pattern recognition, etc.

The topics of this book continue to be discussed in forums like the CPAC Summer Institute and the satellite workshops sponsored by CPAC. A successful CPAC sponsored workshop was held in Rome, Italy in March, 2006, and that concept will be continued with other workshops in the USA and Europe that will be co-sponsored by CPAC. Developments in the various areas of micro-instrumentation are embellished in gatherings like the CPAC Summer Institute and CPAC Satellite workshops. Additional value is obtained when synergistic collaborations result from the discussions. The concept of creating a forum where different technical disciplines from industry, academia, and government come together is important to moving a technical field ahead. This is particularly true in the micro-instrumentation field, which has the ability to affect many aspects of a product life cycle – from the discovery, through development, and into production.

Subject Index

a

Ab2   *see* anti-antibodies
Ab*   *see* anti-ovalbumin
absorption, phase transfer processes   158
absorption cycle heat pump   163
acetic acid   451
– Raman spectra   214
acetic anhydride   452
(S)-2-acetyl tetrahydrofuran
– optical purity   105
– selectivity   105
– synthesis   105
acidity, catalyst characterization   354
acids
– acetic   214, 451
– acrylic   110
– amino acids   288–289
– 2,4-dihydroxybenzoic   114
– nitric   463–465
– phenyl boronic acid synthesis   100–101
– sulfuric   463
acoustic bubble shaking   139
acoustic forces, fluid manipulation   169–170
acrylic acid   110
acrylic acid ester   110
acrylonitrile   110–111
active channels, SPR   246
active micromixers   136–139
actuated valves, electric field   169
acute cardiac infarction   107
adsorption
– microchannel   150
– phase transfer processes   149–150
advanced connection system, micromixer   458
AIMS   *see* automated integrated microreactor system
air-based analyzer, Phased Heater Array Structure for Enhanced Detection (PHASED)   223–227

air sampling, MGA   229
alarm rate   234–237
"all-at-once" contacting, solvent-free processes   124
all-glass LC microchip   293
amides, racemization   439
amine addition, α,β-unsaturated compounds   110–111
α-amino acids   438
amino acids, derivatization   288–289
ammonia oxidation
– AIMS   401–403
– reaction data   402
– sample chromatogram   402
amplification, DNA   295
analysis
– on-line   41
– PCA   215–216, 311–312
analytical chemistry, CPAC summer institute   9–15
analytical loop, fast loop–analytical loop strategy   31
analytical profile   4
analytical technology
– high throughput   70
– microseparation devices   258–297
– potential applications   295–297
– *see also* micro-analytical technology
analytical tools, PAT   7–8
analyzer
– air-based   223–227
– development costs   348
– hydrogen-based   228–229
– *see also* ultra MAG
analyzer systems, MEMS   230–231
anhydride, acetic   452
aniline   117
– in deuterated nitromethane   243
ANSI/ISA standard 76.00.02   16, 316, 324–325

Micro Instrumentation for High Throughput Experimentation and Process Intensification – a Tool for PAT
Edited by M. V. Koch, K. M. VandenBussche and R. W. Chrisman
Copyright © 2007 WILEY-VCH Verlag GmbH & Co. KGaA, Weinheim
ISBN: 978-3-527-31425-6

– fluid ports   325
– mechanical footprint   325
anti-*botulinum* antibody   250
anti-ovalbumin (Ab*)   283
anti-*Y. pestis* antibody   251
antibiotic drug, quinolone   106
antibody
– anti-*botulinum*   250
– anti-*Y. pestis*   251
approach length   196
array structure, heater   223–227
assay system   435
assembly, µMS³ system   333
automated hand-held SPR, modular fluidics   247–248
automated integrated microreactor system (AIMS)   371–373
– ammonia oxidation   401–403
– automation controllers   394
– automation evaluation   396–397
– comparison to MARS   392–396
– hardware evaluation   396–397
– methane oxidation   398–400
– process control loops   397
– reaction performance   397–403
– safety features   388
– sample chromatogram   398
– temperature controllers   397
– temperature sensors   396
– testing   391–403
automated reaction optimization   459–461
automation, testing procedures   386
automation controllers, AIMS/MARS   394
automation evaluation, AIMS   396–397
autosampler tray, moving   461
autothermal reforming   61
average concentration, optical path   199
axial dispersion   410
– 1-dimensional model   421
– laminar pipe flow   415–416
azo pigment Yellow 12
– glossiness   103
– synthesis   102–103

b

bacteria
– *Clostridium botulinum*   250
– Gram negative/positive   106
– *Yersinia pestis*   251
balance, mechanical energy   183–186
ballprobe, Raman spectroscopy   214, 217, 219–220
banding patterns, changing gravity   170
basic building block, µMS³ system   332

basic fast loop, Intraflow System   335
basic flow component boss   327
bath, thermostatic   460–462
bead packing   273
– microchannel   272
beads, suspended   168
benzaldehyde, in deuterated nitromethane   243
bill of materials, MPC configurator   331
biological processes, non-reactor microcomponents   165–166
biopharmaceutical production, micro-scale device applications   296
biotechnology, industrial   296
block and bleed valve system   31
blood substitute, velocity   165
board
– heater driver circuit   377
– reactor   374–375
– temperature controller   376
body force driven processes   167–170
bonding, catalyst characterization   354
boss, basic flow component   327
bottleneck, high throughput research   73
box enclosure   28
branched polyimide membrane channels   160
bromination
– *m*-nitrotoluene   118
– side-chain   119
– thiophene   119–120
bubble shaking, acoustic   139
building block   125
– basic   332
bus, New platform for Sampling and Sensor Initiative (NeSSI)   340–344
butterfly effect   202
bypass channel, microreactor   294

c

C_6–C_{10} alkanes, preconcentration/separation   231
CAE   *see* capillary array electrophoresis
capex   *see* capital expenditure
capillary array electrophoresis (CAE)   154–155, 157
capillary electrochromatography (CEC), chip-based   261–263
capillary electrophoresis (CE)
– chip-based   261–263
– continuous sampling   281–291
– DNA amplification/digestion chip   295
capillary gel electrophoresis (CGE), DNA analysis   156

capital expenditure   43
capital intensification   50
carbonyl compounds, α,β-unsaturated
   110–111
cardiac infarction, acute   107
case studies
– micro instrumentation   79–468
– process intensification   57–63
cast stabilization, silicon micromixer   455
catalyst deactivation, reactor classification
   49
catalysts   44
– gas phase processes   353–361
– heterogeneous reactions   76
– screening   70
– selectivity   361
catalytic distillation   48
catalytic materials, intrinsic kinetics   134
catalytic steam reforming   61
caterpillar micromixer-reactor   110
caterpillar microstructured mixer-reactors
   88–90
CE   see capillary electrophoresis
CEC   see capillary electrochromatography
cell testing, microchannel systems   165
cells, surface plasmon resonance (SPR)   251
center for process analytical chemistry
   (CPAC)
– biannual meetings   317
– 2006 and future meetings   474
– high throughput experimentation   16–19
– NeSSI   315, 322–323
– process intensification   16–19
– refractive index optical detection method
   279
– summer institute   9–15
– summer institute 2004/2005   209
CFC   see chlorofluorocarbons
CFD   see computational fluid dynamics
CGE   see capillary gel electrophoresis
channel-on-a-circle design, electrical heat
   exchanger   95
24-channel SPR, modular fluidics   247–248
channels
– diffusion   201
– manifold/substrate   330
– orthogonal capacity   234–238
chaotic flow micromixers   142
characteristic mixing time analysis   412
chassis, microreactor systems   371–372
chemical exposure   387
chemical research, specialty   74–75
chemical synthesis, flow (micro)reactors
   442–446

chemical vapor deposition (CVD)   274
– microchannel   275
chemical warfare agents (CWA), detection
   222–223, 229
chemistry
– combinatorial   67
– novel   124–125
– traditional   124
chip-based
– CE   261–263
– CEC   261–263
– electrophoresis   262
– HPLC   269–270
– LC   264–280
chlorofluorocarbons (CFC), consumption
   162
chromatogram
– ammonia oxidation   402
– methane oxidation   398
chromatography
– microchannel   149
– process analytical systems   33–38
– see also gas chromatography,
   high pressure liquid chromatography,
   liquid chromatography
chrome grating, GLRS   306
Circor Tech, μMS^3 system   330–333
circuit boards, heater driver   377
clam shell heater   376
Clariant/Frankfurt, phenyl boronic acid
   synthesis   100–101
classification, reactor   49
clip-in flowcell, SPR   249
Clostridium botulinum toxin A   250
coke deposition   359
coking   354
collocated monolith support structure
   (COMOSS) column   263–264
columns
– GC   34
– high pressure liquid chromatography
   (HPLC)   272
– microstripping   54
combinatorial chemistry
– data quality   72–73
– IUPAC definition   67
COMOSS   see collocated monolith support
   structure
component collection, microchannel
   adsorption   150
computational fluid dynamics (CFD)   181,
   189, 205
– microreactor process safety   387
– passive micromixers   140–142

– *see also* software
concentrated sulfuric acid   463
concentration
– optical path   199
– profile at inlet/outlet   410
– steady-state profile   414
configurator layout screen, MPC system   331
connection system
– advanced   458
– Omnifit   457
– optimized   456
– U-shaped legs   456
connectivity architecture   319
connector, slip-fit   336
contacting, "all-at-once"   124
contactor, gas–liquid   161
contaminants, surface plasmon resonance (SPR)   250
contamination, cross   461
contamination-free sampling device   462
continuous microreactor, characterization and optimization   211–219
continuous reaction monitoring   213–218
continuous sampling
– CE chips   281–291
– interface   288
– LC chips   292–293
contraction flow   194–195
– geometry   189
contractions
– pressure drop   189–191
– velocity rearrangement   195
controllers
– automation   394
– mass flow   395
– temperature   397
– temperature controller board   376
convection
– heat exchange   133
– heat transfer   56
– microcomponent flow characterization   198–199
– relative   200
convection-diffusion-reaction equations   206
conventional reactor types, specific interfacial area   53
conversion, oxygen   399
conversion profiles, fractional   414
cooler, Vortec   30
copper microcoil   242
costs
– methanol production   60

– microreactor technology   86
– NeSSI   348, 350
counter-flow heat exchanger-reactors   94
Covion/Frankfurt, production-oriented LED materials development   107
CPAC   *see* center for process analytical chemistry
cream manufacturing plant   98
creatinin clearance, microchannel dialyzer   165
cross-boundary research, microreactor technology   85
cross-contamination, sampling system   461
cross-flow heat exchanger-reactors   94
cross-section, microchannel   266
crystal radius, effective   360
CVD   *see* chemical vapor deposition
CWA   *see* chemical warfare agents
cycle heat pump
– absorption   163
– vapor compression   163
cyclododecatriene, hydrogenation   55
cyclohexanol, extraction   152
cyclone micromixers   141

d

data-acquisition   81
data quality, combinatorial chemistry   72–73
deactivation, catalyst   49
Dean vortex   142
decarboxylation, thermal   115
decarboxylation rate   114
Dechema, MicroChemTec Group   17
deep reactive ion etching (DRIE)   228
dehydrogenation
– ethylbenzene   48
– process intensification   46–47
dendrimers   309–310
deposition, coke   359
derivatization, amino acids   288–289
design, microreactors systems   371–378
desorber, fractal-based   159
desorption, phase transfer processes   158
detachment process, microchannel system   145
detection, chemical warfare agents (CWA)   222–223, 229
– explosives-related compound (ERC)   229
– GLRS   307
– micro gas analyzer (MGA)   232–233
– toxic industrial compounds (TIC)   229
detectors, LC   35
deuterated nitromethane   243
development cycle   209, 472

device technology, micro-scale   295
dialyzer, microchannel   165
dielectric constant, EOF   434
dielectric spectroscopy   253–256
– comparison with other techniques   253
– direct sensing   256
– electrode patterns   254–255
– excitation frequency   256
– imaging capability   255
– non-contact mode   254
– PAT   253
– pre-concentrators   256
– signal strength control   255
– spatial wavelength control   254
DieMate
– electrical interconnect   367
– fluidic interconnect   368–369
– manifold fabrication   370
– manifold heating   370
– manifold insulation   370
– socket   368
diffusion
– gas phase processes   354–360
– microcomponent flow   196–205
– mixing   198
– rectangular channel   201
– relative   200
– T-sensor   202–203
– see also convection-diffusion-reaction equations
diffusion constant   361
diffusion tools, gas phase processes   359–360
digestion, DNA   295
2,4-dihydroxybenzoic acid   114
dimensionless mechanical energy balance   184–193
– manifolds   192–193
dimethyl acetals, synthesis   445
dimethylamine   111
dinitrophenol-lysine, SPR detection   249
direct sensing, dielectric spectroscopy   256
dispersion, axial   410, 415–416, 421
disposable syringe   457
distillation
– catalytic   48
– phase transfer processes   160–161
distributed production
– hydrogen   60–63
– methanol   57–59
DNA analysis, capillary gel electrophoresis (CGE)   156
double layer, electrical (EDL)   137, 167
Dow Chemical Company   25

DRIE   228
driver circuit boards, heater   377
droplets
– fractal-based desorber   159
– moving droplet mixers   143
drugs, quinolone antibiotic drug   106
dry etching, isotropic   274
DSM Fine Chemicals/Linz (Austria), high-value polymer production operation   109

e
edge   203
Edisonian research   71
EDL   *see* electrical double layer
effective crystal radius   360
electric field actuated valves   169
electric field mixing   137
electric fields   167–169
electrical double layer (EDL)   137, 167
electrical heat exchanger, channel-on-a-circle design   95
electrical interconnect, DieMate   367
electrical wiring, AIMS/MARS   394
electrode patterns, dielectric spectroscopy   254–255
electrokinetic instability induced mixing   137
electrokinetically-driven plug flow   260
electromagnets, fluid manipulation   169
electroosmotic flow (EOF)   155, 167–168, 290, 432–434, 444
– glass microreactor   433
– hydrodynamic pumping   432
– microseparation devices   259–261
– peptide synthesis   437
– solvent properties   434
– velocity profile   433
electropherogram
– mixing channel   285
– multichannel   282
– multiple-injection   286
electrophoresis, CAE   154–155, 157
– CGE   156
– chip-based   262
– microchannel   154–157
electrophoretic purification, peptides   438
elevated process temperatures   387
emission spectra, trichloroethane   234
emulsification, microchannel   143–147
emulsions, liquid–liquid mass transfer   52
enclosure
– box   28
– purged   26
energy balance   184–193

- laminar flow   186
- turbulent flow   183

enhanced detection, air-based analyzer   223–227

enthalpy
- methanol synthesis   59
- Michael addition   112

entry length   196
- microcomponent flow   194–195

EOF   see electroosmotic flow

equaling-up, microstructured mixer-reactors   92

equations
- convection-diffusion-reaction   206
- Hagen–Poiseuille   415
- Stokes   187

ERC   see explosives-related compound

ester, acrylic acid   110

esterification, methanol   215, 217–218

etching
- DRIE   228
- dry   274
- etched-through silicon mixers   456–457

ethyl cyanoacetate   444

ethylbenzene, dehydrogenation   48

evaluation
- automation   396–397
- experimental   419–424
- hardware   396–397

evanescent field, GLRS   306

excitation frequency, dielectric spectroscopy   256

exhaust failure, fume hood   388

expansions, pressure drop   189–191

experimental evaluation, liquid phase processes   419–424

experimental methods, microreactors   389–391

experimental protocol, microreactors   390

explosive regime, routes into   125

explosives-related compound (ERC), detection   229

external resistance, mass transfer   45

extraction
- cyclohexanol   152
- phase transfer processes   151–153

f

failure, reactor   395

falling film microreactor   116

false alarm rate
- CWA detection   222
- metrics   234–237

false positive fraction (FPF)   235–237

- MDD   236

fast loop, basic   335

fast loop–analytical loop strategy   31

feed
- premixed/unmixed   414
- preparation   418

feed gas
- mixing board   374
- primary manifold   373

feed system, liquid phase processes   417

FEF   see fringing electric field sensor

Fenix laser marker, LAMIMS   356

field mixing, electric   137

Fieldbus, Foundation   319, 343–345

filtration, process analytical systems   30

fine chemistry
- industrial microreactor process development   100–106
- industrial production   107–109
- microreactor plant   96–99

fitting, Intraflow System   335–336

flammability   387

flat top experiment   420, 422–424

flexible sample line   291

flow
- chaotic   142
- contraction flow geometry   189
- convection   198–199
- disturbances   187–188
- laminar   186, 411, 415–416
- periodic switching   136
- planar   197
- plug flow criterion   416
- slow   182
- three-phase   153
- turbulent   183

flow characterization, microcomponents   181–206

flow component boss, basic   327

flow components, MPC system   330

flow guiding   168

flow rate
- caterpillar mixers   90
- microreactor   434
- platelet mixers   92–93

flow reactor   75
- chemical synthesis   442–446

flow tubesets, μMS^3 system   332

flowcell, clip-in   249

fluid manipulation
- acoustic forces   169–170
- body force driven processes   167–170
- electric fields   167–169
- electromagnets   169

- EOF   167–168
- high pressure liquid chromatography (HPLC)   278–279
- ultrasonic forces   169–170
fluid ports, ANSI/ISA standard 76.00.02   325
fluidics
- DieMate interconnect   368–369
- hardware platform   see new platform for sampling and sensor initiative (NeSSI)
- microchip interface   278
- modular   247–248
fluorescein isothiocyanate   287, 289
fluoropolymer mobile monoliths, in situ polymerization   277
focus design, multi-lamination mixer   140
Forschungszentrum Karlsruhe (FZK), high-value polymer production operation   109
Foundation Fieldbus   319, 343–345
Fourier number (Fo), micromixers   135–136
Fourier-transform infrared spectroscopy (FTIR)   24
- on-line spectroscopy   36
- process environment   5
FPF   see false positive fraction
fractal-based droplet desorber   159
fractal network-type design   134
fraction collector, Gilson   460
fractional conversion profiles, simulated   414
free radicals, polymerization   121–123
frequency, excitation   256
Fresenius
- amine addition to α,β-unsaturated compounds   110–111
- thiopene bromination   119–120
friction factor, slow flows   182
frictional loss coefficient
- laminar flow (K_L)   188–190, 193–194
- manifolds (K_C)   193
- turbulent flow   188
fringing electric field (FEF) sensor   253–254
fritless packing, microchannel   272
frits, silica   445
fuel cell   56
- wicking phase separator   148
fuel processor, pilot scale   99
fume hood exhaust failure   388
functional chemistry, industrial microreactor process development   100–106
FZK   see Forschungszentrum Karlsruhe

g

Gans   see Rayleigh–Gans approximation

gas analyzer
- MEMS   230–231
- ultra MAG   222–238
gas chromatography (GC)
- columns   34
- early systems   24
- micro GC/MS gas analyzer   224
- microreactor experimental methods   391
- on-line   419
- process analytical systems   33–34
gas flow manifold, AIMS/MARS   393
gas–liquid contactor, spinning micro-disk   161
gas–liquid mass transfer, process intensification   53
gas phase
- heterogeneous catalyst reactions   76
- mass transfer   51
- reactions   363–404
gas phase processes
- catalysts   353–361
- diffusion   354–360
- diffusion tools   359–360
- kinetic tools   356–358
- reaction mechanisms   354–360
- reactivity   354–360
- reactivity testing   355
gas–solid systems, mass transfer   54–55
gas stick   321
GC   see gas chromatography
gemifloxacin   106
generation I NeSSI   323–338
generation II NeSSI   339–345
generation III NeSSI   340
Gilson fraction collector   460
glass lined reactor   422
glass microreactor
- EOF   433
- hydrodynamic pumping   432
glossiness, azo pigment Yellow 12   103
GLRS   see grating light reflection spectroscopy
glued-in hypodermic needles, silicon micromixer   455
Gram negative/positive bacteria   106
grating, chrome   306
grating light reflection spectroscopy (GLRS)
- integration onto microchip   280
- monitoring of heterogeneous matrices   305–313
- results and discussion   309–312
- theory   307–308
gravity changes, banding patterns   170
Grignard reagent   105

h

Hagen–Poiseuille equation 415
hand-held SPR, automated 247–248
hardware evaluation, AIMS 396–397
hardware system 378–380
hazardous elements 125
hazardous locations, NeSSI 342
hazards, process 387
heat exchange, convection 133
heat exchanger-reactors
– counter-flow 94
– cross-flow 94
– microstructured 94–95
heat exchangers
– commercial applications 133
– electrical 95
– heavy-pilot 95
– non-reactor micro-components 132–134
– specific surface area 56
heat pump 162–164
– cycle 163
heat transfer
– convection 56
– laminar pipe flow 411
– process intensification 56
heater
– array structure 223–227
– clam shell 376
– transport tube 376
heavy-pilot heat exchanger 95
hemodialysis, non-reactor micro-components 165–166
herringbone micromixer, staggered 142
heterogeneous catalyst reactions, gas phase 76
heterogeneous matrices, grating light reflection spectroscopy 305–313
HFC see hydrofluorocarbons
HGP see Human Genome Project
high-p, T processing 124
high pressure liquid chromatography (HPLC)
– automation 459–461
– chip-based 269
– chromatogram 453–454
– columns 272
– fluid handling 278–279
– industrial applications 296–297
– microseparation 263–264
– nitration 451–454
high pressure ratings, NeSSI systems 338
high throughput analytical techniques 70
high throughput experimentation, CPAC summer institute 16–19
high throughput research 67–77
– description of terms 67
– extraction of information 71–72
– process development as bottleneck 73
high-value polymer intermediate product 109
HPLC see high pressure liquid chromatography
Human Genome Project (HGP) 154–155
human-machine interface (HMI) 319, 378–379, 381–383, 396–397
hybrid autothermal technology 63
hydraulic diameter, length to hydraulic diameter ratio 132
hydrodynamic pumping
– EOF 432
– glass microreactor 432
hydrofluorocarbons (HFC), consumption 162
hydrogen, distributed production 60–63
hydrogen-based analyzer, Sandia 228–229
hydrogen peroxide, synthesis 104
hydrogenation
– cyclododecatriene 55
– nitrobenzene 115–117
hypodermic needles, glued-in 455

i

Idemitsu Kosan/Chiba (Japan) 108
IEEE 1451 344–345
IG Chem/Daejeon, quinolone antibiotic drug intermediate synthesis 106
imaging capability, dielectric spectroscopy 255
IMM see Institut für Mikrotechnik Mainz GmbH
immersion probe, Raman spectroscopy 213
impedance spectroscopy 253
improved reactor sampling, NeSSI components 219
IMRET see International Conference on Microreaction Technology
IMS see ion mobility spectroscopy
IMVT see Institut für Mikroverfahrenstechnik, Karlsruhe
in situ analysis, NeSSI 320
in situ polymerization, fluoropolymer mobile monoliths 277
in vitro testing, microchannel membrane oxygenator 164
industrial applications, high pressure liquid chromatography (HPLC) 296–297
industrial biotechnology, micro-scale device applications 296

industrial pilot plant, radical polymerization 108
industrial process development, microreactors 100–106
industrial production, fine chemistry 107–109
industrial use, laboratory-scale microreactor developments 110–120
inflation process, microchannel system 145
informing power, IMS and MS analytical techniques 238
injection sequence, LC 270
injector loop, sizing 417
injectors, teflon 422
inlet concentration profile 410
inlet tee, reactor 412–414
instability induced mixing, electrokinetic 137
Institut für Mikrotechnik Mainz (IMM) GmbH 108
- (S)-2-acetyl tetrahydrofuran synthesis 105
- amine addition to α,β-unsaturated compounds 110–111
- azo pigment Yellow 12 synthesis 102–103
- free-radical polymerization 121–123
- hydrogen peroxide synthesis 104
- Kolbe–Schmitt synthesis 112–114
- nitrobenzene hydrogenation 115–117
- nitroglycerine production plant 107
- m-nitrotoluene bromination 118
- phenyl boronic acid synthesis 100–101
- production-oriented LED materials development 107
- quinolone antibiotic drug intermediate synthesis 106
- thiopene bromination 119–120
Institut für Mikroverfahrenstechnik (IMVT), Karlsruhe, high-value polymer production operation 109
instrument house 29
instrument interface, surface-mount 347
instrumentation
- NeSSI 347
- surface plasmon resonance (SPR) 247
insulated DieMate 370
integrated microreactor system, gas phase reactions 363–404
integrated production, valves 277
integrated valves production, *in situ* polymerization 277
integration, multiple sensing elements 349
interdigital micromixer 101
interface
- continuous sampling 288

- fluidic 278
- liquid 307
- process 30
- process monitoring 383–385
- "real world" 283
- surface-mount 347
- *see also* human-machine interface
interfacial area, reactor types 53
interfacing, microreactor technology 85
interlock, system 388
intermediates
- high-value polymers 109
- surface 358
internal resistance, mass transfer 45
International Conference on Microreaction Technology (IMRET) 365
International Student Games 333
Intraflow System 334–337
- basic fast loop 335
- fitting 335–336
intrinsic kinetics, catalytic materials 134
Intrinsically safe operating region 343
ion etching, deep reactive (DRIE) 228
ion mobility spectroscopy (IMS), informing power 238
ISA Expo 2005 333
ISA standard 76.00.02 *see* ANSI/ISA standard 76.00.02
isometric view, reactor board 375
isothermal reaction 411
isothermal separation 232
isotopes, xenon 233
isotropic dry etching 274
IUPAC definition, combinatorial chemistry 67

j
jet *see* colliding jet mixers

k
keystone effect 272
kinetic tools, gas phase processes 356–358
kinetics
- catalyst characterization 354
- intrinsic 134
- reaction 287, 425–426
knife edge, T-sensor 203
Knoevenagel reaction
- reaction manifold 443
- solid-supported 443
- solution-phase 442
Kolbe–Schmitt synthesis 112–114
- flow scheme 113–114

l

laboratory applications, NeSSI   346–350
laboratory instrument, conversion to process instrument   4
laboratory modules, process development   462
laboratory nitrations, orienting   451–453
laboratory optimization   452–453
laboratory-range microreactor plants   96–97
laboratory scale   452–465
– microreactor process developments   110–120
LabView   344, 381–383
– AIMS safety features   388
LAMIMS   see laser activated membrane introduction mass spectrometry
laminar flow, mechanical energy balance   186
laminar pipe flow
– axial dispersion   415–416
– heat transfer   411
– plug flow criterion   416
– pressure drop   415
large-scale manufacture   446
laser activated membrane introduction mass spectrometry (LAMIMS)   355–356
laser marker, Fenix   356
latex polystyrene   170
LC   see liquid chromatography
legs, U-shaped   456
length to hydraulic diameter ratio   132
lens, spherical   213
LIGA   see lithography and electroforming
light emitting diode (LED), materials development   107
liquid chromatography (LC)
– all-glass microchip   293
– chip-based   264–280
– continuous sampling   292–293
– detectors   35
– injection sequence   270
– process analytical systems   35
– rhodamine dye   269
liquid interface, microchannel   307
liquid–liquid mass transfer, mixing and emulsions   52
liquid phase processes   407–427
– feed system   417
– system capabilities   417–426
– system design   409–417
– two-dimensional simulation   413–414
liquids
– non-miscible   53
– see also spinning micro-disk gas–liquid contactor
lithography and electroforming (LIGA)   229
loop
– fast loop–analytical loop strategy   31
– injector   417
– Intraflow System   335
– process control   397
lysine   see dinitrophenol-lysine

m

magnetic resonance imaging (MRI)   241–244
malononitrile   444
manifold
– dimensionless mechanical energy balance   192–193
– fabrication   370
– gas flow   393
– heating   370
– insulation   370
– primary feed gas   373
– reactor feed gas   374
manifold channels, MPC   330
manufacturing plant
– cream   98
– methyl methacrylate (MMA)   108
mass flow controllers, AIMS/MARS   395
mass spectrometry (MS)
– informing power   238
– LAMIMS   355–356
– micro GC/MS gas analyzer   224
– on-line   37
– xenon isotopes spectrum   233
mass transfer
– gas–liquid   53
– gas phase   51
– gas–solid systems   54–55
– internal and external resistances   45
– liquid–liquid   52
materials of construction effects   422–423
matrices, heterogeneous   305–313
MCPT   see micro-chemical process technology
MDD   see micro-discharge device
measurements, NeSSI   347
mechanical energy balance
– dimensionless   184–193
– laminar flow   186
– turbulent flow   183
– viscous dissipation   185–186
mechanical footprint, ANSI/ISA standard 76.00.02   325
membranes
– channels   160

- emulsification   144
- oxygenator   164
- piezoelectric   137
- reactors   47
- ultrasound   137

MEMS   *see* micro-electrical mechanical systems

metallic sites, catalyst characterization   354

metallization, microreactor die   369

methane oxidation
- AIMS   398–400
- reaction data   399
- sample chromatogram   398

methanol
- alternative flow scheme   58
- conventional synthesis   57
- conversion of syngas   57–59
- distributed production   57–59
- esterification   215, 217–218
- production costs   60
- Raman spectra   214

methyl acetate, reaction data   214

methyl methacrylate (MMA), pilot manufacture plant   108

MGA   *see* micro gas analyzer

Michael addition   110–111
- flow scheme   111
- microreactor   439
- reaction enthalpies   112

micro-analytical technology
- dielectric spectroscopy   253–256
- micro gas analyzer (MGA)   222–238
- nuclear magnetic resonance (NMR)   241–244
- on-line Raman spectroscopy   211–219
- selected developments   209–313
- surface plasmon resonance (SPR)   246–251

micro-chemical process technology (MCPT)   108

micro-component development, non-reactor   131–171

micro-discharge device (MDD)   227–228, 233–234
- false positive fraction   236

micro-disk, gas–liquid contactor   161

micro-electrical mechanical systems (MEMS)   7, 133, 366–367
- gas analyzer systems   230–231

micro gas analyzer (MGA)   222–238
- air sampling   229
- detection   232–233
- false alarm rate metrics   234–237
- performance goals   223

- preconcentration   230
- preliminary results   230–237
- separation   231

micro-instrumentation   473
- background   3–6
- basic concepts   1–78
- case studies   79–468
- technology developments   79–468

micro modular substrate sampling system (μMS^3)   330–333
- assembly steps   333
- basic building block   332
- flow tubesets   332

micro-structured unit operations   49–56

micro total analysis systems (μTAS)   135, 260–261

microchannel chromatography   149

microchannel dialyzer   165

microchannel electrophoresis   154–157

microchannel emulsification   143–147
- comparison with membrane emulsification   144

microchannel heat exchanger, commercial applications   133

microchannel membrane oxygenator, *in vitro* testing   164

microchannel system
- cell testing   165
- detachment process   145
- inflation process   145
- "straight through"   146

microchannels
- beads   272
- chemical vapor deposition (CVD)   275
- fritless packing   272
- liquid interface   307
- three-phase flow   153
- trapezoidal cross-section   266

MicroChemTec   17

microchips
- all-glass   293
- fluidic interface   278
- GLRS integration   280
- socket   363

microcoil, copper   242

microcomponent flow   181–206
- contraction flows   194–195
- convection   198–199
- diffusion   196–205
- entry lengths   194–195
- pressure drop   182–183
- reactor system   205

microflow techniques   38

microgravity environment   148

microimpellers 138
micromixer-reactor, caterpillar 110
micromixers
– active 136–139
– advanced connection system 458
– chaotic flow 142
– colliding jet 142
– cyclone 141
– interdigital 101
– moving droplet 143
– multilamination 123
– passive 139–143
– recirculation 142
– silicon 454–455
– split and recombination 141
– staggered herringbone 142
– structured packing 142
– T-shaped 137
– T type 139
– ultrasound 138
– Y type 139
microporous silica frits 445
microreactor for automated reaction optimization (MICROTAURUS) 454–458
– setup 460
microreactor plants
– (S)-2-acetyl tetrahydrofuran synthesis 105
– amine addition to α,β-unsaturated compounds 110–111
– azo pigment Yellow 12 synthesis 103
– fine-chemical 96–99
– fine chemistry 96–97
– free-radical polymerization 122
– hydrogen peroxide synthesis 104
– Kolbe–Schmitt synthesis 113–114
– laboratory-range 96–97
– nitrobenzene hydrogenation 116–117
– m-nitrotoluene bromination 118
– phenyl boronic acid synthesis 101
– pilot-range 98–99
– quinolone antibiotic drug intermediate synthesis 106
– table-top size 96
– thiopene bromination 120
microreactor process safety, computational fluid dynamics (CFD) 387
microreactor technology
– costs 86
– cross-boundary research 85
– evolution 366
– fundamental sciences 86
– interfacing 85

microreactors
– bypass channel 294
– chemical synthesis 436–441
– concepts 85–126
– coupling to microseparation devices 281–294
– die 369
– DNA amplification/digestion 295
– experimental methods 389–391
– experimental protocol 390
– falling film 116
– flow rate 434
– heater power profile 403
– hydrodynamic pumping 432
– industrial process development 100–106
– integrated system 363–404
– laboratory-scale process developments 110–120
– Michael addition 439
– micro-scale device applications 297
– novel chemistry 124–125
– packaging 367–370
– packaging hardware 367
– performance increasement 124
– process development 75
– process monitoring 378–386
– process safety 386–388
– processing 85–126
– scaleup piping diagram 372
– system control 378–386
– system design 371–378
– testing procedures 378
– upscale 449–466
microscale device technology
– biopharmaceutical production 296
– industrial biotechnology 296
– microreactors 297
– process development 297
– relevance 295
microscale equipment, surface-to-volume ratio 38
microscale reaction characterization 76
microseparation
– coupling to microreactors 281–294
– historical perspectives 259–280
– HPLC 263–264
– PAT 258–297
– process stream sampling 281–294
microstripping column 54
microstructured heat exchanger-reactors 94–95
microstructured mixer-reactors 88–95
– caterpillar 88–90
– equaling-up 92

– platelets 91
– starlam 91–93
MICROTAURUS see microreactor for automated reaction optimization
microvalve operation, AIMS/MARS 395
mini-trickle bed 104
miniaturized reactor types, specific interfacial area 53
mixer-reactors, microstructured 88–95
mixers
– multi-laminating 140
– recirculation 142
– serpentine 204
mixing
– diffusion 198
– efficiency 52
– electric field 137
– electrokinetic instability induced 137
– liquid–liquid mass transfer 52
– non-reactor micro-components 135–143
– reactor inlet tee 412–414
mixing board, feed gas 374
mixing cascades, parallel
mixing channel, electropherogram 285
mixing time analysis 412
mixing variance, T-sensor 203
MMA see methyl methacrylate
µMS³ see micro modular substrate sampling system
modeling
– axial dispersion 421
– multiscale 45
modular fluidics, automated hand-held SPR 247–248
modular platform components (MPC) system 325–329
– configurator layout screen 331
– 2-port toggle valve 328
– 5-position assembly 329
modules, laboratory 462
monitoring
– refractive index 246
– see also continuous reaction monitoring
monolayer, self-assembled 152
moving autosampler tray 461
moving droplet mixers 143
MPC see modular platform components
MRI see magnetic resonance imaging
MS see mass spectrometry
µTAS see micro total analysis systems
multi-laminating mixers 140
– focus design 140
multiscale technology 88
multichannel electropherogram 282

multilamination micromixer, polydispersity index 123
multiple automated reactor system (MARS) 371, 389–392
– automation controllers 394
– comparison to AIMS 392–396
multiple-injection electropherograms 286
multiple sensing elements, integration 349
multiport rotary valve 33
multiscale modeling 45

n

nanofluidic preconcentration device 157
near-infrared (NIR) spectroscopy 27
NeSSI see new platform for sampling and sensor initiative
new chemistry
– exploration 431–447
– large-scale manufacture 446
new platform for sampling and sensor initiative (NeSSI) 315–352
– ANSI/ISA standard 76.00.02 16
– background 320–322
– bus 340–344
– CPAC 315, 322–323
– electrical layer 339–345
– example systems 318
– generation I 323–338
– generation II 339–345
– generation III 340
– hazardous locations 342
– high pressure ratings 338
– improved reactor sampling 219
– in situ analysis 320
– instrumentation 347
– laboratory applications 346–350
– measurements 347
– overview 316–319
– PAT 320–322
– physical layer 323–338
– platform 19
– "plug-and-play" technology 319–323, 341
– process analytical systems 32
– sensing technologies 349
– top mount component 220
Newtonian fluids 424
NIR see near-infrared
nitration
– high pressure liquid chromatography (HPLC) 451–454
– HPLC chromatogram 453–454
– laboratory scale 451–453, 463–465
– 4-(phenyl)morpholin-3-one 465–466
– upscale 449

nitric acid 463–465
nitrobenzene, hydrogenation 115–117
nitroglycerine, production plant 107
nitromethane, deuterated 243
m-nitrotoluene, bromination 118
nitroxide mediated (NMP) radical polymerization 121
NMP see nitroxide mediated polymerization
NMR see nuclear magnetic resonance
noise see signal-to-noise ratio
non-contact mode, dielectric spectroscopy 254
non-miscible liquids 53
non-reactor microcomponents
– biological processes 165–166
– development 131–171
– hemodialysis 165–166
novel chemistry, microreactors 124–125
novel systems, new chemistry exploration 431–447
nuclear magnetic resonance (NMR) 241–244
– signal-to-noise ratio 241

o

OCC see orthogonal measurement channels
Omnifit connection system, disposable syringe 458
on-line analysis 41
on-line GC analysis 419
on-line MS, process analytical systems 37
on-line Raman spectroscopy, continuous microreactor 211–219
on-line spectroscopy 36
operating procedures, standard (SOP) 387, 390
operating region, intrinsically safe 343
operational temperature, radical side-chain bromination 119
optical path, average concentration 199
optical purity, (S)-2-acetyl tetrahydrofuran 105
optical sampling 212
optimization
– continuous microreactor 211–219
– laboratory 452–453
– reaction 439
optimized connection system 456
optimized micromixer, advanced connection system 458
organics, small 249
orthogonal channel capacity (OCC) 222–223 234–238
outlet concentration profile 410
ovalbumin–antiovalbumin complex 283
oxidation
– ammonia 401–403
– methane 398–400
– partial 61
oxygen
– conversion data 399
– sensor 347
oxygenator, microchannel membrane 164

p

Pacific Northwest National Laboratory (PNNL) 143, 148
– wicking phase separator 132
packing, structured 142
parallel mixing cascades
Parker-Hannifin Intraflow System 334–337
Parker integrated conditioning system (PICS) 334–335
partial oxidation 61
passive micromixers 139–143
– CFD 140–142
PAT see process analytical technology
path, optical 199
path forward 469–475
PCA see principle components analysis
PCR see polymerase chain reaction
PDMS see poly(dimethylsiloxane)
Peclet number (Pe) 199–200
– micromixers 135–136
PEM fuel cell, wicking phase separator 148
peptides
– electrophoretic purification 438
– synthesis 437–438
performance, AIMS 397–403
periodic flow switching 136
pervaporation, phase transfer processes 159
pharmaceutical processes, process analytical technology (PAT) 475
pharmaceutical research 74–75
phase separator
– non-reactor micro-components 148
– wicking 132, 148
phase transfer processes
– absorption 158
– adsorption 149–150
– desorption 158
– distillation 160–161
– extraction 151–153
– microchannel electrophoresis 154–157
– non-reactor micro-components 149–164

Subject Index

- pervaporation 159
Phased Heater Array Structure for Enhanced Detection (PHASED)
- air-based analyzer 223–227
- preconcentration 230
- separation 230
phenyl boronic acid 100–102
4-(phenyl)morpholin-3-one, nitration 465–466
photocleavable SAM 152
PI *see* process intensification
PICS *see* Parker integrated conditioning system
piezoelectric membranes 137
pigment, Yellow 12 102–103
pilot plant
- methyl methacrylate (MMA) 108
- radical polymerization 108
- upscale 466
pilot-range
- microreactor plants 98–99
- microstructured mixer-reactors 88–95
pilotscale, fuel processor 99
pipe flow, laminar 411, 415–416
- Reynolds number (Re) 182–183
piping diagram, scaleup microreactor system 372
planar flow, velocity profile 197
plant, world-scale 39
plant-upgrading 85
platelets, microstructured mixer-reactors 91
"plug-and-play" technology, NeSSI 319–323, 341
plug flow
- electrokinetically-driven 260
- laminar pipe flow 416
- pressure-driven 260
PNNL *see* Pacific Northwest National Laboratory
polarity index, EOF 434
poly(dimethylsiloxane) (PDMS)
- biocompatible 287
- CEC 263, 289–290
- PHASED analyzer 226
poly(dimethylsiloxane) chip 268, 276
polydispersity index
- multilamination micromixer 123
- tube reactor 123
polyimide membrane channels, branched 160
polymerase chain reaction (PCR) 157, 258, 294
polymeric semiconductors 107

polymerization
- free radicals 121–123
- nitroxide mediated 121
- radical 108
polymers, high-value 109
polystyrene, latex 170
porosity, catalyst characterization 354
port, fluid 325
2-port toggle valve, MPC system 328
power profile, microreactor heaters 403
preconcentration
- C_6–C_{10} alkanes 231
- dielectric spectroscopy 256
- micro gas analyzer (MGA) 230
- nanofluidic device 157
- PHASED 230
premixed feeds, simulated conversion profiles 414
preparation, feed 418
pressure connector, slip-fit 336
pressure-driven plug flow 260
pressure drop
- contractions 189–191
- expansions 189–191
- flow disturbances 187–188
- laminar pipe flow 415
- microcomponent flow 182–183
pressure ratings, NeSSI systems 338
pretreatment, samples 296
primary feed gas manifold 373
principle components analysis (PCA) 215–216, 311–312
process analytical chemistry, CPAC summer institute 9–15
process analytical systems
- chromatography 33–38
- developments since 1980 25–27
- early developments 23–24
- evolution 23
- filtration 30
- GC 33–34
- general reviews 32
- LC 35
- NeSSI 32, 319
- on-line MS 37
- on-line spectroscopy 36
- sampling systems 28–31
process analytical technology (PAT)
- analytical tools 7–8
- biological field 475
- dielectric spectroscopy 253
- microseparation devices 258–297
- NeSSI 320–322
- pharmaceutical processes 475

– potential applications 295–297
process automation, AIMS/MARS 393
process control, liquid phase processes 419
process control loops, AIMS 397
process development
– (S)-2-acetyl tetrahydrofuran synthesis 105
– amine addition to α,β-unsaturated compounds 110
– azo pigment Yellow 12 synthesis 102
– free-radical polymerization 121–123
– high throughput research 73–74
– hydrogen peroxide synthesis 104
– industrial microreactors 100–106
– Kolbe–Schmitt synthesis 112
– laboratory modules 462
– laboratory-scale 110–120
– micro-scale device applications 297
– microreactors 75
– nitrobenzene hydrogenation 115
– m-nitrotoluene bromination 118
– phenyl boronic acid synthesis 100
– quinolone antibiotic drug intermediate synthesis 106
– thiopene bromination 119
process hazards microreactors 387
process instrument
– conversion to laboratory instrument 4
– transportation 5
process intensification (PI) 43–64
– case studies 57–63
– commercial dehydrogenation process 46–47
– CPAC summer institute 16–19
– gas–liquid mass transfer 53
– gas phase mass transfer 51
– heat transfer 56
– liquid–liquid mass transfer 52
– micro-structured unit operations 49–56
– reaction engineering 44–48
– scope and definitions 43
process interface 30
process monitoring
– interface 383–385
– microreactors 378–386
– software 381–382
process safety, microreactors 386–388
process stream sampling, microseparation devices 281–294
process temperatures, elevated 387
processing solution
– (S)-2-acetyl tetrahydrofuran synthesis 105
– amine addition to α,β-unsaturated compounds 110–111

– azo pigment Yellow 12 synthesis 103
– free-radical polymerization 122
– hydrogen peroxide synthesis 104
– Kolbe–Schmitt synthesis 113–114
– nitrobenzene hydrogenation 116–117
– m-nitrotoluene bromination 118
– phenyl boronic acid synthesis 101
– quinolone antibiotic drug intermediate synthesis 106
– thiopene bromination 120
production-oriented development, LED materials 107
production proteins, surface plasmon resonance (SPR) 250
production range, microstructured mixer-reactors 88–95
protocol
– simplified 125
– tailoring 124
prototype
– reaction optimizer 460
– silicon micromixer 454–455
pumping
– heat pump systems 162–164
– hydrodynamic 432
purged enclosure 26
purification, electrophoretic 438
pyrazole, synthesis 436

q

quantitative measurements, reaction kinetics 425–426
quinolone antibiotic drug, synthesis of intermediates 106

r

racemization
– amides 439
– α-amino acids 438
radicals
– free 121–123
– polymerization 108
– side-chain bromination 119
Raman, acetic acid spectra 214
Raman effect 211
Raman spectroscopy, ballprobe 214, 217, 219–220
– immersion probe 213
– methanol spectra 214
– methyl acetate spectra 214
– on-line 211–219
Rayleigh–Gans approximation 308
reaction data
– ammonia oxidation 402

- methane oxidation   399
- methyl acetate   214
reaction engineering, process intensification   44–48
reaction enthalpy, Michael addition   112
reaction kinetics
- plot   287
- quantitative measurements   425–426
reaction manifold, Knoevenagel reaction   443
reaction mechanisms, gas phase processes   354–360
reaction monitoring, continuous   213–218
reaction optimization   439, 460
reaction performance, AIMS   397–403
reaction profile   6
reaction time constant, reactor classification   49
reactions
- gas phase   363–404
- heterogeneous catalyst   76
- Knoevenagel   442
- microscale reaction characterization   76
- Wittig   440
- see also convection-diffusion-reaction equations
reactive ion etching, DRIE   228
reactivity
- gas phase processes   354–360
- testing   355
reactor board   374–375
- isometric view   375
reactors
- catalyst deactivation   49
- failure   395
- feed gas manifold   374
- flow   442–446
- flow reactor   75
- glass lined   422
- improved sampling   219
- inlet tee   412–414
- liquid phase processes   418
- membrane   47
- microcomponent flow   205
- microstructured heat exchanger-reactors   94–95
- microstructured mixer-reactors   88–95
- reaction time constant   49
- stainless steel   420
- temperature scanning (TSR)   356
- tube   123
reagents
- Grignard   105
- toxic   437

- unstable   437
real-time (RT) analysis   6
"real world" interface   283
receiver operating characteristic (ROC)   234–238
recirculation mixers   142
recombination   see split and recombination micromixers
rectangular channel, diffusion   201
Redwood mass flow controllers   395
Redwood microvalve operation   395
reflection   see grating light reflection spectroscopy
reforming   61
refractive index, monitoring   246
- optical detection method   279
relative convection   200
relative diffusion   200
replacement, AIMS/MARS   395
research
- Edisonian   71
- high throughput   see high throughput research
- pharmaceutical   74–75
- process concepts   68–69
residence time loop   460
resistance, mass transfer   45
resorcinol   113–114
Reynolds number (Re)
- micromixers   135–136
- pipe flow   182–183
rheometer, MRI   244
rhodamine dye, LC   269
robotics feed preparation   418
ROC   see receiver operating characteristic
rotary valve, multiport   33
run time (RT) engine   384

s
safe operating region   343
safety
- microreactors   386–388
- NeSSI   350
SAM   see self-assembled monolayer, sensor actuator manager
sample pretreatment   296
sampling
- air   229
- contamination-free sampling device   462
- optical   212
- process stream   281–294
- reactor   219
- sampling system   28–31
Sandia hydrogen based analyzer   228–229

sapphire, spherical immersion lens 213
scaledown ratio 50
scaleout issues, microstructured mixer-reactors 88–95
scaleup *see* upscale
screening, catalyst 70
secondary amine, addition to α,β-unsaturated compounds 110–111
segregation index
– caterpillar mixers 90
– platelet mixers 92–93
selectivity
– (S)-2-acetyl tetrahydrofuran 105
– catalysts 361
self-assembled monolayer (SAM), photocleavable 152
Semiconductor Equipment and Materials Institute (SEMI) 321–322
semiconductors, polymeric 107
sensing
– direct 256
– hardware platform *see* new platform for sampling and sensor initiative (NeSSI)
– multiple elements 349
– technologies 349
sensor actuator manager (SAM) 319, 339, 345–346
sensors
– costs 348
– diffusion in T-sensor 202–203
– oxygen 347
– Spreeta sensor elements 247
– SPR 246–251
– temperature 396
separation
– C_6–C_{10} alkanes 232
– isothermal 232
– micro gas analyzer (MGA) 231
– PHASED 230
separator, phase 132, 148
serpentine mixer 204
– variance 205
shear rate, MRI 244
side-chain bromination, radical 119
signal strength control, dielectric spectroscopy 255
signal-to-noise ratio, NMR 241
silica frits, microporous 445
silicon micromixer
– etched-through 456–457
– first prototype 454
– with connector 455
simplified protocols, microreactor performance increasement 125

simulation, two-dimensional 413–414
sizing, injector loop 417
SK Corporation/Daejeon, (S)-2-acetyl tetrahydrofuran synthesis 105
slip-fit pressure connector 336
slow flows, friction factor 182
small organics, analysis 249
SMART *see* styrene monomer advanced reheating technology
socket
– DieMate 368
– microchip 363
software
– Comsol Multiphysics 181
– FEMLAB 181, 271
– LabView 344, 381–383, 388
– process monitoring 381–382
– *see also* computational fluid dynamics
solid systems, mass transfer in gas–solid systems 54–55
solvent-free processes 124
solvent viscosity (η) 168
SOP *see* standard operating procedures
SP76 *see* ANSI/ISA standard 76.00.02
spatial wavelength control, dielectric spectroscopy 254
specialty chemical research 74–75
specific interfacial area
– conventional reactor types 53
– miniaturized reactor types 53
specific surface area, heat exchangers 56
spectroscopy
– dielectric 253–256
– FTIR 5, 24, 36
– grating light reflection 305–313
– IMS 238
– NIR 27
– nuclear magnetic resonance (NMR) 241–244
– Raman 211–219
– *see also* dielectric spectroscopy, grating light reflection spectroscopy, mass spectrometry, nuclear magnetic resonance
spherical lens, sapphire 213
spinning micro-disk, gas–liquid contactor 161
split and recombination micromixers 141
– YM-1 structure 141
spores, surface plasmon resonance (SPR) 251
SPR *see* surface plasmon resonance
Spreeta sensor elements 247
stabilization, cast 455

staggered herringbone micromixer 142
stainless steel reactor 420
standard operating procedures (SOP) 387, 390
starlam microstructured mixer-reactors 91–93
steady-state concentration profile, tracer 414
steam reforming, catalytic 61
stereochemistry 440–441
stilbenes 441
Stokes equation 187
"straight through" microchannel system 146
stream profile, suspended beads 168
stream sampling, microseparation devices 281–294
Strouhal number (St), micromixers 135–136
structured packing micromixers 142
styrene monomer advanced reheating technology (SMART) 48
substrate channels, MPC system 330
sulfuric acid, concentrated 463
surface area, specific 56
surface intermediates 358
surface-mount instrument interfaces 347
surface plasmon resonance (SPR)
– active channels 246
– automated hand-held 24-channel 247–248
– cells 251
– clip-in flowcell 249
– contaminants 250
– dinitrophenol-lysine 249
– fundamentals 246
– instrumentation 247
– production proteins 250
– sensors 246–251
– specific applications 249–251
– spores 251
– viruses 251
surface-to-volume ratio, microscale equipment 38
suspended beads, stream profile 168
Swagelok modular platform components MPC system 325–329
syngas, conversion into methanol 57–59
synthesis
– (S)-2-acetyl tetrahydrofuran 105
– azo pigment Yellow 12 102–103
– dimethyl acetals 445
– flow reactors 442–446
– hydrogen peroxide 104

– methanol 57
– peptides 437–438
– phenyl boronic acid 100–101
– pyrazole 436
syringe, disposable 457
system capabilities, liquid phase processes 417–426
system chassis, microreactors 371–372
system control, microreactors 378–386
system design
– liquid phase processes 409–417
– microreactors 371–378
system hardware, control 378–380
system interlock 388
system size 392–393

t

T-sensor, diffusion 202–203
– knife edge 203
– mixing variance 203
T-shaped micromixer 137
T-type micromixers 139
table-top size, microreactor plant 96
tailoring, protocols 124
target mass, variability 425
TCD see thermal conductivity detector
technology developments, micro instrumentation 79–468
TEDS see transducer electronic data sheets
tee, reactor inlet 412–414
teflon
– feed preparation 418
– injector 422
– tubing 241, 278, 424
temperature controller
– AIMS 397
– board 376
temperature–enthalpy plot, methanol synthesis 59
temperature scanning reactor (TSR) 356
temperature sensors, AIMS 396
temperature swing adsorption (TSA) 54, 149–151
testing
– AIMS 391–403
– automation 386
– *in vitro* 164
– microreactors 378
Texas Instruments Spreeta sensor elements 247
thermal conductivity detector (TCD)
– AIMS 398–399
– ultra MGA 230, 233–234, 236
thermal decarboxylation 115

thermostatic bath 460–462
thiophene, bromination 119–120
three-phase flow, microchannels 153
TIC *see* toxic industrial compounds
time constant, reaction 49
toggle valve, 2-port 328
toluene yield, dependence on reactor temperature 357
tools
– analytical 7–8
– diffusion 359–360
– kinetic 356–358
top mount component, NeSSI 220
toxic industrial compounds (TIC), detection 229
toxic reagents, *in situ* synthesis 437
toxins, *Clostridium botulinum* toxin A 250
tracer, steady-state concentration profile 414
traditional chemistry, limitations of reactors 124
transducer, ultrasound micromixer 138
transducer electronic data sheets (TEDS) 345
transfer *see* heat transfer, mass transfer, phase transfer processes
transport tube heating 376
trapezoidal cross-section, microchannel 266
tray, autosampler 461
trial and error discovery 71
trichloroethane emission spectra 234
Trust Chem/Hangzhou 102–103
TSA *see* temperature swing adsorption
TSR *see* temperature scanning reactor
tube, transport 376
tube reactor, polydispersity index 123
tubesets, flow 332
turbulent flow
– frictional loss coefficient 188
– mechanical energy balance 183
two-dimensional simulation 413–414
– liquid phase processes 413–414

u

U-shaped legs, connection system 456
ULP *see* University Louis Pasteur Strasbourg
ultra MAG 222–238
ultrasonic forces, fluid manipulation 169–170
ultrasound membranes 137
ultrasound micromixer, transducer 138
unit operations, micro-structured 49–56
Universal Oil Products (UOP)/Chicago, hydrogen peroxide synthesis 104
University College London, nitrobenzene hydrogenation 115–117
University Louis Pasteur (ULP)/Strasbourg, free-radical polymerization 121–123
unmixed feeds, simulated conversion profiles 414
α,β-unsaturated compounds, amine addition 110–111
unstable reagents, *in situ* synthesis 437
UOP *see* Universal Oil Products
upscale 43
– in larger laboratory scale 462–465
– microreactor die 369
– microreactors 449–466
– microtechnology to world-scale production 82
– nitration 449
– pilot plant 466
– piping diagram 372
urea clearance, microchannel dialyzer 165
US Food and Drug Agency (FDA) 8, 10–11

v

valves
– actuated 169
– block and bleed system 31
– integrated production 277
– 2-port toggle 328
– rotary 33
vapor compression cycle heat pump 163
variance
– mixing 203
– serpentine mixer 205
– target mass 425
velocity profile
– EOF 433
– MRI 244
– planar flow 197
velocity rearrangement, contractions 195
versatile micro-analytical systems (VMAS) 10
viruses, surface plasmon resonance (SPR) 251
viscosity
– EOF 434
– MRI 244
viscous dissipation, mechanical energy balance 185–186
VMAS *see* versatile micro-analytical systems
Vortec coolers 30
vortex, Dean 142

w

wavelength control, dielectric spectroscopy 254
web enabled technology 319
wicking phase separator
– PEM fuel cell 148
– PNLL 132
wiring, electrical 394
Wittig reaction 440
world-scale plant 39
world-scale production, microtechnology scaleup 82

x

xenon isotopes, mass spectrum 233
Xi'an Huian Industrial Group/Xi'an, nitroglycerine production plant 107

y

Y-type micromixers 139
Yersinia pestis 251
yield, toluene 357
yield pattern 55
YM-1 structure, split and recombination micromixers 141

z

zeta-potential (ζ) 168